Coastal Ecosystems of the Tropics - Adaptive Management

Velmurugan Ayyam • Swarnam Palanivel
Sivaperuman Chandrakasan

Coastal Ecosystems of the Tropics - Adaptive Management

Springer

Velmurugan Ayyam
ICAR-Central Island Agricultural
Research Institute
Port Blair, Andaman and Nicobar Islands,
India

Swarnam Palanivel
ICAR-Central Island Agricultural
Research Institute
Port Blair, Andaman and Nicobar Islands,
India

Sivaperuman Chandrakasan
Zoological Survey of India – ANRC
Port Blair, Andaman and Nicobar Islands,
India

ISBN 978-981-13-8928-3 ISBN 978-981-13-8926-9 (eBook)
https://doi.org/10.1007/978-981-13-8926-9

This Springer imprint is published by the registered company Springer Nature Singapore Pte Ltd.
The registered company address is: 152 Beach Road, #21-01/04 Gateway East, Singapore 189721, Singapore

Foreword

Tropical coastal environment represents one of the most dynamic, vital interfaces at the boundary between land and sea. Coastal areas include several natural and managed ecosystems which are dependent on the land-sea interconnection and dynamic flow of energy and matter. On the other hand, the coastal region has long been under stress from over-exploitation and mismanagement of resources by human population. Furthermore, the looming spectre of sea level rise associated with the effect of global warming presents a new and potentially far more dangerous threat to this region. Certainly, this necessitates appropriate coastal zone management strategies to conserve and derive sustainable benefit from the coastal ecosystems. However, still there are critical gaps in our understanding of the functioning and utilization of these natural ecosystems.

In this context, it gives me great pleasure to introduce the book *Coastal Ecosystems of the Tropics: Adaptive Management* authored by Dr. V. Ayyam, Dr. S. Palanivel, and Dr. S. Chandrakasan. I am also happy to note that the senior author is also an alumnus of **the Carbon Management and Sequestration Center, The Ohio State University, Columbus Ohio, USA**. The contents of this book are of great interest to professionals working on coastal ecosystems of the tropical region. This book collates and synthesizes information and case studies on coastal ecosystems. Thus, large habitat contiguity and diversity in the tropical region will benefit researchers in the field as well as the coastal resource managers.

I commend the authors for their exemplary efforts in bringing out this publication with 25 technical chapters under five different themes concerning tropical coastal ecosystems. The opening chapter provides very good introduction to the outline of the book followed by the status of natural resources. Similarly, the chapters addressing the theme of enhancing the productivity and conservation of coastal ecosystems reflect rational and logical analysis with case studies on legal and institutional frameworks and participatory coastal area management, which are essential to conserving and managing these unique ecosystems. I therefore congratulate the authors and the publisher for publishing this book of topical significance, and I am sure that it will serve as a vital reference material.

Yours sincerely,

Rattan Lal

Distinguished University Professor
of Soil Science, Director, Carbon Management
and Sequestration Centre, College of Food,
Agricultural, and Environmental Sciences
The Ohio State University
Columbus, OH, USA
22 February 2019

Preface

The coastal areas are one of the most dynamic natural systems where the three main components of our planet—the hydrosphere, the lithosphere, and the atmosphere—meet and interact, forming several interconnected systems. The major ecosystems are mangroves, coral reefs, seagrass beds, rocky coasts, mud or salt flats, sandy or dune systems, lagoons, estuaries, and salt marshes, whereas agro-ecosystem in the coastal areas is human-managed system and its surroundings are highly influenced by human interest. These coastal ecosystems are diverse in function and form and dynamic, and the boundaries between various components of the systems are often diffused.

As a result of this uncertainty and human influence, the coastal areas are defined in different ways depending on the need, purpose, scale, and availability of data for delineation. This book follows the definitions adopted in the Millennium Ecosystem Assessment (MEA) and the IPCC to define and understand different coastal ecosystems particularly in the tropics. As per the MEA, the inland extent of coastal ecosystems is defined as the line where land-based influences dominate up to a maximum of 100 km from the coastline or 50-meter elevation (whichever is closer to the sea) and with the outward extent as the 50-meter-depth contour. Marine ecosystems begin at the low water mark and encompass the high seas and deepwater habitats. The Fourth Assessment Report of the Intergovernmental Panel on Climate Change considers coastal systems as the interacting low-lying areas and shallow coastal waters, including their human components. This helped us to compile information from diverse sources and understand the tropical coastal areas for its adaptive management.

Coastal ecosystems are among the most productive because they are enriched by land-based nutrients and nutrients that well up into the coastal waters from deeper levels of the ocean. Coastal ecosystems are repositories of biological diversity and provide a wide range of goods and services. In general, it provides provisioning, regulating, and sociocultural services to coastal communities. In other words, they sustain economies and provide livelihood to millions of people through fisheries, agriculture, ports, tourism, and other industries. This is more pertinent to the tropical region where nearly one third of the total population is concentrated within 100 km from the coast. The ever-increasing human population and strive for economic growth put severe pressure on the natural resources of the tropical coastal areas. As a result, the coastal ecosystems are highly modified or influenced by human activities which are further aggravated by the natural threats.

In this context, literatures suggested that coastal areas are affected by two main types of influences, terrestrial and marine, that are external to these areas. Terrestrial influences include land use changes, exploitations, and all the consequences of changing hydrological regimes which are mostly anthropogenic in nature. Marine influences are mostly natural phenomena such as storms, cyclones, tsunamis, sea surges, and ocean currents. However, the coastal and marine ecosystem goods and services in the tropical areas are not properly documented and highly undervalued. This resulted in unsustainable patterns of resource exploitation, highly degraded ecosystems, and inadequate conservation measures.

In recent times, advances in environmental and ecological studies of the coastal areas have shed light on the intricacies and values of tropical coastal ecosystems. This brought worldwide attention on the vulnerability and interconnections of these coastal ecosystems. The academia and the resource managers have well acknowledged that the coastal ecosystems perform several invaluable functions and offer several tangible and intangible services and commodities to humanity. Furthermore, the policymakers are also getting conscious of the importance of the coastal ecosystems for sustainable development of the humankind. Therefore, it is high time that 'coastal ecosystems of the tropical region' needs to be properly understood to appreciate its value to the human society and the environment; and adaptive management should receive top priority for its sustainable use. This requires an attempt to consolidate the works of researchers from diverse fields concerning tropical coastal areas and to present them comprehensively for the benefit of various stakeholders. We made efforts for the last 2 years to accomplish the task and bring out this publication. We have presented the nature of coastal ecosystems and various aspects of adaptive management in 25 chapters with case examples under 5 major themes in a systematic manner, keeping in mind students, researchers, and general readers at large.

In fact, the theme and presentation style were highly influenced by Prof. R. Lal, director, CMASC-OSU, Columbus, USA, ever since the senior author visited his laboratory. We owe our sincere gratitude to him for his inspiration and suggestions for future research in coastal and island ecosystems. We made this book by souring information from the published works of several researchers across the world, and we are very grateful to all of them. We sincerely thank all our colleagues and research scholars at our institute, reviewers of this proposal, publication team of Springer, and all our family members for their overwhelming support and encouragement. We hope that this book would serve as a reference base on tropical coastal ecosystems management and facilitate further deliberations on specific issues to bring in a sustainable future for the coastal areas of the tropics.

Port Blair
India

V. Ayyam
S. Palanivel
S. Chandrakasan

Contents

About the Authors

Velmurugan Ayyam is working as a Principal Scientist in ICAR-Central Island Agricultural Research Institute, Port Blair, India. He has vast experience of working in the tropical region and has expertise in natural resource management, study of biogeochemical cycles, and climate change impact assessment in agro-ecosystems using remote sensing and GIS. He has worked in the World Bank-funded land degradation management and water harvesting in the tsunami-affected areas to restore the productivity of coastal areas and improve the livelihood of poor people. He has successfully implemented several developmental projects involving technology dissemination for the benefit of coastal communities and tribal people. He was a Visiting Scientist to Ohio State University, Columbus, USA. He has also worked in collaborative educational programme with ITC, the Netherlands, and successfully guided several Ph.D., M.Tech, and P.G. Diploma students in different aspects of natural resource management. He has published several peer-reviewed research articles in reputed research journals and has authored several books.

Swarnam Palanivel is working as a Principal Scientist in ICAR-Central Island Agricultural Research Institute, Port Blair, India. She has vast experience of working in the tropical islands and has proficiency in integrated farming systems, natural resource management, crop modelling, and study of impact of agricultural practices on the island ecosystems. She has been working in resource characterization and optimization, farming system models for disadvantaged areas, and restorative measures for enhancing farm productivity in island ecosystem. She has successfully developed methods for faster recycling of organic wastes, utilization of native microbial resources, and assessment of island agro-ecosystem. She has successfully implemented several developmental projects involving technology dissemination for the benefit of tribal people. She has published more than 50 peer-reviewed research articles and has authored several books.

Sivaperuman Chandrakasan is working as Scientist-D and Officer in Charge of Zoological Survey of India—Regional Centre, Port Blair. He received his Master's degree in wildlife biology and his Doctorate degree in ecology of wetland birds in the Vembanad-Kole Ramsar site from the Kerala Forest Research Institute, Kerala, and Forest Research Institute, Deemed University, Dehra Dun. He has been extensively involved in field surveys in different parts of India. His research interests are biodiversity assessment and documentation, impact assessment, policy analysis, and ecological study of marine and coastal animals. He has published more than 200 research papers in national and international journals and newsletters. He also authored/edited more than 25 books published by respected national and international publishers. He has participated in 36th Indian Scientific Expedition to Antarctica during 2016–2017 and carried out studies on the species abundance and distribution of birds and mammals in Antarctica. He has participated and presented research papers at 60 national and international seminars and symposia. He is a life member of various scientific societies in India and abroad.

Part I

Overview

Coastal Regions of the Tropics: An Introduction 1

Abstract

Tropical coastal environment represents one of the most dynamic and vital interfaces on Earth, at the boundary between land and sea. It encompasses some of the most diverse and productive habitats. These habitats include natural ecosystems, managed ecosystems besides major urban centres. The existence of these ecosystems is dependent on the land-sea interconnection and dynamic flow of energy and matter. At the same time, the coastal region has long been under stress from over-exploitation and mismanagement. The increasing pace of human population and developmental activities in the tropical coastal region has altered the functionality of coastal ecosystems and endangered several flora and fauna that threaten the livelihood of people who depend on them. In addition, the looming spectre of sea level rise associated with the effect of global warming presents a new and potentially far more dangerous threat to this region. This necessitates suitable coastal zone management plan to conserve and derive sustainable benefit from the coastal ecosystems. With this background an overview of tropical coastal countries, its demographic features, natural resources, coastal ecosystems, and its services to the human society are discussed in this chapter. Brief account on effect of human activities and climate change on coastal region sourced from different literatures provides useful information to the researchers, students, and policymakers.

Keywords

Tropical coast · Ecosystems and services · Food production · Climate change

V. Ayyam et al., *Coastal Ecosystems of the Tropics - Adaptive Management*,
https://doi.org/10.1007/978-981-13-8926-9_1

1.1 Introduction

Throughout the human history, coastal regions have been centres of human activity due to its diverse resources and ecosystem services. This region supports 37% of global population (as per 2011 census) though they represent only 21% of the available land. Coastal region comprises of wide diversity of ecosystems providing shelter to diverse flora and fauna. It provides livelihood through agriculture, fisheries, ports, tourism, and other industrial activity; hence, coasts are of great ecological and socio-economic importance. Coastal region is one of the most dynamic natural systems where the hydrosphere, the lithosphere, and the atmosphere meet and interact, forming interconnected systems (Michel and Pandya 2010). It constitutes a transition zone where land and fresh water meet saline water and across which the effects of land on the ocean, and vice versa, are transferred and modified. This region has become very critical where climate change adaptation has assumed importance as it provides habitat for several endangered flora and fauna.

Due to its ecological, socio-economic significance, tropical coastal regions deserve further elaboration in the context of sustainable management. This region consists of several important and distinctive ecosystems as it is located in the interface between marine and terrestrial which has characteristics of climatic regime. However human activities have modified or interfered with the functioning of several important natural ecosystems to support agro-ecosystems and urban centres. Marine, estuary, and coastal wetlands often benefit from flows of nutrients from the land and also from ocean upwelling which brings nutrient-rich water to the surface. They thus tend to have unique coastal ecosystems, support a rich biological diversity and productivity, and contain a valuable assortment of natural resources. Examples of such habitats are estuarine, coral reefs, coastal mangrove forests and other wetlands, tidal flats, and seagrass beds, which also provide essential nursery and feeding areas for many coastal and oceanic aquatic species (FAO 1998a). However, in the last few decades, pressure exerted by population explosion and climate change impacts have been altering and interfering with the normal functioning of these ecosystems in a much faster pace which seriously affect the biodiversity, habitat suitability, and ecosystem services. Worldwide, nearly 50% of salt marshes, 35% of mangroves, 30% of coral reefs, and 29% of seagrasses are either lost or degraded (Valiela et al. 2001; MEA 2005; UNEP 2006; FAO 2007).

The Earth Summit held in Rio de Janeiro, Brazil, in 1992 marking the twentieth anniversary of the Stockholm Conference had placed environmental concerns firmly before the global community. With the increasing realization of the direct link between economic development, poverty, and natural resource degradation, the summit recognized that economic development should go along with environmental conservation. The summit yielded five major instruments signed by world leaders which included Agenda 21 which is considered as a blueprint for sustainable development. Section 1.1 of the preamble states '*No nation can achieve this on its own; but together we can – in a global partnership for sustainable development*' (MEA 2005). This is very pertinent to the tropical coastal ecosystems which are in need of solution to the burgeoning population, demand for development, and concern for environment and biodiversity.

In essence, it is increasingly realized that many of these problems pertaining to the tropical coastal regions are directly related to its geographical settings, anthropogenic activities, and climate change in recent times. Sustainable use of coastal ecosystems is therefore critical for continued economic development of the coastal region in the light of global climate change and population explosion (UNEP 2006). Although climate change and associated events are well studied and documented scientific subject, our understanding of the link between climate change-coastal ecosystems-livelihood securities to achieve sustainable development and climate change adaptation in the tropical coastal region is limited by the lack of comprehensive information. These critical details are covered in various chapters, and as a first step this chapter attempts to provide an overview of these details in a comprehensive manner.

1.2 Tropical Region

The word '*tropical*' specifically means places near the equator. This includes all the areas of the Earth where the sun reaches a point directly overhead at least once a year. The word 'tropics' comes from Greek word *tropos*, meaning 'turn', because the apparent position of the Sun moves between the two tropics within a year. Geographically the tropics are a region of the Earth surrounding the Equator and are delimited in latitude by the Tropic of Cancer in the Northern Hemisphere at 23°26′13.2″ (or 23.437°) N and the Tropic of Capricorn in the Southern Hemisphere at 23°26′13.2″ (or 23.437°) S. These latitudes correspond to the axial tilt of the Earth. This region is also referred to as the 'tropical zone' and the 'torrid zone'. The highlighted region in Fig. 1.1 depicts the tropical regions in the world. The tropics comprise 40% of the Earth's surface area

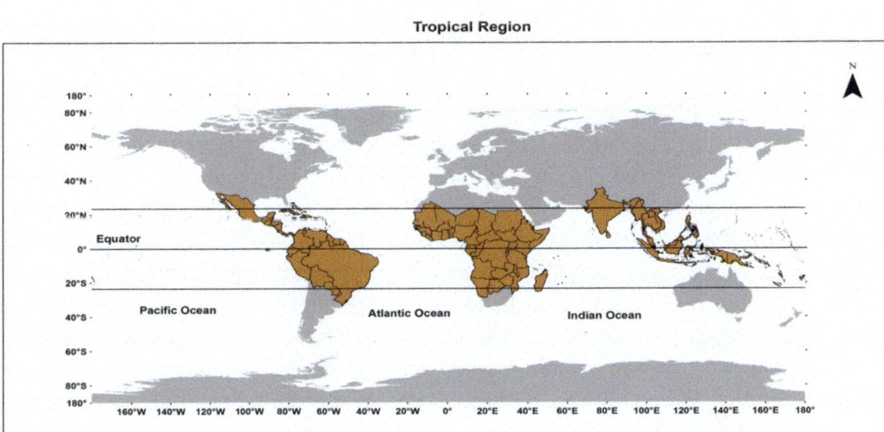

Fig. 1.1 Tropical regions of the world

and contain 36% of the Earth's landmass. The tropics are distinguished from the other climatic and biotic regions of Earth, which are the middle latitudes and the polar regions on either side of the equatorial zone.

Much of the equatorial belt within the tropical climate zone experiences hot and humid weather due to active vertical uplift or convection of air and resultant heavy rainfall. During certain periods in a year, thunderstorms would occur every day. There is less difference between maximum and minimum temperature throughout the year. This region is also known to contain different land forms and varying topographies. These conditions are favourable for different life forms to thrive and proliferate. This is reflected in the form of wide biodiversity in flora and fauna as seen in this region. Many of the tropical ecosystems are very productive and provide suitable habitat for several endangered flora and fauna besides valuable ecosystem services to the human society. Consequently the region has become home to 40% of the world population as of 2014, and this figure is projected to reach 50% by the late 2030s (NGE 2017).

1.3 Nature of Coastal Region in the Tropics

1.3.1 What Constitute the Coastal Region?

The tropical region is endowed with heterogeneous landforms and variations in regional climatic pattern (excluding the general climatic trend) which provide macro-relief of high plateau, open valleys, rolling upland, fertile plains, swampy lowlands, and barren areas. In this, the coastal region is most important because it includes land and marine resources, variety of land forms which directly supports nearly 20% of the total population of the tropical countries. It consists of several critical ecosystems supporting wide variety of flora and fauna. Coastal ecosystems are areas where land and water join to create an environment with a distinct structure, diversity, and flow of energy (FAO 1998a). Both land and marine systems are getting modified or affected by process/agents originating in opposing place. They include salt marshes, mangroves, wetlands, estuaries, and bays and are home to many different types of plants and animals. In this region most often there is dynamic equilibrium between depositional and erosive forces, which results in the formation of different coasts. As the region is rich in resources, major human activities are aimed at deriving their livelihood. Most importantly aquaculture and agricultural activities are highly managed systems and affect the natural ecosystems. The coastal plain land is used for cultivation of crops, animal grazing, and fisheries activities. The river system originating deep inside the hinterland or away from the coastal areas flows into the sea with sediments and nutrient load. The estuaries are important breeding ground for several aquatic species. In other dimension this region facilitates the interactions between various natural processes and human activities which are very important factors for the development of this region. At the same time, this resource-rich region supporting high population density is vulnerable to natural forces and anthropogenic activities. Besides, coasts are continually changing because of the dynamic interaction between the oceans and the land (Michel and Pandya 2010).

1.3.2 Coastal Formation

As discussed in the previous section, the coast is recognized where the sea and land meet with each other, and it is very dynamic. In other words coast are formed at the continental margins as a result of balance between erosive and depositional forces. Continental margins are of two types: active margins where the edge of a continent happens to be at the edge of an oceanic plate (the west coast of South America) and inactive margins where the transition from continental lithosphere to oceanic lithosphere is within a plate rather than at a plate edge (the Atlantic). Coastal areas are therefore characterized by the vertical accretion of nearshore land. This depends on several factors, viz. sediment supply from rivers or from the sea, the width of the shelf or the proximity of a submarine canyon through which currents remove sediments, and the strength of longshore currents and incidence of cyclones, both of which transport and redistribute sediments along the coast. Sedimentation is the major geological activity that shapes coasts, but human-induced modifications are having an increasing impact on coastal morphology (FAO 1998a).

In addition subduction and upliftment play a predominant role in shaping the coastal region. Nevertheless, for management purposes, a variety of landward and seaward boundaries, ranging from fairly narrow and precise ones to much broader and more nebulous ones, have been utilized around the world (FAO 1998a). Management boundaries are pragmatic, being influenced by the geographic scope of relevant management concerns. For these reasons it is expected that the boundaries of a coastal area may change over time for management purposes, as the issues to be faced become more extensive or complex and require more far-ranging solutions.

1.3.3 Defining Coastal Areas

In common parlance 'coast' is the interface or transition area between land and sea, including large inland lakes. Coastal areas are diverse in function and form, are dynamic, and do not lend themselves well to definition by strict spatial boundaries. Unlike watersheds, there are no exact natural boundaries that unambiguously delineate coastal areas (FAO 1998a). Because of this dynamic nature of interactions and different use and modification by human, the boundaries between various components of the coastal region are often diffused. As a result of this uncertainty, the coastal regions are defined in different ways depending on the need, purpose, scale, and availability of data for delineation. In order to understand the coastal areas and its proper management, certain relevant terms should be well understood which are used in this publication, viz. coastal line, coastal zone, coastal area, coastal region, and coastal plains (Fig. 1.2).

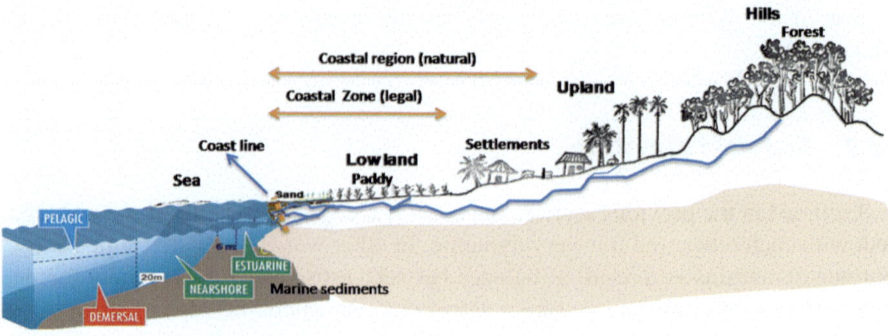

Fig. 1.2 Schematic representation of coastal areas

- A coastline or seashore is the area where land meets the sea or ocean or a line that forms the boundary between the land and the ocean or a lake (*The Merriam-Webster Dictionary*). A precise line that can be called a coastline cannot be determined due to changes in tidal level. In many countries it is described with reference to the high tide line.
- The term 'coastal zone' would refer to the geographic area defined by the enabling legislation for coastal management by any country. This recognizes tidal line and takes into account the land use and land cover of that area. As per the definition adopted by the World Resources Institute, coastal zone includes the intertidal and subtidal areas on and above the continental shelf (to a depth of 200 meters) and immediately adjacent lands. The definition of coastal ecosystems is based on their physical characteristics (their proximity to the coast) rather than only a distinct set of biological features (Burke et al. 2001).
- Whereas 'coastal area' would be used more broadly to refer to the geographic area along the coast that has not yet been defined as a zone for management purposes, it can be used to denote anything along the coast in broader sense without any specific reference to the land forms. Both the terms 'coast' and 'coastal area' often used interchangeably to describe a geographic location or region along the land-ocean interface.
- 'Region' also denotes geographical area, but within the region certain geographical features will have uniformity or largely homogenous. For example, within a climatic region rainfall, temperature would be fairly uniform throughout. A region would be distinguishable from the nearby region where the distribution of the same feature would be different. In the same context, 'coastal region' is understood which is located between marine and terrestrial region or at the interface where various biogeochemical processes are taking place which are different from the inland or ocean.
- Another important term often used by agricultural scientist and planner is 'coastal plain'. Geographically it is a depositional land form, fairly flat terrain lying between the sea and mountain or plateau. This can vary in width and length according to the terrain or geographical extent, for example, East coast plain and West coast plains of India.

Even though these terms are fairly defined, the boundaries of the relevant management area can and usually do change over time without regard to the enabling legislation. The multifaceted approach to the management of coastal resources has become known as integrated coastal management (ICM). In ICM guidelines the term 'coastal area' is preferred to 'coastal zone' to refer to the geographic entity covered by an integrated coastal management plan.

Therefore, more realistically, a combination of distance-to-coast, elevation data, and dominant processes would be preferable along with the dominant process of coastal region to delineate diffused boundaries. Here also different countries have adopted different criteria particularly distance and elevation. For example, in India, 500-m distance from the high tide line or 50-m elevation from the MSL is taken for demarcating the coastal zone. This is strictly regulated and called coastal regulation zone. In South Africa it is 50-km zone where the CZM is applied. In Sri Lanka, the coastal zone is defined as encompassing 1 km seawards of the mean low waterline and 300-m landwards of the mean high waterline, extending to a maximum of 2 km inland in the case of rivers, lagoons, or estuaries (FAO 1998a). Adopting different criteria for defining what constitute coastal region may lead to mismatch in data compilation and interpretations which may result in misunderstanding and disagreements. In order to overcome the problem, this publication follows the definitions of coastal zones adopted in two key publications as given below:

- The Millennium Ecosystem Assessment (2005) defines coastal region as 'the inland extent of coastal ecosystems is defined as the line where land-based influences dominate up to a maximum of 100 kilometers from the coastline or 50-meter elevation whichever is closer to the sea, and with the outward extent as the 50-meter depth contour. Further the marine ecosystems begin at the low water mark and encompass the high seas and deepwater habitats'.
- For the Fourth Assessment Report of the Intergovernmental Panel on Climate Change (IPCC, 2007), 'coastal systems are considered as the interacting low-lying areas and shallow coastal waters, including their human components that includes adjoining coastal lowlands, which have often developed through sedimentation during the Holocene (past 10,000 years), but excludes the continental shelf and ocean margins and inland seas'. Functionally the coastal systems also form part of the larger marine ecosystems that include coasts and open ocean areas.

1.3.4 Climate

Climate of the tropics is the single most important factor influencing different coastal ecosystems as a whole. Figure 1.3 shows different climatic regimes, viz. equatorial, monsoon, and tropical savanna climate of tropical region. Figure 1.4 shows the zonally averaged monthly precipitation of tropical region. The tropics receive more precipitation than higher latitudes. There is good distribution of rainfall in the tropical zone spread across 8 months in a year. In general the tropical

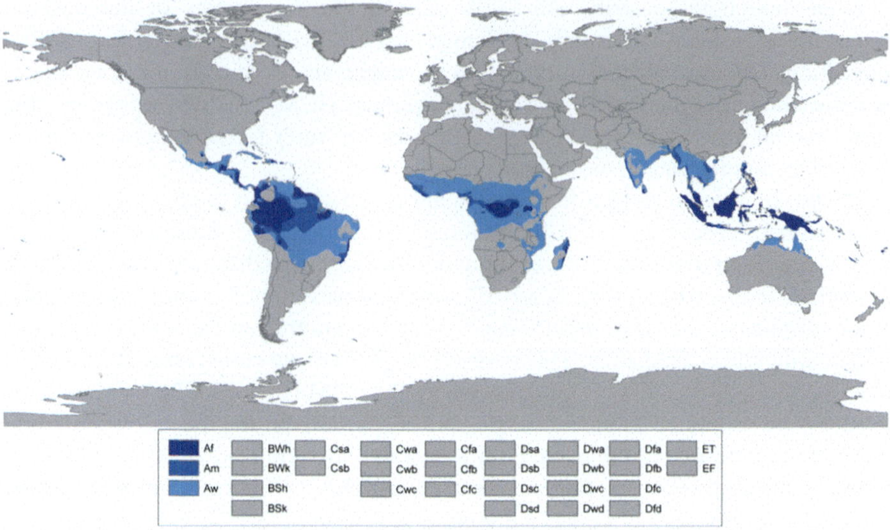

Fig. 1.3 Distribution of tropical climatic regimes. **Af**: equatorial climate. **Am**: monsoon climate. **Aw**: tropical savanna climate

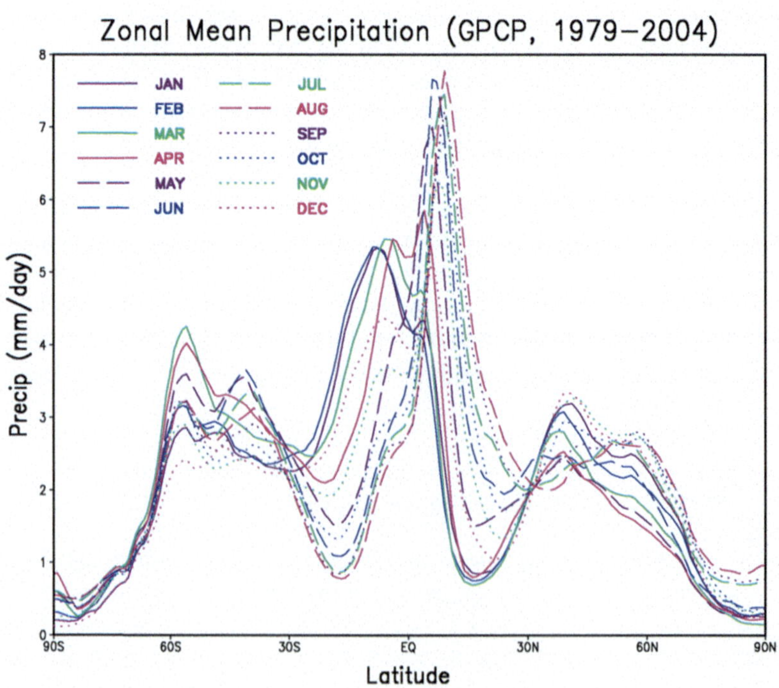

Fig. 1.4 Zonal monthly mean precipitation (GPCP, 1979–2004)

region receives more rainfall during July-Aug-Sep. This has profound influence on the water resources, river flow, and salinity of estuaries and coastal water. The precipitation maximum, which follows the solar equator through the year, is under the rising branch of the Hadley circulation; the subtropical minima are under the descending branch and cause the desert areas.

1.4 Coastal Countries: Demographic Features and Resource Use

Coasts are of great ecological and socio-economic significance; it provides livelihoods through fisheries, agriculture, mineral, bioresources, ports, tourism, and other industries. Therefore, these areas have been centres of human settlement and have become among the most populated regions particularly the tropical coastal regions. If coastal region is defined in its broader sense as 'within 100 km of a coastline', then it comprises 21% of all land but is currently occupied by 2.6 billion people which is 37% of humanity based on 2011 demographic data. If coast is considered as only a narrow natural strip along the land-marine interface, then it is home for nearly 1.36 billion world population which is 20% of all humanity, but it is just 7% of all land, at densities that average twice as great as in inland areas. This led to the urbanization and development of nearly half the world's major cities located within 50 km of a coast (MEA 2006).

The population of Latin America and the Caribbean is even more littoral. The region's coastal states have a collective population of around 610 million, full three-quarters of whom live within 200 kilometres of a coast. The majority of the Caribbean Basin's 200 million permanent residents (including over 20 million people living in 99 coastal counties along the US Gulf Coast) live on or near the seashore. The resident population is swelled every year by the influx of some 100 million tourists, nearly all of whom end up on the region's beaches. In Southeast Asia, Indonesia and Vietnam are two typical examples of Asia's population shift from the hinterlands to coastal areas. Of Indonesia's population of 200 million, 130 million live on the main island of Java, on just 7% of the country's total geographical area, most of them in rapidly growing towns and cities. Similarly majority of Vietnam's population is almost coastal. Not to left behind, over the past two decades, Africa's coastal cities — centres of trade and commerce —have been growing by 4% a year or more, drawing people inexorably out of the countryside. This demographic pattern has great impact on the coastal ecosystems, resources, and its future status. Unless proper CZM plan through legislation is not implemented, this will accelerate the coastal degradation, loss of biodiversity, and ecosystem services, all are directly linked with the demographic features of the tropical coastal areas.

Countries located in the tropics having coast are grouped according to their geographical location. Coastal countries with at least 50% of the total geographical area falling within the coastal region are considered as tropical coastal countries. Tropical countries having coast are grouped based on their continents and listed in Table 1.1. Among the continents, Africa is having more coastal areas followed by

Table 1.1 Continent-wise list of tropical coastal countries

North America	**South America**	**West Africa**
Mexico	Bolivia	Benin
Central America	Brazil	Côte d'Ivoire
Belize	Colombia	The Gambia
Costa Rica	Ecuador	Ghana
El Salvador	French Guiana	Guinea
Guatemala	Galápagos Islands	Guinea-Bissau
Honduras	Guyana	Liberia
Nicaragua	Peru	Mauritania
Panama	Suriname	Nigeria
Caribbean	Venezuela	Saint Helena
Anguilla	Suriname	São tomé and Principe
Antigua and Barbuda	**Central Africa**	Senegal
Bahamas	Angola	Sierra Leone
Barbados	Cameroon	Togo
British Virgin Islands	Congo	**Southeast Asia**
Cayman Islands	Democratic Republic of Congo	Brunei
Cuba	Equatorial Guinea	Burma (Myanmar)
Dominican Republic	Gabon	Cambodia
Grenada	**East Africa**	East Timor
Guadeloupe	Comoros	Indonesia
Haiti	Djibouti	Malaysia
Jamaica	Kenya	Philippines
Puerto Rico	Madagascar	Singapore
Saint Kits and Nevis	Mauritius	Thailand
Saint Lucia	Mayotte	Vietnam
St. Vincent and the Grenadines	Mozambique	**South Asia**
Trinidad and Tobago	Reunion	India
	Seychelles	Sri Lanka
	Somalia	Bangladesh
	Tanzania	Maldives

Southeast Asia. Among the tropical countries, Indonesia is having the largest coastal area. Predictably fisheries, tourism, and agriculture are the three principal activities providing livelihood to the coastal population and constitute almost more than 30% of the GDP of the tropical coastal countries. The contribution could be even higher for Caribbean and some Polynesian countries. In summary total coastal population and area together have profound influence on the resource use pattern and the status of coastal ecosystems. Consequently these populations are very much vulnerable to the coastal disasters, at the same time highly dependent on the services provided by the coastal ecosystems. Therefore, it is vital for these countries to conserve these valuable ecosystems for their own interest and implement climate change adaptation strategies to reduce its vulnerability.

1.5 Ecosystems and Biodiversity

The tropical coastal areas have unique, naturally variable ecosystems, including tropical rainforests, freshwater lakes and streams, salt marshes and mudflats, mangrove and coastal littoral forests, seagrass, and fringing and offshore coral reefs (SPREP 2012). The tropical region ecosystems have high species turnover and an unusual richness of endemic terrestrial and freshwater species (Kinch et al. 2010). The Pacific Islands, within the Coral Triangle region, are broadly considered the centre of highest marine biodiversity on the planet (Veron et al. 2009). Marine species richness tapers off towards the eastern islands of Polynesia, with proportionally increasing endemism in some taxa. Though biodiversity is not distributed uniformly throughout the globe, its conservation is a shared responsibility of human beings in the world. This is because the biodiversity is linked with the human wellbeing and part of ecosystems which are of human concern.

Owing to several favourable natural factors, in general, coastal ecosystems are repositories of biological diversity and provide a wide range of goods and services. These ecosystems cover large surface areas in the shallow tropical coastal seascape but have suffered from serious human accelerated degradation, especially in the last few decades. Part of their diversity, productivity, and functioning seems to be based on their juxtaposition (Michel and Pandya 2010). Ecosystem services can be broadly grouped into three types:

- Provisioning—e.g., food species, water for agricultural and industrial use, timber, fibres, fuel, and genes
- Regulating—e.g., climate regulation; influencing hydrological flows and cycles; regulation of erosion; removal of excess nutrients and wastes; and mitigation/amelioration of natural hazards such as floods, storm surges, landslides, and high winds
- Cultural and religious—e.g., recreational, aesthetic, educational, and scientific opportunities and spiritual and symbolic values

The current widespread decline of regulating services of coastal ecosystem is most worrying as, without these, the other two types of services are not possible. At the same time, anthropogenic pressures are leading to the disappearance, fragmentation, degradation, and outright destruction of habitats (UNEP 2006). Several studies have indicated that this global decrease in coastal ecosystems is known to affect at least three critical ecosystem services, viz.:

- The number of viable (noncollapsed) fisheries decreased by 33%
- The provision of nursery habitats such as oyster reefs, seagrass beds, and wetlands have declined by 69%.
- Filtering and detoxification services provided by suspension feeders, submerged vegetation, and wetlands have also decreased by 63%.

Consequently the loss of biodiversity, ecosystem functions, and coastal vegetation would also contribute to biological invasions, declining water quality, and decreased coastal protection from flooding and storm events (Braatz et al. 2007; Cochard et al. 2008; Koch 2009). Loss or reduction of ecosystem goods and services or loss of biodiversity beyond certain limits can impair the natural functioning of ecosystems. For example, sand mining of coastal areas will severely affect breeding and population of sea turtles. Reduction of ecological resilience which is the natural capacity to recover or revitalize from damage or disturbance beyond certain levels may lead to ecosystem collapse (UNEP 2006).

1.6 Natural Resources and Food Production System

Soils of the tropics are very diverse, their diversity being at least as large as that of the temperate zone. In general, the rates of organic matter decomposition are higher, and therefore it is more difficult to maintain organic matter levels in the tropical as compared to temperate soils; there is no difference in quality and effectiveness of humus in tropical and temperate soils. In the coastal areas due to deposition of soil particles, coastal alluvial soils are formed, and the river deltas are intensively cultivated. The main soil-related constraints to plant production in the tropical region (inland and elevated terrain in the coast) are high rate of weathering and leaching of nutrients resulting in soil acidity, low base saturation, and low phosphorus availability. Further in the low-lying coastal areas, water logging and salinity are the major constraints limiting agricultural production in the coastal soils. It should be noted here that these generalizations are only broad indications; a wide range in soil properties exist among soils of any given order. Furthermore, local conditions and management practices can have a significant effect on the soil's physical, chemical, and biological properties. Another important natural resource essential for the coastal ecosystems and humans is the availability of fresh water. This precious resource is affected by over-exploitation and seawater intrusion. Industrial- and tourism-related activities severely constraint the water resources and contaminate the fresh water. In the tropical region over the last two decades, nearly 30–40% of water resources are either depleted or contaminated. The condition is becoming more challenging for the tropical small island nations due to climate change and population explosion.

While overall agricultural growth is undoubtedly an effective engine for economic development and food security, the form that this growth takes has a bearing on its effectiveness to climate change adaptation. Therefore, the challenge before these countries is to identify specific agricultural development needs and opportunities. This identification and resource allocation process can be facilitated by analysing farming systems as practised in these countries in order to develop an understanding of local factors and linkages (FAO and WB 2001). These resources normally include different types of land, various water sources, and access to common property resources. Based on the difference in the resources and production systems, several farming systems can be recognized in these tropical countries with specific and general constraints. Further, these countries are also grouped into different developing regions, viz. sub-Saharan Africa, South Asia, East Asia and Pacific, Latin America, and Caribbean, to organize the analysis of farming types, constraints, climate change, diversification, and adaption options. Other than agriculture, aquaculture activities are more important which provide livelihood but seriously affect some of the natural ecosystems. Aquaculture accounts for 52% of mangrove loss globally, with shrimp farming alone accounting for 38% (Barbier and Cox 2003). More details in this regard can be seen in Chap. 15 'Alternative Farming Systems for Diversification and Conservation of Agro-biodiversity' of this book.

1.7 Climate Change and Natural Disasters

Although the tropical region is endowed with good amount of rainfall, coastal alluvium, deltas, natural resources, and diverse habitat, agro-ecosystem production is severely affected by geophysical and biological form of natural disasters due to its proximity to ocean and physiographic features. On an average worldwide, the present decade experienced 910 natural catastrophes in a year. Out of which 93% have been identified as weather-related disasters which had huge impact on the output from agro-ecosystem. An insight into the different categories of natural disasters showed that 45% were classified as meteorological in nature (storms), 36% were hydrological (floods), 12% were climatological (heat waves, droughts, etc.), and 7% were geophysical (earthquakes and volcanic eruptions). The economic loss across tropical countries by the major disasters causing the greatest damage and losses to the agriculture sector reviewed between 2003 and 2015 showed that tropical cyclones are the major type accounting for billions of dollar loss (Fig. 1.5). East African drought ($ 10.5 B) followed by Mexican flood ($ 3 B) recorded the largest economic loss in recent disaster history (FAO 2015). Further it is seen that disasters

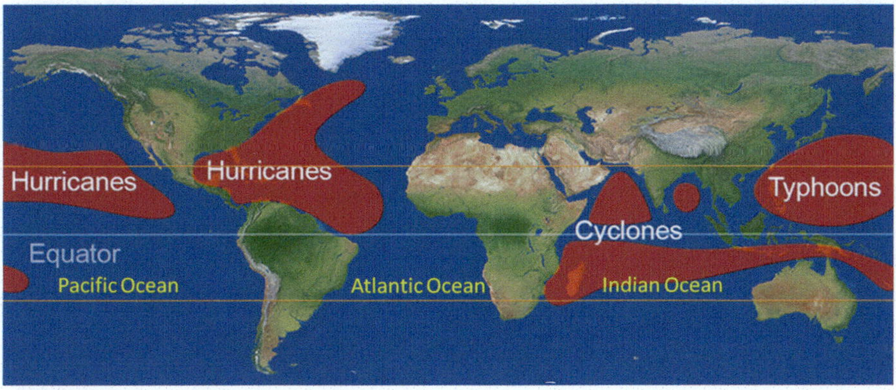

Fig. 1.5 Distribution of tropical cyclones

are normal features in almost all the tropical region (islands not visible) and are extensive and larger magnitude.

The IPCC (2007) report indicated sea level rise of 0.6 meters or more by 2100 and an increase of sea surface temperature by 3 °C, but recent work suggests that these may be underestimates. It is important to note that there is significant regional variation in the coastal impacts of sea level rise. Sea levels can also vary naturally through the geological processes of subsidence and uplift or through human processes such as extraction of water, oil, and gas. Similarly, shoreline retreat, flooding, and saltwater intrusion can occur through natural phenomena such as changes in ocean currents, transport of sediments to coasts, and wind patterns.

Climate change is projected to affect climate variability, increasing the frequency and severity of storm/tidal surges, tropical cyclones, hurricanes, etc. These phenomena also cause coastal flooding, erosion, saltwater intrusion into fresh waterways, salinization of soils, and destruction of coastal infrastructure. The effect will be exacerbated by increasing human-induced pressures on coastal areas.

By the 2080s, many millions more people than today are projected to experience floods every year due to sea level rise. In coastal areas, sea level rise will exacerbate water resource constraints due to increased salinization of groundwater supplies. The numbers affected will be largest in the densely populated and low-lying mega deltas of Asia and Africa, while small islands are especially vulnerable.

Global warming will also increase sea surface temperatures as surface waters absorb heat from higher air temperatures. Such altered temperature regimes can significantly affect the reproduction and survival of species unless they can adapt quickly enough (Nicholls and Nicholls 2008). For example, coral bleaching is associated with increase in sea surface temperature. Studies of several types of ecosystems, including coral reefs, kelp forests, and oceans, show that pressures from human activities can bring about dramatic changes where functioning is severely impaired (Scheffer et al. 2001).

The key challenge in addressing the climate threat to coastal region is timely adaptation to global warming. Nevertheless, serious socio-economic, technical, political, and ecological problems hamper the planning and implementation of adaptive strategies. The Asian and African regions of the Indian Ocean share characteristics that make them more likely to be affected by climate change. Moreover, climate change tends to disproportionately affect the more vulnerable segments of society, such as the poor and the marginalized, as they are already living in locations that have high vulnerability (Turner et al. 2003). At the same time, many of the developing countries of this region have comparatively less adaptive capacity given the speed with which climate change is taking place. The costs of adaptive responses will be highly site-specific within a country but will be greater in low-income economies in coastal zones of developing countries (Michel and Pandya 2010). Resource diversion will be a hard task as national priority is mostly towards poverty alleviation and ensuring food security.

1.8 Problems and Management of Coastal Regions

The Indian Ocean tsunami of December 2004 is a case in point. Mangroves and coral reefs buffered the impact of the waves that hit the coastlines. Assessments of the environmental vulnerabilities of coastal areas after the tsunami highlighted the importance of maintaining the integrity of coastal ecosystems. Coasts are affected by two main types of influences, terrestrial and marine, that are considered external to the coastal zone. Terrestrial influences are mostly anthropogenic in nature such as land use changes, changing hydrological regimes, and nutrient loading from sediment transport, runoff, etc. The increasing pace of human development activities in the tropical coastal region has also strained environmental resources and produced escalating economic and sociocultural impacts (Mimura 2006). In contrast, marine influences are mostly natural phenomena such as weather events (storms and cyclones), tsunamis, and wave patterns and coastal and ocean currents that affect the processes of nutrient, material, and heat transfer and mediate geomorphological changes.

These are called drivers and grouped into direct and indirect. Their impacts contribute to widespread trends on local, regional, and global scales and across different socio-economic strata. The problem of coastal region is further complicated as these drivers do not operate singly but form an interacting and often synergistic complex. Consequently the coastal ecosystems already affected by anthropogenic activities may in some cases prove unable to withstand the additional effects of climate change and may suffer irreversible loss of function and ecosystem services (Tompkins and Adger 2003). An overview of problems facing the coastal region caused by anthropogenic and natural causes is given in Fig. 1.6.

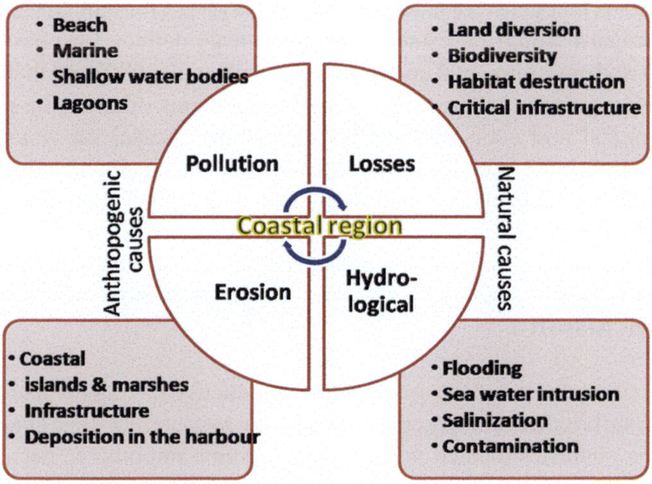

Fig. 1.6 An overview of problems facing the coastal region

In general, irrespective of causes, the problems are grouped into four major types, viz. pollution, losses, erosion, and hydrological regime change. For example, a coastal development is considered as potential hazard which is something that affects the natural environment by man-made products. It doesn't happen in isolation, but both anthropogenic and natural causes interact, get modified, and affect the coastal region which means the function and services of different ecosystems. Along with climate change events, these impacts such as increased pollution and marine traffic – all of which lead to loss of sensitive habitats and put increased pressure on coastal and marine biodiversity. This results in loss of endangered species, changing the status of species, invasive species, out-migration, etc.

Another very important problem affecting the agro-ecosystem productivity in the coastal region is salinity and its associated factors, like waterlogging and/or drought exaggerated by climate change. The increase in saline areas has been directly attributed to both water and soil salinity problems. In the coastal areas, inundation of low-lying areas by sea water and seawater intrusion into the freshwater aquifers due to over-exploitation of groundwater resources contribute to the coastal salinization.

It is evidenced from the foregoing discussion on the problems facing the coastal regions that it is becoming more and more difficult to manage any one particular coastal natural resource/ecosystem or enhance one economic sector in the absence of a comprehensive, integrated framework for policy planning and management. This necessitates the need for a comprehensive integrated coastal zone management programme to provide solution for the best long-term and sustained use of coastal natural resources and for perpetual maintenance of the most beneficial natural environment (FAO 1998b). At present available information suggest that enough is known of coastal sea ecological processes and resource values and also of human impacts, to enable the local and national governments to start integrated coastal zone management programme now to sustain the functional integrity of coastal resource systems that generate natural goods and services for human welfare. A first step in this regard would be establishing interim guidelines for coastal resource management and environmental conservation (Clark et al. 1980) and then improving upon them after gaining experiences. Other key elements of an action plan include further ecological and economic studies on valuing coastal ecosystem services, improving institutional and legal frameworks for management, controlling and regulating destructive economic activities, and developing ecological restoration options (Barbier et al. 2011).

1.9 Conclusion

Coastal regions across the globe are dynamic, productive having diverse ecosystems but highly fragile when compared to the inland systems. In this context coastal regions of the countries located in the tropics assume importance because of high productivity of its ecosystems, concentration of population, exploitation of renewable and non-renewable natural resources, pollution of coastal waters, and spurt in recreational activities. This region is highly influenced by anthropogenic and natural forces, and there is dynamic interaction between the ocean and land; as a result it is continuously changing or getting modified affecting its dependant biodiversity. Agro-ecosystems comprising primarily crop production, livestock rearing, and fisheries are highly modified systems, providing food and livelihood security to one third of tropical populations. Several regional and global-scale studies have proved that climate change has emerged as a major and global change driver that interacts with and exacerbates the impact of already existing anthropogenic drivers. Together it has become a major challenge and seriously affects several coastal ecosystems. Sometimes the changes are irreversible which triggers ecological consequences.

Efforts have been made to address these issues, but it is realized that there is paucity of comprehensive information on biodiversity and its surrounding physical environment of tropical coast although climate change is a much debated and widely documented scientific topic in the present context. Hence, preparation and implementation of integrated coastal zone management plan and suitable regulations are indispensable. This requires several spatial information on land use, land forms, sea level, coastline, ecosystems and its services, ecologically sensitive areas, human-managed production systems, infrastructures, etc. Still there are certain critical information gaps with respect to the interaction of ecosystems and its link with climate change; therefore certain uncertainty exists in our understating of the present coastal ecosystem and required adaptation measures. Under such circumstances we have little choice but to attempt to reduce the exposure to climate change risk of the critical ecosystems and the dependant populations so as to ensure its sustainable use, build climate resilience where possible, and salvage the coastal ecosystems to prevent further degradation and extinction of its unique biodiversity.

References

Barbier EB, Cox M (2003) Does economic development lead to mangrove loss? A cross-country analysis. Contemp Econ Policy 21:418–432

Barbier EB, Hacker SD, Kennedy C, Koch EW, Stier AC, Silliman BR (2011) The value of estuarine and coastal ecosystem services. Ecol Monogr 81(2):169–193

Braatz S, Fortuna S, Broadhead J, Leslie R (2007) Coastal protection in the aftermath of the Indian Ocean Tsunami. What role for forests and trees? Proceedings of the Regional Technical Workshop, Khao Lak, Thailand, 28–31, August 2006. FAO, Bangkok

Burke L, Kura Y, Kassem K, Revenga C, Spalding M, McAllister D (2001) Pilot analysis of global ecosystems: coastal ecosystems. World Resources Institute, Washington, DC

Clark JR, Banta JS, Zinn JA (1980) Coastal environmental management. Federal Instittute Admin. FLA-4, 161 p

Cochard R, Ranamukhaarachchi SL, Shivakotib GP, Shipin OV, Edwards PJ, Seeland KT (2008) The 2004 tsunami in Aceh and Southern Thailand: a review on coastal ecosystems, wave hazards and vulnerability. Perspect Plant Ecol Evol Syst 10:3–40

FAO (1998a) Issues, perspectives, policy and planning processes for integrated coastal area management. Rural development through entrepreneurship. Food and Agricultural Organization of the United Nations, Rome, Italy http://www.fao.org/docrep/ W8440e/W8440e02.htm. Accessed on 26 June 2018

FAO (1998b) Integrated coastal area management and agriculture, forestry and fisheries. FAO guidelines, Food and Agriculture Organization of the United Nations, Rome. http://www.fao.org/docrep/W8440e/W8440e00.htm. Accessed on 26 June 2018
FAO (2007) The worlds mangroves 1980–2005, FAO forestry paper 153. Food and Agricultural Organization of the United Nations, Rome
FAO (2015) The impact of natural hazards and disasters on agriculture and food security and nutrition. FAO guidelines, Food and Agriculture Organization of the United Nations, Rome, p 15
FAO and World Bank (2001) Farming systems and poverty- improving farmers' livelihoods in a changing world. FAO and World Bank, Rome
IPCC (2007) Coastal systems and low-lying areas. In: Parry et al (eds) Climate change 2007: impacts, adaptation and vulnerability, Contribution of Working Group II to the Fourth Assessment Report of the Intergovernmental Panel on Climate Change. Cambridge University Press, Cambridge, p 318
Kinch J, Anderson P, Richards E, Talouli A, Vieux C, Peteru C, Suaesi T (2010) Outlook report on the state of marine biodiversity in the Pacific Island region. Secretariat of the Pacific Regional Environment Programme, Apia
Koch EW (2009) Non-linearity in ecosystem services: temporal and spatial variability in coastal protection. Front Ecol Environ 7:29–37
MEA (2005) Ecosystems and human well-being: current state and trends. Coastal systems, Millennium Ecosystem Assessment. Island Press, Washington, DC
MEA (2006) Marine and coastal ecosystems and human well-being: synthesis, millennium ecosystem assessment, United Nations Environment Programme, p 40
Michel D, Pandya A (2010) Coastal zones and climate change. The Henry L. Stimson Center, Washington, DC. 20036
Mimura N (2006) State of the environment in the Asia and Pacific coastal zones and effects of global change. In: Harvey N (ed) Global change and integrated coastal management: The Asia-Pacific Region. Springer, Dordrecht, pp 17–38
NGE (2017) Tropics, national geographic encyclopedia. National geographic society. https://www.nationalgeographic.org/encyclopedia/tropics/. Retrieved 2017-06-26
Nicholls H, Nicholls R (2008) Global sea-level rise and coastal vulnerability. Sustain Sci 3(1):1–33
Scheffer M, Carpenter S, Foley JA, Folke C, Walkerk B (2001) Catastrophic shifts in ecosystems. Nature 413:591–596
Tompkins EL, Adger WN (2003) Building resilience to climate change through adaptive management of natural resources, Tyndall Center Working Paper No. 27. University of East Anglia, Norwich
Turner BL, Kasperson RE, Matson PA, McCarthy JJ, Corell RW et al (2003) A framework for vulnerability analysis in sustainability science. Proc Natl Acad Sci 100(14):8074–8079. https://doi.org/10.1073/pnas.1231335100
UNEP (2006) Marine and coastal ecosystems and human wellbeing: a synthesis report based on the findings of the Millennium Ecosystem Assessment. United Nations Environment Programme, Nairobi
Valiela I, Bowen JL, York JK (2001) Mangrove forests: one of the world's threatened major tropical environments. Bioscience 51:807–815
Veron JEN, DeVantier LM, Turak E, Green AL, Kininmonth S, Stafford-Smith M, Peterson N (2009) Delineating the coral triangle. Galaxea 11:91–100

Coastal Ecosystems and Services

2

Abstract

Coastal areas occupy one of the most dynamic interfaces on Earth, at the boundary between land and sea, and they support some of the most diverse and productive habitats. The physical environment of the coastal region is highly variable within relatively small spatial extent. The interactions between biological component and their surroundings are more dynamic through flow of matter and energy which makes the coastal ecosystems very unique than the terrestrial ecosystems. The most important coastal ecosystems are mangrove forests, coral reefs, other wetlands, seagrass beds, marshes, rocky and sandy beaches, estuarine, and tidal flats which also provide essential nursery and feeding areas for many coastal and oceanic aquatic species. These ecosystems provide provisioning, protective, and regulating services. Besides these regions are highly influenced by the presence of agro-ecosystem which is highly managed and aimed at providing economic benefit and livelihood to coastal communities. The coastal ecosystems are affected by terrestrial and marine influences. Loss or reduction of ecosystem goods and services or loss of biodiversity beyond certain limits can impair the natural functioning of ecosystems. Therefore, it is vital to conserve these valuable ecosystems and should be managed for sustainable economic benefit.

Keywords

Coastal habitats · Marsh · Seagrass · Agro-ecosystem · Economic value · Ecosystem services

V. Ayyam et al., *Coastal Ecosystems of the Tropics - Adaptive Management*,
https://doi.org/10.1007/978-981-13-8926-9_2

2.1 Introduction

The interface or transitional area between land and sea is known as coastal areas. Such interface may be gradual or with sharp boundaries which largely depends on the nature of land form, type of coast among other features. In coastal areas land and water meet each other to create an environment with a distinct structure, diversity, and flow of energy. Such a diverse environment, land form, and coast geomorphology besides climatic factors manifest in the form of diverse but most complex coastal ecological systems on Earth. The major ecosystems include mangroves, wetlands, corals, marshes, seagrass, estuaries, bays, etc., which are home to different types of plants and animals and provide valuable services. As a result coastal areas have become centres of human economic activity with ever-increasing population density. Coastal ecosystems are not only economically valuable for agriculture, fisheries, forest-based products, commerce, industry, navigation, and recreation but also include some of the most ecologically important environments which provide valuable ecosystem services. As they are natural systems, it tends to establish as a contiguous patch in any favourable conditions until altered by human activities or natural phenomena.

Although the coastal areas are interface zone, for management purposes, a variety of landward and seaward boundaries, ranging from fairly narrow and precise ones to much broader and more indefinable, are in practice around the world. This means that boundaries of a coastal area may change over time as the issues to be addressed become more extensive or complex. Moreover, the natural ecosystems have its own boundary irrespective of human interests. Therefore, in order to have clear understanding of the coastal region and the ecosystems, distinction should be made between the terms 'coastal zone' and 'coastal area'. The term 'coastal zone' would refer to the geographic area defined by the enabling legislation for coastal management, while 'coastal area' would be used more broadly to refer to the geographic area along the coast which are natural landscape (FAO 1998). It is deliberated in several international conventions and agreed by all coastal countries that coastal region includes inshore waters, intertidal areas, and extensive tracts of contiguous land (Chua 1986; Clark 1992; Boelaert-Suominen and Cullinan 1994). Therefore all the natural systems occurring in these areas are considered as coastal ecosystems.

In this context, defining and delineating the boundaries of coastal ecosystem in its natural extent rather than fixed distance from the coastline are very essential to conserve them, derive continued benefit, and continue the scientific study to understand the effect of human activities and global changes. This is more important because most of the coastal ecosystems are very sensitive to changes in the environment, and there is concern that some areas are now struggling to maintain their diversity due to human activity, the introduction of non-native species, and other factors. With these backgrounds this chapter attempts to highlight different tropical coastal ecosystems with reference to its nature, function, and services.

2.2 Coastal Ecosystems

In simple terms an ecosystem is all the organisms living in a place together and interacting with their environment and among themselves. In other words ecosystem has living and nonliving components interact through flow of matter such as water, nutrient cycles, and energy. An ecosystem can be huge, such as a large forest, or it can be small, such as a mud puddle or a single bush. In the context of coastal ecosystem, it is the community of interacting organisms and their physical environment in the coastal areas. The physical environment is highly variable within relatively small spatial extent. For example, in the coastal areas from highly saline shallow water to terrestrial condition, it is available within a distance of 25 m which may be interspaced with rock shore and sandy beaches. This is one of the reasons for more dynamic interactions through flow of matter and energy which make the coastal ecosystems very unique than the terrestrial ecosystems.

2.2.1 Nature of Coastal Environment

There is no single, comprehensive definition for coastal zone as it is practically variable and changes with the management objectives. Some authors have referred to it as 'that part of the land most affected by its proximity to the sea and that part of the ocean most affected by its proximity to the land' (Hinrichsen 1998). The PAGE study defined coastal regions to be the intertidal and subtidal areas on and above the continental shelf to a depth of 200-meter areas routinely inundated by saltwater and immediately adjacent lands (Burke et al. 2001). The Millennium Ecosystem Assessment (2005a, b) defines coastal region using distance from the coast and elevation. According to this the inland extent of coastal ecosystems is where land-based influences dominate up to a maximum of 100 km from the coastline or 50-m elevation whichever is closer to the sea and with the outward extent as the 50-m depth contour. Further the marine ecosystems begin at the low-water mark and encompass the high seas and deepwater habitats.

The world's coastal regions are subdivided by physical rather than only on biological characteristics; however it includes a wide array of terrestrial, nearshore, intertidal, benthic, and pelagic marine environments. In these environments salinity, depth of water, and depth of penetration of sunlight significantly vary particularly in the wet part of the coastal region. Such diverse habitats often coexist and are dynamic systems; therefore, it is difficult to identify exact locations and extent or delineate clear boundaries between them. Examples of such communities are shown in Table 2.1. In all the definitions of coastal areas, one thing is common, that is, coastal ecosystems are areas where land and water join to create an environment with a distinct structure, diversity, and flow of energy. They include mangroves, wetlands, salt marshes, estuaries, and bays and are home to many different types of plants and animals (Michel and Pandya 2010). In other words, this region consists of several important and distinctive ecosystems as it is located in the interface between marine and terrestrial which has characteristics of climatic regime. But

Table 2.1 Coastal environment and habitats

Coastal environment	Characters of different coastal habitats
Terrestrial	Land-based process dominates, abundant sunlight, moist to waterlogged condition
	Dunes, cliffs, beach, freshwater wetland, river floodplains, agricultural landscape, urban
Nearshore	Interface of land and sea, abundant sunlight, more often under the influence sea waves
	Rocky shore, sandy shore
Intertidal	This is intertidal zone where periodical flushing by sea waves occur, restricted sunlight below the water level
	Estuaries, mangroves, deltas, lagoons, mudflats, salt marshes, salt pans, saltwater wetland, marinas, aquaculture beds
Benthic	This is soft-bottom environments above the continental shelf, restricted sunlight below the water level, deposition of materials eroded from the land
	Seagrass beds, kelp forests, and coral reefs
Pelagic	Open waters above the continental shelf, sunlight limits the growth of photosynthetic organism at depth, salinity and sea surface temperature dominate the distribution of organism
	Freestanding fish and shoals, plankton blooms

habitat is a smaller proportion when compared to the ecosystem as the habitat of an organism is the place where it lives within an ecosystem. Several populations can share the same habitat and it provides food, water, shelter, and space.

2.2.2 Recognition of Different Coastal Ecosystems

As discussed earlier, the coastal region occupies one of the most dynamic interfaces on Earth, at the boundary between land and sea. It has distinct climate regime, level of anthropogenic activities, and ecosystem richness, among other factors. The habitats and features along the world's coastline are highly varied—from the flat, coastal plains to the mangrove and coral reef-lined shores, to the rugged, rocky coastline. Thus the coastal areas tend to have unique ecosystems, support a rich biological diversity and productivity and contain a valuable assortment of natural resources. Examples of such habitats are coastal mangrove forests and other wetlands, rocky and sandy beaches, estuarine, coral reefs, tidal flats, and seagrass beds. These habitats provide essential nursery and feeding areas for many coastal, oceanic, and other aquatic species (FAO 1998). This necessitates delineation of these ecosystems with descriptive attributes to provide baseline information and reference points for assessing the condition of the ecosystem's goods and services.

In practice, delineation/recognition of different ecosystems is rather difficult task than simply describing them based on certain conditions. Because coastal ecosystems can encompass a wide range of environmental conditions over short distance particularly salinity (from fresh to hypersaline) and energy (from sheltered wetlands to energetic wave-washed shorelines). At a much coarser geographical scale, there is a spectrum of climate types—from tropical to polar—with concomitant broad-scale differences in biogeophysical processes and features. They are also a major factor in the vulnerability and resilience of an area to a particular pressure either natural or anthropogenic. As a result, the distinct physical environment, diverse biological systems, their complex interactions, and dynamic flow of energy and matter make the coastal ecosystems unique and dynamic. Based on these considerations, mangroves, seagrass meadows, lagoons, estuaries, rock shore, sandy beach, flats, and salt marshes are identified as major coastal ecosystems. Agro-ecosystem is also a part of coastal areas, but as it is completely under the human management, it is dealt separately. Nevertheless, agro-ecosystem also comes under the influence of coastal processes and hazards. Similarly the runoff from the agricultural field and human activities within agro-ecosystem affect other coastal ecosystems. Further the presence and consideration of coastal communities as part of coastal systems (human element) make the system more dynamic and intimately linked.

2.3 Ecosystem Services

Coastal ecosystems occur where the land meets the sea, that is, the interface region. They extend along more than 1.6 million km of coastline in the coastal countries. These ecosystems include a diverse set of habitat types, encompassing both terrestrial and marine habitat. Coastal areas are home to approximately one third of the world's population, and almost 40% of the world lives within 100 km of the coast. Therefore, it is important to recognize the many different values of the ecosystems which in turn help to understand the role of ecosystems in climate change adaptation.

Ecosystem services, in the context of coastal ecosystem, are the benefits that the society receives from the ecosystems such as regulatory, provisioning, and cultural services (Fig. 2.1). They provide a pragmatic framework for managing ecological sustainability and strongly influence human wellbeing. Ecosystems provide provisioning services such as food, water, fuel wood, forage, timber, medicine, etc. For example, coastal ecosystems support many of the world's poorest communities, who rely on the provisioning services of these systems for their food supply and livelihoods. Coral reefs, mangroves, and other ecosystems are important for fisheries and fish nurseries, which provide people with a key source of protein as well as livelihood opportunities. Regulatory services are other very important benefits of ecosystems which include recycling of water and chemicals, mitigation of floods, pollination of crops, and cleansing of the atmosphere. It also provides cultural services such as recreation, tourism, and aesthetic and spiritual benefits (MEA 2005a, b).

All of these services depend on ecosystem processes that are sometimes known as supporting services. Supporting services are the fundamental ecological processes that control the structure and functioning of ecosystems. These processes include biogeochemical cycles, diversity maintenance, and disturbance cycles. It is now

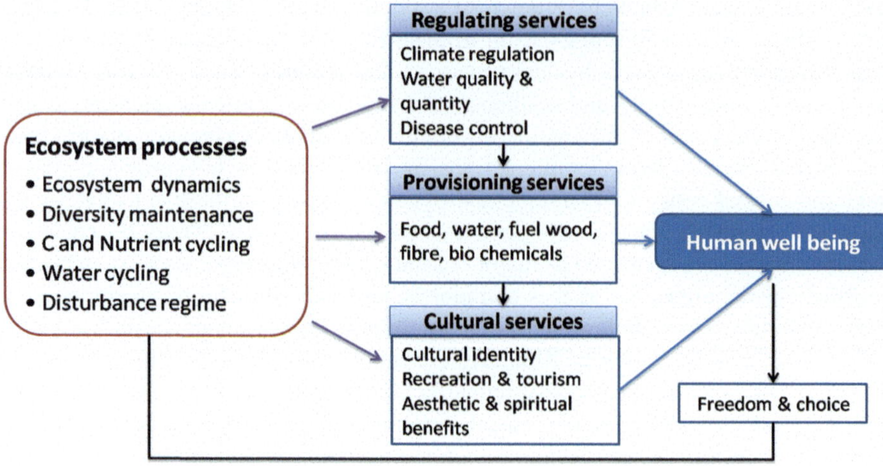

Fig. 2.1 Relationship among ecosystem process, services, and human wellbeing

clear that unless these underlying properties are properly maintained, other services that are more directly recognized and valued by the society cannot be sustained.

Valuation and importance of ecosystem services arise because of overuse or misuse of resources that alter the functioning of ecosystems and the services they provide. The extent and loss of these natural habitat types serve as a proxy condition indicator for many of the ecosystem services and values that are otherwise difficult to quantify. Land use change can reduce the amount of water infiltrating into the groundwater or results in water pollution. Degradation and loss of wetlands can expose communities to increase damage from floods and storm surges (Kates et al. 2006). Introduction and invasion of non-native species due to human activities cause enormous damage to living resources and threaten human wellbeing as well. Human activities also affect ecosystem services through changes in the atmosphere, hydrological systems, and climate. At the same time, management decisions often involve choices that reflect trade-offs among ecosystem services. An important step in ecosystem management is to assess potential impacts of decisions on multiple ecosystem services. This is most challenging task, given the huge number of services provided by ecosystems and uncertainties in their responses to a particular action.

2.4 Major Coastal Ecosystem Types

The coastal ecosystems display important characteristics specific to the region in spite of some common types of stresses exerted on coastal regions in the tropics. Though there are several studies on the interaction of biological system with its surroundings, still a lot many things are not documented especially in coastal and marine environments. At many places organisms and the environmental factors that determine their survival are distributed as gradients that blend into one another at the edges of the space they occupy. Further many indirect effects of global change on ecosystem functioning will unfold themselves gradually rather than as readily apparent losses of ecosystem integrity. For these reasons the characteristics of ecosystems vary over time and space. These aspects are manifested in various coastal ecosystems.

Any ecosystem study is incomplete without the inclusion of human component. Humans are part of the ecosystem of any place on this planet. They and their constructed systems/managed ecosystem have to be included in any analysis of ecosystem. Therefore, agro-ecosystems specific to the coastal areas are also discussed in this chapter. The major coastal ecosystems, their characters, and significance are discussed in the following subsections.

2.4.1 Mangrove Ecosystem

2.4.1.1 Structure and Composition

Mangrove ecosystem represents one of the most productive natural systems (plant communities and marshy conditions) capable of producing a wide range of goods and services for coastal environments and society as a whole. In general, mangrove denotes a group of salt-tolerant plant species that occur in the tropical and subtropical intertidal estuarine regions, sheltered coastlines, and creeks (Fig. 2.2). Mangrove communities comprise both biotic and abiotic components. The biotic component includes plants, animals, and microbial organisms that are highly adapted to intertidal environmental conditions (Jayatissa et al. 2002). Generally two categories of plants are recognized in mangroves, namely, 'plants which are restricted to mangrove habitats' (tropical intertidal habitats) and 'plants which are not restricted to mangrove habitats' (Polidoro et al., 2010). These are known as 'true mangroves' and 'mangrove associates', respectively.

Often it is observed that mangrove vegetation is gradually merged with salt marsh, seashore, freshwater marsh, or other terrestrial vegetations depending on the terrain. Globally, a total of 69 species in 27 genera belonging to 20 families are

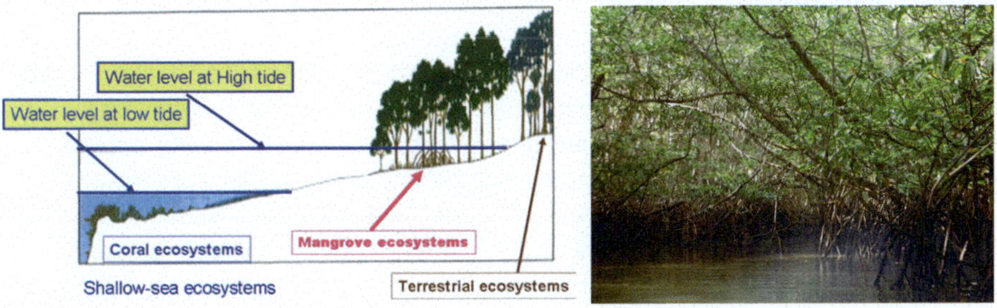

Fig. 2.2 Occurrence of mangroves in the intertidal zones

Fig. 2.3 Mangrove communities in the Sundarbans delta

considered as true mangrove species (Duke 1992; Kathiresan and Bringham 2001). Major families are Rhizophoraceae, Avicenniaceae, Acanthaceae, Meliaceae, and Sonneratiaceae. True mangroves generally possess some special adaptations and are exclusively seen in mangrove areas. They are morphologically, physiologically, and reproductively adapted to saline, waterlogged, and anaerobic conditions. Morphologically, true mangroves are sclerophyllous broadleaved halophytic plants with some special adaptations like specialized roots (stilt roots and pneumatophores), viviparous germination, etc. Sometimes it is very difficult to separate the plant species and habitat as in the case of Sundarbans (India and Bangladesh coast) where mangrove flourishes in small islands interspersed with sea water of varying salinities (Fig. 2.3).

2.4.1.2 Distribution

In general, mangroves are found in the intertidal region between 30° North and South of the equator with some notable extensions. About 112 countries and territories have mangroves within their borders. An estimate by Spalding et al. (1997) concluded that mangroves are distributed along approximately one-quarter of the world's tropical coastlines, covering an estimated surface area of 181,000 km^2. Four countries, viz. Indonesia, Brazil, Nigeria, and Australia, account for about 41% of all mangroves showing skewed distribution. The comprehensive assessment of mangrove area worldwide has now fallen below 15 million hectares, down from 19.8 million ha in 1980 (FAO 2003). From the available resources, it is concluded that nearly 50% of the world's mangrove forests have been lost (Kelleher et al. 1995). Indeed, extensive losses from the original distribution, particularly in the last 50 years, include an estimated 83.7% of mangroves in Thailand and 67% in Panama during the 1980s. Although the net trend is clearly downward, in some regions, mangrove area is actually increasing as a result of plantation forestry and small amounts of natural regeneration (Spalding et al. 1997).

2.4.1.3 Significance of Mangrove Ecosystem

Mangrove habitats exhibit community assemblage of specific species among mangroves and associations with some non-mangrove species as well. Mangrove is a unique environment linking land and the sea and sustains ecosystems containing a variety of plants, providing habitat for a wide range of flora and fauna. Most importantly mangroves provide habitats for a large number of molluscs, crustaceans, birds, insects, monkeys, and reptiles. It is also rich in associated flora occurring in difficult conditions which includes bacteria, fungi, algae, seagrasses, lichen, and other vegetation. In general the significance of mangrove ecosystem is grouped into three categories based on the basic services it offers, viz. provisioning, protective, and ecosystem services (Fig. 2.4). Variety of fishes, shell fish, prawns, crustaceans, and other marine species are abundantly found in the mangrove ecosystem which are source of food for many including coastal communities. Mangrove roots trap the sediments washed from the land. There are specific microbial communities living in the saline mangrove sediments which help in decomposition and release of nutrients. Mangroves help in coastal protection acting as barriers against devastating tropical storms and waves. Mixed mangrove species along with other measures in the

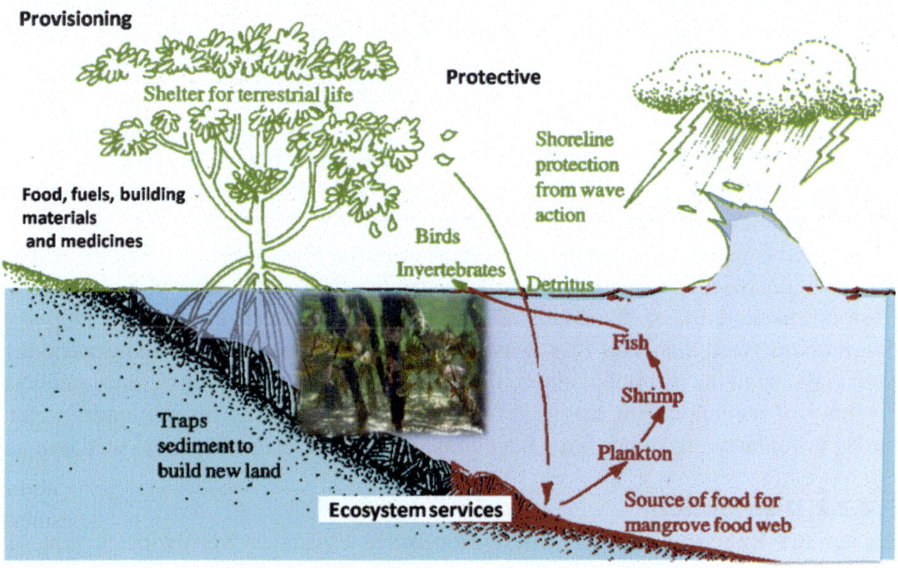

Fig. 2.4 Significance of mangrove ecosystem

seafront acts as a bio-shield against the surging sea waves and tsunamis. In addition, mangroves have tremendous ecological and social values. Other mangrove ecosystem services include the filtering and trapping of pollutants and the stabilization of coastal land by trapping sediment.

2.4.2 Wetlands (Other Than Mangroves)

Unlike mangroves, other wetland types (marshes, swamps, and peatlands) are less clearly defined and documented due to various reasons. Ramsar Convention of 1971 defines the term wetlands as 'areas of marsh, fen, peat land or water, whether natural or artificial; standing or flowing; fresh, brackish or salt, including areas of marine water, the depth of which at low tide does not exceed six meters' (Navid 1989). In addition, it is difficult to distinguish *coastal* wetlands from *freshwater* wetlands without a combination of remote sensing and ground-based survey. A broad definition of wetlands used by Ramsar Convention, and which is internationally accepted, also encompasses reef flats and seagrass beds in coastal waters (Davies and Claridge 1993). More details on wetland ecosystem is provided separately (Chap. 6) in this book.

2.4.2.1 Wetland Environment

Study of geological and biogeographical history of our planet showed that wetlands had continuously evolved in time and space. Coastal wetlands developed along passive margin coasts with low-gradient coastal plains and wide continental shelves. At several places along the tropical coast, it has been observed that the combination of low hydraulic energy and gentle slope normally provides an ideal setting for the development of wetlands. The formation of low-lying coastal plains has been in response to the pattern of relative sea level change. In general, as per the internationally accepted definition of wetlands, it must have one or more of the three attributes (Cowardin et al. 1979), viz.: (i) the land should support predominantly hydrophytes at least periodically, (ii) the substrate should be predominantly undrained hydric soil, and (iii) the substrate should be saturated with water or covered by shallow water.

Wetland ecosystems are some of the most important biodiverse areas in the world and provide essential habitats for many species (Fig. 2.5). The global Ramsar Convention network of 'Wetlands of International Importance' strongly supports unique biodiversity, species (e.g. water birds, amphibians, and wetland-dependant mammals such as hippopotamus, manatees, and river dolphins), and genetic diversity of wetlands. Among the different species inhabiting the wetlands, water birds particularly migratory birds have more striking feature of tropical wetlands.

2.4.2.2 Distribution

As per the Ramsar Convention definition-based estimates worldwide, extent of wetlands is 12.8 million km^2. Coastal wetlands are more extensive in the wetter parts of the tropics. However, Mitsch and Gosselink (2007) estimated the extent of

Fig. 2.5 Coastal wetlands with high biodiversity

the world's wetlands slightly low at seven to ten million km^2 (about 5–8% of land surface area and 1.37–1.96% of the total surface area of Earth). There is lack of comprehensive global information to document historical changes in coastal wetlands. Where national data do exist, however, the habitat loss is often very high. For example, some 46% of Indonesia's peat swamps and as much as 98% of Vietnam's are believed to have been lost (Mackinnon 1997).

2.4.2.3 Significance of Wetland Ecosystem

The Millennium Ecosystem Assessment (MEA 2005a, b) states that the ecosystem goods and services provided by wetlands are extremely valuable to people all over the world. Wetlands harbour a large biodiversity highly disproportionate to their areal extent. They provide ecosystem services, viz. provisioning, regulating, supportive, and cultural, that are critical to the entire life on the Earth. These ecosystem services result from the interactions between different biodiversity components and their abiotic variables as noted earlier (Fig. 2.6). Another most important service provided by wetlands is related to water-based ecosystem services, such as food (fish, prawn, rice, and many other plants), wastewater purification, hydrological regulation of floods and droughts, carbon sequestration and climate regulation, storm protection, and erosion control. Wetlands enhance aesthetics and support a wide range of livelihoods besides various cultural/recreational activities (Gopal 2015). In addition, wetlands hold important spiritual values for some cultures particularly in the tropics.

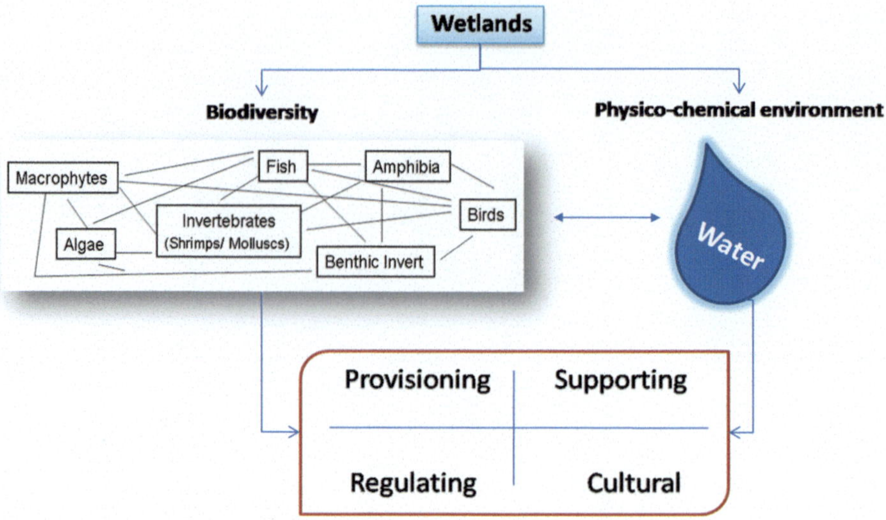

Fig. 2.6 Ecological significance of wetland ecosystems

2.4.3 Coral Reef Ecosystem

2.4.3.1 Ecosystem Components

In coral reef ecosystem, corals build reef in shallow waters which are inhabitated by large variety of marine biota having a close relationship among themselves and the surroundings. Corals are two-layered, tiny invertebrate animal called *polyps* that live in groups and are related to jellyfish and sea anemones. Corals belong to the phylum Coelenterata which is primitive, occurring from low tidemark to up to a depth of 6000 m. In coral ecosystem each polyp is like a fluid-filled bag with a ring of tentacles surrounding its mouth and looks like a tiny anemone (Fig. 2.7). The individual polyps within a colony are linked by living tissues and can share their food with other individuals (Allen and Steene 1994). In some corals, the polyp extracts calcium carbonate from the sea and secretes it as a cup of calcium carbonate from the bottom half of its body. These cups provide anchorage for the polyps, but when threatened, the polyp can retreat into the safety of the hard cup. When the calcium carbonate cups of many billions of these polyps fuse together, they form coral reefs (Veron 2000). This limestone structure sometimes goes up to 1300-m thick from the surface to its base on volcanic rock (Enewetak Atoll) or sometimes more than 2000-km long (Great Barrier Reef).

There are two types of corals, viz. reef forming corals called ‘hermatypic’ and non-reef-building corals known as ‘ahermatypic’. The hermatypic corals mostly live in shallower waters and depths less than 30 m. They are having their ability to resist hydrodynamic stretch during their rising up to the surface of the sea. They also have symbiotic zooxanthellae in their endodermal cells (gastrodermal cells) which help to segregate higher amount of calcium carbonate. Ahermatypic lacks

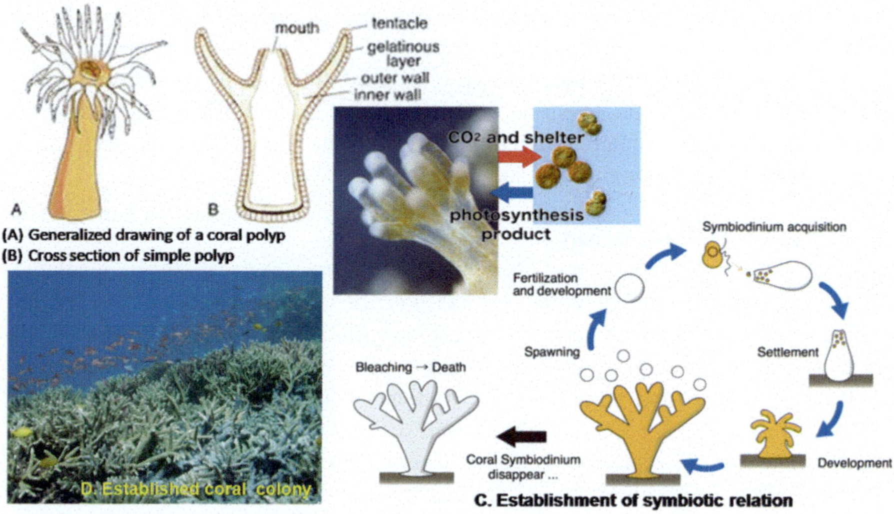

Fig. 2.7 Coral polyp, symbiotic relationship, and established coral

zooxanthellae in their tissue. They are smaller in size, grow more slowly, and do not provide significant contribution to the reef formation (Spalding et al. 2001). However, there are exceptions, available for both the classifications. Many solitary ahermatypic corals (*Fungia*, *Diaseris*, *Cycloseries*, and *Heteropsammia*) live in shallow waters and contain zooxanthellae but does not helping for reef-building mechanisms. Similarly, hermatypes (*Tubastrea micranthus*) without zooxanthellae contribute significantly for the formation of reefs. The hermatypic corals thrive mostly in the intertropical Indo-Pacific regions where the waters are clean in nature with light, temperature, salinity, and nutrient level.

The more prominent members of the coral reef community and associated tropical marine habitats are as follows: sponges, scleractinian or stony corals, soft corals (alcyonaceans), sea fans, sea whips (gorgonaceans), Polychaeta worms, wide varieties of bivalve and gastropod molluscs, octopus, squid, spiny lobsters, crabs, sea urchins, sea stars, sea cucumbers, crinoids, fishes, and brittle stars. The bacterial and viral diseases are very common among these communities.

2.4.3.2 Coral Reef Environment

The coral reef environment is dynamic and highly productive ecosystem. It occurs in shallow water to the extent of sunlight penetration. Although the salinity of sea water varies significantly, coral reef can't tolerate much of sediment load (turbidity) which reduces light penetration. The reef system alone produces limestone at the rate of 400–2000 tons per hectare per year (Chave et al. 1972). The amount of deposition of calcium carbonate in this process has been estimated roughly half the calcium that enters the sea throughout the world in a year (Smith 1978). Each deposited atom of calcium would fix a molecule of CO_2 and resultant of gross fixation of CO_2 on the order of 700 billion kg carbon per year (Birkeland 1996). The balance between deposition and destruction of calcium is mainly carried over by the interaction of physical and biological controls. The biotic mainly concerns with production of calcium carbonate through biological activities of coral polyp which in turn depends upon photosynthetic activities by zooxanthellae. Similarly the destruction produced in calcium carbonates has been dissolved due to the respiration which releases the carbon dioxide and increases the pH.

2.4.3.3 Global Distribution

Information on the extent and distribution of coral reefs is fairly available. Worldwide, there are an estimated 255,000 km^2 of shallow coral reefs, with more than 90% of that area in the Indo-Pacific region (Spalding and Grenfell 1997). Only less than 8% of the world's reefs are found in the Caribbean and Atlantic. Among the tropical islands and island nations, Indonesia and the Philippines constitute 76,000 km^2 reef areas which are almost half of the reef area in the tropical region. In general, coral reef degradation is a more serious problem than outright reduction in coral reef area on a global basis. The reef area has been significantly reduced in some parts of the world through land reclamation and coral mining. Additionally, as increasing numbers of coral reefs become weakened from coral bleaching, coral diseases, and other stresses caused by human activities and global warming, mortality is likely to increase. When reefs do not recover, the reef will eventually erode and as a result there will be a loss in coral reef area. However, these changes can be documented with and quantified only if the changes occur in large-scale means at measurable scale.

2.4.3.4 Significance

Coral ecosystems not only provide economic benefit by various activities, but its environmental protection service is highly valuable. It has bio-shield characters which protect the coastal areas. The gigantic structures of limestone are called coral reefs have a very thin layer of living organic materials. This thin layer produced the inorganic materials which accumulate over a time period and form the reefs. The most important benefits of coral ecosystems are:

- Being the most biologically productive ecosystems, it has high potential for fisheries. Out of 100 million tons of fisheries production in the world, nine million tons per year of fishes arrived from the coral reef environment (Smith 1978; Munro 1984). Tens of millions of people living in the tropics dependent on their livelihoods on coral reef environment (Salvat 1992). Number of small-scale fishery operation dependent on only coral reef fisheries harvest (Wells and Hanna 1992).
- Lime is extracted from coral and used for construction and other purposes. However, large-scale mining of coral/over-exploitation is causing damage to the entire coral reef.
- The coral reef has greatest diversity per hectare of any ecosystem in the ocean (Fig. 2.8). It is the habitat for innumerable reef fishes and other marine life. It provides protection to juveniles.

Lagoon and barrier reef, Lakshadweep, India — Branching coral — Anemone fish in coral surroundings — Star fish and coral biodiversity

Fig. 2.8 Coral ecosystem in tropical waters

- Coral reef protects the coast and islands and in some places islands are formed over the coral bed.
- The shallow water corals also fix the carbon through zooxanthella which is known as blue carbon and thus help to reduce atmospheric carbon level.

2.4.4 Seagrass Ecosystem

2.4.4.1 Seagrass Environment

Seagrasses are type of submerged aquatic vegetation evolved from terrestrial plants over a period of time. These plants have become specialized to live in the marine environment with specific adaptations. Like terrestrial plants, seagrasses also have leaves, roots, conducting tissues, flowers, and seeds and manufacture their own food by photosynthesis (Fig. 2.9). However, unlike terrestrial plants, seagrasses do not possess the strong, supportive stems and trunks required to overcome the force of gravity on land. Alternatively seagrass blades are supported by the natural buoyancy of water, remaining flexible when exposed to waves and currents (Diane 1986). Seagrasses are different from sea macroalgae which strongly attaches with the hard substratum, whereas seagrasses possess true roots that help the plant to hold in place and also specialized for extracting minerals and other nutrients from the sediment

Fig. 2.9 Seagrass bed in the tropical ocean

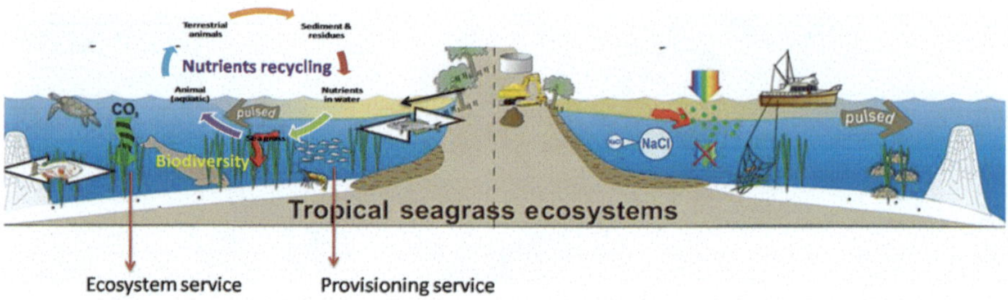

Fig. 2.10 Schematic representation of tropical seagrass ecosystems

(Fletcher and Fletcher 1995). It is not found in intertidal sediments where normally mangroves will occupy.

Seagrass has very high rate of carbon assimilation and thereby total biomass production. *Halodule wrightii* has an estimated annual production (as measured in grams of carbon per square meter) of 182–730 g/C/m^{-2}; *Syringodium filiforme* has an estimated annual production of 292–1095 g/C/m^{-2}; and *Thalassia testudinum* has an estimated annual production of 329–5840 g/C/m^{-2}. Seagrass has very good blade elongation which will support its high photosynthetic ability. The blade elongation in seagrasses averages 2–5 mm per day in *Thalassia testudinum*, 8.5 mm in *Syringodium filiforme*, and as much as 3.1 mm in *Halodule wrightii*. In the Indian River Lagoon, *Halodule wrightii* has been shown to produce one new leaf every 9 days during spring—the season of highest productivity (Robert 1982).

Seagrass habitat offers food, shelter, and essential nursery areas to commercial and recreational fishery species and to the myriad invertebrates that are produced within or migrate to seagrasses. The complexity of seagrass habitat gets increased when several species of seagrasses grow together, their leaves concealing juvenile fish, smaller finfish, and benthic invertebrates such as crustaceans, bivalves, echinoderms, and other groups. Juvenile stages of many fish species spend their early days in the relative safety and protective environment created by seagrasses in the tropical waters. Schematic representation of tropical seagrass ecosystems is given in Fig. 2.10. Additionally, seagrasses provide both habitat and protection to the faunal organisms living within the substratum as seagrass rhizomes intermingle to form dense networks of underground runners that deter predators from digging in faunal prey from the substratum.

Seagrass meadows also help dampen the effects of strong currents, providing protection to fish and invertebrates while also preventing the scouring of bottom areas. Finally, seagrasses provide attachment sites to small macroalgae and epiphytic organisms such as sponges, bryozoans, forams, and other taxa that use seagrasses as habitat. A number of studies have found epiphytes to be highly productive components of seagrass habitats (Tomasko and Lapointe 1991), with epiphytes in some systems accounting for as high as 30% of ecosystem productivity and more than 30% of the total above ground biomass (Morgan and Kitting 1984, Heijs 1984). Seagrass epiphytes also contribute to food webs, either directly via organisms grazing on seagrasses or indirectly following the deaths of epiphytes, which then enter the food web as a detritus carbon source (Kitting et al. 1984).

2.4.4.2 Distribution

It was difficult to estimate the worldwide extent and changes in seagrass beds till the advent of modern tools to map this subtidal ecosystem. Historically, most seagrass habitat loss has been the result of degrading water quality primarily caused by high nutrient and sediment loadings caused by human activities (NOAA 1999). Direct damage from vessels, dredging, and trawling are other activities that have significantly harmed many seagrass beds in the tropical water. Compilation of data from the available resources and other indirect methods showed that the magnitude of loss in this ecosystem is believed to be high (Global Seagrass Survey 1999).

2.4.4.3 Significance

Seagrass ecosystem is very unique and vital for the tropical coastal areas. It performs a variety of functions within ecosystems and has tremendous economic and ecological values. The high level of productivity, structural complexity, and biodiversity of tropical seagrass beds is recorded by several researchers. The ecosystem services, processes, and functions for seagrass ecosystem are given in Table 2.2. Within seagrass communities, a single acre of seagrass can produce over 10 tons of leaves per year. This vast biomass provides food, habitat, and nursery areas for a myriad of adult and juvenile vertebrates and invertebrates. For example, in the western part of Indian Ocean, abundant and widespread fish species are associated with seagrass beds which belong to the families Apogonidae, Blenniidae, Centriscidae, Gerreidae, Gobiidae, Labridae, Lethrinidae, Lutjanidae, Monacanthidae, Scaridae, Scorpaenidae, Siganidae, Syngnathidae, and Teraponidae. Studies revealed that seagrass beds support a variety of benthic, demersal, and pelagic organisms including many fish and shellfish species (Gullström et al. 2002). Seagrass roots bind the bottom strata and prevent from further disturbance and take CO_2 from the water during photosynthesis, thereby rendering valuable service to the other marine organisms. Further, seagrasses are vital for the survival of several endangered marine mammal species and other large faunal communities. Higher sediments and nutrients washed from the nearby terrestrial area could adversely affect the seagrasses.

Table 2.2 Ecosystem services, processes, and functions for seagrass ecosystem

Ecosystem services	Ecosystem process and functions	Important controlling components
Raw materials and food	Biological productivity and diversity	Vegetation types and density, habitant quality
Coastal protection	Attenuate and /or dissipates sea waves	Wave height and length, water depth above canopy, seagrass bed size, and distance from shore. Wind climate, beach slope, seagrass species and density, reproductive stage
Erosion control	Provide sediment stabilization and soil retention in vegetation root structure, physical obstruction	Sea level rise, subsidence, tidal stage, wave climate, coastal geomorphology, seagrass species and density
Water purification	Provides nutrient and pollution uptake, as well as retention, particle deposition	Seagrass species and density, nutrient load, water residence time, hydrodynamic conditions, light availability
Maintenance of fisheries	Provides suitable reproductive habitat and nursery grounds, sheltered living space for several marine life	Seagrass species and density, habitat quality, food sources, hydrodynamic conditions
Carbon sequestration	Generates biogeochemical activity, sedimentation, biological productivity	Seagrass species and density, water depth, light availability, burial rates, biomass export

Though there are several studies on the benefits of seagrasses, corals, and mangroves, many of the important benefits of seagrass beds and sand dunes and beaches could not been assessed properly. Even for coral reefs, marshes, and mangroves, important ecological services have yet to be valued reliably, such as cross-ecosystem nutrient transfer (seagrasses and coral reefs), erosion control (marshes), and pollution control (mangroves) (Barbier et al. 2011). Hence, many of the ecosystem services are only approximation or qualitative.

2.4.5 Marshes

Marshes, estuaries, and mangrove forests are part of unique wetland ecosystems in semi-sheltered areas near the ocean coastline. This section describes about the marsh ecosystem with a focus on its biodiversity and ecological significance. Marshes as part of large wetland system are dealt separately in Chap. 6. In general a marsh is a wetland that is dominated by herbaceous vegetations rather than woody plant species. Marshes can often be found at the edges of lakes and streams, in the coastal areas (in the context of this chapter) where they form a transition between the aquatic and terrestrial ecosystems. They are often dominated by grasses and other herbaceous vegetations. On the other hand, swamps are wetlands dominated by trees and mire that have accumulated deposits of acidic peat.

2.4.5.1 Marsh Environment

Marshes provide habitat for many species of plants, animals, and insects that are adapted to live in flooded/marshy conditions. The plants growing in the marshes gradually evolved to survive in wet mud/waterlogged situations with low oxygen levels. In some places the water level may vary with season, but the ground remains fully saturated. Many of these plants have aerenchyma channels within the stem (air conducting cells) that allow air to move from the leaves to the roots. As a consequence, there are several symbiotic associations exists with many microbes which help the plant to survive in such situations and in turn benefited from the oxygen and other growth substances secreted by the roots. Marsh plants also tend to have rhizomes for underground storage and reproduction.

Most marsh plants flourish in the environment; after they fall, they begin to decay and are distributed within the same marsh or into other marshes and mudflats where they become the first level of the food chain. Microscopic organisms like bacteria, small algae, and fungi help decompose the detritus resulting from salt marsh plants. These microorganisms and the remaining decomposing plant material become an ideal source of food for bottom-dwellers in salt marshes like worms, fishes, crabs, and shrimps. The cycle continues when the faeces of the bottom-dwellers is cleaned up by microorganisms. As with many food webs, microorganisms at the most primary level on the food chain are responsible for more than one role.

Similarly aquatic animals have developed special characters or respiration mechanisms which enable them to live with a low amount of oxygen in the water. Some can obtain oxygen from the air instead, while others can live indefinitely in conditions of low oxygen. Marshes provide habitats for many kinds of invertebrates, fish, amphibians, waterfowl, and aquatic mammals. The pH in marshes tends to be neutral to alkaline depending on the flow of sea water, as opposed to bogs, where peat accumulates under more acid conditions.

2.4.5.2 Types of Marshes

Based on the location and salinity, the nature of marshes and the biodiversity it harbours greatly vary. Water and salinity level and its seasonal changes significantly influence the range and scope of animal and plant life that can survive and reproduce in these environments. Based on the physical and biological environment, there are three main types of marsh, viz. salt marshes, freshwater tidal marshes, and freshwater marshes.

Saltwater marshes are found around the world wherever there are sections of protected coastline, in lagoons, estuaries, and on the sheltered side of sandpit. They are located close to the shoreline that it comes under the periodical influence of sea water and sporadic tidal surges during which they are covered with saltwater (Fig. 2.11). In general, salt marshes flourish where the rate of sediment buildup is greater than the rate at which the land level is sinking. Salt marshes are dominated by specially adapted vegetation dominated by salt-tolerant grasses and herbaceous plants (Keddy 2010).

Freshwater tidal marshes are basically a freshwater marsh but are affected by the ocean tides at times. However, the salinity level is low due to flow of fresh water or

Fig. 2.11 Coastal salt marshes

runoff due to heavy rain. Unlike salt marshes, the salinity stress is low here; therefore, the diversity of plants and animals that live in and use freshwater tidal marshes is much higher. At the same time, the most serious threats to this ecosystem are the increasing pollution level and fragmentation of land due to urbanization of its surroundings.

Freshwater marshes occur mostly on the side of freshwater stream, river mouth, and coastal low-lying areas of humid tropics. They are also the most diverse of the three types of marshes. They have the highest primary productivity and support wide variety of flora and fauna. It also attracts migratory birds for nesting due to abundant availability of fishes and other aquatic life.

2.4.5.3 Significance of Marshes

Among coastal ecosystems, marshes provide a high number of valuable benefits to humans, including raw materials and food, coastal protection, erosion control, water purification, maintenance of fisheries, carbon sequestration, and tourism, recreation, education, and research (Barbier et al. 2011). Some of these important values have been estimated (Table 2.3).

Marshes have extremely high levels of biological production and therefore are important in supporting fisheries. Marshes also improve water quality by acting as a sink to filter pollutants and sediment from the water that flows through them (Mitsch and Gosselink 2008). Marshes along with other wetlands are able to absorb water during periods of heavy rainfall and slowly release it into waterways and therefore reduce the magnitude of flooding and maintain base flow in the streams during dry season. Salt marshes are most commonly found in lagoons, estuaries, and on the sheltered side of sandpit. Water currents carry fine particles and nutrients around to the quiet side of the spit, and sediment begins to build up. The marshes located in these areas absorb the excess nutrients from the water running through them before they reach the oceans and estuaries. Further, any stream passing over or through marshes get filtered of sediment load; thus the shallow water ecosystems of the ocean get benefitted.

Table 2.3 Ecosystem services, processes, and functions for marshes

Ecosystem services	Ecosystem processes and functions	Important controlling components	Ecosystem service value
Raw materials and food	Generates biological productivity and diversity	Vegetation type and density, habitat quality, inundation depth, habitat quality, healthy predator populations	Income from livestock grazing and fisheries
Erosion control	Provides sediment stabilization and soil retention by the thick vegetation and root structure	Sea level rise, tidal stage, coastal geomorphology, subsidence, fluvial sediment deposition and load, marsh grass species and density, distance from sea edge	Estimates unavailable
Coastal protection	Attenuates and/or dissipates waves Also reduces physical energy by friction force	Tidal height, wave height and length, water depth in or above canopy, marsh area and width, wind climate, marsh species and density, local geomorphology	Value of possible loss of land and infrastructures within the vulnerable zone
Maintenance of fisheries	Provides suitable nursery grounds and sheltered living space	Marsh grass species and density, marsh quality and area, primary productivity, healthy predator populations	Estimate the economic value of all fisheries catch
Water purification	Favours nutrient and pollution uptake, as well as retention, particle deposition	Marsh grass species and density, marsh quality and area, nutrient and sediment load, water supply and quality, healthy predator populations	Cost savings over traditional waste treatment
Carbon sequestration	Improves biological productivity, stabilizes biogeochemical cycle and sedimentation	Marsh grass species and density, sediment type, primary productivity, healthy predator populations	Amount of incremental carbon stored in soil and biomass
Recreation, education, and research	Provides unique and aesthetic landscape, suitable habitat for biodiversity	Marsh grass species and density, habitat quality and area, prey species availability, healthy predator populations	Mostly qualitative in nature and varies

Modified from Barbier et al. (2011)

2.4.6 Agro-ecosystem

As in the case of natural ecosystem, in agro-ecosystem, there are interactions between biological components and its surroundings as well as flow of energy and matter. However, it is a highly managed ecosystem where humans try to modify or influence each of the natural relationships and flow of energy and matter between different components. Profit or livelihood is one of the motives behind the management of agro-ecosystem as it is highly human centric. In agro-ecosystem maximization of crop and livestock using available resources is carried out with

Fig. 2.12 Coastal agro-ecosystem is modified by natural ecosystems

human management. Due to variation in climate, biotic and abiotic components influenced by human social systems resulted in different farming systems.

2.4.6.1 Resource Characterization

The analysis of situations of individual farming systems practiced in the tropical region has revealed the great diversity in developmental challenges which are grouped into developmental regions. There are dominant systems in a developmental region and within this there are subsystems. As against the crop alone in a specialized farming, mixed farming involving crops, farm animals, and fish suiting to different conditions is normally seen in this region. A typical farm located in the coastal areas can be seen in Fig. 2.12. Defining and characterizing these farming will help to evaluate its economic and ecological sustainability; besides it provides a base to assess its climate resilience. In this context a farming system is defined as a population of individual farm systems that have broadly similar resource bases, enterprise patterns, household livelihoods, and constraints and for which similar development strategies and interventions would be appropriate. Depending on the scale of the analysis, a farming system can encompass a few dozen or many millions of households. This approach, broadly, is based on the available natural resource base and dominant pattern of farm activities and household livelihoods existing in the region.

The delineation of the major farming systems provides a useful framework within which appropriate agricultural development strategies and interventions can be determined with a particular focus on adaptation and addressing food insecurity. This inevitably results in a considerable degree of heterogeneity within any single system. However, the alternative of identifying numerous, discrete, micro-level farming systems in each developing region/island would complicate the interpretation of appropriate regional and global strategic responses and detract from the overall impact of the analysis (FAO and WB 2001).

Table 2.4 Some characteristics of farming systems practiced in tropical islands

Farming system	Principal livelihoods	Constraints	Location
Rice and rice-tree	Wetland/rainfed rice, vegetable, legumes, off-farm activities Rice, maize, legume complimented by banana and coffee	Irrigation water is required during break in monsoon season and for second crop. Limited fodder production, decline in soil fertility and profitability	South Asia, viz. India, Sri Lanka, Andaman Islands, and Madagascar
Highland mixed	Range of cereals, legumes, tubers, mixed plantation, fodder, fodder trees, and livestock	Reduced forest cover, poor soil and water conservation, accelerated erosion, decline in soil fertility, poor adaptive capacity	Indonesia and other South East Asia, Sri Lanka, Caribbean, sub-Saharan islands
Coastal artisanal fishing	Fishing, coconuts, mixed cropping, banana, livestock	Storms, water logging, soil salinity, coarse soil, poor fertility	Pacific Ocean, sub- Saharan islands, S.E Asia, South Asian islands
Coconut based	Coconut, areca nut, other plantations, spices	Monocropping, soil fertility decline, high evaporation, storm, outbreak of pest, mean temperature increase	South Asian islands, Polynesian islands
Lowland rice	Rice, maize, pulse, sugarcane, soybean, oil seeds, aquaculture	High mean temperature, irrigation water, shift in pest and diseases, declining soil fertility	South East Asia, Caribbean
Root tubers	Root, vegetables, fruits, livestock	Soil erosion, declining fertility, moisture stress	Papua New Guinea, Caribbean, South Asian islands
Tree crop mixed	Rubber, oil palm, coconut, coffee, tea, spices, livestock	Low profitability, moisture stress, decline in processing	Indonesian islands, Borneo, South Asian islands
Coastal plantation and mixed farming	Family-owned mixed farming and large-scale plantations produce export crops, tubers, fishing	Salinity and water logging, storms	Caribbean, island of sub-Saharan region, and South Asian islands

Modified from FAO and WB (2001)

2.4.6.2 Farming Systems

Only the major farming systems have, therefore, been identified and then mapped in order to estimate the magnitudes of their populations and resource bases (Table 2.4). The major farming systems are rice and rice-tree, highland mixed, rainfed mixed, coastal artisanal fishing, coconut based, lowland rice, root tubers, tree crop mixed, and coastal plantation and mixed farming. Each of these broad systems is characterized by a typical farm type or household livelihood pattern, although significant subtypes are described where appropriate.

Rice system dominates in the South Asian coastal and islands, while the rice-tree crop subsystem with banana and coffee cultivation complemented by rice, maize, cassava, and legumes is the predominant system in Madagascar and some other sub-Saharan islands. In this region the cattle numbers are relatively low probably influenced by the availability of green fodder. Though there is abundant rainfall, this region faces water shortage during the post-monsoon season (January–March) due to high evapotranspirations. Coastal plantation and mixed farming occupy some of the richest agricultural lands in the region but also include mangrove swamps and isolated areas of tropical forest. There are several challenges to sustainable production particularly resource degradation and climate change adaptation. In the Southeast Asian islands, wetland rice-based farming system is one of the major systems which depend on the monsoon rains supplemented by irrigation. Rice-fish and rice-rice are the prominent subsystems.

Rainfed farming systems in humid areas have high resource potential, characterized by a crop activity or mixed crop-livestock systems, whereas rainfed farming systems practiced in steep and highland areas are mixed crop-livestock systems. In these areas declining soil fertility and water scarcity are emerging as major constraints. Coastal artisanal fishing is often mixed farming systems. On the other hand, Coastal Artisanal Farming Systems often have good access to services, but the underlying resource base varies with locations. The few areas with fertile soil often face serious risks of storms and floods, as occurs around the Bay of Bengal. Many systems in the rainfed region include some tree or multipurpose trees and small livestock, especially goats and poultry.

In order to overcome these challenges and improve upon the adaptation of different farming systems, specific actions and management of natural resources are required. But the constraints of these farming systems are deteriorated further by the natural calamities. A typical example is the Indian Ocean tsunami of December 2004 which inflicted heavy damage to the islands located in the Indian Ocean. It was seen that soil and water resources were severely affected due to seawater intrusion into the areas adjoining the coast in several inhabited islands. In addition, area under agriculture particularly of rice and coastal plantation areas decreased due to permanent submergence and periodic water logging of the coastal areas caused by the subduction of the land and earthquake. In these areas integrated farming system-based resource management strategies proved to be highly sustainable and restorative.

2.5 Conclusion

In coastal areas land and water join to create an environment with a distinct structure, diversity, and flow of energy. These areas include inshore waters, intertidal areas, and extensive tracts of contiguous land. It contains many of the Earth's most complex and diverse ecological systems, productive in both the biological and economic senses. Some of the most important coastal ecosystems are mangroves, coral reef, seagrasses, salt marshes, wetlands, estuaries, and bays. The region also supports agro-ecosystem which is one of the highly human-managed ecosystem providing food and other services in the coastal region.

Coastal ecosystems can encompass a wide range of environmental conditions over short distances, particularly of salinity from fresh to hypersaline and energy from sheltered wetlands to energetic wave-washed shorelines. As a result coastal ecosystems have become repositories of biological diversity, and some of them are very unique and endemic to the region which necessitates conservation. Some of the natural features of the ecosystems provide significant coastal protection, including coral reefs, the most extensive and effective coastal protection structures for the islands and coastal areas, sand and gravel beaches, which function as wave energy sinks, and barrier beaches, which act as natural breakwaters. Coastal dunes form natural buffers and sand repositories, from which sand may be extracted during storms without major shoreline retreat; coastal vegetation particularly mangroves often absorbs wind or wave energy, retarding shoreline erosion. Even the importance of salt marsh as a sea defence and mangroves as a sediment trap has been well recognized.

However, increasing human activities and exploitation of several coastal ecosystems in a unsustainable manner seriously affects the normal functioning of these ecosystems and sometimes causes irreparable damages. In the same context, these economic, social, and environmental benefits are at a risk as the coastal areas are affected by both gradual and recurrent processes such as accretion and erosion and by extreme natural events. This combination of natural and human forces and the uncertainties involved in their origins and impacts presents major challenges to the conservation of coastal ecosystems and its sustainable management for economic benefit.

References

Allen GR, Steene R (1994) Indo-Pacific coral reef field guide. Tropical Reef Research, Singapore, p 378

Barbier EB, Hacker SD, Kennedy C, Koch EW, Adrian C, Stier AC, Silliman BR (2011) The value of estuarine and coastal ecosystem services. Ecol Monogr 81(2):169–193

Birkeland C (1996) Life and death of coral reefs. Chapman and Hall, New York, p 536

Boelaert-Suominen S, Cullinan C (1994) Legal and institutional aspects of integrated coastal area management in national legislation. FAO, Rome. 118 pp

Burke L, Kura Y, Kassem K, Revenga C, Spalding M, Mcallister D (2001) Pilot analysis of global ecosystems, coastal ecosystems. World Resources Institute, Washington, DC. http://www.wri.org/wr2000

Chave KE, Smith SV, Roy KJ (1972) Carbonate production by coral reefs. Mar Geol 12:123–140

Chua TE (1986) Managing ASEAN coastal resources. Tropical coastal area management, vol. 1, no 1. ICLARM, Manila, pp 8–10

Clark JR (1992) Integrated management of coastal zones. FAO fisheries technical paper, no. 327. FAO, Rome, 167p. http://www.fao.org/docrep/003/T0708E/T0708E02.htmch2

Cowardin LM, Carter V, Golet FC, La Roe ET (1979) Classification of wetlands and deepwater habitats of the United States. FWS/OBS-79/31. U. S. Fish and Wildlife Service, Washington, DC, p 103

Davies J, Claridge G (1993) Wetland benefits. The potential for wetlands to support and maintain development. Asian Wetland Bureau, Kuala Lumpur. Gloucester, UK: International Waterfowl and Wetland Research Bureau, and Manomet MA: Wetlands for the Americas

Diane DB (1986) The Indian River Lagoon – seventy years of cumulative impacts. In: Esteves ED, Miller J, Morris J, Hamman R (eds) Proceedings of the conference: managing cumulative effects in Florida Wetlands, Oct 17–19, 1985, New College of University S. Fla., Sarasota, FL. E.S.P. Publ. #38, Omnipress, Madison, WI, pp 193–218

Duke NC (1992) Mangrove floristics and biogeography. In: Robertson AI, Alongi DM (eds) Tropical mangrove ecosystems. American Geophysical Union, Washington, DC, pp 63–100

FAO (1998) Integrated coastal area management and agriculture, forestry and fisheries. FAO guidelines, Food and Agriculture Organization of the United Nations, Rome. http://www.fao.org/docrep/W8440e/W8440e00.htm

FAO (2003) Status and trends in mangrove area extent worldwide. In: Wilkie ML, Fortuna S (eds) Forest resources assessment working paper no. 63. Forest Resources Division, FAO, Rome. http://www.fao.org/docrep/007/j1533e/J1533E00.htm

FAO and World Bank (2001) Farming systems and poverty- improving farmers' livelihoods in a changing world. FAO and World Bank, Rome

Fletcher SW, Fletcher WW (1995) Factors affecting changes in seagrass distribution and diversity patterns in the Indian River Lagoon complex between 1940 and 1992. Bull Mar Sci 57(1):49–58

Global Seagrass Survey (1999) Global seagrass survey results 1997–1998. On-line at http://www.seagrass.unh.edu. Last Updated on 10/23/99. Accessed on 15 Oct 2018

Gopal B (2015) Guidelines for rapid assessment of biodiversity and ecosystem services of wetlands, version 1.0. Asia-Pacific Network for Global Change Research (APN-GCR), Kobe, Japan, and National Institute of Ecology, New Delhi, p 134

Gullström M, Torre Castro M, Bandeira SO, Björk M, Dahlberg M, Nils Kautsky N, Rönnbäck P, Öhman MC (2002) Seagrass ecosystems in the Western Indian Ocean. AMBIO J Hum Environ 31(7):588–596. https://doi.org/10.1579/0044-7447-31.7.588

Heijs FML (1984) Annual biomass and production of epiphytes in three monospecific seagrass communities of *Thalassia hemprichii* (Ehrenb.) Aschers. Aquat Bot 20:195–218

Hinrichsen D (1998) Coastal waters of the world: trends, treats and strategies. Island Press, Washington, DC

Jayatissa LP, Dahdouh-Guebas F, Koedam N (2002) A review of the floral composition and distribution of mangroves in Sri Lanka. Bot J Linn Soc 138(1):29–43

Kates RW, Colten CE, Laska S, Leatherman SP (2006) Reconstruction of New Orleans after Hurricane Katrina: a research perspective. Proc Natl Acad Sci U S A 103:14653–14660

Kathiresan K, Bingham BL (2001) Biology of mangroves and mangrove ecosystems. Adv Mar Biol 40:81–251

Keddy PA (2010) Wetland ecology: principles and conservation, vol 497, 2nd edn. Cambridge University Press, Cambridge

Kelleher G, Bleakley C, Wells S (1995) A global representative system of marine protected areas, vol 1. World Bank, Washington, DC

Kitting L, Christopher BF, Mark DM (1984) Detection of inconspicuous epiphytic algae supporting food webs in seagrass meadows. Oecologia 62:145–149. https://doi.org/10.1007/BF00379006

Mackinnon J (1997) Protected areas systems review of the Indo-Malayan realm. Asian Bureau for Conservation, Canterbury

MEA (2005a) Ecosystems and human well-being: current state and trends, Coastal systems. Millennium ecosystem assessment. Island Press, Washington, DC

MEA (2005b) Online guide to the millennium ecosystem assessment reports. Available via http://www.millenniumassessment.org/en/index.html. Accessed 10 Nov 2017

Michel D, Pandya A (2010) Coastal zones and climate change. The Henry L. Stimson Center, Washington, DC. 20036

Mitsch WJ, Gosselink JG (2007) Wetlands. Wiley, New York, p 582

Mitsch WJ, Gosselink JG (2008) Wetlands. Van Nostrand Reinhold, New York

Morgan MD, Kitting CL (1984) Productivity and utilization of the seagrass *Halodule wrightii* and its attached epiphytes. Limnol Oceanogr 29:1099–1176

Munro JL (1984) Coral reef fisheries and world fish production. ICLARM Newsletter 7:3–4

Navid D (1989) The international law of migratory species: the ramsar convention. Nat Resour J 29:1001–1016

NOAA (1999) Trends in U. S. Coastal Regions 1970–1998. National Oceanic and Atmospheric Administration, 1305 East-West Hwy., 9th Fl., Silver Spring, MD 20910-3281. http://state-of-coast.noaa.gov/natdialog/index.html

Polidoro BA, Carpenter KE, Collins L, Duke NC, Ellison AM et al (2010) The loss of species: mangrove extinction risk and geographic areas of global concern. PLoS One 5(4):1–10

Robert WV (1982) Leaf growth rate of the seagrass *Halodule wrightii* photographically measured in situ. Aquat Bot 12(3):209–218

Salvat B (1992) Coral reefs a challenging ecosystem for human societies. Glob Environ Chang 2:12–18

Smith SV (1978) Coral-reef area and the contribution of reefs to processes and resources in the world's oceans. Nature 273:1149–1160

Spalding MD, Blasco F, Field CD (1997) World mangrove atlas. The International Society for Mangrove Ecosystems, Okinawa

Spalding MD, Grenfell AM (1997) New estimates of global and regional coral reef areas. Coral Reefs 16:225–230

Spalding MD, Ravilious C, Green EP (2001) World atlas of coral reefs. Prepared at the UNEP World Conservation Monitoring Centre. University of California Press, Berkeley, p 432

Tomasko DA, Lapointe BE (1991) Productivity and biomass of *Thalassia testudinum* as related to water column nutrient availability and epiphyte levels: field observations and experimental studies. Mar Ecol Prog Ser 75:9–16

Veron JEN (2000) Corals of the world, vol 3. Australian Institute of Marine Science, p 410

Wells S, Hanna N (1992) The Greenpeace book of coral reef. Sterlings Publishing, New York, p 160

Part II

Status of Natural Resources in the Coastal Ecosystem

Land Resources of the Tropics vis-a-vis the Hinterland

3

Abstract

Land resources are vital for the development of any region which varies from place to place. In the tropical region, high rainfall, intensive solar radiation, luxuriant vegetation, and type of parent materials besides topography are the factors determining the soil formation and its properties. Laterization, humification, and salinization are the major specific pedogenic process in the tropical region. In the coastal tropical regions, alluvial, laterite, and organic soils are the major soils which are highly modified or influenced by human activities. Occurrence of acid sulphate and sandy soils is specific to the tropical coastal areas which are major constraint for the crop production. As a result land degradation of various types occurred causing decline in land productivity. Different soil resource management options are suggested based on the constraints of different locations for the sustainable resource use in the tropical coastal areas.

Keywords

Tropical soils · Assessment · Laterization · Organic soils · Management

3.1 Introduction

Land is indispensable resource for human wellbeing which also provides habitat for innumerable lives on Earth. Land includes not only soil but also water, vegetation, landscape, and microclimate. In general, description of soil resource covers all other components of land as all of them influence soil formation and are interconnected. In other words soil characters are spatially variable because soils found in a particular region are influenced by the vegetation, landscape, climate, and geological characters of that place; all of them are spatially variable. Alternatively, soils of a particular place influence the vegetation, and both are in dynamic equilibrium with the climate of that place. Thus the soil is a complex system and dynamic natural

V. Ayyam et al., *Coastal Ecosystems of the Tropics - Adaptive Management*,
https://doi.org/10.1007/978-981-13-8926-9_3

body. It contains all necessary things required for life such as water, nutrient, energy, and oxygen besides providing habitat for large diversity of living things. Being a component of the lithosphere and biosphere systems, soils provide food, fibre, fodder, and fuelwood for meeting the basic human needs. However, the capacity of a soil to produce is limited, and the limits to production are set by its intrinsic characteristics, agroecological settings, and use and management (FAO 1991).

For supporting food production system to meet the increasing population, the intrinsic characters of soils and the way it is being managed are very important which vary from place to place. At the same time, soil resources are being degraded by human activities besides natural causes resulting in deterioration of soil quality. In this context it is essential to understand the term soil quality which is defined as the capacity of a soil to function, within the limits imposed by the ecosystem, to preserve the biological productivity and environmental quality, and to promote plant, animal, and human health (Doran and Parkin 1994). Going by this definition, it has been observed that the soil resources are being degraded by not only the human activities taking place in the coastal region but even from far-off places. Meanwhile land use change in the coastal areas of the tropics has triggered soil erosion and witnessed decrease in area of quality agricultural land and decrease in forest area.

These facts suggested the importance of understanding and documenting soil resources at finer scale for maintaining soil quality and sustain the production (Scherr and Yadav 1996). Therefore it is imperative to study the soil resources of the coastal region and hinterland for its proper use and conservation. There are three reasons for the degradation of land resources in the coastal areas. Firstly, waterlogging during monsoon season and water scarcity during dry season limit the cultivable area and crop productivity. Secondly, degradation of soil and water occurs due to the phenomena like saline water flooding following breach or overflow of embankments as a result of very high tides or storm waves and presence of shallow brackish groundwater table near the soil surface. Finally, there is unrestricted diversion of agricultural land witnessed in recent times to other land uses due to population growth and developmental needs. The condition is further likely to be aggravated due to the projected sea level rise following global warming. Under such situations, for effectively managing the soil resources, three elements must be considered, viz. information about natural resources, clear management strategy, and participation of all the stakeholders (including local people). Thus in this chapter, we have attempted to highlight soil resources with reference to its formation and process, characterization, and technical options for sustainable management in the tropical coastal areas and hinterland.

3.2 Soil Formation in the Tropics

There is great diversity of soils in tropical and subtropical regions of the world. They range from young fertile volcanic soils to very old, often infertile, red tropical soils. There are sandy soils, almost without any soil profile, and the wet soils of equatorial regions, where water percolates almost continuously to deeper soil layers. These suggest that there are differences in soil processes (e.g. oxidation, reduction, eluviations, leaching, transport, and accumulation). However weathering and soil formation are often quite different in the tropics, because of important differences in soil climate and in biological processes. In the humid tropics and subtropics, weathering is more severe and intense. Processes of soil formation are more active and often continuous. The relative importance of these fundamental reactions, their combinations, and the more complex processes of soil formation determines the ultimate nature of the soils (Buringh 1970). Some typical characteristics of the factors and process of soil formation are briefly described.

3.2.1 Factors of Soil Formation

Soils vary greatly in their nature and extent of development from place to place. Apparently, several factors are responsible for these changes, and some factors are more effective than others in bringing about changes in any given soil. There are five factors of soil formation, viz. climate, organism, relief, parent material, and time called 'CLORPT' (Fig. 3.1). These factors act simultaneously at any point on the surface of the Earth to produce a characteristic soil. Parent materials, relief, and time are considered as passive factors, while climate and biosphere (vegetation and organisms) are termed as active factors. Active factors supply energy for the soil-farming process and drive it towards progressive maturity, whereas the passive factors provide a base on which the active soil-forming factors work or act for the development of soil. Thus human interference or modification of these factors will have consequential effect on the soil characters.

3.2.1.1 Climate

The temperature in the tropical region is generally warm, and the difference between mean maximum and minimum temperature is minimum throughout the year. This region receives relatively higher amount of rainfall than other climatic region. But,

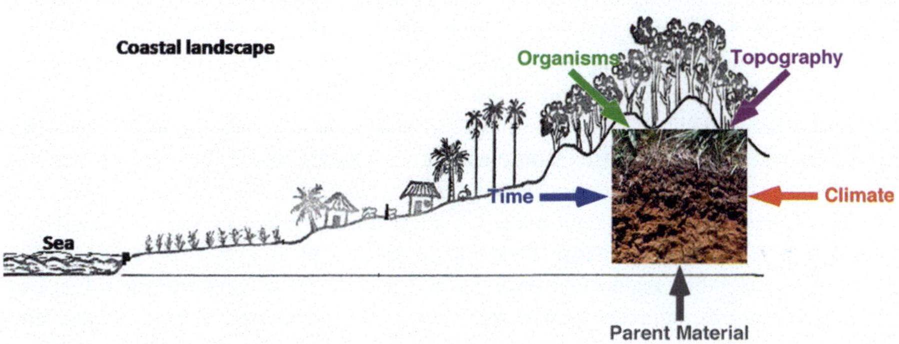

Fig. 3.1 Factors of soil formation

there is very great range in annual precipitation, usually 3000–6000 mm in the humid tropics, while it is more than 6000 mm in some wet equatorial region. Besides these variations, there is also diversity in precipitation between seasons (alternating dry and wet seasons) or during a day. Heavy showers of 5 mm per minute and daily totals of 250 mm do occur occasionally or during cyclonic weather. Unlike rain in temperate regions, rainwater in the tropics is warm and intensive. The relative humidity is very high (75–90%) throughout the year due to intensive solar radiation and copious amount of rainfall. The weather is almost constant in some regions and extremely variable in few others. Many soils in the tropics and subtropics are very old, even tertiary, and their formation is being influenced by changes in climate in the Pleistocene Era, as pluvials and interpluvials. Many old soils therefore are polygenetic (Buringh 1970). In spite of high rainfall received during more than 7 months in a year, this region also experience soil moisture deficit during dry months due to high evapotranspiration.

3.2.1.2 Vegetation

Vegetation of a place also influences the soil formation and the process by way of supplying energy in the form of organic matter and active plant roots. It is a general feature that natural vegetation type is closely related to the climate of any region. For example, it may be a dense tropical forest consisting of wide varieties of plant species, or it may consist of a few sparse grasses in arid regions. For soil formation these differences are quite important because of variation in biomass production or organic matter addition which may vary from almost nothing in a desert to several tons per ha in a tropical forest. Further, the kind of organic material (easily or hardly decomposable) provided by the tropical vegetation and soil moisture are the major determining factors of soil organic matter accumulation. Similarly in agricultural soils, cropping system, intensity of cropping, land management, etc. will have significant impact on the soil status by way of organic matter addition, activity of microbes, and tillage. This is very important because organic matter governs various processes, in particular the biological activity of soil.

3.2.1.3 Time

Soil formation is a very slow process requiring thousands of years to develop a mature pedon. The period taken by a given soil from the stage of weathered rock up to the stage of maturity (fully developed horizons A, B and C) is considered as time. In this context, the time that nature devotes to the formation of soils is termed as pedologic time. This varies from place to place. Some soils in the tropics are extremely old, devoid of bases, and even without weatherable minerals. Others are very young, with soil formation hardly begun. In the humid tropics, where soil formation proceeds almost continuously, soils can be altered in a short period; yet in arid regions, lack of moisture slows down many soil processes, so that changes are hardly detectable, even over several centuries. Based on mineralogical features of soil, weathering happens in five stages, viz. initial, juvenile, virile, senile, and final. In general, ageing is more rapid in warm and humid climate than in cold or arid climate. Similarly soils age faster on flat to gently sloping uplands than on flat lowlands or on steep sloping positions. In a floodplain, soils remain young as long as flood waters inundate them periodically.

3.2.1.4 Parent Material

The parent material (weathered rock from which soil develops) provides the bulk and has a distinct effect on soil. The soil parent material is quite different in the various regions. Often the material, which has undergone some cycles of weathering and soil formation, is transported and deposited again, forming a new but already entirely weathered parent material. An extreme type of parent material is formed by fresh pyroclastic material produced by active volcanoes. Fresh volcanic ash may be transported in the wind over huge distances, rejuvenating old soils on which they are deposited (Buringh 1970).

The parent material determines within broad limits, such physical properties of soil as texture, structure, and water holding capacity. It may affect the downward movement of water for the profile development. Different parent materials affect profile development and produce different soils, especially in the initial stages. Under humid tropical conditions, acid igneous rocks (granite, rhyolite) produce light-textured podzolic soils (Alfisols), while basic alluvium or aeolian materials produce fine to coarse textured soils (Entisols or Inceptisols). Similarly silica-sesquioxide ratio of the soils also differs. The maximum (>2.0) being in the basic or least weathered alluvium- and aeolian-derived soils and the minimum (<2.0) are observed in soils of the tropical regions.

3.2.1.5 Relief/Topography

Erosion, colluviation (movement of material to the foot of slopes by gravity), and sedimentation as induced by variations in topography have affected many soils, particularly in regions with old soils. Another important effect is the relation of altitude to air temperature. In the tropical region, this temperature decreases about 0.6 °C for every 100 m rise in altitude. Thus high mountains in the tropics and subtropics have a kind of temperate climate.

High water tables in tropical and subtropical regions cause the accumulation of soluble salts in the rooting zone of soils. Further, in tropical region particularly in the coastal areas, floods and tides often inundate large areas in river plains, estuaries, and coastal lowlands which are rarely protected by dikes. Due to saturated soil conditions, minerals are present in reduced state, and it also causes hydration of rock/parent materials.

3.2.2 Soil Processes

The differences of tropical and subtropical soils from soils of other region can be ascribed to the special nature of soil-forming factors and process mainly soil climate and biology. At the same time, there are basic processes involved in the soil formation which is applicable to all soils, viz. gain (water, oxygen, organic matter, energy), losses (water, erosion, etc.), transformation (weathering, structural

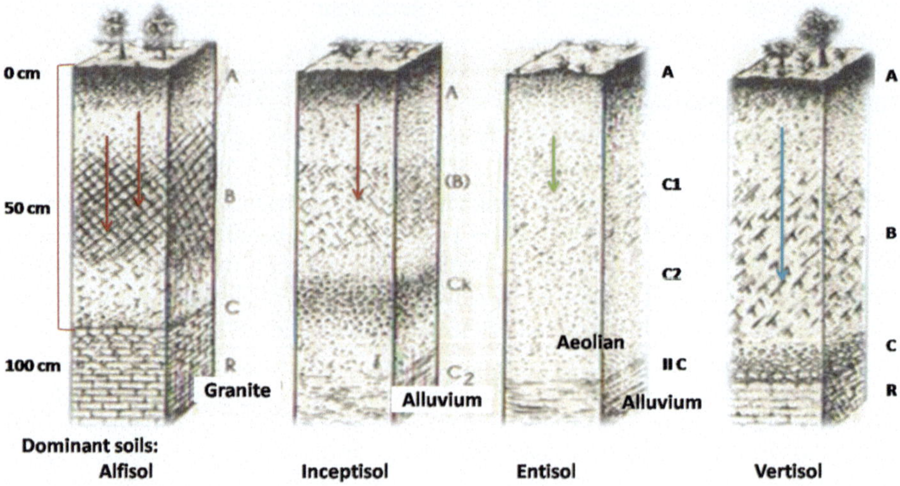

Fig. 3.2 Evolution of soil on different parent materials with horizons

improvement), and translocation (eluviations, illuviation). The soil climate or pedoclimate in the humid and wet tropics encourages rapid soil formation. In other words the soil matures more rapidly than dry or temperature region. Pedogenic process happening in different parent materials under tropical conditions is depicted in Fig. 3.2. In true tropical soils, the difference between the average summer and winter soil temperature (at 50 cm depth) is less than 5 °C, while in the subtropics this difference is more than 5 °C. The biological activity of the soils is closely related to these pedoclimatic classes. Soil temperature, moisture content in relation to the permeability, structure, and water storage capacity are factors that may vary considerably during the year and between different regions. These are the most important physical factors which largely decide the soil properties. The water that penetrates and percolates soils is always warm (24–25 °C) and plenty. Similarly the soil is also much warmer. For these reasons (Buringh 1970):

- More water penetrates deeper in the soil profile resulting in profile homogeneity.
- Ionization of water is four times as high as at 10 °C due to more energy.
- Solution proceeds much quicker.
- The hydrolytic power of the solution is much higher.
- Silica is eight times as soluble hence quickly leached down.
- Less CO_2 penetrates the soil.
- Soil water is less viscous.

These are most important soil environmental conditions which result in specific pedogenic process which imparts distinct soil properties for tropical region.

At the same time, the total amount of water that penetrates soil is less than total precipitation, because of surface runoff and considerable evaporation from the leaves of plants and from soil. Even in flat regions of the tropics, effective rainfall is often less than three quarters of the precipitation because of high evaporation. In the humid tropical regions, wet and dry seasons alternate. The total monthly precipitation in 'dry' months in such areas is less than 60 mm and spread over 3 months. During this dry season, weather proceeds faster than soil formation from parent material, while soil formation takes place when the soil is moist. These conditions are favourable for laterization. It is the process that removes silica and leaves sesquioxides (iron and aluminium oxide and hydroxide) in the soil solum.

Biological processes in the soil depend mainly on soil organic matter, moisture, and temperature. The most intense biological activity is in warm and moist soils of the humid tropics. The vegetation produces much organic matter but is rapidly decomposed. Mineralization of organic matter is most important factors which support dense vegetation. Under natural tropical forest, percentage organic matter is almost constant in the soil due to large addition and fast decomposition. In wet equatorial regions, temporary wet conditions slightly delay decomposition of organic matter. Under such condition, humus accumulates and favours the formation of organic soils. There is possibility for some humification and percolation of humic acids through the soil profile causing some podzolization. Humification is the process of transformation of raw organic matter into humus.

In summary, the pedogenic processes particularly occurring in the tropical regions are complex and dynamic involving many chemical and biological reactions and usually operate simultaneously in a given area. One process may counteract another, or it is possible that two different processes may work simultaneously to produce the same result.

3.3 Important Soil Groups of the Tropics

In the early twentieth century before the comprehensive soil taxonomy, natural resource scientist used to refer 'rubber soils' or 'coffee soils' reflecting a set of properties that were suitable for growing specific crops in the tropics. Likewise, there were terms like 'forest soils' and 'savanna soils', and these also reflect the link between land use or vegetation and soil properties. These descriptions don't completely provide the characters of soils nor useful to provide comprehensive taxonomy. We now know that there are different soils under forest or rubber and linking soils to land use or vegetation has some severe limitations. Nonetheless, such linkages are still made (Hartemink 2004). However, these are indirectly referring to one or two factors of soil formation in the tropics. In the same context, 'tropical soils' is used suggesting that, in the tropics, soils are conspicuous different from soils in other climatic zones. Hence, to understand soils of tropics, all the factors and process involved in making a particular soil have to be described.

Weathering and soil formation are often quite different in the tropics, because of important differences in soil climate and in biological processes. Further, in the humid tropics and subtropics, weathering is more severe and intense. For these reasons there is an extreme diversity of soils in tropical and subtropical regions of the world. They range from young fertile volcanic soils to very old, often infertile, red tropical soils. So the soil types should cover the occurrence, properties, characteristics, genesis, classification, and agricultural evaluation of soils in the tropics and subtropics. In the tropics laterite soils, alluvial soils, Podzols of the lowlands, organic soils and soils formed from volcanic eruption, and sandy soils (formed in coastal sand dunes) are the major types (Buringh 1970). Glimpses of soil profile of different soils are given in Fig. 3.3. This shows only the general description and view. The detailed properties are discussed in the following subsections.

Fig. 3.3 A glimpse of soil profiles of different tropical soils

3.3.1 Laterite Soils

The term 'laterite' is derived from the word '*later*' meaning brick or tile. Laterite soils are formed by the process of laterization. In this process silica is removed from the upper layers leaving sesquioxides to concentrate in the solum. Laterite soils occur in wet tropical regions, e.g. Indonesia, India, Southeast Asia, Central Africa, Brazil, and various islands of Hawaii and Cuba.

3.3.1.1 Factors Favouring Laterization

The process occurs in the following conditions:

- It occurs in warm and humid tropical climate with more than 2000 mm rainfall and continuous high temperature (>25 °C) throughout the year. Such conditions cause intense leaching and favour rapid decomposition of parent rock. The soluble products of weathering are continuously being dissolved and leached to deeper groundwater stream by the gravitational force.
- The natural vegetation of the laterite soils is dense tropical rainforest, with deep-rooting vegetation, producing much organic matter. But the organic matter is quickly oxidized due to the prevailing high temperature. Further organic material is also used by several soil organisms, and consequently the soils remain low in humus.
- The topography is characterized by an undulating and rolling terrain in old land surfaces, leading to a good natural drainage of the soils, because without the natural internal drainage, no laterite soils would have formed. In flat areas and depressions, the water table may reach the solum, so that plinthite has formed in the subsoil.
- The parent materials should have sufficient iron-bearing ferro-magnesium minerals such as pyroxene, amphiboles, biotite, and chlorite. By weathering they release iron which combines with oxygen to form oxides.

3.3.1.2 The Process of Formation

The high temperature, intense leaching due to more rainfall, sloppy land, and basic kind of parent materials all favour the removal of silica and accumulation of sesquioxides. During the initial phase of weathering, soluble basic ions such as Ca, Mg, K, and Na are quickly released from the parent materials. Similarly bases present in the organic combination are also released. These basic ions have high mobility in the soil profile due to which the soil pH increases. Under such basic conditions, the silica is also leached out as the solubility of quartz and amorphous silica increases with increasing temperature. The sesquioxides (iron and aluminium oxide and hydroxides) are left behind as these are more stable under such conditions. Consequently in the mature/residual soils due to removal of basic cations and accumulation of sesquioxide, the soil pH becomes acidic.

In humid tropical climate, the hydrated form of iron oxide ($Fe_2O_3.H_2O$) being unstable gets dehydrated to Fe_2O_3 in short pedologic period. With time, the soil deepens, and the primary minerals are decomposed, and the iron in oxidized form coats the clay, silt, or sand particles and imparts the characteristics red colour to such soils. At the same time, the aluminium released in the process hydrolyses in a water solution to form aluminium hydroxide and hydrogen ions. The aluminium oxide, like iron oxides, imparts grey coatings to the clay, silt, or sand particles. The shades of grey colour depend upon the amount of aluminium oxide present in the soil environment (Buringh 1970).

The high groundwater in the coastal and tropical areas may also favour the supply of sesquioxides. In the wet period, the water containing sesquioxides in reduced (soluble) form accumulates. In dry period the water table falls, and oxidizing conditions return which stabilize sesquioxides in such a way that these are not readily mobilized by further wet period. Eventually this leads to the formation of a soil horizon dominated by sesquioxides. These soils are grouped under Oxisols (laterites) and Ultisols (lateritic).

3.3.1.3 Soil Character and Agricultural Suitability

In general these soils are very old, deeply weathered red to yellow clayey soils almost uniform throughout the profile or without distinct horizons. These soils are nonplastic and non-cohesive and have granular structure, normally drained and aerated. These soils are low in cation exchange capacity and low in fertility. This is due to lack of organic matter and hydrous nature of clay. Such soils need organic matter addition and more fertilization for growing agricultural crops. These soils are having high phosphorus fixing capacity (as iron and aluminium phosphate), due to which phosphorus become unavailable to crop. Soil crusting is a major problem for seed germination. If properly managed, these soils are suitable for plantations of coconut, coffee, banana, pineapple, and areca nut with sufficient nutrient supply.

3.3.2 Alluvial Soils

Alluvial soils are very important for agriculture and used for other construction activities. They occur all over the world in river plains, estuaries, deltas, and low coastal regions. In coastal areas these soils are being degraded not only by agricultural activities but also by industrial pollution and urbanization. Alluvial soils are azonal, usually defined as young soils of recent and subrecent sedimentary deposits, without horizons or with only weak ones. Although alluvial soils are relatively young, there are still some processes like sedimentation, gleying, etc. that influence the soil conditions.

3.3.2.1 Processes of Soil Formation

In wet and humid regions, rainfall is high or very high and vegetation is dense. The flat or lowland of the terrain makes most of the alluvial soil region wet with a high water table, permanently or during specific periods. Such soils are called hydromorphic soils. Young sediments deposited under water contain much more water than normal soils. During maturity the physical characteristics of the clays change. The chemical process is mainly oxidation and decarbonation. On drying, air penetrates the soil through the many cracks; organic matter oxidizes rapidly. The bluish-grey of the reduced mud-clay changes to brownish by oxidation of iron compounds.

A fluctuating water table causes alternate oxidation and reduction in soil layers above the groundwater. Bluish-grey reduction (gley) and reddish oxidation (rust) mottles form. The rust spots, mainly orange-brown in temperate regions, are often more reddish or yellowish in the tropics, through warmer soils and some dehydration. Rust occurs particularly along cracks and root channels; sometimes there is a tendency to form concretions in the bottom layers.

When soils are poorly drained, very wet, or submerged, the decomposition of organic matter produced by the dense vegetation is slow and stagnates. A partly decomposed organic layer is formed on the surface of the mineral soil. Sometimes organic matter is formed during deposition, and a mixture of organic matter and clay, containing much water, forms. Such soils are called peat soil, have low load-bearing capacity and a weak surface, and cannot carry grazing cattle or farm machinery.

3.3.2.2 Factors of Alluvial Soil Formation

The climate does not have an important influence, except for the leaching (decarbonation, desalination) of the young soils. Vegetation is an important factor, firstly during the geological process of sedimentation (decreasing the velocity of flood water) and secondly by producing organic matter, transpiring water (ripening), perforating the soil, and encouraging microbes. Relief differences are usually limited to microrelief. In shallow depressions with poor drainage, organic matter or salts accumulate. Hydrology is another main factor. Groundwater, floods, and the effect of tides influence oxidation and reduction.

Alluvial soils consist mainly of kaolinitic clays, quartz, and highly resistant minerals, low in natural fertility. Some large rivers have a catchment area entirely in the equatorial region (Congo, Amazon); others rise in an equatorial region but end in the subtropics (Nile) or rise in temperate regions and end in the subtropics (Tigris, Euphrates) or in the tropics (Ganges, Mekong). The parent material of the alluvial soils deposited by the rivers is obviously quite different. Marine clays often are more alike. They are richer, often illitic. However they are poor in calcium carbonate because of the higher, more uniform temperature of sea water and the thorough leaching in high rainfall.

3.3.2.3 Soil Character and Agricultural Suitability

Soils are very young often stratified mineral soils in recent or subrecent sedimentary material, which has been carried in suspension, transported and deposited by water. They have a heterogeneous mineral composition and are permanently or seasonally wet, often with groundwater influence. They form flats in the lowest parts of a landscape and have hardly any profile, (A) C soils. Alluvial soils are mostly grouped under Entisol and Inceptisol as per US soil taxonomy. Such soils form a special type of *cambic* subsurface horizon and therefore often belong to the order Inceptisols and the suborder *Aquepts*. In the order Entisols, the suborder *Fluvents* is the typical suborder for alluvial soils without clear characteristics of wetness.

Alluvial soils of river plains and coastal areas are intensively cultivated areas with high inputs. But in some regions, the productivity of alluvial soils is low, partly because of high acidity, through floods and inundation. These low-lying areas are not protected by dikes, and hence artificial provision to drain the land is very much required. Very often the period of cultivation (length of growing season) for some waterlogged alluvial soils is limited. In some regions salinization or alkalization is a problem. In many coastal areas, soils are unripened clay soils with high sulphate content that become acid sulphate soils on reclamation. Suitable crops for alluvial soils are rice, sugarcane, and banana, and on better drained sites, maize, wheat, cocoa, coffee, and citrus are grown.

3.3.3 Some Other Tropical and Subtropical Soils

Besides laterite and alluvial soils discussed in the preceding section, there are some other groups of soils, e.g. lowland and mountain region Podzols, organic soils in mountain regions, and the peat or bog soils in coastal flats, in swamps, and on mountain plateaus. Such soils only occur regionally. Most are less important for agriculture, except the low mountain soils developed in volcanic parent material as in Indonesia (Java, Sumatra) and Hawaii.

3.3.3.1 Lowland Tropical Podzols

The soil-forming process is podzolization because leaching of bases and sesquioxides at high water temperature (22–26 °C) causes the proportion of silica to increase. Bases, sesquioxides (especially iron), and soluble organic compounds (fulvic acids) are leached in various complex combinations to a great depth by high rainfall or a fluctuating water table. Very often there is a layer of unchanged organic material; the pH is always very low, about 4.0–4.5. A slight slope and high permeability encourage lateral movement of water. The parent material of inland Podzol soils is probably repeatedly decomposed, altered, eroded, transported, and redeposited. In coastal areas the parent material consists of poor quartz sands of former beach ridges (Buringh 1970).

The agricultural value of these Podzols is low because of the low nutrient status, the high permeability and low water storage capacity, the low field capacity, the acidity, and the low content of organic matter and clay. In the future some of these soils in a favourable topographical position could probably be improved by manuring, green manure, and if necessary by supplementary irrigation. Lowland Tropical Podzols cover 7 million hectares in various tropical countries, mainly in limited inland areas or patches and only exceptionally (Surinam) in slightly larger regions. They frequently occur in poor marine coastal quartz sands in older beach ridges (Klinge 1968).

3.3.3.2 Organic Soils

Organic soils are formed in tropical areas with extreme hydromorphic conditions. For example, in coastal regions, swampy inland depressions and in the mountains with high rainfall and stagnant water are favourable for organic soil formation. Such stagnant water can also occur over an impermeable rock such as hard lava on Hawaii Island where a blanket of organic soil is formed in the dense forest. The main process in their formation is the accumulation of peat which means that organic matter is produced by growth of vegetation faster than it is decomposed in the soil. Sediment or parent material/rock underlying the peat is usually little altered by weathering. There may be some leaching or formation of gley minerals such as pyrite or siderite, but most of the weatherable minerals and structures of the parent materials are retained. Depending upon the clay content, they contain organic matter ranging from 20 to 30 percent. These soils are grouped under Histosols with four suborder based on degree of decomposition as per US Soil Survey staff.

In some tropical coastal regions, the peat layer may be several metres thick. The peat can be in various stages of humification depending on the soil, climate, and vegetation conditions. In Africa, various papyrus and phragmites swamps occur locally. Large forest peat areas occur in some tropical coastal regions where such conditions favour, e.g. Kalimantan, Sumatra, and West Irian (Indonesia) and the Western Ghats and the Andaman Islands of India (Fig. 3.4). Organic soils form the order Histosols; such soils must have a high content of organic matter. The organic soils of the tropics are not important for agriculture, because they are wet and hydromorphic, and occur mainly in small areas. When drained, they oxidize rapidly and subside tremendously. Various organic soils in coastal regions, especially in the brackish zone, may be transformed into acid sulphate soils when drained.

3.3.3.3 Tropical Sandy Soils

Light-textured sandy soils are ubiquitous throughout the tropics and constitute an important soil resource on which millions are dependent upon for their livelihoods. Sandy soils are characterized by less than 18% clay and more than 68% sand in the first 100 cm of the solum (Valentin and Bresson 1992). But due to their inherent nature, they present unique sustainability and environmental challenges to resource managers. Sandy soils are characterized by the predominance of rigid coarse

Fig. 3.4 Occurrence of organic soils in the coastal areas of Southeast Asia

particles that are associated with small amounts of clay minerals and humus. Thus they tend to have low cation exchange capacity. Sandy soils are characterized by a lack of structure or it is weakly developed. These physical attributes contribute to the significant spatial and temporal variability. For these reasons, these soils have low buffering capacity to changes in physical, chemical, and biological predisposes them to accelerated rates of degradation.

Physical, chemical, and biological characteristics of sandy soils often act as a severe limitation in crop production. Further to elaborate, their sandy nature, low organic carbon content, high hydraulic conductivity rates, low nutrient and water supply capacity, limited buffering capacity, and inadequate biological diversity invariable necessitate high levels of external inputs. Other kinds of problems arise, when more intensive agriculture is initiated. Agricultural activities such as pesticide mixing and tank rinsing and storage of manure, fertilizer, and fuels may pose many risks on sandy soils. Handling agrichemicals requires extra precautions on such soils due to their rapid contamination of groundwater and aquifers.

Sandy soils in the tropics are generally subjected to a cycle of wetting and drying associated with seasonality. In this respect small changes in composition lead to significant differences of physical properties. One of the major soil characteristics to be taken into account is the size distribution of the sand grains. If fine sand induces greater porosity, water retention, and resistance to penetration than coarse sand, they exhibit lower permeability. While silty sands are more compact than sandy soils, most silt particles occupy the voids between sand grains, thereby reducing porosity and consequently permeability. An increase in the silt-sand ratio would also result in a decrease in the porosity (Agrawal 1991). This shows that sandy soils highlight large differences in soil behaviour associated with minor changes in intrinsic soil properties that cannot be entirely attributed to natural heterogeneity. Improved sandy soil characterization would lead to a better understanding of processes of soil change, classification of relevant factors into a hierarchical system, and appropriate management recommendation that would enhance the sustainable utilization of these soils.

3.4 Land Degradation

Soil is a three-dimensional natural body on Earth's surface that is essential to numerous ecosystem functions including production of biomass and net primary productivity (NPP), moderation of climate, purification of water, biodegradation of pollutants, storage of water and plant nutrients, and recycling of elements (Lal 2009). At the same time, soil is being degraded by human activities resulting in reduction of its capacity to perform ecosystem functions. In other words, soil degradation refers to reduction of soil's current or potential capacity to perform ecosystem functions, notably the production of food, feed, and fibre as a result of one or more degradation processes. Principal soil degradation processes include physical (e.g. decline in soil structure, crusting, compaction, accelerated erosion), chemical (e.g. nutrient depletion, elemental imbalance, acidification, salinization), and biological (e.g. depletion of soil organic matter, reduction in the activity, and species diversity of soil microorganisms) (Lal 1993). Depletion of natural resources and increasing competition for limited soil and water resources particularly in the tropical coastal areas have been related to resource/soil degradation.

3.4.1 Global Assessment

The Global Land Assessment of Degradation (GLASOD) study estimated that nearly 8.7 billion hectares of agricultural land, pasture, forest, and woodland of 2 billion hectares (22.5%) have been degraded since mid of the twentieth century. Some 3.5% of the total area has been degraded so severely that it can be reversible only through costly engineering measures, if at all. Just over 10% has been moderately degraded and is reversible only through significant on-farm investments. Another nearly 9% is lightly degraded and easily reversible through good land husbandry practices (Oldeman et al. 1991).

The most important on-farm effects of land degradation are declining potential yields. The threat of degradation may also be reflected in the need to use a higher level of inputs in order to maintain yields. Serious degradation sometimes leads to temporary or permanent abandonment of some plots. In other cases, degradation induces farmers to convert land to lower-value uses; for example, less-demanding cassava may be substituted for maize, fallow periods lengthened, cropland converted to grazing land, or grazing lands converted to shrubs or forests.

3.4.2 Hotspot Areas of Degradation

Although human activities accelerated the land degradation (both soil and water), there are some 'hotspots' where the land degradation poses a significant threat to food security for large numbers of poor people, local economic activity, and important environmental products and services. However, significant differences exist in the patterns of land use among the regions, particularly where intensive agricultural systems are used. There are thus notable differences between the problems highlighted in the different regions. Priority concerns for the future will increasingly be related to environmental damage in Latin America (deforestation, water issues, chemical pollution, seawater intrusion) and dryland and hillside regions where levels of poverty are high. In Africa land degradation will continue to be linked to food security problems and to the lack of adoption technologies for sustainable agricultural intensification in areas of rapid population increase. In South and West Asia, the principal degradation problems are those that pose challenge to irrigated agriculture (salinization) and the livelihood of the poor in dryland areas (devegetation and soil fertility degradation). In the richer areas of East and Southeast Asia, environmental concerns particularly coastal deforestation will also come to the fore and water issues. In the poorer areas of this region, threats to food supply from stagnating yields in irrigated areas, loss of land to invasive species, and the challenges of managing lower-fertility and sloping lands are likely to be more important (Scherr and Yadav 1996).

3.5 Management of Tropical Soils

Soil management is an integral part of land management and tends to focus on differences in soil types and soil characteristics to define specific interventions that are aimed to enhance the soil quality for a particular land use. In the coastal areas, some specific soil management practices are needed to protect and conserve the precious soil resources. An effective response to land degradation at local levels calls for improving the incentives for farmers to care for their land and improving their access to the knowledge and inputs required for proper care. Proper management of soils require thorough understanding of the soil resource of a place, identification of the constraints, and choosing right options for amelioration/correction to improve the soil productivity. This also requires policy supports and involvement of all the stakeholders and supporting organizations because soil degradation is no more a localized issue as it has link with the global food production system.

3.5.1 Status of Soil Resources and Management Options

The adverse effects of human activities on soil and the ensuing degradation affect the food production systems and ecosystem services emanating from the soil environment. This can be alleviated through strategies involving soil restoration based on ecological principles alongside crop management. While the adverse effects of soil degradation on food and livelihood security can be buffered to certain extent by crop management, there is no economically viable alternative to soil quality restoration than the integrated resource management in the coastal areas. Furthermore, there should be multidisciplinary approach where soil scientists have to work with crop scientist to improve nutrient capture from degraded soil by the genetic manipulation of crop plants (Hirsch and Sussman 1999) along with measures to improve the soil quality. In view of the increasing demand for land besides food production and improvements in its nutritional quality, there is a need for change in the context of agricultural science particularly land management (Evans 2005; Brklacich et al. 1991). At the same time, there are significant variations in soil degradation and challenges to crop production among different soil groups and regions which demand multidimensional approach to the problems. Table 3.1 highlights some of the soil-related problems normally encountered in the tropical region and suggested management options to address the challenges in the context of improving the soil productivity and enhancing the food production.

Conversely it is equally important to understand how the preferred management option is sustainable in addressing both the environmental concerns and human food demand, diffuse and minimize pollution from agricultural practices, predict changes in crop productivity over time, and adapt to ecological systems of changing societal needs (Ewert et al. 2005; Giloli and Baumgärtner 2007). These sustainable and efficient practices should address global environmental impacts of climate change and developmental needs of human population. Further, sustainable practices of soil management must be fine-tuned to site-specific needs and the growing

Table 3.1 Global soil resource problems and their management options

Sl. no.	Soil resource/problem	Management options
1	Soil resources are not equally distributed among biomes and geographic regions	Choose land use and farming system on the basis of climatic, physiographic, and hydrologic parameters with due consideration for local aspirations
2	Most soils are prone to degradation by land misuse and soil mismanagement particularly during tillage operations	Select cropping systems, tillage methods, water conservation, and nutrient management options on the basis of soil quality and desired output suitable for the land
3	Soil erosion and erosion-induced degradation depend on 'how' rather than 'what' crops are grown	Adopt conservation agriculture, mulch farming, cover cropping, contour hedges of perennials, and control grazing (mainly in the sloppy land of the tropics)
4	Susceptibility to soil degradation increases with increase in mean annual temperature and decrease in precipitation	Identify management systems with low cropping and grazing intensity (e.g. drought-tolerant agroforestry model) and based on water harvesting, groundwater recharge, and multiple use of scarce water resources
5	Soil resilience depends on inherent physical, chemical, and biological properties and processes	Identify land use and soil management practices that will maintain and enhance soil's ability to recover from anthropogenic and natural perturbations (positive C and elemental/nutrient balance)
6	Processes of soil degradation operate at a faster rate than those of restoration	Identify key soil properties and processes and understand their critical/threshold levels to avoid irreversible soil degradation
7	Soils are a nonrenewable resource over the human time scale	Choose preventative measures for erosion, salinization, and pollution over restorative inputs
8	Soil structure depends on volume, stability, and continuity of retention and transmission pores	Promote green manuring and crop residue incorporation and activity of earthworms, include cover crops with a deep tap root system, and use compost and organic amendments
9	Optimal levels of soil physical properties and processes are important to the effectiveness of chemical and biological properties and processes	Improve soil structure and optimize soil temperature and moisture regimes to enhance use efficiency of fertilizers and realize the benefits of biological N fixation, nutrient mobilization, and mycorrhizal inoculation
10	Soil productivity is constrained by the weakest parameter/link (e.g. PAW, micronutrients, SOC concentration, rooting depth)	Use INM to replace macro elements and micronutrients harvested in crops and animal products and adopt micronutrient-dense varieties

Modified from Lal (2008)

aspirations of rapidly increasing populations in developing countries. Ecologically restored and judiciously managed, global soil resources are adequate to meet the essential needs of the present and future populations. Therefore, there is a need for a paradigm shift in land husbandry (Gowing and Palmer 2008) and principles and practices of soil management.

3.5.2 Sandy Soils

As discussed earlier, sandy or skeletal soils found in the coastal region and the hinterland of the tropics have high proportion of drainage pores, less clay, and low water holding capacity. Therefore, water and dissolved substances are rapidly lost to deeper layers in the soil or translocated to groundwater (Osunbitan et al. 2005). These soils have been referred to as droughty soils and also nutrient deficient. Higher doses of fertilizers are sometimes recommended to counteract the low fertility and the inability of the soils to retain nutrients. Over-fertilization with nitrogen frequently leads to contamination of the groundwater; a high concentration of nitrate in drinking water is a health hazard particularly to the very young and the very old. A nutrient management plan, based on leaching losses and retention ability of the soil, should be developed for its management. Wind erosion may carry applied fertilizers to water bodies. It is hence important to have residue enhancing crop rotations, cover cropping, reduced tillage, shelter belts, and even grassed waterways. Sandy soils have poor structure, and therefore it is difficult to maintain a reasonable ground cover without appropriate land management strategy.

It is long been recognized that enhancing the organic matter content is key to alleviating the soil moisture and nutrient retention problems of such soils. Conventional agronomic practices have not been successful or the systems have not been sustainable, and this presents one of the greatest and immediate research challenges for the use of these soils. In such soils grasses that are drought tolerant, energy plantation and fodder crops having the capacity to produce high amounts of biomass can be grown. In the coastal areas, natural vegetation should be encouraged to stabilize the sand dunes, while efforts should continue to prevent further land degradation. In addition, rainwater harvesting should be part of the land management strategy to provide irrigation water to the crops under such dry conditions.

3.5.3 Waterlogged and Saline Soils

In the coastal lowlands, waterlogging and salinity are twin problems for improving crop production. Waterlogging is linked to high rainfall, drainage congestion, rising of groundwater, and low elevation, while salinity is related to seawater intrusion, rising of saline groundwater, and capillary rise of salts when evaporation exceeds precipitation during the dry season. In the coastal region, many times waterlogged condition gets aggravated when high tide coincides with heavy rainfall. During rainy season in the coastal areas, water table rises up to the surface resulting in ponding of water in the depression and surface. It also results in accumulation of several toxic compounds, soluble form of aluminium, iron, and other organic compounds.

Similarly the increase in salt accumulation is reflected as crop failure and gradual appearance of wild salt-tolerant plants, and the soil salinity is measured by ECe of soil. Salts including sodium accumulate and increase at the soil exchange sites by replacing other ions. As human-induced processes, this may occur mainly through incorrect planning and management of irrigation schemes. Sometimes acid-saline soils occur due to the presence of sulphate and/or release of iron and aluminium from weathering. Pumping of groundwater for urban and industrial use is a major emerging cause. In most places in the coastal areas, salinization occurs along with waterlogging causing severe degradation and lowering the productivity of land. The coastal lowlands, where surface water stagnation occurs, need proper land shaping and provision of surface drains. Land shaping is discussed separately as it is most important to manage soil salinity and acid-saline soil as well. Some of the important management measures are:

- Install drainage wherever gradient is available and clean the existing drainage network.
- Use crop rotations and include deep-rooted crops such as alfalfa, clover, etc.
- Grow crops suited to wetter soil conditions or crops that are planted later in the growing season. For example, rice, soybeans, winter wheat, pasture, or agroforestry.
- Use disease-resistant/disease-tolerant crop varieties in such areas.
- Use tillage carefully to expose soil to the air for evaporation and soil warming.
- Provide diversion channels and one-way sluice gates to prevent the entry of water from nearby hills or sea water
- Encourage earthworm populations for macropore development, by leaving residue on the soil surface wherever possible.
- Use bio-drainage methods to remove some excess moisture from the soil.

3.5.4 Acid Sulphate Soils

This is one of the serious limitations, in some places, along the coastal areas. Acid sulphate soils are formed and deposited in areas that are, or once were, coastal environments. Coastal acid sulphate soils are commonly found in mangrove forests, saltmarsh, floodplains, and salt- and freshwater wetlands. This can also result from the drainage of coastal swamps. Acid sulphate soil is referred to the soils that contain metal sulphides and consequent low pH (<4.5). In an undisturbed and waterlogged state, these soils may pose no or low risk. However, when disturbed or exposed to oxygen, acid sulphate soils undergo a chemical reaction known as oxidation. Oxidation produces sulfuric acid, and as a result the soil pH is lowered. Sometimes these soils may contain soluble salts with high EC, when these soils contain both acidity and salinity called acid-saline soils (pH <5.5, EC >4.0). Human activities may increase their formation and accumulation; as a result, water bodies become brown to dark reddish colour due the presence of iron. These soils should be kept under wet condition to minimize acid formation and use land shaping techniques to remove the toxics from the soil surface. Use poultry manure, agricultural lime, and organic matter to reduce the effect of soil acidity on crop plants.

3.5.5 Policy Support

Management of coastal land resources in the context of enhancing the production depends on several factors. Among other things appropriate policy support is very essential to bring the desired changes in sustainable use of this precious resource. Based on the lessons learned from past successes and failures in managing land resources in the tropics, the following policy recommendations are suggested.

- Develop more flexible and participatory planning systems for sustainable use of land resources.
- Provide for suitable legislation to prevent land degradation and conservation of soil resources by classifying or notifying hotspot areas.
- In the same context, it is highly essential to improve information systems for land management.
- Promote investment in land improvement and conservation activities.
- Modify property rights to encourage long-term land investments and improvement.
- Support local organizations to manage local resources.
- Promote alternative choices of land use if it helps to improve the land quality.
- Develop international and national network for cooperation and information interchange with reference to land management strategies and methods.
- Encourage rural income growth and income diversification to minimize land degradation.
- Reduce discrimination against marginal regions, farmers, and other deprived who have a stake at the land improvement and encourage public investment.
- Increase research and technology development for land improvement.

3.6 Conclusion

Land encompasses soil, water, microclimate, vegetation, and landscape. Soil resource assessment includes characters, constraints, and its potential for sustainable use, while the soil-forming factors and processes get reflected in the assessment. Further, all these elements are interlinked and influence the characters of the soil found in the coastal tropical region. Therefore, in general, soil resource assessment and its sustainable plan are considered as land management. Global-level assessment of soil resources revealed that in the coastal region taxonomically, Entisols, Inceptisols, Alfisols, and Histosols are found having different potential and constraints.

Beginning from the second half of the twentieth century, human activities have accelerated the land degradation in addition to the natural hazards and constraints. Further expansion of cultivation into areas of fragile soils or critical habitats for biodiversity preservation can lead to significant environmental deterioration unless carefully managed. Areas with rapidly increasing domestic and industrial demands for water and soil along with poor agricultural land management could threaten water availability, land quality, and human health.

Land degradation particularly human accelerated is one of the most obvious factors influencing the agriculture, peoples economics, health and wellbeing, ecosystem, and hence livelihood of world population. With adoption of proven management options, global soil resources are adequate enough to meet food and nutritional needs of the present and future population. Further, land form-specific soil and water conservation measures need to be followed in the coastal ecosystems where the terrain is undulating, involving mechanical measures as well as vegetative barriers. International cooperation is very essential to study the climate change impact on the coastal land resources and suggest policy options.

References

Agrawal RP (1991) Water management in sandy soils by compaction. Soil Tillage Res 19:121–130

Brklacich M, Bryant CR, Smit B (1991) Review and appraisal of concepts of sustainable food production systems. Environ Manag 15:1–14

Buringh P (1970) Introduction to the study of soils in tropical and subtropical regions. Agricultural University of the Netherlands, Centre for Agricultural Publishing and Documentation, Wageningen, p 99

Doran JW, Parkin TB (1994) Defining and assessing soil quality. In: Doran JW, Coleman DF, Bezdicek DF, Stewart BA (eds) Defining soil quality for a sustainable environment, Special Publication, vol 35. Soil Science Society of America, Madison, pp 3–21

Evans LT (2005) The changing context of agricultural science. J Agric Sci 143:7–10

Ewert F, Rounsevelle MDA, Reginster I, Metzger MG, Leemans R (2005) Future scenarios of agricultural land use. I. Estimating changes in crop productivity. Agric Ecosyst Environ 107:101–116

FAO (1991) Guidelines: land evaluation for extensive grazing, FAO Soils Bull.5 8. FAO, Rome

Giloli G, Baumgärtner J (2007) Adaptive ecosystem sustainability enhancement in Sub-Saharan Africa. EcoHealth 4:428–444

Gowing JW, Palmer M (2008) Sustainable agricultural development in Sub-Saharan Africa: the case for a paradigm shift in land husbandry. Soil Use Manag 24:92–99

Hartemink AE (2004) Soils of the tropics. Geoderma 123:373–375

Hirsch RE, Sussman MR (1999) Improving nutrient capture from soil by the genetic manipulation of crop plants. Trends Biotech 17:356–361

Klinge H (1968) Report on tropical podzols, 1st draft. FAO, Rome

Lal R (1993) Tillage effects on soil degradation, soil resilience, soil quality and sustainability. Soil Tillage Res 27:1–7

Lal R (2008) Laws of sustainable soil management. Agron Sustain Dev 29:7–9

Lal R (2009) Soil degradation as a reason for inadequate human nutrition. Food Secur 1:45–57

Oldeman L, Hakkeling R, Sombroek W (1991) World map of the status of human- induced soil degradation. GLASOD-ISRIC, Wageningen

Osunbitan JA, Oyedede DJ, Adekalu KO (2005) Tillage effects on bulk density, hydraulic conductivity and strength of a loamy sand soil in southwestern Nigeria. Soil Tillage Res 82:57–64

Scherr SJ, Yadav S (1996) Land degradation in the developing world: implications for food, agriculture, and the environment to 2020, Food, Agriculture, and the Environment Discussion Paper 14. International Food Policy Research Institute, Washington, DC, pp 20036–23006

Valentin C, Bresson LM (1992) Morphology, genesis and classification of soil crusts in loamy and sandy soils. Geoderma 55:225–245

Coastal Floral Diversity and Its Significance

4

Abstract

There is a wide diversity of flora which occupies different tropical coastal ecosystems in the tropics and performs certain important ecological functions. They provide environmental services to man in the form of buffer zones against storm surges, reduced beach erosion, and recreation areas, among others. Some of these plants are also used in ethnomedicinal practices by the local people. However, coastal ecosystems particularly mangroves, wetlands, corals, sand dunes, and marshes are severely threatened all over the tropics by human activities due to their location near the ocean, ideal spots for tourism, their resource endowments, and over-exploitations. Therefore documenting and understanding different flora of coastal ecosystems, their ecological importance, and economic significance that inhabit these regions are extremely important for their judicious use and conservation. Our aim in this chapter is to describe what we know about various aspects of the vegetation of these tropical coastal areas. We also discuss variations in species diversity, number, and extent across different tropical coastal countries and their significance particularly mangroves, sand dune, and marshy vegetations.

Keywords

Coastal ecosystems · Endangered plants · Ethnomedicinal value · Dune vegetations

4.1 Introduction

The transitional area between land and sea is generally known as coastal areas. Due to the transitional nature, an environment with a distinct structure, diversity, and flow of energy exists as the land and ocean meet each other in different ways. Such diverse environment is manifested in the form of most complex coastal ecological

V. Ayyam et al., *Coastal Ecosystems of the Tropics - Adaptive Management*,
https://doi.org/10.1007/978-981-13-8926-9_4

systems on Earth (FAO 1998). The major ecosystems of these areas are mangroves, wetlands, corals, marshes, sea grass, estuaries, bays, and sand dunes which are home to different types of plants and animals besides providing valuable ecosystem services animals (Michel and Pandya 2010). As they are natural systems, it tends to establish as a contiguous patch in any favourable conditions until altered by human activities or natural phenomena.

All these ecosystems consist of different kinds of flora, viz. grasses, small plants, shrubs, bushes, and trees. At the same time, the distribution of these floras and its species varies among different coastal ecosystems. They exhibit remarkable morphological and genetic diversity as they tend to adapt to the different ecosystems. Based on the physical condition, they manifest as small plants, creepers, climbers, shrubs, and trees with different root and shoot modifications. They are valuable to the natural ecosystems as well as for the coastal populations. They not only are the source of food and livelihood to the coastal communities but also provide habitat for the wild animals. As a result coastal areas have become centres of human economic activity with ever-increasing population density.

In this context, it is very essential to document different flora and its distribution in different coastal ecosystems rather than studying the floral distribution. This will help to conserve, derive continued benefit, and continue the scientific study to understand the effect of human activities and global changes. This is more important because most of the coastal ecosystems are very sensitive to changes in the environment to which the flora is a dynamic part, and there is concern that some areas are now struggling to maintain their diversity due to human activity, the introduction of non-native species, and other factors. With these backgrounds this chapter attempts to provide information on different flora of various tropical coastal ecosystems with reference to taxonomical details, benefits, functions, and its conservation.

4.2 Importance of Coastal Flora

The floral diversity of the coastal areas is significant and very unique primarily due to its adaptation and existence in different coastal ecosystem. They are vital component of coastal biodiversity and its various interconnections. In other words biodiversity is the backbone of sustainability of the ecosystem and its continued services to humans. Coastal floral biodiversity plays an important role in maintaining ecosystem productivity, stability, sustainability, and other ecosystem services that are essential for human wellbeing (Kappelle et al. 1999). Some of the plant communities particularly mangroves and littoral vegetations act as a bio-shield against any calamities originating in the sea. Due to the unique adaptation, as in the case of dune grasses and mangroves, coastal flora as a community helps to conserve the coastal ecosystem itself. Other arguments in favour of biodiversity for investment in conservation of biodiversity at global, national, regional, and local levels include the possible discovery of new medicines for humans developed from native plants with the active support from the ethnomedicinal practitioners not yet discovered by scientists. It has been proved that for majority of the crop production problems of biotic origin, the solution was found mostly in the wild relatives of cultivated species in the form of genes for resistant or novel product.

The analysis of data pertaining to the coastal flora points to the fact that our knowledge about the coastal flora and its diversity in the tropical region has significant gaps. The lack of comprehensive information hampers society's ability either to understand the magnitude of the problem or to prevent further losses. Therefore it is moral obligations on the part of human beings to support the co-living things of the Mother Earth, and humans have no right to cause the extinction of this coastal flora across their habitat (Devall 2006). In spite of deriving innumerable benefit from the coastal biodiversity, human activities are one of the major causes of biodiversity loss. In the last century, several species has become extinct, and many others are threatened because of unjust anthropogenic activities. At the same time, climate change aggravated by human-induced greenhouse gas emission is fastly becoming unfavourable for many of the coastal ecosystems resulting in ecological consequences. Therefore, a balance approach between economic benefit through exploitation and conservation of coastal floral diversity is needed.

4.3 Distribution of Coastal Flora

4.3.1 Biodiversity of Coastal Flora

Biodiversity particularly floral diversity provides the basis for life on Earth, including human life, and is the key for safeguarding the wealth of the world for future generations. In general, biodiversity is assessed with reference to genes, species, and ecosystems and to make comparisons based on quantitative values ascribed to one or more of these levels. But the distribution of flora and its biodiversity in the coastal areas of the tropics is not uniform. There are areas on Earth which supports specific species/particular taxa, more biodiversity and number of species than other areas due to natural endowments and favourable physical conditions in a particular ecosystem. For example, only mangrove species can thrive under marshy conditions; on the other hand, grasses are found in sand dunes.

The tropical coastal region is very rich in floral biodiversity which are unique than other regions of the world. In order to be able to conserve tropical biodiversity, it is advantageous to know species distribution and abundance. It is well documented that the tropical biodiversity constitutes more than half of the world's species (Wilson 1988) which probably exceed more than that as far as marine biodiversity of the tropical waters is concerned. Nevertheless the species diversity is not equally spread over the total tropical area but is more or less clumped in centres of high diversity, and these are sometimes characterized by high levels of endemism (Gentry 1986; Myers 1990).

On the other hand, the tropical island species appear to have a patchy in its spatial distribution often limited by the geographical isolation of the islands. The vast expanse of ocean around the islands limits its spread, and the natural selection continued for the island conditions resulting in high level of endemism as observed today. Furthermore, the ecology of the majority of tropical island species is poorly known, and this is especially true of the largest taxonomic group. It is widely accepted that the biodiversity present in tropical forests and coral ecosystem represents an enormous potential resource of species and genetic variation which may benefit humanity (Balic and Mendelsoon 1991; ICBP 1992).

4.3.2 Status of Coastal Flora

The tropical coastal region has diverse land forms and ecosystems which are reflected in its floral diversity and distribution. A summary of coastal flora of major tropical nations is given in Table 4.1. There is wide diversity in total flora and threat to its survival across the tropical region. In general, the islands are rich in vascular plants with high endemism due to its geographical isolation and unique physical conditions. The Indian Ocean region has the highest number of vascular plants with endemism; for that reasons the Western Ghats (India) Sri Lanka is recognized as one of the global biodiversity hotspots (Chua 1986). Another important floral diversity hotspot is found in the Caribbean region. Haiti, Cuba, and Jamaica together contribute more than half of the regional flora and its diversity. At the same time Jamaica and Cuba are having the largest number of endangered/threatened species. Though Brazil, Belize, and Mexico have more floral population, it is very difficult to separate out only the coastal species. Many of the plants in these regions have universal distribution, but still a large number of coastal flora are endangered due to human activities. The Pacific region island nations called coral triangle are known for its richest marine biodiversity (Allen 2007) and Southeast Asia for its terrestrial plant diversity. At the same time, this region has the highest number of endangered or threatened species which are very unique to the region.

Most rainforest islands are in the Pacific, but there are island nations with significant floral diversity that are located in the Indian Ocean (the Andaman, Nicobar, Comoro, Mascarene, and Seychelles islands), a few in the Atlantic (Principe, São Tomé, and Annobon), and many in the Caribbean. The total area of these rainforests in the tropical coastal region is comparatively small but rich in species than their continental counterparts and high rate of endemism. At the same time, this coastal flora is also highly endangered (Primack 2006).

4.3.3 Flora of Different Coastal Ecosystems: A Case Example of Southeast Asia

The plants of coastal areas are grouped together in communities, or plant formations, that owe their special physiognomy to habitat conditions found at the site, to the kinds of species that have managed to arrive and become established locally, and to the length of time these species have been growing together, free from major disturbance. As can be seen from Table 4.2, the coastal area of Southeast Asia is

Table 4.1 Reported flora and its status in some tropical coastal countries

Name	Reported vascular plants	Endangered/IUCN red listed	References
Caribbean region			
Antigua and Barbuda	1113	04	Daltry (2007) www.earthendangered.com
Cuba	6015	42	www.cubanaturaleza.org www.earthendangered.com
Bahamas	1124	07	BMMRO (2009) www.earthendangered.com
Grenada	1068	02	GoG (2014)
Guyana	1182	18	FAO (2005)
Haiti	4685	22	www.earthendangered.com
Jamaica	2746	186	GoJ (2010), www.earthendangered.com
American region			
Belize	4423	22	Meerman (2005)
Mexico		283	http://*earthsendangered.com/* search-regions3.asp
Brazil	–	404	http://*earthsendangered.com/* search-regions3.asp
Asian region			
India	18,000	247	BSI (2011)
Bangladesh	–	486	Irfanullah (2011)
Thailand	–	**90**	http://*earthsendangered.com/* search-regions3.asp
Myanmar	–	**37**	http://*earthsendangered.com/* search-regions3.asp
Philippines	–	**217**	http://*earthsendangered.com/* search-regions3.asp
Papua New Guinea	–	**149**	http://*earthsendangered.com/* search-regions3.asp
Indian Ocean islands			
Sri Lanka	3771	280	BDS/MoE and DNBG (2012)
Mauritius	691	86	NRCBD (2015)
Maldives	583		http://*earthsendangered.com/* search-regions3.asp
Seychelles	707	44	http://*earthsendangered.com/* search-regions3.asp
Madagascar		**284**	http://*earthsendangered.com/* search-regions3.asp
Pacific islands			
Solomon	7000	13	http://*earthsendangered.com/* search-regions3.asp
Fiji	324	63	https://chmfiji.wordpress.com/about/ overview-of-fijis-biodiversity
Palau	1200	05	Kitalong (2010)

Table 4.2 Total flora recorded in Southeast Asian countries across different coastal ecosystems

Name of the country	Coastline (km)	Strictly coastal flora Sandy coast	Rocky coast	Mangrove	Flora of the coastal areas[a]
Singapore	193	219	79	254	456
Malaysia	4675	314	114	328	702
Brunei	161	142	51	181	307
Indonesia	54,716	328	108	332	709
Timor-leste	706	121	55	120	204
Thailand	3219	269	80	271	556
Burma	1930	193	68	197	372
Cambodia	443	179	59	175	318
Vietnam	3444	174	70	226	478
Philippines	36,289	147	89	238	510

CIA (2008), Wong (2005)
[a]Includes near the sea vegetation and coastal flora

characterized by environmental heterogeneity that is caused in part by geomorphological diversity and geological process. Flora of this region is found on a variety of land forms, and the plants are greatly influenced by oceanicity. The resulting scenario is a mosaic of vegetations on different ecosystem types.

Studies revealed that coastal vascular plants are generally very widespread. Most of them are distributed in the Asian/African/Australian category (517 coastal species sensu lato or 414 coastal species sensu stricto). Ninety-eight species sensu lato or 78 species sensu stricto are pantropical because coastal species are usually water-dispersed which is effective over long periods of time. A number of such species are native to America but naturalized in Southeast Asian region. For example, *Passiflora foetida* from South America and *Argemone mexicana* from the Caribbean region have arrived here. Interestingly there are 55 coastal species sensu lato or 37 coastal species sensu stricto that have a limited distribution. Some of them are endemic, such as *Eulalia ridleyi* that can only be found east coasts of Malaysia, while others are limited to one or two areas such as *Dracaena maingayi*, which can be found in Peninsular Malaysia and Singapore. Such coastal species with limited distribution might be more prone to extinction from habitat destruction as a result of rapid coastal development.

In this region, a total of 876 species from 133 families across 7 vascular plant phyla were found to be coastal. Of those, 192 were placed in the 'near the sea' category. Hence, only 684 species were recorded from sandy coasts, rocky coasts, and mangrove forests with 392 species found in sandy coasts, 129 species in rocky coasts, and 380 species in mangrove forests. The latter 3 numbers do not add up to 684 as some species are found in more than 1 habitat. Across the region, mangrove ecosystem has large number of flora, while rocky coast has less due to the difference in favourable growing conditions. Among the countries of this region, Indonesia has the highest number of coastal species (709 species sensu lato) followed by Malaysia (702 species sensu lato), while Timor-Leste has the lowest number of species (204 species sensu lato).

4.4 Coastal Flora of Ecological Significance

4.4.1 Coastal Zonation

Tropical coastal areas are characterized by environmental heterogeneity that is primarily due to geomorphological diversity and different in ocean-land surface interactions. This is manifested in the form of different ecosystems such as mangroves, sand dunes, wetlands, rock shore, etc. that is inclusive of variety of flora and fauna. Near the ocean, plant communities on sandy deposits usually reveal a sharp zonation pattern subdivided into an outer pioneer zone and inner dense coastal undergrowth. Plant communities farther inland do not necessarily follow a linear sequence across the dune field or coastal plain. However, they are located along a gradient of increasing community complexity, that is, greater species richness and cover, taller stature, and greater biomass. Therefore, the best approach to presenting these vegetation types is perhaps along the natural gradient from the beach towards inland. At the same time, it is generally observed that there is not only a gradient of factors resulting from the presence of the ocean (salt spray, wind velocity, seawater intrusion, sea surges) but also a temporal gradient in the age of the substrate, which increases in this direction. In contrast, on broad coastal plains, this gradient may not be continuous in sea-inland directions which are also subjected to human modification. Therefore conservation management should include all these zones irrespective of its areas so that all the coastal flora and fauna are conserved (Boelaert-Suominen and Cullinan 1994).

4.4.2 Coastal Floral Communities

Due to variations in the physical, climatological, and geological conditions, different plants and their communities are observed along the coastal areas. In the evolutionary process, these vegetations not only adapted to different conditions but also assumed significant ecological role. This includes mangrove and non-mangrove species, salinity-tolerant woody species, tree/shrubs with high economic values, and non-mangroves with high ecological significance. Different flora of coastal ecosystems and their ecological significance are summarized in Table 4.3.

Mangrove communities comprise unique plant species that form the critical interface between terrestrial, estuarine, and nearshore marine ecosystems in tropical and subtropical regions. They grow only in mangrove environment and do not extend into terrestrial plant community and are morphologically, physiologically, and reproductively adapted to saline, waterlogged, and anaerobic conditions. It is often associated with another group of plant communities known as mangrove associates. Mangrove ecosystem represents one of the most productive natural capital capable of producing a wide range of goods and services for coastal environments and communities and society as a whole. When sustainably managed, they can support the livelihoods of millions of coastal people while storing globally

Table 4.3 Important coastal flora of the tropical region and its characters

Name of the species	Characters and its use
Mangrove species	
Avicennia spp.	10 m height with 35cm diameter, multiple branches, strong stem, aerial roots (pneumatophores), dense populated withstand salinity and wind
Sonneratia spp.	20 m height with 20 cm diameter, pneumatophores present dense population, withstand salinity
Rhizophora spp.	25 m height with 30 cm diameter, pneumatophores present, aerial roots and buttresses are present withstand the salinity
Non-mangrove shelter belts	
Casuarina equisetifolia Forst.	35 m with 50 cm diameter, low branching, straight stem, fast growing tolerant to wind and salinity, old trees form buttresses, controls soil erosion
Manilkara littoralis	32 m with 130cm diameter straight strong stem, wider root spread with buttresses, highly tolerant to salinity and wind
Ficus spp.	20 m height 60 cm diameter, tolerant to wind and the aerial roots for structural support
Thespesia populneoides	10 m height with 30 cm diameter withstand high wind and salinity, deep-rooted fast-growing trees
Hibiscus tiliaceus L.	10 m height with 30 cm diameter, common coastal plant, stem mostly lying on the ground, curved branches withstand heavy wind and salinity, deep rooted and fast growing
Barringtonia spp.	12 m height with 25 cm diameter, close proximity to the sea, deep-rooted short-boled stem, withstand salinity
Saline tolerant, littoral	
Pongamia pinnata L.	25 m height with 80 cm diameter, multiple branches deep rooted, and withstand salinity and wind
Azadirachta indica A. Juss.	20m height 45cm diameter deep-rooted strong stem withstand salinity and wind
Morinda citrifolia	10 m height with 20 cm diameter, deep-rooted saline-tolerant bush-shaped branched stem
Cassia fistula L.	12 m height with 18 cm diameter, strong branched bole salinity tolerant deep rooted
Sandy saline areas	
Chrysobalanus icaco (beach plum)	They grow in the sandy areas which is saline as well due to the intrusion of sea water but not marshy to accommodate mangrove vegetations
Dalbergia ecastaphyllum (coin vine)	
Jacquinia armillaris	
Secondary protective layer with high economic benefits	
Cocos nucifera L.	30 m height with 30 cm diameter, straight single bole, profuse lateral roots tolerant to salinity and mild wind
Anacardium occidentale L.	14 m height with 30 cm diameter, irregularly shaped stem with multiple branches, deep rooted, and withstand salinity and wind
Syzygium spp.	30 m height with 40 cm diameter, strong stem with deep roots having multiple branches, salt and wind tolerant

(continued)

Table 4.3 (continued)

Name of the species	Characters and its use
Pandanus tectorius	20 m height with 15 cm diameter, moderate growth rate, grown in dense population having prop root near the base, roots are adventitious and often branched, the roots form a pyramidal tract to hold the stem, trunk is short, and branches withstand the wind and salinity
Non-mangrove with ecological values	
Ipomoea spp.	Typical vine covered with entire coastal area and control the sand erosion, fast growing and each nodes of the vine having dense roots
Ipomoea pes-caprae (creeper)	Creepers cover the coastal areas, protect sand dunes and sandy beaches
Sporobolus virginicus (creeper)	
Scaevola plumieri (creeper)	
Remirea maritima	Scrub vegetation of the sandy coast
Jacquinia armillaris	
Vriesea neoglutinosa (scrub)	
Clusia hilariana (*shrub*)	
Nypa fruticans	The stem or trunk of the palm is under the mud, dense population the palm leaves length is around 9 m. the stem branches at intervals to form individual clumps of large withstand swampy coastline often with mangroves
Allagoptera arenaria	Stemless palm
Bixa orellana L.	6 m height with 12 cm diameter, multiple branched stem, deep rooted, and withstand heavy wind and salinity
Bambusa arundinacea (Retz.)	30 m height, clumps having many numbers of culms with mat like dense fibrous roots around the clump to hold the soil, withstand heavy wind and salinity, grown tall and straight poles

significant quantities of carbon and reducing the vulnerability of coastal regions to storm surges and other threats.

Mangrove plants characteristically range from trees (*Avicennia*, *Rhizophora*), to shrubs (species of *Aegiceras*, *Aegialitis*, *Pemphis*, and *Conocarpus*), to the trunkless palm (*Nypa fruticans*), and ground fern (*Acrostichum*). Trees and shrubs might be columnar and erect (*Pelliciera rhizophorae*, *Bruguiera parviflora*), spreading or sprawling (*Acanthus* spp., *Scyphiphora hydrophyllacea*), and multiple-stemmed (*Ceriops decandra*). Growth form might also vary within the same species having both an erect tree form and low tangled thicket forms (*Lumnitzera littorea* and *Rhizophora*). In general, edge plants (both waterfront and landward) have more lower limbs and foliage, and their stems are laterally sprawling and sinuous, rather than erect and straight. Some species typically form combined closed canopies as in the case of *Avicennia marina*, *Rhizophora apiculata*, *Bruguiera parviflora*,

Bruguiera gymnorhiza, *Camptostemon schultzii*, and *Xylocarpus* spp., while others are commonly found as undercanopy plants beneath the closed canopy (e.g. *Aegiceras*, *Cynometra*, *Acanthus*, *Acrostichum*, and *Ceriops decandra*).

Another important plant community is found in sand dunes/beach areas all over the tropics. This vegetation is probably the best known of all the plant communities found on sandy shores. These plants form a dense cover on sheltered beaches where storm surges are uncommon, and there is some degree of protection from strong winds (*Ipomoea* spp., *Sporobolus virginicus*, *Scaevola plumieri*). On unprotected beaches, subject to the full force of the waves and prevailing winds, the low shifting foredunes are only partially fixed by plants. In either case, this vegetation is inevitably washed by the sea at one time or another. Sometimes extensive areas of sandy coastal plains are covered by open scrub vegetation that may lie behind the coastal thicket described above or farther inland. The presence of this vegetation is closely related to the topography of the strand plain. Where series of parallel beach ridges alternate with depressions in a regular pattern, dry open scrub vegetation occurs at the higher elevations, that is, on the crest of the beach ridges, while wet open scrub lies in depressions (*Jacquinia armillaris*, *Vriesea neoglutinosa*, *Clusia hilariana*)

4.5 Coastal Flora and Human Wellbeing

A large majority of people from developing countries and tropical world still depend on traditional medicine either fully or partially for primary healthcare needs; a major portion of these involves the use of medicinal plants (Bennerman et al. 1983). Often majority of the tribes from the tropical world depend on the indigenous medicinal plants. At present around 70% of the drugs used as medicines are modelled on natural compounds, yet these are derived from only around 250 plant species, when only less than 0.1% of the plant species have been examined for their medicinal value (Dobson Andrew 1998). In the rural and tribal lives of tropical world, medicinal plant sector has traditionally occupied an important position in the socio-economic position in the sociocultural and spiritual and medicinal arena.

The Caribbean and Central American region is known for its rich floral diversity and ethnomedicinal plants with high degree of endemism (Table 4.4). During the European expansion/colonial period, some of these plants are even taken to other parts of the world for commercial exploitation. Mexico recorded the highest number of vascular plants and medicinal plants with high degree of endemism. This is followed by Cuba with similar trend. Some of the Caribbean region plants having high commercial value are introduced by European colonial powers during the eighteenth to nineteenth centuries.

The practice of ethnomedicine is a complex multidisciplinary system constituting the use of plants, spirituality, and the natural environment and has been the source of healing for people for millennia. A summary of different medicinal plants and its uses is given in Table 4.5. The ethnomedicinal uses of plants are one of the most successful criteria used by the pharmaceutical industry in finding new

Table 4.4 Summary of number of medicinal plants and endemism reported from Caribbean and Central American region

Caribbean countries	Number of vascular plant species	Number of medicinal plants	% of endemism
Belize	4423	64	19
Jamaica	2746	116	31
Haiti	4685	161	30
Mexico	30,000	5000	52
Nicaragua	5796–7000	229	19
Panama	8500–9000	265	19
Cuba	6015	101	53
Puerto Rico	2128	172	10
Dominican Republic	4685	200	30

Source: Wendy Torres-Avilez et al. (2015)

Table 4.5 Details of some medicinal plants of Indian Ocean region

Sl. no.	Scientific name	Family	Habit	Parts used	Reported in aliment
1.	*Centranthera indica*	Scrophulariaceae	Herb	Whole plant	Diuretic
2.	*Crotalaria retusa*	Fabaceae	Herb	Leaves and flowers	Fever, cough,
3.	*Helianthus argophyllus*	Asteraceae	Herb	Seed, leaves	Astringent, diuretic
4.	*Justicia betonica*	Acanthaceae	Herb	Leaves	Diarrhoea, dysentery
5.	*Gloriosa superba*	Liliaceae	Herb	Seed, tubers	Snakebite, ulcers, arthritis, cholera, colic, kidney problems, typhus, itching, leprosy, bruises, sprains, haemorrhoids, cancer rheumatism
6.	*Clitoria ternatea*	Fabaceae	Herb	Flower, seed, leaves	Arthritis, vertigo, and brain weakness
7.	*Aloe vera*	Asphodelaceae	Herb	Whole plant	Burns, cuts, stings, bruises and rashes to welts, itching, blisters, infections, and abrasions
8.	*Alpinia calcarata*	Zingiberaceae	Shrub	Rhizomes	Bronchitis, cough, respiratory ailments, diabetics, asthma, and arthritis

(continued)

Table 4.5 (continued)

Sl. no.	Scientific name	Family	Habit	Parts used	Reported in aliment
9.	*Zingiber zerumbet*	Zingiberaceae	Shrub	Rhizome	Appetizers
10.	*Ardisia humilis*	Myrsinaceae	Shrub	Roots, leaves, fruit	Diarrhoea, bruise, dysmenorrhoea, gout, mental disorder, rheumatic arthritis, rheumatism, skin sore
11.	*Woodfordia fruticosa*	Lythraceae	Shrub	Flower, leaves	Diarrhoea, piles, and dysentery
12.	*Stachytarpheta indica*	Verbenaceae	Shrub	Leaves	Diabetes and fever
13.	*Melastoma malabathricum*	Melastomataceae	Shrub	Whole plant	Diarrhoea, dysentery, haemorrhoids, cuts and wounds, toothache, and stomach ache
14.	*Abutilon heterotrichum*	Malvaceae	Undershrub	Leaves	Wound
15.	*Catharanthus roseus*	Apocynaceae	Shrub	Leaves	Dysentery and diarrhoea, leukaemia, and cancer
16.	*Vitex negundo*	Verbenaceae	Shrub	Leaves, Roots	Vata and Kapha
17.	*Barleria lupulina*	Acanthaceae	Shrub	Leaves	Swellings, bleeding wounds, and boils
18.	*Abroma augusta*	Sterculiaceae	Shrub	Roots	Uterine tonic, uterine stimulant, anodyne
19.	*Cassia auriculata*	Caesalpiniaceae	Shrub	Leaves	Fevers, diabetes, diseases of urinary system, and constipation
20.	*Crinum asiaticum*	Amaryllidaceae	Shrub	Bulb and leaves	Inflamed joints and sprains
21.	*Datura metel*	Solanaceae	Shrub	Leaves, seed	Asthma, bone setting, antiasthmatic, antispasmodic, hypnotic, and narcotic
22.	*Furcaria surattensis*	Malvaceae	Climber	Leaf and stem	Venereal sores and urethritis
23.	*Solanum trilobatum*	Solanaceae	Creeper	Whole plant	Cough and cold
24.	*Barringtonia acutangula*	Lecythidaceae	Tree	Leaves, fruits, and roots	Jaundice, liver disorders, stomach disorders, leprosy, and splenic disorders
25.	*Pagiantha dichotoma*	Apocynaceae	Tree	Leaves and bark	Wounds, eye infections, toothache, fistula, snake bites, centipede bites
26.	*Cinchona*	Rubiaceae	Tree	Bark	Malaria

(continued)

Table 4.5 (continued)

Sl. no.	Scientific name	Family	Habit	Parts used	Reported in aliment
27.	*Myristica fragrans*	Myristicaceae	Tree		Diarrhoea, nausea, stomach spasms and pain, and intestinal gas
28.	*Sterculia villosa*	Sterculiaceae	Medium tree	Gum, root,	Rheumatism, diuretic, cooling, and aphrodisiac
29.	*Santalum album*	Santalaceae	Tree	Seed and mace	Patches on the skin, inflammation, itching, and eczema
30.	*Ficus racemosa*	Moraceae	Tree	Bark, root	Diabetes, liver disorders, diarrhoea, inflammatory conditions, haemorrhoids, respiratory, and urinary diseases
31.	*Averrhoa carambola*	Oxalidaceae	Tree	Fruit	Laxative for intestinal tract
32.	*Calotropis gigantea*	Asclepiadaceae	Small tree	Root, stem, and leaves	Syphilis, boils, inflammation (swelling), epilepsy, hysteria, fever, muscular spasm
33.	*Adenanthera pavonina*	Mimosaceae	Tree	Leaves, seed, and roots	Inflammation
34.	*Mesua ferrea*	Clusiaceae	Tree	Flower, leaves	Fever, vomiting, urinary tract disorders, migraine
35.	*Terminalia arjuna*	Combretaceae	Tree	Fruit	Fever, vomiting, urinary tract disorders
36.	*Terminalia chebula*	Combretaceae	Tree	Fruit	High cholesterol and digestive disorders, including both diarrhoea and constipation and indigestion
37.	*Bauhinia purpurea*	Caesalpiniaceae	Tree	Leaves, bark	Astringent in diarrhoea, ulcers, laxative
38.	*Cassia fistula*	Caesalpiniaceae	Tree	Flower, leaves, bark	Antiseptic, ulcers, and wounds
39.	*Saraca indica*	Caesalpiniaceae	Tree	Bark	Gynaecological
40.	*Phyllanthus emblica*	Euphorbiaceae	Tree	Fruit	Dietary source of vitamin C, amino acids, and minerals
41.	*Punica granatum*	Punicaceae	Tree	Fruit	Cancer, osteoarthritis

Compiled from Dagar et al. (1991), Parthasarathy (1999), WHO (1998), Guebas et al. (2006), and Ranjith Mahindapala (2004); http://bit.ly/gadgets_cheap

therapeutic agents for the various fields of biomedicine. Some outstanding medicinal drugs which have been developed from the ethnomedicinal uses of plants include vinblastine and vincristine from *Catharanthus roseus* (the periwinkle) used for treating acute lymphoma, acute leukaemias, etc.; reserpine from *Rauwolfia serpentina* (Indian snake root) used for treating hypertension; aspirin from *Salix purpurea* used for treating inflammation, pain, and thrombosis; and quinine from *Cinchona pubescens* (cinchona) used for treating malaria.

At present about 80% of the world's population rely predominantly on plants and plant extracts for healthcare. In addition, of the top 150 proprietary drugs used, 57% contain at least 1 major active compound currently or once derived from plants (Grifo and Rosenthal 1997). According to data released by the World Health Organization (WHO), ethnomedicine has maintained its popularity in all regions of the developing world, and its use is rapidly expanding in the industrialized countries, for example, in China traditional herbal preparation account for 30–50% of the total medicinal consumption. In Ghana, Mali, Nigeria, and Zambia, the first-line treatment for 60% of children with malaria is the use of herbal medicine.

In general, most of the ethnomedicinal plants are used for curing diuretic, fever, cough, respiratory problems, stomach problems, diabetics, improving immunity, healing of wounds, and antiseptic purposes. Most importantly nearly 70% of people living with HIV/AIDS use traditional medicine prepared from plants.

4.6 Conclusion

Coastal areas of the tropics contain many of the Earth's most complex and diverse ecological systems, productive in both the biological and economic senses. These areas are ecologically diverse comprising inshore waters, intertidal areas, and extensive tracts of contiguous land. Coastal zones are not only economically valuable but also include some of the most ecologically important environments and functions. Some of the most important coastal ecosystems are mangroves, coral reef, sea grasses, salt marshes, wetlands, estuaries, and bays. These ecosystems extend in different geomorphological conditions in the coastal areas. Consequently the physical conditions, along with the prevailing regional climate, greatly influence the structure and function of different plant communities besides its distribution (Fig. 4.1).

These vegetations provide protection against sea surges, sea level rise, erosion, habitat, and livelihood to millions of coastal people. The role of mangroves together with other littoral vegetations as a bio-shield deserves special mention. Similarly some dune grasses play a vital role in dune stabilization in the coastal areas. Coastal vegetations have been used in ethnomedicinal practices for a long time.

However, the coastal plant biodiversity is being depleted by the loss of habitat, fragmentation of habitat, over-exploitation of resources, human-sponsored ecosystems, climatic changes, pollution-invasive exotic spices, diseases, shifting cultivation, poaching of wildlife, etc. Estimates of species loss rates suggest that, unless current trends are reversed, from one quarter to one half of the Earth's species will become extinct in the next 30 years. Many of the tropical countries/islands offer a unique opportunity to conserve biodiversity that is increasingly under threat in the respective regions.

Dense mangrove in the sea front — Sand grass and Ipomea sp.

Littoral vegetation with strong poles — Coconut plantations

Casurina based coastal shelter belt — Pandanus with dense foliage in the sea front

Fig. 4.1 Different coastal vegetations and their ecological importance in the tropical region

References

Allen GR (2007) Conservation hotspots of biodiversity and endemism for Indo-Pacific coral reef fishes. Aquat Conserv Mar Freshwat Ecosyst 18(5):541–556. https://doi.org/10.1002/aqc.880

Balic MJ, Mendelsoon R (1991) Assessing the economic value of traditional medicines from tropical rain forests. Conserv Biol 6(1):128–130

BDS/MoE and DNBG (2012) The National Red list of Sri Lanka. In: Weerakoon DK, Wijesundara S (eds) Conservation status of the fauna and flora. Ministry of Environment, GoS, Colombo, pp. Viii+476

Bennerman R, Burton J, Chen WC (1983) Traditional medicine and health care coverage. World Health Organisation, Geneva

BMMRO (2009) Bahamas Marine Mammal Research Organisation. Annual report to The Bahamas Department of Marine Resources. Govt. of Bahamas

Boelaert-Suominen S, Cullinan C (1994) Legal and institutional aspects of integrated coastal area management in national legislation. FAO, Rome. 118 pp

BSI (2011) The botanical survey of India. Annual report 2009–10. Ministry of Environment and Forests, Government of India, New Delhi

Central Intelligence Agency (2008) Field listing– coastline. The world fact book. Central Intelligence Agency, Washington, DC. https://www.cia.gov/library/publications/ the-world-factbook/fields/2060.html. Accessed 26 Mar 2017

Chua TE (1986) Managing ASEAN coastal resources. Tropical Coastal Area Management, ICLARM, Manila, 1(1): 8–10

Dagar JC, Mongia AD, Bandyopadhyay AK (1991) Mangrove of Andaman and Nicobar Islands. Oxford and IBH Publishing Co. Pvt. Ltd, New Delhi

Daltry JC (2007) An introduction to the herpeto fauna of Antigua, Barbuda and Redonda, with some conservation recommendations. Appl Herpetol 4:97–130

Devall B (2006) Conservation of biodiversity: opportunities and challenges. Hum Ecol Rev 13(1):60–75

Dobson Andrew P (1998) Conservation and biodiversity. Scientific American Library, New York, pp 231–253

FAO (1998) In: Food and Agriculture Organization of the United Nations (ed) Integrated coastal area management *and* agriculture, forestry and fisheries. FAO guidelines, Rome. http://www.fao.org/docrep/W8440e/W8440e00.htm

FAO (2005) Global forest resources assessment 2005. Food and Agriculture Organization, Rome. http://www.fao.org/forestry/country/20807/en/guy. Accessed 20 May 2018

Gentry AI (1986) Endemism in tropical versus temperate plant communities. In: Soulé ME (ed) Conservation biology. The science of scarcity and diversity. Sinauer Associates, Sunderland, pp 286–303

GoG (2014) Fifth national report to the convention on biodiversity. Government of Grenada, Grenada, p 75

GOJ (2010) NEPA. Fourth national report for Jamaica to the convention on biological diversity. www.earthendangered.com

Grifo F, Rosenthal J (1997) Biodiversity and human health. Island Press, Washington DC

Guebas D, Koedam FN, Danielsen F, Sorensen MK, Olwig MF, Selvam V, Parish F (2006) Coastal vegetation and the Asian tsunami. Science 311:37–38

ICBP (1992) Putting biodiversity on the map priority areas for global conservations. International Council for Bird Preservation, Cambridge, p 90

Irfanullah HM (2011) Conserving threatened plants of Bangladesh: miles to go before we start? Bangladesh J Plant Taxon 18(1):81–91

Kappelle M, Van Vuuren MMI, Baas P (1999) Effects of climate change on biodiversity: a review and identification of key research issues. Biodivers Conserv 8:1383–1397

Kitalong (2010) The Republic of Palau statewide assessment of forest resources and resource strategy. Bureau of Agriculture, Republic of Palau

Meerman JC (2005) Compilation of information on biodiversity in Belize. Forest Department, Belize, p 56

Michel D, Pandya A (2010) Coastal zones and climate change. The Henry L. Stimson Center, Washington, DC, p 20036

Myers N (1990) The biological challenge: 'hot-spots' in tropical forests. Environmentalist 10:243–256

NRCBD (2015) Fifth national report on the convention on biological diversity. Ministry of Agro-Industry and Food Security, Republic of Mauritius, p 87

Parthasarathy N (1999) Tree diversity and distribution in undisturbed and human-impacted sites of tropical wet evergreen forest in southern Western Ghats, India. Biodivers Conserv 8:1365–1381

Primack RB (2006) Essentials of conservation biology, 4th edn. Sinauer Associates, Sunderland

Torres-Avilez W, Méndez-González M, Durán-García R, Boulogne I, Germosén-Robineau L (2015) Medicinal plant knowledge in Caribbean Basin: a comparative study of Afro-Caribbean, Amerindian and Mestizo communities. J Ethnobiol Ethnomed 11:18. https://doi.org/10.1186/s13002-015-0008-4. Accessed on 15th April 2018

WHO (1998) Medicinal plants in the South Pacific, World Health Organisation regional publications, Series No.19. pp 253

Wilson EO (1988) The current status of tropical biodiversity. In: Wilson EO (ed) Biodiversity. National Academy Press, Washington, DC, pp 1–18

Wong PP (2005) The coastal environment of Southeast Asia. In: Gupta A (ed) The physical geography of Southeast Asia. Oxford University Press, Oxford, pp 80–93

Coastal Fauna and Human Perturbation 5

Abstract

Coastal areas include inshore waters, intertidal areas, and extensive tracts of contiguous land providing diverse habitats for innumerable flora and fauna. The tropical coastal areas are not only economically valuable for fisheries, commerce, navigation, and recreation but also include some of the ecologically most important and unique fauna. They represent almost all the animal phyla which directly or indirectly depend on this region wholly or during part of their life cycle. They provide livelihood to millions of coastal communities. However, the combination of natural and human forces presents major challenges for conservation of faunal communities. It has been observed that beginning from the twentieth century, the level of contamination and type and amount of pollutants have steadily increased in response to the increase in industries, tourism activities, and urbanization. Besides, these areas also receive pollutants from faraway places through drainage systems. Such situations pose severe threat to the survival, proliferation, and distribution of coastal faunal communities. This chapter presents coastal fauna and its diversity in different coastal ecosystems of the tropics and the effect of human activities on them with some case examples.

Keywords

Faunal diversity · Tropical region · Coastal pollution · Urbanization · Impacts

5.1 Introduction

Coastal areas contain many of the Earth's complex and diverse ecological systems, productive in both biological and economic senses. Coastal ecosystems are areas where land and water meet each other in different ways creating an environment with a distinct structure, diversity, and flow of energy. They are mangroves, salt marshes, sandy coast, rock shores, wetlands, estuaries, and bays. Such diverse

V. Ayyam et al., *Coastal Ecosystems of the Tropics - Adaptive Management*,
https://doi.org/10.1007/978-981-13-8926-9_5

environment is home to different types of plants and animals; in addition, it provides valuable ecosystem services (Michel and Pandya 2010).

At the landscape level, all these coastal ecosystems provide suitable habitat and sustain different kinds of faunal communities which inhabit the surface, subsurface, wetlands, shallow water, estuaries, and plant canopies. At the same time, there is greater diversity in population, species composition, and its vulnerability/threat among the tropical coastal countries. It is often observed that these coastal fauna exhibit remarkable morphological and genetic diversity as they tend to adapt to the different ecosystems and the changing environment. Greater than the sum of its parts, coastal areas with its diverse ecosystems sustain major portions of global biodiversity. They are valuable to the functioning of the natural ecosystems as well as for the coastal populations (Daniels et al. 1991).

With over one third of the world's population living within 50 km of the coast, human influence on coastal biodiversity and its ecosystems is inevitable, making the study of this areas particularly tropical region even more important. At the same time over the years, the level of contamination and type and amount of pollutant reaching the coastal areas have been increasing at an alarming rate as they harbour several industries, support tourism activities, and give space for urbanization besides the coastal region receive pollutants even from faraway places through drainage systems. In addition, human interference of the natural system including climate drivers is affecting the coastal ecosystem through the feedback mechanism. It affects several critical habitats in the coastal areas. The combined effect of population growth, race for economic development, and the environmental change is threatening the coastal ecosystems to which diverse and several unique fauna forms integral part (Creel 2003).

In this context, it is essential to understand different fauna, its spatial distribution, and their status in different coastal habitats. This is more pertinent to the tropical coastal ecosystems because most of them are very sensitive to changes in the environment to which the flora and fauna are a dynamic part. Further, it is observed that some of these critical habitat areas are now struggling to maintain their diversity due to human activity, the introduction of non-native species, and other factors. This will help to conserve, derive sustainable benefits, continue the scientific study in their natural environment, and understand the effect of human activities and global changes on their survival and distribution. With these backgrounds, we have made an attempt to provide information on different coastal fauna of various tropical ecosystems with reference to their population, threat, diversity, and the effects of human activities with case examples.

5.2 Faunal Diversity and Population Status

The tropical biodiversity constitutes more than half of the world's species (Wilson 1988) which probably exceed more than that as far as coastal biodiversity is concerned. But the spatial distribution of biodiversity is not uniform. There are areas which supports more biodiversity and number of species than other areas due to natural endowments and favourable physical conditions. This is observed in the tropical coastal region which is not only rich but unique in faunal diversity, supported by diverse habitats than other regions of the world. In order to facilitate the conservation of these faunal diversity and its habitats, it is vital to know species distribution and abundance. Nevertheless the species diversity is not equally spread over the total tropical area but is more or less clumped in centres of high diversity, and these are sometimes characterized by high levels of endemism (Gentry 1986; Myers 1990). Such areas are called biodiversity hotspots (Myers et al. 2000).

At the same time, it is often observed that many of the species of the tropical island appear to be patchy in its spatial distribution limited by the geographical isolation of the islands. The vast expanse of ocean around the islands limits its spread, and the natural selection continued for the island conditions resulting in high level of endemism as observed today. It is widely accepted that the biodiversity present in mangroves and coral ecosystem represents an enormous potential resource of species and genetic variation which may benefit humanity (Balick and Mendelsohn 1992; ICBP 1992).

The faunal diversity of coastal countries and islands located in the tropical region is summarized in Table 5.1. From this it is observed that the region is rich in fishes, mammals, corals, and birds with high endemism due to the reasons discussed earlier. In the Indian Ocean region, India harbours large number of fishes, mammals, and birds. Similarly many of the island nations in this region have large number of fishes, mammals, and corals due to favourable physical conditions. Another important tropical biodiversity hotspot is the Caribbean region. Haiti, Cuba, and Jamaica together contribute more than half of the regional biodiversity. At the same time, Jamaica and Cuba are having the largest number of endangered/threatened species due to over-exploitation and human interference. The Pacific region islands called Coral Triangle are known for its richest marine biodiversity in the world particularly the coral ecosystem. It is regarded as the global epicentre of marine biodiversity; therefore, it has become the global priority for conservation (Allen 2007). At the same time, this region has the highest number of endangered or threatened species which are very unique to the coastal ecosystem (Fig. 5.1). In general majority of the islands and coastal population derive their livelihood from this, but over-exploitation, aggressive fishing practices, and pollution of the coastal areas are responsible for the threatened status of some of these faunal communities.

5.3 Fauna of Sand Dune and Beach

In general, it is observed that coastal areas particularly sand dunes and active beaches are rich in unique faunal diversity. The floral communities existing in these areas provide shelter to the coastal faunal communities. The interrelationships between the coastal fauna and its surrounding physical and biological environments are very intricate. Some of the most important fauna of these areas are discussed in the following subsections.

Table 5.1 Reported faunal population and endangered/IUCN Red Listed species

Country/territory	Reported species				IUCN Red Listed/endangered species		References
	Birds	Mammals	Fish	Reptile	Mammals	Birds	
Caribbean region							
Antigua and Barbuda	182	17	400	26	–	–	Daltry (2007)
							www.earthendangered.com
Bahamas	57	4	248	53	06	13	Buden (1975) and BMMRO (2009)
Belize	574	152	669	126	05	08	Meerman (2005)
Cuba	350	42	963	121	16	18	www.cubanaturaleza.org
							www.earthendangered.com
Grenada	150	22	317	13	–	09	GoG (2014)
Guyana	814	225	467	179	05	07	FAO (2005)
Haiti	245	25	184	217	05	14	www.earthendangered.com
Jamaica	106	26	1500	43	05	15	GoJ (2010)
American region							
Mexico					128	95	http://earthsendangered.com/search-regions3.asp
Brazil					103	229	http://earthsendangered.com/search-regions3.asp
Atlantic region							
Guinea-Bissau	282	132	337	–	15	14	Catarino (2004)
Sao Tome and Principe	49	10	33	16	–	–	NBSAP II (2015)
Indian Ocean region							
India	500	423	3022	526	116	136	ZSI (2015)
Bangladesh	–	–	–	–	138	566	
Sri Lanka	482	91	916	183	15	11	BDS/MoE and DNBG (2012)

Mauritius	28	5	493	17	03	11	NRCBD (2015)
Maldives	167	–	1090	–			Anderson and Baldock (2001)
Madagascar	–	–	–	–	92	52	http://earthsendangered.com/search-regions3.asp
Seychelles	65	6	–	22	02	20	Government of Seychelles (2014)
Southeast Asia							
Thailand					81	130	http://earthsendangered.com/search-regions3.asp
Myanmar					60	110	Gomez (1990)
Philippines					55	129	
Papua New Guinea					52	86	
Pacific islands							
Solomon	163	53	420	80	19	38	http://earthsendangered.com/search-regions3.asp
Fiji	162	06	575	27	05	13	https://chmfiji.wordpress.com/overviewoffijis-biodiversity
Palau	171	8	1387	44	02	05	Kitalong (2010)

Pacific Golden-Plover *Pluvialis fulva*

Greater Sand Plover *Charadrius leschenaultii*

Common Red shank *Tringa totanus*

Common Sand piper *Actitis hypoleucos*

Madagascar Humblot's heron

Kelp Gull (Brazil) *Larus dominicanus*

Fig. 5.1 Some of the shorebirds of tropical coastal and island region

5.3.1 Birds

Coastal areas support an amazing variety of bird species from small sand pipers to large pelicans (Fig. 5.1). Many birds spend large portions of their lives over the water for food, but they all must eventually come out of the air to find a mate and reproduce. The sand, rocks, pebbles, and plants of the beaches serve as breeding grounds for many species of bird. Therefore, preservation of beaches as nesting sites is crucial for the protection of bird species. When beaches are developed, polluted, or destroyed, habitat critical to the life cycle of birds is also destroyed. Beach sand mining is a major mechanism for habitat destruction. This necessitates the study of bird population, its habitat, and human impact on them. Such studies provide useful information for basic and applied ecology and identifying priority areas for conservation (Peterson et al. 2000; Pankaj and Manchi 2013).

5.3.2 Crabs

Crabs are soft-bodied animals with hard exoskeleton for protection which inhabit varied coastal environment including the clay marshes. It is an economically and culturally important species present in many tropical countries and coastal areas. Crab habitat varies; some species live exclusively in the ocean in a range of water depths with a preference for shallow waters, while some species live entirely on land. Some land species have structures that function as primitive lungs. There are over 6500 known crab species making them an extremely diverse group of animals. Crabs, as a group, are scavengers with some species eating just about anything: algae, molluscs, worms, crustaceans, fungi, and other organic material. Habitat loss is a major threat to many crab species as well as invasive and non-native species. Beach development, nourishment, and dredging produce fine sediment which is harmful to crabs as well.

5.3.3 Sea Turtles

Sea turtles are found in every ocean of the world, but there are only seven species. All of them regarded as threatened, and five of them are endangered or critically endangered. Among their adaptations are their highly evolved front flippers, ability to hold their breath for long periods of time, and hard shell. Sea turtles use beaches as nesting sites for incubation of their eggs. Without nesting sites, it would be impossible for sea turtles to complete their life cycle. They spend much of their time eating seagrasses, and when the females are able to reproduce, they return to the beach to lay their eggs. In most cases, the females return to the beach where they were born using the Earth's magnetic field to navigate. Once on the beach, they lay large clutches of eggs in shallow pits of sand where the young turtles mature. It takes around 2 months for the young turtles to mature. When the turtles finally hatch, they must crawl to the ocean quickly to avoid predation by birds, other animals, and human threats. There are numerous human threats to sea turtles. The major threat is illegal poaching and harvesting of turtle eggs and shells for crafts work. Sometimes they get entangled in the fishing nets. Beach sand mining also represents a major threat to sea turtles (UNEP 2006).

5.3.4 Seashells

Marine molluscs such as snails, bivalves, and chitons have hard exoskeleton made up of calcium carbonate. It serves to protect and support their body which is secreted by the mantle, a skin-like tissue in the mollusc's body wall. In general there are six major classes of molluscs that have shells, viz.:

- Gastropoda—snails, slugs, and limpets
- Bivalvia—clams, oysters, and mussels
- Polyplacophora—chitons
- Cephalopoda—squid, octopus, and nautilus
- Scaphopoda—tusk-shaped shells
- Monoplacophora

They inhabit wet to dry sand in the beach and offshore areas. Seashells are an important part of biological and geological beach processes as well as an important part of human culture. For example, they provide materials for birds' nests, a home, or attachment surface for algae, seagrass, sponges, and a host of other microorganisms. Fish use them to hide from predators, and hermit crabs use them as temporary shelters. They are greatly affected by the sand and coral mining.

5.4 Human Impact

5.4.1 Effect on Faunal Diversity

Among the components of biodiversity, coastal realm consist of many important biodiversity-rich ecological units like coastal wetlands, estuaries, coral reefs, mangroves, sand dunes, lagoons, and seagrass and seaweed ecosystems. Diversity of flora and fauna of these habitats has an inevitable and collective role in the existence, maintenance, and wellbeing of the larger coastal and marine zones. In addition, they exist in an intricate relation with flow of energy and matter between them. However, human activities have inflicted heavy damage to the structure and functioning of the coastal biodiversity in general and fauna which are unique in particular. Reduction of a single species can affect the energy flow in the respective food web, and later it reflects on the entire ecosystem. In other words over-exploitation of a single species or community will reduce its population below its minimum population required for survival. Thus the gradual decline of a species to its extinction results in reduction of faunal diversity which might have serious and long-term effects on the biosphere. With the increasing need for seafood, catch of fish species is inevitable, but consequently it passes thorough different phases such as commercial exploitation, over-exploitation, and finally extinction of species. Similarly construction of aquaculture ponds in the coastal areas significantly affected the wild faunal population in these areas primarily by habitat loss and over-exploitation (Shahidul Islam and Tanaka 2004). More than 70% of the mangroves were lost in Kenya, Liberia, the Philippines, and Puerto Rico.

The activities related to oil and gas exploration and exploitation severely affected marine, wetland, and coastal fauna at various stages of development such as exploration, site access, site preparation, drilling, production, pipeline installation, spill control and cleanup, and site closure. The loss of wetland area in deltaic regions decreases the value of the estuarine zone as a nursery ground for aquatic animals and also results in huge loss of biodiversity. Thus, the loss of faunal diversity in coastal ecosystems is influenced by habitat loss, habitat fragmentation, edge effect, pollution, biomagnification, eutrophication, introduction of cultivable species, and invasive species through ballast water. Other anthropogenic stressors like anchoring of boats in coral reef areas, sewage water pollution from the urban areas, tourism pressure in the pristine beaches, harvest of ornamental shells for sale, and destructive fishing using explosives also account for biodiversity loss in the most diverse coastal environment.

5.4.2 Pollution

The major sources of coastal and marine pollution come from both land (through urbanization, industrialization, agriculture, mining, etc.) and sea (through dredging, mining, dumping, and shipping). More than 50% of population of Bangladesh, North Africa, and many of the Pacific island nations lives in coastal areas causing heavy damage to coastal resources through various activities. The coastal regions in the tropics are increasingly urbanized, and 14 out of 17 megacities are located in coastal region, and 11 of these large cities are located in Asia. The rapid growth of the coastal urban centres in tropical coasts of Asia, Africa, and Latin America and the fast development of the coastal tourism sector produce vast quantities of pollution from untreated domestic sewage, posing a threat to the nearshore habitats particularly coral reefs, estuaries, and mangrove forests covering 25% of world's tropical coastline. Further, the increased solid waste generation with inadequate disposal system results in release of nutrients and toxic chemicals to the environment. Non-point sources of pollution from urban areas include suspended solids, oil and grease, organic chemicals, pathogens, litter, and heavy metals which are brought down by runoff to the coastal areas from urban and even off-farm sources. Table 5.2 summarizes impacts of various human activities on fauna of different tropical coastal countries.

It is estimated that in the Gulf of Thailand, nearly 659–34,376 tonnes per annum of municipal and industrial wastes are dumped (Cheevaporn and Menasveta 2003). Gomez (1990) reported that over 80% of sewage originating from urban slums in

Table 5.2 Human impact on fauna of different coastal ecosystems in the tropics

Causes	Effects	Region	References
Industrial	Stock decimation of fishes	Bangladesh	Alam et al. (1998)
Pollution	Pollution endangered and threatened (Ganges River dolphins, *Platanista gangetica*)	Bangladesh	Smith et al. (1998)
Pollution	Damage to mollusc producing beds in estuaries and bays	North and Central America and Europe	MacKenzie and Burrell (1997)
Pollution	Diminishing yields of American oysters (*Crassostrea virginica*)	Coastal lagoons of Mexico	Marin et al. (1997)
Pollution	Loss of nesting habitats and population decline of leatherback turtle	Malaysia	–
Pollution due to the Gulf War 1991	Interruptions in the life cycles, morbidity, emigration, and recruitment collapse of penaeid shrimp, *Penaeus semisulcatus* and supporting fisheries	Arabian Gulf	Mathews et al. (1993)
Industrial, commercial, agricultural, and domestic pollution	Massive fish kills and polluted water	Laguna de Bay, Philippines	Oledan (2001)
Pollution from industrial waste discharge, mining, pesticides, and oil residues and spills	Stock declines of fish and other commercially important aquatic organisms; changes in lake ecosystem and water quality	African Great Lakes	Reeves et al. (1997)
Plastic debris (anthropogenic debris)	Endangered green sea turtle (*Chelonia mydas*) was found with plastic debris in their digestive tracts	South of Brazil coast	Bugoni et al. (2001)
Pollution from intensive ship scrapping activities, sewage disposal, and antifouling paints	Bioaccumulation of butyltins and resulting butyltin pollution in fishes	India, Bangladesh, Thailand, Indonesia, Vietnam, Taiwan, Australia, Papua New Guinea, and the Solomon Islands	Kannan et al. (1995)
Intense persecution by farmers and coconut plant owners	Seychelles parakeet (*Psittacula wardi*)	Indian Ocean islands	Kundu et al. (2012) and Anil et al. (2014)
Plastics	*S. magellanicus* stomach analysis. Sixteen percent contaminated with plastic pieces, elastic bands, and pellets which were associated with squids. Items ranged from 2.5 to 14.5 cm. Plastics related with fishing	Brazil	Azevedo and Schiller (1991)

(continued)

Table 5.2 (continued)

Causes	Effects	Region	References
Debris—all	*C. mydas*, *C. caretta*, *D. coriacea*. Stomach and oesophagus analysis. Sixty percent of the *C. mydas* contaminated (highest rate in the literature). Mostly plastic bags and nylon ropes. External injuries related with fishing	Beach, Brazil	Bugoni et al. (2001)
Light sticks—debris	*Artemia* sp., *Nauplii*, and eggs exposed to light stick contents and elutriate. Both presented hazardous potential	Beach, Brazil	Ihara et al. (2005)
Plastics	*R. lalandii*. Plastic rings (36–42 mm) around gills or mouth. Related to fishing activities. Possibly compromising ventilation and feeding	Ocean, Brazil	Sazima et al. (2002)
Plastics	Sharks and seals Plastic rings, related to fishing activities on sharks and marine mammals	Ocean, Brazil	Santos (2006)
Pollutants (concentrations of heavy metals—Hg, Cd, Pb, and Cu (ppm wet weight))	Skipjack tuna (Hg, 0.160)	Northern Indian Ocean	Kureishy et al. (1983)
	Katsuwonus pelamis (Cd, 0.93)		
	Linnaeus (Pb, 1.19)		Sen Gupta et al. (1989)
	Bluefin tuna (Hg, 0.120)		
	Mackerel tuna (Hg, 0.13)		
	Solandri (*Cuvier*) (Cd, 0.43)		
	Dolphinfish (Hg, 0.01–0.042)		
	Coryphaena hippurus (Cd, 1.19)		

Southeast Asia goes into the sea untreated. In Indonesia, organic pollution emissions into the South China Sea have more than doubled between 1980 and 1993 (Chou 2009). Sewage effluents entering coastal waters contain a diverse array of polluting agents including pathogens, organic substances, heavy metals, and trace elements posing direct and indirect threats to coastal ecosystem. The heavy metals like lead, copper, chromium, arsenic, and cadmium concentration of 0.1–0.7 mg l^{-1} were found in the sewage wastes. It adversely affects the normal functioning and survival of several coastal fauna. The sewage also contains harmful human pathogens such as *Escherichia coli*, *Mycobacterium tuberculosis*, *Vibrio cholera*, *Citrobacter freundii*, *Salmonella* sp., etc., and they were also found to occur in marine mammals (Grillo et al. 2001). Decaying organic matter and nutrients in sewage may lead to eutrophication resulting in alterations in ecosystem structure (Shahidul Islam and Tanaka 2004).

Oil pollution is a major threat, and huge volume of oil is added into the coastal and marine environment from various sources like drainage, routine maintenance, smoke, natural seeps, oil spills, onshore drilling, etc. (Ventikos and Sotiropoulos

2014; Nwilo and Badejo 2005). The oil pollution is high in the Straits of Malacca, the South China Sea, the Gulf of Thailand, and the Jakarta-Manila and Singapore-Bangkok routes which seriously affected the coastal and coral reef fauna (Lu 2003; Todd et al. 2010). Further, heavy metals have serious effects on terrestrial and aquatic ecosystems, which increase physiological health risks (Davies 1978; Pandey 2012). Major impacts on the aquatic organisms are inhibition of recruitment, fertilization, and development in various marine organisms including corals and crustaceans.

Agriculture particularly intensive practices in the coastal plains and deltas is one of the major contributors to coastal pollution accounting about 50% of the total pollution source of surface water. Further, agricultural activities pollute the coastal region by means of soil erosion leading to siltation, nutrient enrichment, pesticide contamination, and agro-industrial pollution observed in the Coastal Asia, the Caribbean region, and Small Island Developing States (SIDS) in the tropical region (Eyre 1990; Shahidul Islam and Tanaka 2004).

DDT is still used for malarial control in the Philippines, Thailand, Papua New Guinea, and India though many countries have banned these compounds. Several reports documented the presence of these compounds in marine waters and animals. The bioaccumulation and biomagnification of pesticides can interfere with hormonal, reproductive, and nervous systems in molluscs, crustaceans, fishes, shorebirds, marine mammals, as well as humans (Todd et al. 2010). Other important toxics are butyltin compounds which are used as antifouling paint on ship hulls also and also contribute to marine pollution proving detrimental to living organisms. They are known as endocrine disruptors, causing ovarian spermatogenesis in oysters and imposex in gastropods and bivalves. Occurrence of imposex has been recorded in many Southeast Asian countries which causes decline and species extinction in some natural populations (Cadee et al. 1995).

5.5 Climate Change and Other Stresses

Climate change is one of the most important human-induced stressors on coastal ecosystems and faunal diversity having large-scale effects such as survival, breeding, and migration. Increased sea surface temperature, sea level rise, tropical cyclones, sea surges, salinization, etc. are linked to the global climate change and have negative effect on the coastal faunal population both directly and indirectly. Halpern et al. (2007) have demonstrated the staggering extent of multiple stressors on oceans with 40% of the world's oceans already heavily impacted by human activities and no area of the global oceans left unaffected by human influence. Fishing impacts marine ecosystems through overfishing and destruction of habitat and by catch which will get aggravated by the projected global climate change (Worm et al. 2006). Together these stressors affect the resilience of ecosystems, thereby increasing the vulnerability of coastal fauna to climate change.

But natural systems exhibit some degree of resilience to these changes. But, there is often a threshold beyond which an altered ecosystem cannot recover from many kinds of disturbances. In order to overcome these stresses, protected areas have been created to conserve the flora, fauna, and the entire ecosystem in biosphere reserves. But protection works best as a conservation tool if the area remains protected for the foreseeable future and the active involvement of the local population. Contrary to this concept, under climate change, species for which a particular protected area was established may no longer survive there. As the climate change impacts are uncertain, therefore, monitoring provides important information on which future management decisions can be made.

5.6 Conclusion

Coastal areas of the tropics contain many complex and diverse ecological systems extending in different geomorphological conditions. Compilation of studies pertaining to the tropical coastal ecosystems revealed that there is wide diversity of fauna in corals, rock shore, sandy, estuaries, salt marsh, wetland, and mangrove ecosystems. It is estimated that 90% of the world's fish production is dependent on coastal areas including coral environment at some part of their life cycle. Mangroves represent most important habitat for large number of juveniles inhabiting the coastal waters. Coastal sand deposits forms part of the critical habitat required for the survival of many terrestrial and marine fauna. In addition, coastal areas support large numbers of migratory and non-migratory waterfowl and shorebirds and endangered reptiles, such as turtles and alligators. The presence of high species density and occurrence of many threatened species emphasize the need for continued protection and preservation of the tropical coastal areas.

However, human activities have resulted in damage to the structure and functioning of the coastal biodiversity in general and fauna in particular. Pollution is seriously affecting the faunal population, diversity, and distribution. The major sources of coastal pollution come from both land (through urbanization, industrialization, agriculture, mining, etc.) and sea (through dredging, mining, dumping, and shipping). Heavy metals and pesticide residues are reported to interfere the normal body functioning of several fauna. The activities related to oil and gas exploration affected the coastal fauna by way of inhibition of recruitment, fertilization, and development in various marine organisms including corals and crustaceans. At the same time, over-exploitation of commercially important species has resulted in gradual decline of that species and sometimes to its extinction. Therefore it is highly essential to take protective measures to conserve this faunal diversity and minimize or regulate such human activities which are causing pollution in the coastal areas.

References

Alam MGM, Jahan N, Mazid MA (1998) Impact of aquatic pollution and its effect on fisheries in Bangladesh. Mer Tokyo 36(1):23–37

Allen GR (2007) Conservation hotspots of biodiversity and endemism for Indo-Pacific coral reef fishes. Aquat Conserv Mar Freshwat Ecosyst 18(5):541–556. https://doi.org/10.1002/aqc.880

Anderson RC, Baldock M (2001) New records of birds from the Maldives, with notes on other species. Forktail 17:67–73

Anil MNV, Kumari K, Wate SR (2014) Loss of biodiversity and conservation strategies: an outlook of Indian scenario. Asian J Conserv Biol 3(2):105–114

Azevedo TR, Schiller A (1991) Notes on the diet and the ingestion of plastic material by the magellanic penguin Spheniscus magellanicus on Santa Catarina Island and mainland (Brazil). Unit Doc Zool Seru Ethol Psichol Anim Univ Liege Belgium Rapp 457:1–8

Balick MJ, Mendelsohn R (1992) Assessing the economic value of traditional medicines from tropical rain forests. Conserv Biol 6(1):128–130

BDS/MoE, DNBG (2012) The national red list of Sri Lanka. In: Weerakoon DK, Wijesundara S (eds) Conservation status of the fauna and flora. Ministry of Environment, GoS, Colombo, pp Viii–V476

BMMRO (2009) Bahamas Marine Mammal Research Organisation. Annual report to The Bahamas Department of Marine Resources, Govt. of Bahamas

Buden DW (1975) A taxonomic zoogeographic appraisal of the big-eared bat in *(Macrotus waterhousii Gray)* on the West Indies. J Mammal 56:758–769

Bugoni L, Krause L, Petry MV (2001) Marine debris and human impacts on sea turtles in Southern Brazil. Mar Pollut Bull 42:1330–1334

Cadee GC, Boon JP, Fischer CV, Mensink BP, Tjabbes CC (1995) Why the whelk *Buccinum undatum* has become extinct in the Dutch Wadden Sea. Neth J Sea Res 34:337–339

Catarino L (2004) Fitogeografia da Guiné-Bissau. Dissertaçãodoutor. Universidade Técnica, Lisboa

Cheevaporn V, Menasveta P (2003) Water pollution and habitat degradation in the Gulf of Thailand. Mar Pollut Bull 47(1–6):43–51

Chou LM (2009) The East Asian seas state of the marine environment report. UNEP EAS/RCU, Bangkok

Creel L (2003) Ripple effects: population and coastal regions, making the link. Population reference Bureau, Washington, DC. https://www.prb.org/wp-ontent/uploads/2003/09/RippleEffectsEng.pdf. Accessed on 16 July 2018

Daltry JC (2007) An introduction to the herpeto fauna of Antigua, Barbuda and Redonda, with some conservation recommendations. Appl Herpetol 4:97–130

Daniels RJR, Hegde M, Joshi NV, Gadgil M (1991) Assigning conservation value: a case study from India. Conserv Biol 5(4):464–475

Davies AG (1978) Pollution studies with marine plankton; part II. Heavy metals. Adv Mar Biol 15:381–508

Eyre LA (1990) Forestry and watershed management, mineral development planning and production national strategy consultation workshop, 25–27 April 1990, Kingston, Jamaica

FAO (2005) Global forest resources assessment 2005. Food and Agriculture Organization, Rome. http://www.fao.org/forestry/country/20807/en/guy. Accessed on 24th Aug 2018

Gentry AI (1986) Endemism in tropical versus temperate plant communities. In: Soulé ME (ed) Conservation biology. The science 0/ scarcity and diversity. Sinauer Associates, Sunderland, pp 286–303

GoG (2014) Fifth national report to the convention on biodiversity. Government of Grenada, Grenada, p 75

GOJ (2010) NEPA. Fourth National Report for Jamaica to the Convention on Biological Diversity. www.earthendangered.com

Gomez ED (1990) State of the marine environment in the East Asian Seas Region. UNEP Regional Seas Reports and Studies No. 126

Government of Seychelles (2014) Fifth national report to the United Nations Convention on Biological Diversity. Ministry of Environment and Energy, Victoria

Grillo V, Parsons ECM, Shrimpton JH (2001) A review of sewage pollution and cetaceans: a Scottish perspective. Paper presented to the Scientific Committee at the 53rd Meeting of the International Whaling Commission, 3–16 July 2001, London

Halpern BS, Selkoe KS, Micheli F, Kappel CV (2007) Evaluating and ranking the vulnerability of global marine ecosystems to anthropogenic threats. Conserv Biol 21(5):1301–1315

ICBP (1992) Putting biodiversity on the map priority areas for global conservations. International Council for Bird Preservation, Cambridge, UK, p 90

Ihara PM, Ivar do Sul JA, Filgueras AS, Portz LC, Filmann G, Mathiensen A (2005) Avaliacao da toxicidade do conteudo de atratores luminosos ao microcrustaceo planctonico Artemia sp. In: Proceedings of the Congresso da Sociedade de Quinica e Toxicologia Ambiental (SETAC), Santigo, Chile

Kannan K, Tanabe S, Iwata H, Tatsukawa R (1995) Butyltins in muscle and liver of fish collected from certain Asian and Oceanian countries. Environ Pollut 90(3):279–290

Kitalong (2010) The Republic of Palau Statewide assessment of forest resources and resource strategy. Bureau of Agriculture, Republic of Palau

Kundu S, Jones CG, Prys-Jones RP, Groombridge JJ (2012) The evolution of the Indian Ocean parrots (Psittaciformes): extinction adaptive radiation and eustacy. Mol Phylogenet Evol 62(1):296–305

Kureishy TW, Sanzgiri S, George MD, Braganca AM (1983) Mercury, cadmium and lead in different tissues of fishes from the Andaman Sea. Indian J Mar Sci 12:60–63

Lu J (2003) Marine oil spill detection, statistics and mapping with ERS SAR imagery in South-east Asia. Int J Remote Sens 24:3013–3032

MacKenzie CL Jr, Burrell VG Jr (1997) Trends and status of molluscan fisheries in North and Central America and Europe: a synopsis. NOAA Tech Rep NMFS 0(127):1–14

Marin ML, Baeza L, Zapata PO, Gold BG (1997) Trace metals in the American oyster, Crassostrea virginica, and sediments from the coastal lagoons Mecoacan, Carmen and Machona, Tabasco, Mexico. Chemosphere 34(11):2437–2450

Mathews CP, Kedidi S, Fita NI, Yahya A, Rasheed K (1993) Preliminary assessment of the effects of the 1991 Gulf War on Saudi Arabian prawn stocks. Mar Pollut Bull 27:251–271

Meerman JC (2005) Compilation of information on Biodiversity in Belize. Forest Department, Belize, p 56

Michel D, Pandya A (2010) Coastal zones and climate change. The Henry L. Stimson Center, Washington, DC, p 20036

Myers N (1990) The biological challenge: 'hot-spots' in tropical forests. Environmentalist 10:243–256

Myers N, Mittermeier RA, Mittermeier CG, Fonseca GAB, Kent J (2000) Biodiversity hotspots for conservation priorities. Nature 403:853–858

NBSAP II (2015) National biodiversity strategy and action plan 2015–2020. National Commission for Monitoring and Evaluation, Ministry of Infrastructure, Natural Resources and Environment, São Tomé and Príncipe, p 122

NRCBD (2015) Fifth national report on the convention on biological diversity. Ministry of Agro-Industry and Food Security, Republic of Mauritius, p 87

Nwilo PC, Badejo OT (2005) Oil spill problems and management in the Niger Delta. International Oil Spill Conference, Miami

Oledan MTT (2001) Challenges and opportunities in watershed management for Laguna de Bay (Philippines). Lakes Reserv Res Manag 6(3):243–246

Pandey VC (2012) Phytoremediation of heavy metals from fly ash pond by *Azolla caroliniana*. Ecotoxicol Environ Safe 82:8–12

Pankaj K, Manchi S (2013) Avifaunal records from Chalis Ek, North Andaman Island: insights into distribution of dome Andaman Island Birds. Check List 9(1):034–041. www.checklist.org.br. Accessed on 20 Aug 2018

Peterson AT, Ball LG, Brady KW (2000) Distribution of the birds of the Philippines: biogeography and conservation priorities. Bird Conserv Int 10(2):149–167

Reeves RR, Wang JY, Leatherwood S (1997) The finless porpoise, *Neophocaena phocaenoides* (G. Cuvier, 1829): a summary of current knowledge and recommendations for conservation action. Asian Mar Biol 14:111–143

Santos IR (2006) Tubaoes de coleira. Cienc Hoje 38:54–55

Sazima I, Gadig O, Namora RC, Motta FS (2002) Plastic debris collars on juvenile 606 carcharhinid sharks (*Rhizoprionodon lalandii*) in southwest Atlantic. Mar Pollut 607 Bull 44:1149–1151

Sen Gupta R, Naik S, Varadachari VVR (1989) Environmental pollution. In: Bourdeau P, Haines JA, Klein W, Krishna Murti CR (eds) Coastal areas of India. Ecotoxicology and climate, SCOPE. Published by Wiley

Shahidul Islam M, Tanaka M (2004) Impacts of pollution on coastal and marine ecosystems including coastal and marine fisheries and approach for management: a review and synthesis. Mar Pollut Bull 48:624–649

Smith BD, Haque AKMA, Hossain MS, Khan A (1998) River dolphins in Bangladesh: conservation and the effects of water development. Environ Manag 22(3):323–335

Todd AP, Ong X, Chou LM (2010) Impacts of pollution on marine life in South east Asia. Biodivers Conserv 19:1063–1082

UNEP (2006) Marine and coastal ecosystems and human wellbeing: a synthesis report based on the findings of the Millennium Ecosystem Assessment. United Nations Environment Programme, Nairobi

Ventikos NP, Sotiropoulos FS (2014) Disutility analysis of oil spills: graphs and trends. Mar Pollut Bull 81:116–123. https://doi.org/10.1016/j.marpolbul.2014.02.007

Wilson EO (1988) The current status of tropical biodiversity. In: Wilson EO (ed) Biodiversity. National Academy Press, Washington, DC, pp 1–18

Worm B, Barbier EB, Beaumont N, Duffy JE, Folke C, Halpern BS et al (2006) Impacts of biodiversity loss on ocean ecosystem services. Science 314:787–790

ZSI (2015) In: Chandra K, Sheela S, Das D (eds) Animal discoveries new species and new records. Zoological Survey of India Report, Government of India, Ministry of Environment, Forest and Climate Change, p 152

6 Coastal Wetlands: Status and Strategies for Development

Abstract

Wetlands, a major feature of coastal landscape, are among the most important ecosystems harbouring biodiversity on Earth. They are among the most productive ecosystems in the world, which have for thousands of years provided human communities with provisioning, supporting, regulative services, and cultural services. They played a critical role in maintaining global biodiversity in the coastal areas partly through their rich productivity, which helps to support food chains, and partly through provision of habitats for specially adapted plant and animal species. In spite of its pivotal role in providing niche habitat and ecosystem services, they are under constant pressure from anthropogenic activities besides other natural forces. In brief, this chapter discusses the meaning and significance of wetlands with reference to tropical coastal areas, their formation and distribution, biodiversity, and issues of threats impacting wetlands and wetland conservation strategies.

Keywords

Wetland types · Biodiversity · Mangroves · International conventions · Conservation

6.1 Understanding Wetlands

6.1.1 Definition of Wetlands

Wetlands are among the most productive ecosystems in the world, which have for thousands of years supplied human communities with food, drinking water, building materials, and countless other benefits besides vital ecosystem services. They play a critical role in maintaining global biodiversity, partly through their rich productivity, which helps to support food chains, and partly through provision of

V. Ayyam et al., *Coastal Ecosystems of the Tropics - Adaptive Management*,
https://doi.org/10.1007/978-981-13-8926-9_6

habitats for specially adapted plant and animal species. Owing to the global importance, the 'Convention on Wetlands of International Importance', popularly known as the Ramsar Convention of 1971 (RCS 2013), has globally come into force for conservation and wise use of wet or watery lands. This intergovernmental treaty defines the term wetlands as 'areas of marsh, fen, peat land or water, whether natural or artificial; standing or flowing; fresh, brackish or salt, including areas of marine water, the depth of which at low tide does not exceed six meters' (Navid 1989). It includes all lakes and rivers, underground aquifers, swamps and marshes, wet grasslands, peatlands, oases, estuaries, deltas and tidal flats, mangroves and other coastal areas, coral reefs, and various man-made wetlands. However, as far as marine/ coastal areas are concerned, the term 'wetland' is applicable to only those areas where depth of water does not exceed 6 m at low tide (Anderson 2012).

Deep insight into the definition shows that wetlands are transition zone as they exhibit characteristics of both aquatic and terrestrial systems of varying degrees. In this context, Baldassarre and Bolen (1994) stated that wetlands represent transitional zones between upland and aquatic ecosystems, and therefore they exhibit some characteristics of each. The United States Fish and Wildlife Service's (Cowardin et al. 1979) definition emphasizes the transitional characteristics of wetlands as 'wetlands are the lands transitional between terrestrial and aquatic systems where the water table is usually at or near the surface or the land is covered by shallow water'. In general, wetlands must have one or more of the following three attributes: (1) at least periodically, the land supports predominantly hydrophytes; (2) the substrate is predominantly undrained hydric soil; and (3) the substrate is saturated with water or covered by shallow water at some time during the growing season of each year (Cowardin et al. 1979).

6.1.2 Types of Coastal Wetlands

Wetlands are very important component of the coastal ecosystem which range from marine to freshwater system (Fig. 6.1). It is very unique in some places both marine

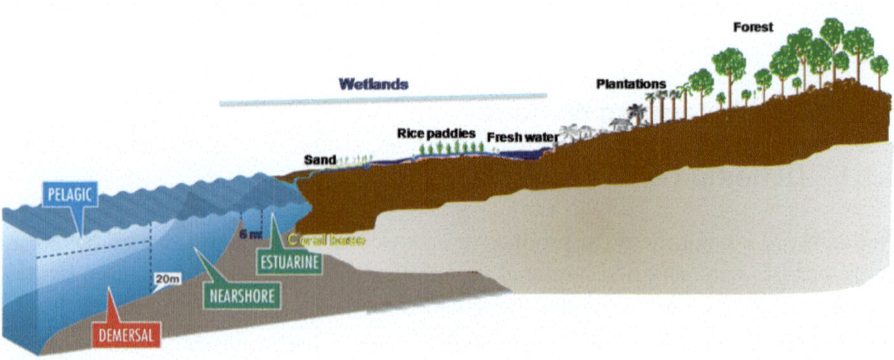

Fig. 6.1 Schematic diagram showing wetlands in a typical coastal landscape

and fresh water mix together depending on the terrain, season, tide, etc., at various point of time. Consequently the characteristics of such wetland habitats will also vary which normally harbours unique and endemic species. The coastal wetlands regulate water quantity, groundwater recharge, floods, and at times the impacts of storms. Most importantly, wetlands help in erosion control and sediment transport, thereby contributing to land formation and increasing resilience to storms (ten Brink et al. 2013).

Matthews and Fung (1987) gave distribution of wetlands in the world by climatic zones. They estimated 2.7 million km^2 wetland area in polar/boreal zone, 0.7 million km^2 in temperate zone, and 1.9 million km^2 wetland area in the tropical/subtropical zone. In addition, they estimated 1.5 million km^2 of rice paddies. It is very difficult to estimate the wetland only in the coastal areas as it gradually intermingles with the inland, and most of the freshwater bodies originate in the highland flow through the coastal areas towards the sea. There were many attempts to characterize the wetlands found in the coastal areas using remote sensing data (Fig. 6.2) either using digital classifier or by visual interpretation. This helped to identify the spatial distribution of different wetlands. A summary of important wetlands found in the coastal areas are given in Table 6.1. Under broad considerations, there are two types of wetlands based on spatial location, viz. coastal and inland. Wetlands can further be divided based on their origin or formation as natural or man-made (artificial). Wetlands are also categorized as saline (>45,000 μMhos/cm^3), brackish (500–45,000 μMhos/cm^3), or fresh (<500 μMhos/cm^3) based on salinity levels (Stewart and Kantrud 1971). In addition, there are some basic wetland types based on ecological viewpoints which are:

- Marshes—seasonally or permanently inundated wetlands dominated by emergent herbaceous vegetation adapted to saturated soil conditions
- Salt marshes—halophyte-dominated marshes bordering saltwater bodies characterized by tidal or nontidal water level fluctuations
- Swamps—wetlands dominated by trees or shrubs (e.g. mangrove swamp)
- Peatlands—a general term for wetlands having the tendency of accumulating partially decayed vegetation
- Bogs—a type of peatland having no noticeable incoming or outgoing flow of water and supporting acidophilic mosses
- Fens—a type of peatland that receive some drainage from the mineral soils in environs and usually has marsh-like vegetation

However, there are numerous other types of wetlands across the globe due to variations in physicochemical and biological characteristics. It is difficult to describe and document all of them.

a

b

Fig. 6.2 Some coastal wetlands as seen through satellite images. (**a**) Sinuous channels and tidal wetlands South Andaman, India. (**b**) Coastal wetlands—mangroves, estuarine, intertidal marshes, and marine in the Sunderban delta (India and Bangladesh)

Table 6.1 Marine/coastal wetlands as per Ramsar convention

Wetland type	Subtypes	Remarks
Natural		
A. Permanent shallow marine waters	Sea bays and straights	In most cases <6 m at low tide
B. Marine subtidal aquatic beds	Kelp bed	Penetration of sunlight up to this bed is very critical
	Seagrass bed	
	Tropical marine meadows	
C. Coral reefs	Coral reef system up to 6 m	Found in tropical shallow water and having richest biodiversity
D. Rocky marine shore	Rocky offshore islands	These are erosional features of sea waves, animal of this habitat having typical adaptation features
	Sea cliffs	
E. Sand, shingle, or pebble shore	Sandy bars and spits	Indicates depositional features
	Sandy islets	
	Dune systems	
	Humid dune slacks	
F. Estuarine waters	Permanent water of estuaries	Confluences of fresh and sea water with different salinity gradients
	Estuarine systems of deltas	
G. Intertidal mud, sand, or salt flats	Mudflats	Deposition of mud/sand, mix of marine and terrestrial condition
H. Intertidal marshes	Salt marshes	Varying physical conditions and water accumulation in the marine coastal interface
	Salt meadows	
	Saltings	
	Tidal brackish/fresh water	
I. Intertidal forested wetlands	Mangrove swamps	Vegetation having special adaptation features and provide habitat for large biodiversity
	Nipah swamps	
	Tidal freshwater swamps	
J. Lagoons	Costal brackish/saline lagoons	Brackish to saline lagoons having narrow connection with sea
	Coastal freshwater lagoons	
Zk(a). Karst land	Surface and subterranean hydrological systems	Marine/coastal
Human-made		
A. Aquaculture ponds	Fish/shrimps ponds	Used for culturing of aquatic foods
B. Ponds (generally <8 ha)	Farm ponds	Used for water storage and fisheries
	Small tanks	
C. Seasonally flooded agricultural land	Intensively managed or grazed pasture	Under human management with less biodiversity
	Canals and drainage channels	

(continued)

Table 6.1 (continued)

Wetland type	Subtypes	Remarks
D. Irrigated land	Irrigation channels	Under human management with less biodiversity
	Rice fields	
E. Salt exploitation site	Saltpans	Under human management with less biodiversity
	Salines	
F. Water storage areas	Reservoirs/barrages	Used for storage of water
	Dams/impoundments (>8 ha)	
G. Excavations	Gravel/brick/clay pits	
H. Wastewater treatment area	Sewage farms	Water treatment system not a typical wetland
	Settling ponds	
Zk(c). Karst	Surface and subterranean hydrological systems	Human-made

Some of these wetland types can be seen in Annexure 1)

6.1.3 Importance

Wetlands are unique in many ways, of which their hydrologic characteristics and roles as ecotones between terrestrial and aquatic systems are notable. At the same time, wetland management requires systematic understanding of the various aspects of the system, balanced with legal, institutional, and economic realities and expertise from disciplines of biology, ecology, and hydrology. Wetlands are closely connected with humankind particularly in the coastal ecosystem than any other ecosystems, for their intimate role in human sociocultural development and settlement across the globe. Consequently recent few decades witnessed advances in environmental and ecological studies which have shed light on the intricacies of wetlands: their values are being recognized, and wetland protection is considered imperative in many parts of the world. Policy makers are also getting conscious of the importance of the wetlands for sustainable development of the humankind.

Wetlands harbour a large biodiversity highly disproportionate to their areal extent and provide ecosystem services that are critical to the entire life on Earth. These ecosystem services result from the interactions between different biodiversity components and their abiotic variables as noted earlier (Fig. 6.3). Wetlands provide water and water-related ecosystem services, such as food (fish, prawn, rice, and many other plants), wastewater purification, hydrological regulation of floods and droughts, carbon sequestration and climate regulation, storm protection, and erosion control. Wetlands enhance aesthetics and support a wide range of livelihoods besides various cultural/recreational activities (Gopal 2015). In addition, wetlands may hold important spiritual values for some cultures. Thus, wetlands contribute to human wellbeing, cultural identity, and economy.

The tropical coastal areas are bestowed with vast expanses of coastally sensitive ecosystems, which include coral reefs, mangrove forests, and seagrass beds with

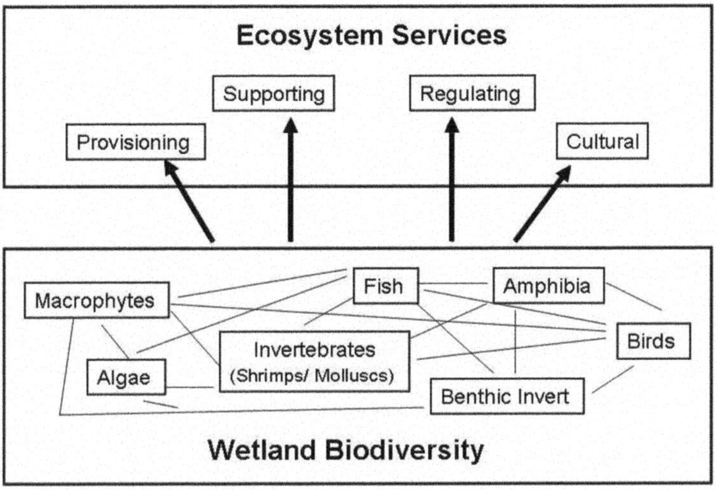

Fig. 6.3 Linkages between wetland biodiversity and ecosystem services

rich marine biodiversity. Mangroves and corals protect shorelines and are an integral part of contributing to the fishery industry. They act as nursery grounds to fish seed and juveniles. Apart from raising the shores of the islands and protecting them against sea level rise in coral reefs, mangroves also act as a barrier and natural breakwater to small islands and hence protect them against storm surges and tsunamis. They also provide shelter to fish stock and help support the fishery industry hence producing revenue for local population. Coral reefs are the main habitat for variety of marine life, and together with lagoons, it attracts revenue in the field of tourism through recreational activities (Dam Roy et al. 2017). The freshwater wetlands are vital not only to provide drinking water but also for maintaining the salt balance in the groundwater by pushing back the saline sea water. Numerous aquaculture ponds scattered along the coast provide livelihood and food security to small farmers. Hence, they are significant in sustaining the economy of coastal and small island states.

On the other hand, wetlands are disappearing at alarming rates across the globe, apparently since they were largely considered synonymous with wastelands to be filled up, drained, occupied, and diverted for other human needs, disregarding the vital ecological services these ecosystems offer and the ecological goods we derive from them. A more recent and global issue that concerns both coastal and inland wetlands is the climate change. In addition, release of chemical-laden wastes by industries, dumping of solid wastes, and excessive use of agrochemicals that ultimately reach the wetlands have emphasized a need for strategies for the conservation, sustainable utilization, and management of wetlands.

6.2 Development and Distribution of Coastal Wetlands

In the modern world, the number of wetlands is declining, wetland area is shrinking, and wetland diversity is decreasing, mainly due to human's unsustainable developmental activities. Therefore, it is important to understand the development, historic and current abundance, distribution, and diversity of wetlands existing on our planet.

6.2.1 Development of Coastal Wetlands

Our understanding of the functioning of the coastal wetlands has grown significantly over the past few decades, but still lots of gap remain particularly with reference to its spatial distribution. Yet the scientific world gained insight into the roles of geomorphic process, hydrologic dynamics, biotic feedbacks, and disturbance factors in creating a variety of coastal wetland ecosystems across climatic gradients. The variety is understood predominantly as a result of different physical processes which both facilitate and constraint a diversity of plant and animal communities.

Study of geological and biogeographical history of our planet reveals that wetlands continuously evolved in time and space. Coastal wetlands best develop along passive margin coasts with low-gradient coastal plains and wide continental shelves. At several places along the tropical coast, it has been observed that the combination of low hydraulic energy and gentle slope normally provides an ideal setting for the development of wetlands. The formation of low-lying coastal plains has been in response to the pattern of relative sea level change. It is also known that major coast modifiers such as tsunamis and storm surges act more decisively on higher relief coasts than low coasts wetlands. Therefore, coincidence of sea level with suitable topography is critical. All these facts support the conclusion that there were several factors that have favoured the development of these extensive coastal wetlands over the past 7000 years. Different components of wetlands exchange mass and energy with each other which are mostly synergistic (Fig. 6.4). In general, the coastal wetlands are areas that have combined physical sources and biological process to develop structure that continues to take benefit from the natural energy inputs.

Global distribution of wetlands has some general feature particularly in the coastal region. Coastal wetlands are more extensive in the wetter parts of the tropics. There are several reasons to explain this kind of distribution. First, deltas are more extensive at the mouth of rivers in high rainfall regions because greater sediment loads are eroded from the catchment and carried to the coast. Second, inundation is more common and widespread, both because of the river-fed flood waters and also as a result of direct precipitation.

Tropical coastal wetlands are mostly composed of mangroves and swamps in the low-lying coastal plains followed by intertidal flats of various types and corals. Mangrove characterizes the upper intertidal zone on many low energy tropical

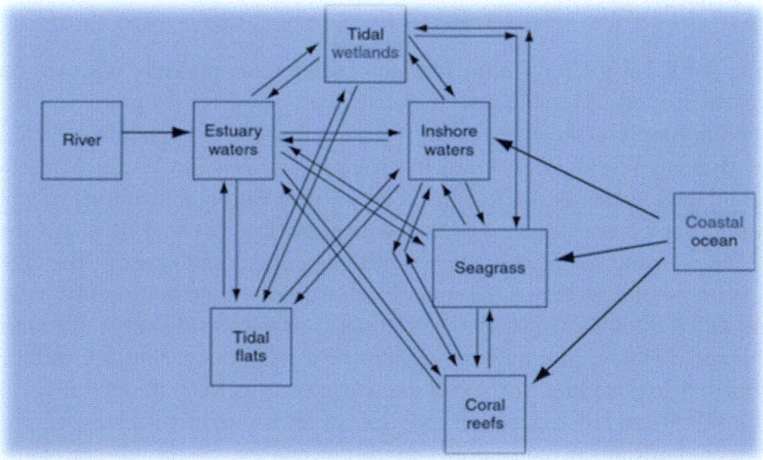

Fig. 6.4 Energy and mass flows in a coastal ecosystem

coasts, often with salt marsh and associated wetlands that can form landward of such halophytic vegetation (Alongi 2008). Sedimentation in these wetlands is a function of the type of substrate and the extent to which sediment sources are derived from inland catchments, the nearshore or in situ production of organic matter. Study of mangrove sediments has shown that sea level has been the primary force controlling the way in which the wetlands have developed. Species diversity is relatively less in the biogeographical zones of Atlantic Coast, in contrast to Indo-West Pacific zones centred on Southeast Asia and Northern Australia. In the Caribbean region, the pattern of decelerating sea level rise has resulted in a transgressive sequence of sediments that has only recently changed to progradation of mangrove wetlands at few locations. In southeast Asia and Northern Australia, after the stabilization of sea level 7000 years ago, extensive coastal plains have been built, nature of wetland on the landward side was controlled by climate, extensive salt flats occurred in arid and semiarid areas, and the peat swamp occurred on the wettest areas.

Tidal salt marshes are found in coastal areas in the middle and high latitudes. They are common wherever accumulation of sediment is equal to or greater than the rate of land subsidence and where there is adequate protection from destructive waves and storms. Salt marshes are predominantly intertidal and found in areas that are at least occasionally inundated at high tide but not flooded during low tide. The upper and lower boundaries of salt marshes are usually set by the tidal range. While their physiognomy is much simpler than a bottomland hardwood forest, the vegetation of tidal salt marshes does vary across salinity and flooding gradients and provides habitat for a variety of plants, animals, and microbes that are adapted to deal with the stresses of this environment (Mitsch and Gosselink 2000).

6.2.2 Current Status

Following the Ramsar Convention and increased human pressure imposed on wetland ecosystems, considerable attention has been given for promotion of wetland conservation. Based on the investigation by a number of organizations such as Asian Wetland Bureau, World Wide Fund for Nature, IUCN (The World Conservation Union), and the International Council for Bird Preservation, the number of wetlands of high importance is estimated as 947 with the total area of 120 Mha (Van Zon 2004). This is scattered all over the world both inland and coastal. But the main interest is on coastal wetlands particularly on tropics which is difficult to estimate and compile. Wetlands exist in all the continents of the world except in Antarctica.

In the early 1980s, Maltby and Turner (1983) estimated that 8.6 million km^2 (more than 6.4% of the land surface of the world) was wetland. Later, Finlayson and Davidson (1999) included nearshore marine areas as one of the categories of wetlands as per the Ramsar Convention definition and estimated that the worldwide extent of wetlands is 12.8 million km^2 (8.6% of land surface area and 2.51% of total surface area of Earth). However, based on several other previous studies, Mitsch and Gosselink (2007) estimated the extent of the world's wetlands as 7–10 million km^2 (about 5–8% of land surface area and 1.37–1.96% of the total surface area of Earth). Finlayson and Davidson (1999) have given regional minimum estimates for wetland area. Neotropics constitutes 32% of global wetland area (4,149,170 km^2) followed by North America (2,415,740 km^2) and Eastern Europe (2,292,170 km^2). The largest continents, Asia (2,042,450 km^2) and Africa (1,246,860 km^2), constitute only 16% and 10%, respectively, of the total wetland area of the world (Fig. 6.5).

These are global distribution of wetlands; however, the type of wetland varies from place to place (Fig. 6.6). The exact estimate of different types particularly in the tropical coastal areas is very difficult as it is predominantly influenced by sea waves and climatic factors. In the coastal areas, wetlands predominantly include mangroves, salt marshes, tidal flats, and seagrasses, and in the seafront corals are the major components in tropical world. They are found in all continents; however cliffs and rocky shores are the only coasts with minimal wetlands. Worldwide

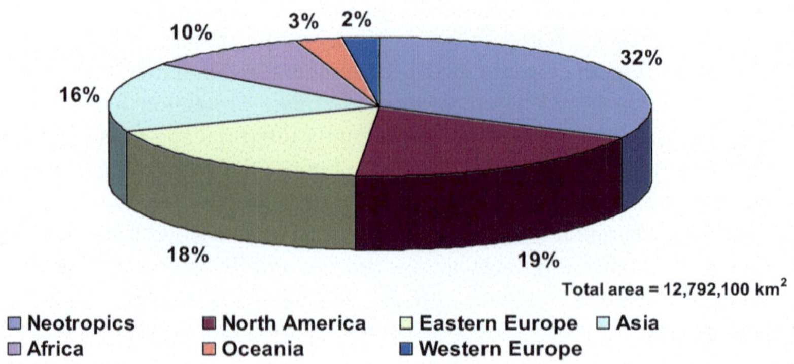

Fig. 6.5 Global distribution of wetlands

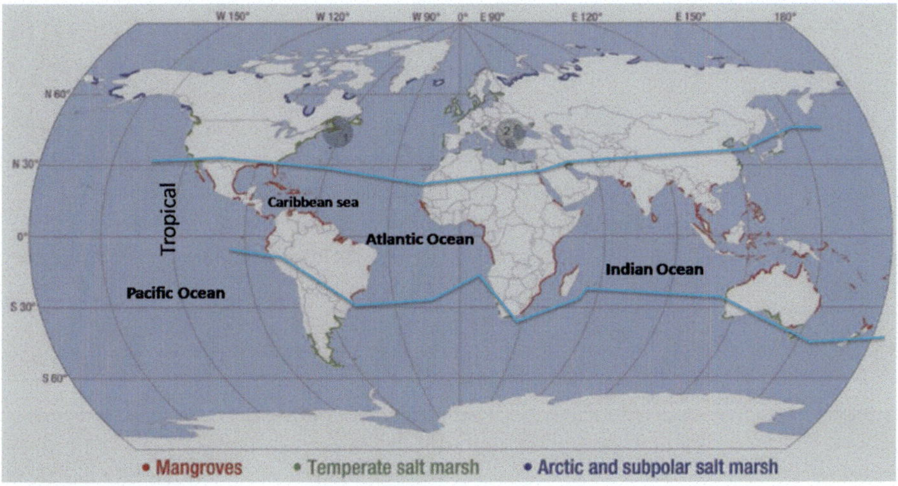

Fig. 6.6 Major coastal wetland types in different region

mangroves cover about 230,000 km^2 area but are restricted to mangrove belt in the tropics. Salt marshes cover a larger area, but in tropics it is found in the upper intertidal areas and also landward of mangroves. Tidal flats and freshwater coastal wetlands are likely to occupy nearly 300,000 km^2. Another important diversity-rich marine wetland ecosystem is represented by coral reef. The shallow coral reefs aerial extend is 284,300 km^2 in the world ocean (Venkatraman and Satyanaryana 2012). Majority of the world coral reefs occurs in the Indo-Pacific region centred on the waters of the Philippines, Indonesia, Malaysia, and Papua New Guinea. Less than 8% of the world's reefs are found in the Caribbean and Atlantic.

6.3 Biodiversity and Wetlands

A major goal of the Convention on Biological Diversity (1992) is to conserve biological diversity, world over. Evaluation of the species status is possible only if detailed inventories are available for monitoring the species in different habitats at required intervals. Wetlands are productive aquatic systems that support a variety of life forms. The biodiversity in wetlands has to be documented to plan for conservation mainly because of their rapid decline due to various natural and anthropogenic stressors (Krishnan et al. 2017). However, biological data are complex (e.g. hierarchical structures of gene ontology), heterogeneous (composed of different types), and highly dynamic (revision and addition) (Ozsoyoglu et al. 2006), thus offering challenges in choosing technologies and designing database schema for data reliability. At the same time, biodiversity is inextricably linked with the goods and services provided by ecosystems. The Millennium Ecosystem Assessment (2005) concluded that the ecosystem goods and services provided by wetlands are extremely

Fig. 6.7 Wetland birds found in Indian Ocean tropical islands

valuable to people all over the world. Hence, complete documentation of the biological resources of wetlands is essential.

Wetland ecosystems are some of the most important biodiverse areas the world and provide essential habitats for many species. The global Ramsar Convention network of 'Wetlands of International Importance' comprises over 2000 sites covering over 1.9 million km^2 (15% of estimated global wetland area), supports unique biodiversity (coral reefs, peatlands, freshwater lakes, marshes, and mangroves), species (e.g. waterbirds, amphibians, and wetland-dependant mammals such as hippopotamus, manatees, and river dolphins), and genetic diversity.

Among the different species inhabiting the wetlands, waterbirds, particularly migratory birds, are more striking feature of wetlands. In the South Asian region, some birds depend on wetlands almost totally for breeding, nesting, feeding, and shelter or during their annual cycles (Figs. 6.7 and 6.8). Birds that need functional access to a wetland or wetland products during their life cycle can be called 'wetland-dependent'. Many birds that inhabit intertidal habitats are migrants and travel annually along the Central Asian Flyway, which extends from central Siberia

Pacific Golden Plover : *Pluvialis fulva* Greater Sand Plover: *Charadrius leschenaultii*

Common Redshank: *Tringa totanus* Wood Sandpiper: *Tringa glareola*

Fig. 6.8 Shore birds found in Indian Ocean tropical islands

through the Himalayas to the Indian subcontinent. During peak annual migration periods, hundreds of thousands of birds migrating along the Central Asian Flyway descend upon the coastal wetlands of India in search of refuge and food. The important migratory birds utilizing the coastal wetlands are ducks, shore birds, gulls, terns, and flamingos.

Mangroves are one of the world's dominant coastal ecosystems comprised chiefly of flowering trees and shrubs uniquely adapted to marine and estuarine tidal conditions. It is largely confined to the regions between 30° North and South of the equator, with notable extensions beyond this to the north in Bermuda (32°20′N) and Japan (31°22′N) and to the south in Australia, New Zealand, and the east coast of South Africa (Fig. 6.9). Mangroves have broader ranges along the warmer eastern coastlines of the Americas and Africa than along the cooler western coastlines. The widespread occurrence of mangrove vegetation and the floristic divergence between the 'old' and the 'new' world can only be explained by geological events (Spalding 1997).

Mangroves are a diverse group of predominantly tropical trees and shrubs which are functionally diverse and complex. They also provide essential structure and

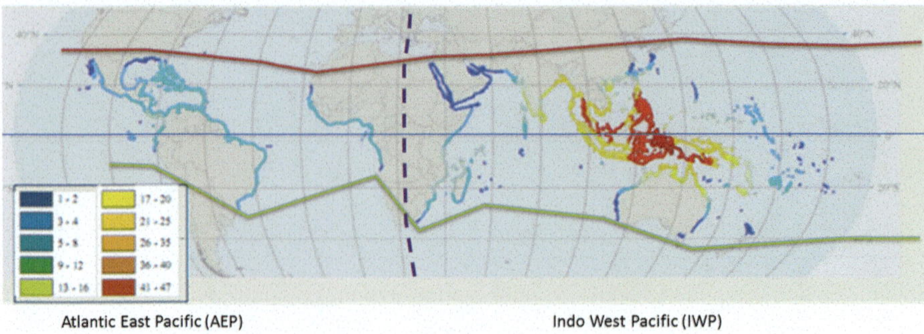

Fig. 6.9 Distribution of mangroves and species diversity with limit of its distribution

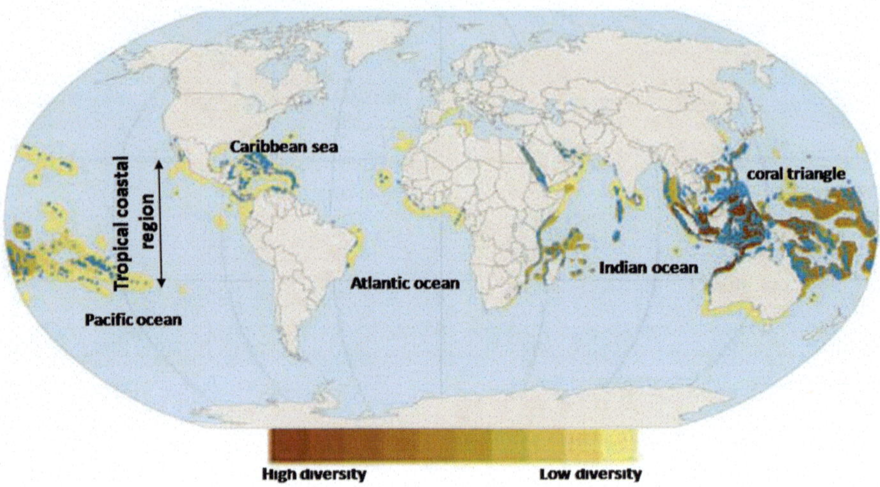

Fig. 6.10 Coral distribution and diversity in the tropical region. (Modified from: Spalding et al. 2001)

habitat for a host of marine and intertidal species (Duke et al. 1998). There are about 77 species belonging to 21 flowering plant genera found in mangroves; there is a relatively high degree of specialization for the tidal wetland habitat. The only plant families comprised exclusively of mangrove taxa are Avicenniaceae and Sonneratiaceae, although this continues to be debated (Kathiresan and Bingham 2001). According to Duke et al. (2007), among the 21 flowering plant genera found in mangroves, there is a relatively high degree of specialization for the tidal wetland habitat.

Coral reefs of the tropical ocean represent 1.2% of continental shelf area and 0.09% of total area of World Ocean. It occurs in the tropical ocean, and among the tropical nations, Indonesia and the Philippines constitute 76,000 km^2 reef area which is almost half of the reef area in the tropical region. It is also seen that the diversity and density is much higher in Indo-Pacific region called coral triangle (Fig. 6.10).

The more prominent members of the coral reef community and associated tropical marine habitats are as follows: sponges, scleractinian or stony corals, soft corals (alcyonaceans), sea fans, sea whips (gorgonaceans), polychaeta worms, wide varieties of bivalve and gastropod molluscs, octopus, squid, spiny lobsters, crabs, sea urchins, sea stars, sea cucumbers, crinoids, fishes, and brittle stars. Coral reef system is one of the most productive systems among the wetland and other ecosystems. The standing stocks of coral reef fishes are about 30–40 times higher than the Southeast Asia, the Mediterranean, or other temperate regions standing stocks of demersal fishing grounds (Birkeland 1996). The estimation of reef fishes for Atlantic is 160 mt km^2 and in Pacific 93–239 mt km^2 (Williams and Hatcher 1983).

6.4 Challenges to the Wetland Ecosystems

The major challenges of the coastal region particularly of the tropical countries are loss of land due to subsidence and erosion, overpopulation, communication and connectivity, amplified prices, unemployment, and vulnerability to natural calamities and oceanic changes. Low availability of resources, high costs of living, and high vulnerability due to natural disasters are common features which make it further vulnerable to projected climate change events like sea level and precipitation changes. As a result, the global extent of wetlands is now estimated to have declined between 64% and 71% in the twentieth century, and wetland losses and degradation continue worldwide. Because of wetland losses and degradation, people are deprived of the ecosystem services that wetlands provide. Adverse changes to wetlands, including coral reefs, are estimated to result in more than US$ 20 trillion in losses of ecosystem services annually (Gardner et al. 2015).

6.4.1 Climate Change

Climate change has emerged as one of the main challenges humankind will have to face for many years to come. It is also expected that the extent of vulnerability may differ widely in different regions of the globe, but no region or country is exception to its impacts. It could become a major threat to world food security, ecosystem services, and sociocultural aspects. Developing, least developed nations because of their limited adaptive capacity are more vulnerable. Coastal and island ecosystems in the tropics are intrinsically linked to climate in many ways. Sea level rise is the most important climate-related phenomenon with a major influence on coastlines. It has now been reasonably established that the climate change is happening and the global average temperature has risen by about 0.6 °C over the twentieth century. Globally, the ten warmest years on record all occurred after 1991. Even a temperature rise of 1–2.5 °C will have serious effects, including increase in the severity and frequency of extreme weather, changes in rainfall patterns leading to severe water shortages and/or flooding, melting of glaciers resulting in flooding and soil erosion, and an increased risk of extinction of plant and animal species (Dam Roy and George 2010).

Further, sea level rise will lead to the destruction of coasts worldwide with some islands possibly facing complete inundation. Coastal region is highly dependent on the ocean; the fragile relationship of the environment and human population with the ocean makes the resident population vulnerable to natural calamities.

6.4.2 Extreme Events

Tsunami, tropical cyclones, sea surges, floods, and droughts are the most destructive severe weather and climate extreme events which affect the coastal areas of tropical countries. A classical example of extreme events in South Asia is the tsunami of 26 December 2004. On that day a massive earthquake of magnitude 9.0 struck the coastal area of Northern Sumatra in Indonesia. A number of aftershocks also occurred which triggered tsunamis that affected Indonesia and neighbouring countries in Asia (including India, Malaysia, Maldives, Sri Lanka, and Thailand) and the east coasts of Africa (including Somalia and Yemen), causing serious damage to the coastal areas and small islands. While the final death toll will never be known, an estimated 250,000 persons have perished in the tsunami, majority of them women and children. Millions more have been displaced or rendered homeless. The reports indicated that natural ecological systems such as coral reefs, mangroves, and other wetlands had suffered extensive damages (Ramachandran et al. 2005). It is likely that maximum tropical cyclone wind intensities could increase, by 5–10% by around 2050 (Walsh 2004). Under this scenario, peak precipitation rates are likely to increase by 25% as a result of increases in maximum tropical cyclone wind intensities, which in turn cause higher storm surges. Consequently physical damages might impact the structure and function of coastal ecosystems and their ability to sustain marine life and support livelihood of coastal communities.

6.4.3 Coastal Erosion

The wearing away of land or the removal of beach or dune sediments by wave action, tidal currents, wave currents, or drainage due to natural or anthropogenic causes is a serious issue confronted by the coastal regions. This is a more serious concern for small island nations. While in the mainland, the littoral transport is interfered by coastal structures that include groynes, ports, inlet jetties, etc., the rise in sea water levels also plays a major role in coastal erosion. Over 80% of the land area in the Maldives is less than 1 m above mean sea level. Being so low-lying, the islands of the Maldives are very vulnerable to inundation and beach erosion. Presently, 50% of all inhabited islands and 45% of tourist resorts face varying degrees of beach erosion. Climate change and projected sea level rise would aggravate the present problem of beach erosion in small islands. It is expected that even a 1 m rise in sea level would cause the loss of the entire land area of Maldives. This also affects the coastal wetland and coral ecosystem by way sedimentation. Erosion and sedimentation also reduces light penetration into the benthic layer in the coastal/shallow marine waters affecting the overall productivity.

6.4.4 Population Pressure

Most settlements in the coastal and small islands, with the exception of some of the larger Melanesian and Caribbean islands the prime city or town is hosting the main port, international airport, and centre of government activities. Rapid and unplanned movements of rural and outer-island residents to the major centres are occurring throughout the zones, resulting in deteriorating urban conditions, with pressure on access to urban services required to meet basic needs. High concentrations of people in coastal urban areas create various social, economic, and political stresses and make people more vulnerable to short-term physical and biological hazards such as tropical cyclones and diseases. It also increases their vulnerability to the impacts of climate change and sea level rise besides profound anthropogenic impacts on the coastal ecosystems.

6.4.5 Habitat Destruction

Another larger problem of the coastal region is the challenge from the developmental activities. Coastal and offshore developments including land reclamation, dredging and disposal, construction of ports and jetties, desalination plants, large-scale aquaculture, and 'barotrauma' (impact of human-generated noise on the marine environment) play a major role in the loss of biodiversity. Though these issues are confronted by all coastal environments, the issue of land reclamation and unplanned development is particularly acute for the small island conglomerations. It is also almost inevitable that the ecological systems of coastal and islands, and the functions they perform, will be sensitive to the rate and magnitude of climate change and sea level rise, especially where exacerbated by human activities (e.g. in the case of the small islands in the Pacific, ADB 2004). For instance, analysis of coral reef surveys over three decades has revealed that coral cover across reefs in the Caribbean has declined by 80% in just 30 years, largely as a result of continued pollution, sedimentation, marine diseases, and overfishing (Gardner et al. 2003).

6.4.6 Over-Exploitation

Over-exploitation of resources such as groundwater, fishes and other biological resources from marine as well in the freshwater ecosystem, and soil will lead to ecological consequences and degradation in short span. The most important resource of the coastal region and small islands is the fresh water. Atoll countries and limestone islands have no surface water or streams and are fully reliant on rainfall and groundwater harvesting. Many small islands are experiencing water stress at the current levels of rainfall input, and extraction of groundwater is often outstripping supply.

6.4.7 Pollution of Wetlands

The most dreaded factor inducing loss of biodiversity is environmental pollution which includes land pollution, water pollution, and industrial pollution. Globalization is accelerating development, urbanization, and consumerism across many islands, but increasing liquid, solid, and hazardous wastes undermine ecosystem health and impact on food, water, and health security (SPREP 2014). Pollution comes from land-based sources, including industrial waste, chemical waste, pesticide runoff, sedimentation, urban runoff, plastic debris, as well as nutrient pollution from fertilizer runoff and sewage waste. Many plastics are broken down into small particles, ingested by seabirds and marine animals, bioaccumulating in food webs (Gregory 1999). Larger plastic pieces concentrate in big ocean gyres such as in the northern Pacific Ocean (SPREP 2014). Increased reliance on imported goods has increased ship-based pollution from discharge of sewage and waste oil, disposal of plastic, oil spills, and toxic chemical leaching from antifouling paints. Inaccurate navigation charts and poor practice lead to ship grounding and sinking; with limited options for removal, abandoned ships leach chemicals and other pollutants, sometimes devastating coral reef ecosystems (Work et al. 2008; Kelly et al. 2012).

6.5 Conservation of Wetlands and Habitats

6.5.1 The Strategy

In several countries wetlands are seriously threatened by reclamation through drainage and landfill, pollution, hydrological alterations, and over-exploitation of their natural resources resulting in loss of biodiversity. As discussed earlier these experiences pointed to the fact that conservation of wetlands won't be successful without the active involvement of the local communities and strong government policy support as many of the developing countries the government prioritize would tend towards pro-development. Therefore to attain a balance between sustainable use, societal expectations, and conservation, the following strategies are essential.

1. Developing wetland management plans for ensuring the stand and diversity
2. Zoning or prioritizing wetland ecosystems for proper management and utilization
3. Encouraging community regulation for long-term sustainability
4. Assessing the value of wetlands and its services for efficient planning and prioritization
5. Improving community information
6. Rehabilitation of degraded wetlands and protection for existing for social and environmental benefits

In this context the terms restoration and rehabilitation are used to describe the wetland conservation. The term 'restoration' has been adopted to specifically mean any activity/policy that aims to return system to a preexisting condition, whereas the term 'rehabilitation' is applied more generally and is used to denote any activity including restoration and habitat creation that aims to convert a degraded system to a stable alternative.

6.5.2 Global Initiatives

The global loss of biodiversity and shrinking of wetlands particularly its ecosystem services that has link across several climatic regions (transboundary) incited several regional and international organizations, national governments, scientist, and local communities against species extinction particularly wetland habitats that have been the main driving force for the emergence of the global conservation regime. A number of 'soft' instruments such as consultative conference, declarations, and guidelines and 'hard' instruments such as bilateral and multilateral treaties and conventions have been promulgated over the years that have components for the conservation of wetlands (Table 6.2). They are described below.

- The United Nations conferences have emphasized the need for conservation and protection of wetlands among the member countries especially those in coastal areas. In some cases, given explicit instructions/agreement on the way forward, usually achieved through international conventions and national legislation.
- 2 February 1971, marks the date of the signing of the Convention on Wetlands in the Iranian city of Ramsar on the shores of the Caspian Sea. Originally for waterfowl later extend to cover wetlands comprehensively. Consequently various countries across regions initiated activities to conserve the wetland ecosystem in their territory. The convention helped to protect the biodiversity and to retain the good and services provided by the ecosystems.
- There are a number of regional level instruments that have been evolved in the form of cooperative agreements and action plans focusing on various parts of the environment especially with respect to the conservation initiatives related to coastal ecosystems.
- Conservation policies are broad statements of intent to deal with the environmental issues that aid in framing appropriate legislation within a sovereign state. Broadly it covers forest, water, land, and biodiversity. These policies are framed in consonance with the environmental and social principles, set by the international regulatory instruments. The guiding principles include the ecological sustainability, equity, human rights, and public participation. However, providing incentives for the implementation of the policies and the market-oriented approach has been widely adopted at national and international levels.

6.5.3 Wetland Conservation: A Case Example of South Asia

Sri Lanka is an island nation situated Southeast of India. A massive 68% of the internationally important sites in Sri Lanka were reported to be under moderate or

Table 6.2 Global perspectives and major initiatives on wetland conservation

Initiative/measure	Year and location	Salient features
The United Nations conferences		
The United Nations Conference on the Human Environment	1972, Stockholm, Sweden	Representative samples of natural ecosystems, must be safeguarded for the benefit of present and future generations
United Nations Conference on Environment and Development (The Earth Summit)	1992, Rio de Janeiro, Brazil	Integrated coastal zone management (ICZM) and ecosystem-based approach
World Summit on Sustainable Development (WSSD)	2002, Johannesburg, South Africa	Develop national, regional, and international programmes for halting the loss of marine biodiversity, including coral reefs and wetlands
United Nations Conference on Sustainable Development (Rio+20)	2012, Rio de Janeiro, Brazil	Green economy in the context of sustainable development and poverty eradication
Multilateral agreements		
Ramsar Convention	1971, Ramsar, Iran	Wetland conservation and sustainable use
Conservation of Migratory Species of Wild Animals (CMS)	Also known as Bonn Convention, signed in 1979	Conserve terrestrial, aquatic, and avian migratory species throughout their range
United Nations Convention on the Law of the Sea (UNCLOS)	1994	Conservation and utilization of living marine resources within the exclusive economic zone and high seas
Convention on Biological Diversity (CBD)	Came into force on 29 December 1993	Its main goals are conservation of biological diversity including marine (Jakarta Mandate 1995), sustainable use of its components, and fair and equitable sharing of benefits arising from genetic resources
Convention on International Trade in Endangered Species of Wild Fauna and Flora (CITES)	Came into force on 1 July 1975	The aim of this convention is to ensure that international trade in specimens of wild animals and plants does not threaten their survival and avoids over-exploitation
Regional instruments		
South Asia Co-operative Environment Programme	1982	Established by the governments of South Asian region to promote and support protection, management, and enhancement of the environment in the region
Mangroves for the Future (MFF)	Initiated after December 2004 tsunami	To promote investment in the coastal ecosystem conservation for sustainable development and is co-chaired by IUCN and United Nations Development Programme

(continued)

Table 6.2 (continued)

Initiative/measure	Year and location	Salient features
Instruments of the Convention on Migratory Species (CAF)		The CAF seeks to conserve various habitats that provide abode to migratory waterbirds and important migration routes. Indian Ocean-South-East Asian Marine Turtle Memorandum of Understanding is covered under this
Policies		
Forest conservation	All the countries	Based on the need and international obligation of the national government different laws are enacted
Water use policy		
Land use policy		
Protection of wildlife		

high threat. Therefore, Central Environmental Authority of Sri Lanka (CEA) formulated the Wetland Conservation Project, in 1990 to protect the wetlands of Sri Lanka. Subsequently, the Asian Wetland Directory described 41 wetland sites of international importance in Sri Lanka, covering 274,000 ha in total. The country is signatory to all the international conventions and conferences protecting wetlands and biodiversity. The country has enacted several laws related to forestry, water resources, wildlife, and coastal zone for sustainable utilization (NWP 2000).

The Republic of Maldives is a collection of small islands comprising about 1190 islands grouped into 26 atolls forming a chain of over 820 km in length. The wetlands of Maldives are mostly located within or partly along the coast and are either fresh or brackish water. Of the 1190 Maldivian islands, only 41 islands have wetlands as the coral reef is not classified as wetland. In the Maldives, there are natural depressions or wetlands (locally known as Kulhi) situated within or partly along the coast. These wetlands are either brackish or fresh water, may be linked to lagoons, and almost always support mangroves. The policy context in the Maldives is supportive of wetlands and coral reefs conservation from the standpoint of biodiversity conservation and climate change adaptation. The Strategic Action Plan (SAP; also called National Framework for Development 2009–2013) stresses conserving and sustainably utilizing biological diversity to ensure maximum ecosystem benefits.

The wetland inventory of India prepared by the Space Applications Centre (2011) and published as 'National Wetland Atlas' categorizes the wetlands of India into two: inland and coastal. There are 13,033 coastal wetlands in India which are spread over 4,140,116 ha. These wetlands encompass numerous varieties of flora and fauna and are the hotspots of biodiversity. The country is signatory to all the international conventions and conferences protecting wetlands and biodiversity. Apart from this India has enacted number of legislation protecting coastal wetlands, biodiversity, and wildlife. A comprehensive coastal regulation zone act provides for conservation and sustainable development of entire coastal zone of India. Several coastal wetlands such as Chilika Lake (Fig. 6.11), Kolleru Lake, Pulicat Lake, Gulf of Kachchh (Marine National Park), Gulf of Mannar, and Point Calimere play vital role in protecting biodiversity and providing valuable ecosystem services sometimes

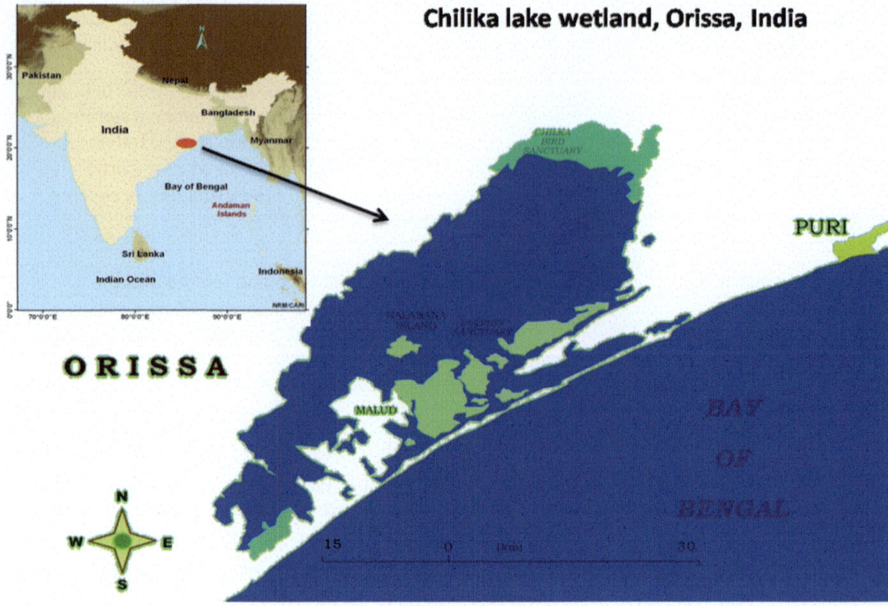

Fig. 6.11 Chilika Lake is the first Ramsar site (1165 km^2) from India where conservation efforts restored the ecologically very important wetland ecosystem

transboundary. Chilika Lake wetland site protects avian (migratory and local), terrestrial and aquatic flora and fauna, and marine animal including seagrasses.

In Bangladesh fisheries and wildlife play a critical role in the economic development, and hence, in order to safeguard these resources, the government has taken several measures to minimize the disruption to the natural aquatic environment in streams and waterways (MoWR 2005). In this line National Environmental Management Action Plan (NEMAP) was formulated. Similar to CRZ of India, Bangladesh framed Coastal Zone Policy (CZP) as a step to conserve Bangladesh's vulnerable coastal zone, which is prone to natural disaster such as cyclone, floods, storm surges, coastal erosion, increase of soil salinity, and coastal pollution. Bangladesh also has developed the Bangladesh Climate Change Strategy and Action Plan (BCCSAP) in 2009.

Similar to these efforts, several other tropical countries have enacted wetland rules in line with international conference and conventions and implemented strategic action plan for conservation of wetlands.

6.6 Conclusion

Tropical coastal wetlands represent one of the most productive ecosystems that contribute to both terrestrial and marine biodiversity. In general, these wetlands are formed as low-lying coastal plains in response to the pattern of relative sea level change. Coastal wetlands can be fresh, saline, and brackish which can also be grouped either as natural or man-made. Wetland is one of the most important ecosystems covering 5–8% of land surface area of the world, but they are under increasing threat by anthropogenic activities. In recent time climate change-associated events have been observed as single most important factor contributing to the deterioration of wetlands in the coastal zone. Further, rapid deterioration and decline in all types of wetlands has occurred in the tropical coastal region particularly in South Asian region due to the rapid human population rise and increased demand for natural resources for food, fuel, and fodder. Decline in size and number of wetlands is happening despite the fact that it provides numerous ecological and socio-economic functions that benefit humankind. Therefore, conservation of wetlands is extremely important to provide sustainable livelihood to the local people and conserve the biodiversity. In this regard, Ramsar Convention is an international effort for wetland conservation that facilitates the framework for national action and international cooperation for conservation and wise use of wetlands and their resources. In addition to this, number of initiatives are taken by United Nations, UNEP, and various national government aimed at protecting wetlands.

Annexure 1: Different Wetland Types Found in the Tropical Coastal Areas

a. Salt marshes

b. Flooded rice paddies

c. Freshwater lake

d. Mangrove

e. Tidal flats

f. Corals and mangrove patch during high tide

g. Rockshore

h. Chennel carrying tidal water

References

Alongi DM (2008) Mangrove forests: resilience, protection from tsunamis and responses to global climate change. Estuar Coast Shelf Sci 76:1–13. https://doi.org/10.1016/j.ecss.2007.08.024

Anderson JT (2012) A defining moment for wetlands: is it a wetland or is it not a wetland? Jalaplavit 3(1):5–9

Asian Development Bank (2004) Impact of climate change: small islands in the Pacific. Asian Development Bank (ADB), Manila

Baldassarre GA, Bolen EG (1994) Waterfowl ecology and management. Wiley, New York, p 609

Birkeland C (1996) Life and death of coral reefs. Chapman and Hall, New York, p 536

CBD (1992) Convention of biological diversity. https://www.cbd.int/doc/legal/cbd-en.pdf. Accessed 26 Nov 2017

Cowardin LM, Carter V, Golet FC, La Roe ET (1979) Classification of wetlands and deepwater habitats of the United States, FWS/OBS-79/31. U. S. Fish and Wildlife Service, Washington, DC, p 103

Dam Roy S, George G (2010) Marine resources of Islands: status and approaches for sustainable exploitation/conservation with special emphasis to Andaman and Nicobar. Indian J Anim Sci 80(4):57–62

Dam Roy S, Krishnan P, Patro S, George G, Velmurugan A, Kiruba Sankar R, Ramachandran P (2017) Wetlands of small island nations in South Asia vis-à-vis the mainland and island groups in India: status and conservation strategies. In: Prusty BAK, Chandra R, Azeez PA (eds) Wetland science. Springer (India) Pvt. Ltd, New Delhi, pp 31–48

Duke NC, Ball MC, Ellison JC (1998) Factors influencing biodiversity and distributional gradients in mangroves. Glob Ecol Biogeogr Lett 7:27–47

Duke NC, Meynecke JO, Dittmann S, Ellison AM, Anger K, Berger U, Cannicci S, Diele K, Ewel KC, Field CD, Koedam N, Lee SY, Marchand C, Nordhaus I, Dahdouh-Guebas F (2007) A world without mangroves? Science 317:41–42

Finlayson CM, Davidson NC (1999) Global review of wetland resources and priorities for wetland inventory, Ramsar Bureau Contract 56. Ramsar Convention Bureau, Gland, p 26

Gardner TA, Cote I, Gill G, Grant A, Watkinson A (2003) Long-term region wide declines in Caribbean corals. Science 301:958–960

Gardner RC, Barchiesi S, Beltrame C, Finlayson CM, Galewski T, Harrison I, Paganini M, Perennou C, Pritchard DE, Rosenqvist A, Walpole M (2015) State of the world's wetlands and their services to people: a compilation of recent analyses, Ramsar Briefing Note no. 7. Ramsar Convention Secretariat, Gland

Gopal B (2015) Guidelines for rapid assessment of biodiversity and ecosystem services of wetlands, version 1.0. Asia-Pacific Network for Global Change Research (APN-GCR)/National Institute of Ecology, Kobe/New Delhi, p 134

Gregory MR (1999) Plastics and South Pacific island shores: environmental implications. Ocean Coast Manage 42:603–615

Kathiresan K, Bingham BL (2001) Biology of mangroves and mangrove ecosystems. Adv Mar Biol 40:81–251

Kelly LG, Barott KL, Dinsdale E, Friedlander AM, Nosrat B, Obura D, Sala E, Sandin SA, Smith JE, Vereij MJA, Williams GJ, Willner D, Rohwer F (2012) Black reefs: iron-induced phase shifts on coral reefs. ISME J 6:838–649

Krishnan P, Samuel D, Sreeraj V, Abhilash CR, Shesdev Patro KR, Sankar R, Purohit M, Dharani G, Lakshmi A, Purvaja R, Ramesh R (2017) Digital repositories for coastal wetland biodiversity in South Asia: a conceptual framework from India. In: Prusty BAK, Chandra R, Azeez PA (eds) Wetland science. Springer (India) Pvt. Ltd., New Delhi, pp 51–66

Maltby E, Turner RE (1983) Wetlands of the world. Geol Mag 55:12–17

Matthews E, Fung I (1987) Methane emissions from natural wetlands: global distribution, area, and environmental characteristics of sources. Glob Chem Cycles 1:61–86

MEA (2005) Online guide to the millennium ecosystem assessment reports. Available via http://www.millenniumassessment.org/en/index.html. Accessed 10 Nov 2017

Mitsch WJ, Gosselink JG (2000) Wetlands, 3rd edn. Wiley, New York

Mitsch WJ, Gosselink JG (2007) Wetlands. Wiley, New York, p 582

MoWR (2005) Coastal zone policy. Ministry of Water Resources, Government of the People's Republic of Bangladesh. http://www.warpo.gov.bd/pdf/czpo_eng.pdf. Accessed on 19 June 2016

Navid D (1989) The international law of migratory species: the Ramsar Convention. Nat Resour J 29:1001–1016

NWP (2000) National wetland policy of Sri Lanka, 2000. http://www.dwc.gov.lk/index.php/en/2015-01-26-04-14-43/wildlife-policy. Accessed on 10 July 2017

Ozsoyoglu ZM, Ozsoyoglu G, Nadeau J (2006) Genomic pathways database and biological data management. Anim Genet 37(s1):v41–v47

Ramachandran S, Anitha S, Balamurugan V, Dharanirajan K, Ezhil Vendhan K, Marie IPD, Senthil Vel A, Sujjahad Hussain I, Udayaraj A (2005) Ecological impact of tsunami on Nicobar Islands (Camorta, Katchal, Nancowry and Trinkat). Curr Sci 89(1):195–200

Ramsar Convention Secretariat (2013) The Ramsar convention manual: a guide to the Convention on Wetlands (Ramsar, Iran, 1971), 6th edn. Ramsar Convention Secretariat, Gland, p 109

SAC (2011) Information brochure – national wetland inventory & assessment, sponsored by ministry of environment and forests. Govt. of India. Space Application Centre –ISRO, Ahmedabad, p 24

Spalding M (1997) The global distribution and status of mangrove ecosystems. International Newsletter of Coastal Management-Intercoast Network, Special edition 1, pp 20–21

Spalding MD, Ravilious C, Green EP (2001) World atlas of coral reefs, Prepared at the UNEP World Conservation Monitoring Centre. University of California Press, Berkeley, p 432

SPREP (2014) State of conservation in Oceania 2013, Regional volume. Secretariat of the Pacific Regional Environment Programme, Apia

Stewart RE Kantrud HA (1971) Classification of natural ponds and lakes in the glaciated prairie region. U.S. Fish and Wildlife Service, Resource Publication 92, Washington, DC, p 26

ten Brink P, Russi D, Farmer A, Badura T, Coates D, Förster J, Kumar R, Davidson N (2013) The economics of ecosystems and biodiversity for water and wetlands. Executive summary. Institute for European Environmental Policy (IEEP), p 13. https://www.ieep.eu

Van Zon JCJ (2004) Wetland conservation and management in Sri Lanka: a status paper. In: Wetland conservation in Sri Lanka. Proceedings of the National Symposium on Wetland Conservation and Management: Sri Lanka. p 75

Venkatraman K, Satyanaryana CH (2012) Coral identification manual. Zoological Survey of India, Kolkata, pp 1–136

Walsh K (2004) Tropical cyclones and climate change: unresolved issues. Clim Res 27:77–83

Williams D, Hatcher AI (1983) Structure of fish communities on outer slopes of inshore, mid-shelf and outer shelf reefs of the Great Barrier Reef. Mar Ecol Prog Ser 10:239–250

Work TM, Aeby GS, Maragos JE (2008) Phase shift from a coral to a corallimorph-dominated reef associated with a shipwreck on Palmyra Atoll. PLoS One 3(8):e2989

Crop Genetic Diversity in the Tropical Coastal Areas

7

Abstract

Rice and coconut are the two most important crops grown in the tropical coastal areas for two different purposes. These two crops require abundant rainfall and hot humid climatic conditions but differ in their soil conditions. Rice is the staple food of more than half the world's population, whereas coconut is the most important plantation crop providing livelihood to millions. Therefore an attempt was made in this chapter to document the information pertaining to rice and coconut genetic diversity with special reference to the coastal areas so as to understand the current status of these crops in the tropical world and its potential use as a variety to improve crop production and as a donor for producing varieties tolerant/resistant to the emerging adverse situations including climate change. Information on collection, conservation, and characterization of rice and coconut germplasm is also provided.

Keywords

Crop diversity · *Oryza* · Coconut palm · Wild relatives · Conservation

7.1 Introduction

Crop production and the ability to meet the emerging challenges largely rely on the genetic diversity of crop plants. Since the very beginning of agriculture during the process of domestication and cultivation of crop plants, a wealth of genetic diversity has been utilized and partly preserved. It is estimated that not even 15% of the potential diversity has been utilized in the world. Thousands of precious allelic variations of traits of economic significance remain unutilized in nearly all crop plants. These potentials can be discovered and effectively used to meet the existing and emerging challenges particularly in the coastal areas of the tropics. This is directly

V. Ayyam et al., *Coastal Ecosystems of the Tropics - Adaptive Management*,
https://doi.org/10.1007/978-981-13-8926-9_7

linked with ensuring livelihood security of local population. Unfortunately, this genetic wealth is being eroded due to neglect and over-exploitation.

Current developmental activities and land use planning in the coastal areas are destroying natural habitats, and improved varieties are replacing native species and landraces, resulting in a reduction of varietal diversity and loss of adaptive capacity. Major crop species of the coastal areas in the tropics are rice, coconut, pulse, and other millets which are of special significance to the agroclimatic conditions of the coastal areas. In order to successfully meet future food requirements and provide sustainable livelihood and adapting to climate change, it is necessary to cope up the continuing genetic erosion and address the issues of genetic conservation and optimum utilization of genetic diversity of important crop plants.

Therefore it is imperative to document the information pertaining to rice and coconut genetic diversity with special reference to the coastal areas so as to understand the current status of these crops in the tropical world and its potential use as a variety to improve crop production and as a donor for producing varieties tolerant/resistant to the emerging adverse situations. This chapter also discusses the role of these crops in cropping systems of coastal regions with a case example from tropical coastal region.

7.2 Importance of Rice and Coconut

Rice and coconut are the two most important crops grown for two different purposes in the tropical coastal region. These two crops require abundant rainfall and hot humid conditions but differing soil conditions. Rice is the staple food of more than half the world's population which is even more for the countries in the tropical Asia. As the humid tropical condition is suitable for the growth of rice, naturally Tropical Asia contributes more than 90% of the world's total rice production. Rice has been cultivated in Asia since prehistoric times, and for generations farmers have maintained thousands of different varieties (Jackson 1995). These landraces along with 22 pan-tropical, wild species of *Oryza* are the genetic foundation for the breeding efforts needed to sustain the productivity of rice. It is a well-established fact that the wild species exhibit tremendous diversity in morphological traits, height, tillering, flowering, growth habit, panicle, leaf, culm, and seed characteristics and adaptation to different habitats and agronomic traits. Besides this, the genetic resources of rice also encompass natural hybrids, commercial and obsolete varieties, breeding lines, and a range of different genetic stocks.

Coconut palm, the other important multipurpose crop of the humid tropics, supports millions of people in coastal and island ecosystems for their livelihood. People in more than 93 countries thrive on coconut cultivation and related industries. It has generally been accepted that the coconut originated in the Indian-Indonesia region and float-distributed itself around the world by riding ocean currents and late by human interventions. Coconut fruit in the wild are light, buoyant, and highly water resistant. Humans played a major role in dispersal of coconut mainly because of its ability to provide different resources like food and water (Chan and Elevitch 2006). In the Hawaiian Islands, the coconut is regarded as a Polynesian introduction, first brought to the islands by early Polynesian voyagers from their homelands in Oceania. They have been found in the Caribbean and the Atlantic coasts of Africa and South America for less than 500 years. Coconut is now almost ubiquitous between 26°N and 26°S, well adapted to the coastal conditions except for the interiors of Africa and South America. Coconut is an important subsistence crop of the humid tropical zones and is a life-supporting species in fragile island and coastal ecosystems.

7.3 Diversity of Rice (*Oryza* Species)

7.3.1 Genetic Diversity

The genus *Oryza* belonging to Gramineae family is having 23 species but is remarkable in its adaptation to diverse ecological conditions. In relation to genera containing other cereals, *Oryza* occupies a distinct phylogenetic position in a separate subfamily, the Ehrhartoideae (Kellogg 2001). The principle morphological characteristics of the genus include rudimentary sterile lemmas, bisexual spikelets, and narrow, linear, herbaceous leaves with scabrous margins (Launert 1965; Tateoka 1964). The changes in basic nomenclature of *Oryza* species occurred with the advancement of genetic and molecular studies beginning from the 1960s (Vaughan and Morishima 2003).

Tateoka (1963) analysed species across the whole genus on the basis of studies carried out in the world's main herbaria and in the field in Asia and Africa. His work clarified the basic groups of species within the genus, and he called these groups as species complexes. Since the 1960s, four new *Oryza* species have been described, *O. meridionalis*, *O. rhizomatis*, *O. indandamanica, and O. neocaledonica*. These are all closely related to previously known species (Table 7.1). Among these species, *O. sativa* is distributed worldwide, while *O. latifolia*, *O. alta*, and *O. grandiglumis* belong to Central and South America. Most of the species belongs to South and South East Asia and associated with wetland habitats. These wild species are reservoirs of many useful genes, particularly for resistance to major biotic and abiotic stresses. However, these wild species are associated with several weedy traits, such as grain shattering, poor plant type, poor grain characteristics, and low seed yield. Besides, several incompatibility barriers limit the transfer of useful genes from wild species into cultivated species, whereas cultivated species are known to give more yield which recorded the maximum number of accessions.

Table 7.1 Distribution of *Oryza* species, chromosome number, genomic composition, and their useful traits

Sl. no.	Species	2 n	Genome	No. of accessions	Distribution	Useful traits
O. sativa complex						
1.	*O. sativa* L.	24	AA	96,564	Worldwide	Cultigen, high yielding
2.	*O. glaberrima* Steud.	24	A^gA^g	1562	West Africa	Cultigen; tolerance to drought, acidity, iron toxicity, P-deficiency; resistance to BB, blast, RYMV, African gall midge, nematodes, weed competitiveness
3.	*O. rufipogon* Griff.	24	AA	858	Tropical and subtropical Asia, tropical Australia	Resistance to BB, blast, BPH, tungro virus; moderately tolerant to Shb, tolerant to aluminium and soil acidity, increased elongation under deepwater; source of CMS and yield enhancing loci
4.	*O. nivara* Sharma and Shastry	24	AA	1260	Tropical and subtropical Asia	Resistance to grassy stunt virus, BB
5.	*O. longistaminata* A. Chev and Roehr	24	$A^l A^l$	203	Africa	Resistance to BB, nematodes, stem borer, drought avoidance
6.	*O. breviligulata* A. Chev. and Roehr.	24	A^gA^g	218	Africa	Resistance to GLH, BB; drought avoidance; tolerance to heat and drought
7.	*O. meridionalis* Ng	24	A^mA^m	56	Tropical Australia	Elongation ability; drought avoidance; tolerance to heat and drought
8.	*O. glumaepatula* Steud.	24	$A^{gp}A^{gp}$	54	South and Central America	Elongation ability; source of CMS; tolerance to heat
O. officinalis complex						
9.	*O. punctata* Kotschy ex Steud.	24, 48	BB, BBCC	71	Africa	Resistance to BPH, BB, zigzag leafhopper; tolerance to heat and drought
10.	*O. minuta* J.S. Presl. ex C.B. Presl.	48	BBCC	63	Philippines and Papua New Guinea	Resistance to BB, blast, BPH, GLH

(continued)

Table 7.1 (continued)

Sl. no.	Species	2 n	Genome	No. of accessions	Distribution	Useful traits
11.	*O. officinalis* Wall ex Watt	24	CC	265	Tropical and subtropical Asia, tropical Australia	Resistance to thrips, BPH, GLH, WPH, BB, stem rot; tolerance to heat
12.	*O. rhizomatis* Vaughan	24	CC	19	Sri Lanka	Drought avoidance, resistance to blast; tolerance to heat
13.	*O. eichingeri* A. Peter	24	CC	30	South Asia and East Africa	Resistance to BPH, WBPH, GLH
14.	*O. latifolia* Desv.	48	CCDD	40	South and Central America	Resistance to BPH, BB, high biomass production
15.	*O. alta* Swallen	48	CCDD	6	South and Central America	Resistance to striped stem borer; high biomass production
16.	*O. grandiglumis* (Doell) Prod.	48	CCDD	10	South and Central America	High biomass production
17.	*O. australiensis* Domin.	24	EE	36	Tropical Australia	Resistance to BPH, BB, blast; drought avoidance; tolerance to heat and drought
O. meyeriana complex						
18.	*O. granulata* Nees and Arn. ex Watt	24	GG	24	South and SouthAsia	Shade tolerance, adaptation to aerobic soil
19.	*O. meyeriana* (Zoll. and (Mor. ex Steud.) Baill.)	24	GG	11	Southeast Asia	Shade tolerance; adaptation to aerobic soil
O. ridleyi complex						
20.	*O. longiglumis* Jansen	48	HHJJ	6	Irian Jaya, Indonesia, and Papua New Guinea	Resistance to blast, BB
21.	*O. ridleyi* Hook. F.	48	HHJJ	15	South Asia	Resistance to blast, BB, tungro virus, stem borer, whorl maggot
Unclassified						
22.	*O. coarctata* Tateoka	48	KKLL	1	Asian coastal area	Tolerance to salinity, stoloniferous

(continued)

Table 7.1 (continued)

Sl. no.	Species	2 n	Genome	No. of accessions	Distribution	Useful traits
23.	*O. brachyantha* A. Chev. and Roehr	24	FF	19	Africa	Resistance to BB, yellow stem borer, leaf folder, whorl maggot; tolerance to laterite soil
24.	*O. schlechteri* Pilger	48	KKLL	1	Papua New Guinea	Stoloniferous
25.	*Leersia perrieri* A. Camus	24	Unknown	1	Africa	Shade tolerance, stoloniferous

Sanchez et al. (2014)

7.3.2 Diversity in Rice Cultivation

Rice is grown under diverse growing conditions and is generally recognized into four major ecosystems (Khush 1984), viz. irrigated, rainfed lowland, upland, and flood-prone. All these conditions are generally prevalent in coastal areas of different tropical coastal countries. In each of the condition, the selection varieties are based on the climate, soil, water availability, and the physiography besides the choice of the farmers.

7.3.2.1 Irrigated Rice

Approximately 55% of the rice-growing area in the world is irrigated and is the most productive rice-growing system though it consumes large quantities of water. Further, 75% of the world rice production comes from irrigated areas, and major portions of the human population from Asia and other tropical regions are fed from irrigated rice. Irrigation water is provided by human intervention by river diversions, reservoirs, channels, and wells. This is a popular form of rice in most of the river deltas in the coastal areas. The area of high-yielding, fertilizer-responsive, irrigated rice has expanded markedly in the last three decades due to high demand for food. Rice yields in irrigated areas have witnessed increase in yield (5–7 ton/ha) in the past 30 years, and there is considerable scope for further yield improvement. The irrigated areas are further divided into irrigated wet season when rainfall is supplemental with irrigation water and irrigated dry season when rainfall is very low and irrigation is the primary source of water supply. Yields during the dry season are higher than during the wet season due to higher incoming solar radiation.

7.3.2.2 Rainfed Lowland Rice

About one-fourth of the world rice area is rainfed lowland practised in most of the tropical countries with limited inputs. Yields average about 2–3 tons per hectare. Rainfed lowlands have a great diversity of growing conditions that vary by amount of rainfall and duration of rainfall, depth of standing water, duration of standing water, flooding frequency, time of flooding, soil type, and topography. Rainfed lowland fields are bunded, and water is impounded, when available, just like irrigated fields. Rainfed lowlands are further subdivided into four categories as the rainfed condition is subject to the performance of monsoon and soil conditions.

- Rainfed shallow favourable, where rainfall is adequate. Short periods of moisture stress may occur. Improved varieties developed for irrigated conditions are grown in such area and yields average around 3 tons.
- Rainfed shallow drought-prone, where rainy period is short (90–110 days) and periods of mild to severe drought stress occur during the growing season. Photoperiod insensitive varieties with short duration and degree of drought tolerance are most suitable.
- Rainfed shallow submergence prone, where rice crop is submerged during periods of heavy rainfall for up to 10 days. Rainy period is generally long, and crop is harvested after the rainy season is over. Photoperiod-sensitive varieties are generally grown.
- Rainfed medium deep, where water accumulates in the fields in low-lying areas and stagnates for 2–5 months because of impeded drainage. Photoperiod-sensitive varieties with tolerance to stagnant water are grown.

7.3.2.3 Upland Rice

Upland or dryland rice is grown under rainfed, naturally well-drained soils in bunded or unbunded fields without surface water accumulation. Some of the upland rice areas are on sloping mountain sides mostly away from the tropical region. Rice is planted under dry conditions just as wheat or maize. Rice varieties are photoperiod insensitive and have deep roots and some level of drought tolerance. The major problem for rice production is that many of the upland soils have low pH and are deficient in nutrients. Yields average about 1.5–2.5 tons per hectare. About 16 million hectares of world rice-growing area are classified as upland having genetically different materials than the lowland rice.

7.3.2.4 Flood-Prone Rice

Flood-prone rice is grown in low-lying lands in river deltas of South and Southeast Asia and coastal lowlands of tropical islands. Standing water depth may vary from 50 cm to more than 3 m due to heavy monsoonal rainfall. However, flooding occurs only during part of the growing season. Fields are unbunded, and rice is often broadcast sown where tall varieties with photoperiod sensitivity are grown. Varieties grown in deeper areas (>100 cm) have elongation and low oxygen tolerance ability. Oxygen moves to the root through leaf stomata also, and there are very specific symbiotic microbial associations with the roots. About 9 million hectares are planted to flood-prone rice of which 3 million fall in the category of deepwater rice. Average yields are around 1.5–2.0 tonnes per hectare.

Table 7.2 Rice accessions collected and conserved at IRRI, Manila

Sl. no.	Country	No. of accessions
1	India	16,013
2	Laos PDR	15,280
3	Thailand	5985
4	Bangladesh	5923
5	Philippines	5515
6	Cambodia	4908
7	Malaysia	4028
8	Myanmar	3335
9	Vietnam	3039
10	Nepal	2545
11	Sri Lanka	2123
12	7 countries (with >1000 and <2000 accessions)	10,241
13	105 countries (<1000 accessions)	11,821
	Total	108,256

7.3.3 Conservation of Rice Germplasm

Conservation of rice germplasm is utmost important to sustain the production and look for beneficial traits for the future use and climate change adaptation. Most countries in Asia maintain huge collections of rice germplasm as it is their main food crop, and the largest are in China, India, Thailand, and Japan (IRRI 1996). In Africa, there are significant collections in Nigeria and Madagascar, while in Latin America, the largest collections are in Brazil, Peru, Cuba, and Ecuador. The details are given in Table 7.2. All these collections conserve landrace varieties as well as breeding materials. Four centres of the Consultative Group on International Agricultural Research (CGIAR), the International Rice Research Institute (IRRI) in the Philippines, the West Africa Rice Development Association (WARDA) in Côte d'Ivoire, the International Institute for Tropical Agriculture (IITA) in Nigeria (on behalf of WARDA), and the International Centre for Tropical Agriculture (CIAT) in Colombia, maintain huge collection of rice genetic diversity. IRRI holds the largest collection in the world, and it is also the most genetically diverse and complete rice collection in the world. The WARDA/IITA and CIAT collections have some specific regional representation besides several breeding materials that are developed at these centres.

7.4 Coconut Biodiversity

Coconut is cultivated all around the tropical belt including natural growth in many of the tropical islands, and its large genetic diversity has long been described at both worldwide and regional levels (Ovasuru 1994; Ashburner et al. 1997; Zizumbo-Villarreal and Piñero 1998). Several efforts have been made worldwide to collect,

Table 7.3 Variants for coconut morphological traits

Trait	Variants
Habit	Tall, dwarf, and intermediate forms
Leaf	Normal, albinism, plicata, small or long leaf/leaflet
Floral	Vegetative transformation (bulbils), midget/hapxanthic, very early to very late flowering/bearing, spicata, secondary spikelets, apocarpy, persistent inflorescences on stem
Sex expression	(Dioecy) male/female, male sterility, hermaphrodite
Fruit	Jelly-like endosperm, fragrant endosperm, edible husk, variable size of fruit (<50 g to 3 kg), colour and husk proportion (20–80%), beak-, ring-, or horn-like structures on fruit

conserve, and utilize the diversity to improve the production and exploit several other useful traits. Attempts to account for this diversity have resulted in various classification systems but were not always easy to obtain a completely consistent system, because some of the traits used in these classifications represented variation between populations, while others were polymorph within a given population (Baudouin and Santos 2005). Therefore it is necessary to take into account intra- and interpopulation variation.

Describing the morphological and agricultural aspects of coconut genetic variation appears to be a complex task and requires the observation of a fairly large number of quantitative traits. Several methods are being developed to assess these variations and identify coconut varieties (Harries 1981). Study of morphometric variation of coconut can generate two kinds of information. The first is an evaluation of the agricultural potential of the cultivars tested; the second is a better understanding of the genetic diversity, in relation to results obtained with molecular markers. Further, using molecular markers, it is now possible to assess the genetic relationships between different coconut cultivars collected across the entire geographic range of the plant.

7.4.1 Morphological Diversity

Studying morphological diversity in coconut involves comparing traits that are measured on different parts of the plant. Some important variants for morphological traits such as habit, leaf, floral, sex expression, and fruit in coconut palm are given in Table 7.3. It is often useful to compare the results of different studies, and this comparison is easier if standardized techniques are used. Such studies have indicated that the variability in local coconut types was reported to be highest in Southeast Asia (Whitehead 1976). However, it has not been possible to establish either a true centre of diversity or centre of origin for the species. These simple but basic factors are of great importance for understanding the extent and distribution of coconut genetic diversity and for locating useful variation. Moreover, any crop improvement programme for coconut requires certain basic information about different accessions/local varieties planned to be used in the breeding programme. Although evaluation of phenotypic traits is direct, cost-effective, and easy to observe, it also may be influenced by environmental factors and other factors (Perera et al. 2003).

7.4.2 Morphotypes

Several authors across the tropical world have reported variations in the size of the coconut fruits and its components in different varieties. They appear to be similar in large collections from different countries which are referred as morphotypes (John and Narayana 1949; Gangolly et al. 1957). In other words this is defined as any of a group of different types of individuals of the same species in a population. The details are summarized in Table 7.4. From the large pool of coconut collections, it has been observed that occurrence of similar morphotypes in various major growing countries is an instance of parallel variations.

7.4.3 Genetic Diversity

The worldwide coconut collections are having veritable gene pools, and it needs to be characterized in terms of their morphological and genetic traits for its further use. As the latter are independent of the environment, it defines precisely the identity of the germplasm collection. Because characterization of genetic resources in coconut is a prerequisite for crop improvement, initiatives in coconut as the breeding efforts are cumbersome in coconut owing to its perennial nature, high heterozygosity, long juvenile phase, requirement of large area, and longer time duration for any evaluation trials. Development of biotechnological tools, drastic improvement in genetic studies, and use of marker-aided selection have revolutionized the plant genetic characterization.

Different types of traits have been used to characterize genetic diversity in coconut populations. A variety of phenotypic traits was used by N'Cho et al. (1993) to describe the diversity of 17 Tall cultivars. In an initial study involving 17 morphological fruit traits, Zizumbo-Villarreal and Pinero (1998) separated Mexican coconut palms into three groups: two groups of Pacific Talls and one group of Atlantic Talls. Fruit traits were used by Ashburner et al. (1997) to describe the structuring of 29 cultivars from the South Pacific (28 Talls and 1 Dwarf) into genetic groups. Phenotypic observations are directly related to agronomic traits, but they are largely subjected to selection and to environmental conditions. It was therefore necessary to develop new tools for characterizing cultivars, independently of their growth habit

Table 7.4 Distinctive characters of coconut morphotypes

Sl. No.	Name of the variety	Character	Country of representation	References
1.	Makapuno	Light curd-like endosperm fills the cavity instead of firm endosperm which is considered a delicacy	Philippines India and Sri Lanka	Copeland (1931) and Jerard et al. (2013)
2.	Klapa Wangi	Fragrant endosperm	Singapore	Burkill (1935)
3.	Coconino	Baby coconut, very dwarf in nature, possibly similar to micro and nana forms	Philippines	Copeland (1931)
4.	Cocos nana Grif	Very small dwarf variety going by the name Maladive coconut also similar to Maldives Micro Tall	South Asia	Trimen (1898)
5.	Spicata	Inflorescence unbranched with no spikelets present, bears with 100 female flowers and 50 male flowers in an inflorescence, produces 10–20 mature fruits/inflorescence	India	Jacob (1941) and Sankaran et al. (2015)
6.	Sam Ramon	High yielding with large nuts, similar types reported from India, Sri Lanka, Malaysia, and Polynesia	Philippines	Copeland (1931)
7.	Pugai	Very dwarf palm with short juvenile period	Philippines	Copeland (1931)
8.	Tembili	Possesses pink endosperm, more commonly known as king coconut, a dwarf form	South Asia	Watt (1889) and Trimen (1898)
9.	Male coconut	The inflorescence bears only male flowers and no female flowers, with 5000 male flowers per inflorescence	–	John and Narayana (1949)
10.	Freaks and abnormalities	Viviparous and suckering forms and teratological abnormalities	–	John and Narayana (1949)
11.	Mangipod	Intermediate between Coconino and Pugai, has very short juvenile phase	–	Copeland (1931)
12.	Niu Leka	Dwarf coconuts and prolific bearer	Fiji	–
13.	King coconut	Also known as Rath Thembili, juice is very sweet, esteemed for culinary purposes	Sri Lanka	–
14.	Nudam	Husk "exceptionally" thin, hardy in nature	Fuji	–
15.	Kaithathali	A tall form, mesocarp soft and fleshy, sometimes eaten, said to be a good antidote for seasickness	India	Patel (1938)

(continued)

Table 7.4 (continued)

Sl. No.	Name of the variety	Character	Country of representation	References
16.	Taban	Similar forms are known as Cayamis in Northern Mindanao and Kalpa Tebu in Java, husk sweet when tender, which can be chewed like sugarcane	Philippines	–
17.	Nasasi	Mesocarp is sweet	Sri Lanka	Trimen (1898)
18.	Klapa Logo	Nuts are edible at the tender-nut stage	Malaysia	–
19.	Coco raisin	Nuts small and spherical, resembling grapes	Seychelles	–
20.	Nu Wa or Cocos suere	Mesocarp sweet and edible like sugarcane	New Caledonia	–
21.	Klapa Angi	Endosperm possesses pleasant smell, used in preparing medicines	Malaysia	–

and associated traits. The most commonly employed tools to study the genetic diversity are molecular markers and biochemical markers.

Different types of molecular markers were developed to describe the diversity of coconut collections and subsequently to identify cultivars or individuals. Each genetic marker corresponded to a DNA sequence located at a precise spot of the genome (or locus), whose polymorphism between individuals was shown using different tools, which generally included enzymatic digestion of DNA and specific or random multiplication of selected sequences (polymerase chain reaction (PCR) amplification). This polymorphism corresponded to variants (or alleles) of the sequence being studied. The variation appears in the form of bands on electrophoresis gel after developing with a radioactive or non-radioactive stain (e.g. silver nitrate or fluorochrome) (Lebrun et al. 2005).

Some of the molecular markers used in diversity analysis of coconut are restriction fragment length polymorphism (Lebrun et al. 1999), randomly amplified polymorphic DNA (RAPD), amplified fragment length polymorphism (Teulat et al. 2000), inverse sequence-tagged repeat (ISTR), and microsatellites or simple sequence repeats (Baudouin and Lebrun 2002).

All these studies detected two major cultivar groups. The resulting groups are summarized in Table 7.5. The first was the Pacific group that was both the most polymorphic and geographically most extensive. It spreads from Southeast Asia to the east coast of Latin America. It includes four subgroups with blurred boundaries, viz. Southeast Asia, Melanesia, Micronesia, and Polynesia. The coconut palms from Panama and Peru formed a distinct fifth subgroup. All dwarfs, irrespective of their geographical origin, form a genetically uniform group that belongs to the Southeast Asia subgroup. The Niu Leka Dwarf is a notable exception as this cross-fertilizing Dwarf originated from the South Pacific. The second was the Indo-Atlantic group that originated from the Indian subcontinent, from where it was subsequently transported to West Africa and the Atlantic coast of Latin America. East Africa is also

Table 7.5 Molecular grouping of different coconut cultivars

Group	Subgroup	Type
Pacific group	A1 Southeast Asian Dwarfs	A1a Malayan type
		A1b Philippine type
	A2 Pacific Dwarfs (and semi-Talls)	
	A3 Southeast Asian Tall	A3a Continental type
		A3b Indonesian type
		A3c Philippine type
	A4 Melanesia	A4a North New Guinea type
		A4b South New Guinea type
		A4c Insular PNG type
		A4d Markham Valley type
		A4e Vanuatu type
	A5 Micronesia	
	A6 Polynesia	
	A7 Panama (and Peru)	
Indo-Atlantic group	B1 Introgression absent to very low	
	B2 Low introgression rate	
	B3 Moderate to high introgression	

populated with coconut palms of the Indo-Atlantic type, though these received an input from cultivars of Southeast Asian origin, resulting from Austronesian migrations to Madagascar. These distinct molecular groups can be described by a code comprising three characters, viz. capital letter for the first level, a digit for the second level, and a lower case letter for the third level.

Biochemical markers, such as isozymes and polyphenols, were first used to describe the diversity of coconut collections. Among the different traits studied in coconut, fruit component traits were found to be the best for discriminating the varieties. Following this, several researchers have used fruit component, viz. allozymes (Geethalakshmi et al. 2004) and polyphenols (Jay et al. 1989; Chempakam and Ratnambal 1993), to evaluate genetic diversity in coconut from various regions. Unlike with morphological traits, biochemical markers do not require measuring different characters from different organs in a full-sized palm. It is enough to take an organ sample (leaflet, root, etc.) to reveal the biochemical identity of the palm.

7.4.4 Conservation of Coconut Biodiversity

Countries that are signatories to the Convention on Biological Diversity endowed with significant amount of genetic and species diversity have a responsibility to the world at large to conserve them and make them available for use. The main objective in any plant genetic resource (PGR) conservation programme is to maintain the highest possible level of genetic variability present across the genepool of a given species or crop. The options for conserving coconuts are dependent on the biological characteristics of the whole plant and its component organs and tissues, as well as on the state of the technology as applied to coconuts. There are two approaches of conservation, viz. ex situ and in situ. It is generally recognized that the two approaches of conservation are both important in the conservation and use of genetic diversity and should be regarded as complementary to each other (Ramanatha Rao 1998).

The Coconut Genetic Resources Network (COGENT) was established in 1992 by the International Plant Genetic Resources Institute (IPGRI) under the aegis of the Consultative Group on International Agricultural Research (CGIAR). It successfully developed and disseminated worldwide the International Coconut Genetic Resources Database (CGRD) containing passport and characterization data and images of 1416 accessions which are conserved by national programmes in 28 sites in 23 countries. Due to climate change, there is a need to identify, collect, and conserve drought-tolerant germplasm, noting that almost all coconuts are grown in rainfed conditions. Consequently COGENT established a multisite International Coconut Gene (ICG) bank in 1995. It comprises four regional gene banks hosted by Indonesia for Southeast and East Asia, Papua New Guinea for the South Pacific, India for South Asia, and Côte d'Ivoire for Africa and the Indian Ocean. Four types of coconut accessions will be conserved in the ICG: (1) major varieties (parents of existing hybrids and advanced generations of selected cultivars), (2) varieties/cultivars threatened with genetic erosion or total loss, (3) varieties/cultivars with special traits/genetic markers, and (4) genotypes being used for current genetic diversity studies using molecular markers.

7.5 Conclusion

Rice and coconut are the two most important crops of the tropical coastal areas grown for two different purposes. Both require hot humid climate but differ in edaphic conditions. Rice is a pan-tropical crop which grows in a wide variety of habitats. Rice has large genetic diversity which is source of several traits useful for increasing the crop production and adapt to climate change. The wild relatives of rice have furnished genes for the hybrid rice revolution and other genes from *Oryza* species with major impact on rice yields and sustainable rice production in most of the tropical countries. Similar studies have shown that tropical coastal areas including the tropical islands are bestowed with abundant diversity in coconut. This diversity enabled it to play a pivotal role in the economic development of the tropical coastal and islands in the last century with early trade and investment in the region largely based on copra. Copra remains the sole source of cash but with the advent of modern-day technologies of value addition has transformed the coconut-based industry and livelihood for millions.

It is observed that rice-based cropping system especially rice-rice dominates the coastal plains and lowlands of island region either under rainfed or irrigated conditions. Plantation of coconut is prominent in island region, but the productivity is very low. Due to low productivity, poor resource use efficiency and lack of diversification alternative cropping systems are required in the rice-based systems. Coconut has become an important component in many kinds of farming systems, transformed from its most common niche in the fragile coastal environment. In such farming systems, it can play a crucial role in providing ecologically sound and sustainable production through longevity and the protection of soil from erosion which is vital for these islands. The coconut palm is also well suited to mixed farming and intercropping systems and has been used this way for millennia. It has the adaptation capabilities to withstand cyclones and heavy winds often experienced in the tropical coastal region.

References

Ashburner GR, Thompson WK, Halloran GM, Foale MA (1997) Fruit component analysis of south Pacific coconut palm populations. Genet Resour Crop Evol 44(4):327–335

Baudouin L, Lebrun P (2002) The development of a microsatellite kit for use with coconuts. IPGRI, Rome. 66p

Baudouin L, Santos G (2005) Morphometric methods of determining diversity in coconut. In: Batugal P, Ramanatha Rao V, Oliver J (eds) Coconut genetic resources. International Plant Genetic Resources Institute – Regional Office for Asia, the Pacific and Oceania (IPGRI-APO), Serdang2005

Burkill IH (1935) A dictionary of the economic products of the Malay Peninsula. Crown Agent, London

Chan E, Elevitch CR (2006) Species profiles for Pacific Island Agroforestry: Cocos nucifera, Traditional Tree. http://agroforestry.net/tti/Cocos-coconut.pdf. Accessed 18 June 2018

Chempakam B, Ratnambal MJ (1993) Variation for leaf polyphenols in coconut cultivars. In: Nair MK, Khan HH, Gopalasundaram P, Bhaskara Rao EVV (eds) Advances in coconut research and development. Oxford and IBH Publishing Co. Pvt. Ltd, New Delhi, pp 51–53

Copeland EB (1931) The coconut, 3rd edn. Macmillan, London

Gangolly SR, Satyabalan K, Pandalai KM (1957) Varieties of coconut. Indian Coconut J 10:3–28

Geethalakshmi P, Niral V, Parthasarathy VA (2004) Allozyme variation in population of dwarf coconut cultivars. J Plant Crop 32:13–15

Harries HC (1981) Germination and taxonomy of coconut. Ann Bot 48:873–883

IRRI (1996) Standard evaluation system for rice, 4th edn. IRRI, Manila

Jackson MT (1995) Protecting the heritage of rice biodiversity. GeoJournal 35:267–274

Jacob KC (1941) A new variety of coconut palm (Cocos nucifera Linn) var. spicata (K.C. Jacob). J Bombay Nat Hist Soc 41:906–907

Jay P, Bourdex R, Potier F, Sanlaville C (1989) Note on polymorphism of coconut leaf polyphenols. Oleagineux 44:151–161

Jerard BA, Damodaran V, Niral V, Samsudeen K, Rajesh MK, Sankaran M (2013) Conservation and utilization of soft endosperm coconut accession from Andaman Islands. J Plant Crop 41(1):14–21

John CM, Narayana GV (1949) Varieties and forms of the coconut Cocos nucifera. Indian Coconut J 2:209–226

Kellogg EA (2001) Evolutionary history of grasses. Plant Physiol 125:1198–1205

Khush GS (1984) Terminology for rice growing environments. In: Terminology of rice growing environments. International Rice Research Institute, Manila, pp 5–10

Launert E (1965) A survey of the genus Leersia in Africa. Senckenb Biol 46:129–153

Lebrun P, Grivet L, Baudouin L (1999) Use of RFLP markers to study the diversity of the coconut palm. In: Oropeza C, Verdeil JL, Ashburner GR, Cardena R, Santamaria JM (eds) Current advances in coconut biotechnology. Kluwer Academic Publ, Dordrecht, pp 73–87

Lebrun P, Berger A, Hodgkin T, Baudouin L (2005) Biochemical and molecular methods for characterizing coconut diversity. In: Batugal P, Ramanatha Rao V, Oliver J (eds) Coconut genetic resources. International Plant Genetic Resources Institute – Regional Office for Asia, the Pacific and Oceania (IPGRI-APO), Serdang2005
N'Cho YP, Sangare A, Bourdeix R, Bonnot F, Baudouin L (1993) Assessment of a few coconut ecotyp es – a biometric approach, study of tall populations. Oleagineux 48(3):121–132
Ovasuru T (1994) Preliminary analysis of coconut (Cocos nucifera L.) germplasm in Papua New Guinea. Aust Cent Int Agric Res Proc 53:33–40
Patel JS (1938) The coconut: a monograph. Government Press, Madras
Perera L, Russell JR, Provan J, Powell W (2003) Studying genetic relationships among coconut varieties/populations using microsatellite markers. Euphytica 132:121–128
Ramanatha Rao V (1998) Strategies for collecting of tropical fruit species germplasm. In: Arora RK, Ramanatha Rao V (eds) Tropical fruits in Asia: diversity, maintenance, conservation and use. Proceedings of the IPGRI/ICAR/UTFANET Regional Training Course on the Conservation and Use of Germplasm of Tropical Fruit in Asia, 19–31 May 1997, IIHR, Hesaraghatta, Bangalore, India. IPGRI South Asia Office, New Delhi, pp 73–78
Sanchez PL, Wing RA, Brar DS (2014) The wild relative of rice: genomes and genomics. In: Zhang Q, Wing RA (eds) Genetics and genomics of rice, Plant Genetics and Genomics: Crops and Models, vol 5. Springer Science+Business Media, New York. https://doi.org/10.1007/978-1-4614-7903-1_2
Sankaran M, Damodaran V, Jerard BA, Abirami K, Roy DS (2015) Multiple Spicata Coconut (MSC): a rare type of coconut in Andaman Islands. Transcriptomics 3:123–128. https://doi.org/10.4172/2329-8936.1000123
Tateoka T (1963) Taxonomic studies of Oryza III. Key to the species and their enumeration. Bot Mag Tokyo 76:165–173
Tateoka T (1964) Notes of some grasses. XVI. Embryo structure of the genus Oryza in relation to their systematics. Am J Bot 51:539–543
Teulat B, Aldam C, Trehin R, Lebrun P, Barker JHA, Arnold GM, Karp A, Baudouin L, Rognon F (2000) An analysis of genetic diversity in coconut (Cocos nucifera L.) population from across the geographical range using sequence tagged microsatellites (SSRs) and AFLPs. Theor Appl Genet 106:411–422
Trimen H (1898) Handbook of the flora of Ceylon 4, Dulau, London, pp 337–338. https://doi.org/10.5962/bhl.title.10864
Vaughan DA, Morishima H (2003) Biosystematics of the genus Oryza. In: Wayne Smith C, Dilday RH (eds) Rice: origin, history, technology and production. Wiley, New York, pp 27–65
Watt G (1889) Dictionary of the economic products of India. Coconut in volume 2, p 418
Whitehead RA (1976) Coconut. In: Simonds NW (ed) Evolution of crop plants. Longman, London, pp 221–225
Zizumbo-Villarreal D, Piñero D (1998) Pattern of morphological variation and diversity of Cocos nucifera (Arecaceae) in Mexico. Am J Bot 85:855–865

Water Resources and the Changing Needs

8

Abstract

Only 3% of the total water on Earth is fresh water, out of which underground water constitute 29% and less than 1% is in the form of lakes and rivers on the Earth's surface. The total freshwater resources in the world are estimated to be in the order of 43,750 km^3 $year^{-1}$, distributed throughout the world. In the coastal areas, fresh groundwater is present as a lens over the sea water. Due to increasing human population, urbanization and intensification of agriculture freshwater demand have been increasing. At the same time industrial and domestic use is growing relative to that for agriculture. Global assessment points to the fact that the demand for water is influenced by population growth, industrial development, and food production besides climate change. Total global freshwater withdrawal is estimated at 3800 km^3 in which 70% for irrigation, with huge variations across and within countries. Since the balance between water demand and water availability has reached critical level in many regions of the world and increased demand for water and food production is likely in the future, a sustainable approach to water resource management has become imperative. This chapter highlights various sources, demand, and supply and management of fresh water with special reference to coastal freshwater resources.

Keywords

Water availability · Demand · Climate change · Coastal aquifers · Renewable sources

8.1 Introduction

Water is the most important resource essential for sustenance of life on Earth and economic development of human society. Nearly 97% of the total water on Earth is in oceans in the form of saline water, and only the remaining 3% is fresh water. Out

V. Ayyam et al., *Coastal Ecosystems of the Tropics - Adaptive Management*,
https://doi.org/10.1007/978-981-13-8926-9_8

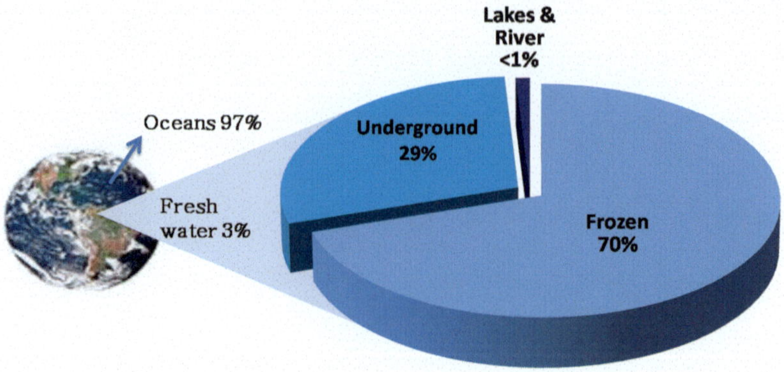

Fig. 8.1 Percent contribution of different sources of water to the global water resources

of this, nearly 70% of fresh water is in the form of ice present in the polar region and higher mountain ranges. Underground water constitutes 27% of fresh water, and only less than 1% is in the form of surface water present in lakes and rivers (Fig. 8.1). Rapid changes in human lifestyle coupled with urbanization and industrialization have created pressure on the limited freshwater resources. Drinking water is becoming a rare commodity due to population explosion increased wastage of water. The surface and the underground water resources are getting exploited to satisfy the rising water demand for household, agriculture, and industrial development (World Resources Institute 2000).

At the same time, most of the surface water resources are seasonal in nature which depend on the rainfall. Constructions of dams and man-made reservoirs are helping to relieve the stress on water demand. At many places, water is being transported from water resource-rich areas to deficit areas. However, the gap between water supply and demand is on the rise with reference to quantity and quality too. Besides, traditional water resources are getting dried because of both natural and man-made activities. The limited groundwater resource is being exploited to meet the demand, while surface water is limited and seasonally available. Further it has been observed that at many places, the groundwater is being contaminated due to man-made activities. Blocking of natural recharge by way of land use change, sand mining, and disturbance to the natural recharge region besides over-exploitation of underground water is disturbing the balance of natural groundwater recharge and groundwater supply (Collin and Melloul 2012). In the coastal region, this resulted in the intrusion of sea water into the freshwater aquifers.

Total global freshwater withdrawals are estimated at 3800 km^3, with 2700 km^3 (or 70%) for irrigation, with huge variations across and within countries. To produce enough food to satisfy human needs takes about 3000 l of water converted from liquid to vapour—about 1 l per calorie. Only about 2–5 l of water are required for drinking. About 80% of agricultural evapotranspiration come directly from rain and about 20% from irrigation. Industrial and domestic use is growing relative to that for agriculture. And water for energy generation—hydropower and thermo cooling—is growing rapidly due to urbanization and industrialization. Not all water withdrawn from available resources is lost; much is available for reuse but often its quality is degraded. In the future more people will require more water for food, fibre, industrial crops, and household needs thereby creates more pressure on the existing freshwater resources. Industrial activities will tend to pollute and disturb the natural hydrological cycle as well. Under such situations, the amount of water per person can be reduced by changing what people consume and how they use water to produce food (IWMI 2007).

Since the balance between water demand and water availability has reached critical levels in many regions of the world and increased demand for water and food production is likely in the future, a sustainable approach to water resource management has become imperative (Mancosu et al. 2015). This chapter highlights concepts of water resources; its availability, human demand, and use of fresh water; the effect of climate change; and other factors on freshwater availability and ways to manage this precious natural resource with reference to the coastal areas.

8.2 Basic Concepts of Water Resources

The concept of water resources comprises qualitative socio-economic and environmental dimensions besides its quantitative and physical aspects. The source of all forms of water either directly or indirectly is precipitation, often used interchangeably with total rainfall. This is part of the global water cycle. However, with reference to water resources, precipitation is considered as gain, evapotranspiration is viewed as loss, and human use including for agriculture is described as demand. Figure 8.2 gives an overview of different water resources beginning from the rainfall. However, precipitation also includes other form of water as well. Part of this water is evapotranspired on reaching ground or moves to the freshwater resources (surface and groundwater). The water can be renewable and nonrenewable, and all the water is not exploitable resource. With respect to the coastal water resources, groundwater is the major source, and it is in dynamic equilibrium with the sea water. Due to increase in demand, majority of the surface flow is utilized in the inland by storing and diverting the flow. Consequently the amount of water that flows into the sea through the coastal region has been drastically reduced limiting its availability to the region. The details are discussed in the following subsections.

8.2.1 Renewable and Nonrenewable

Freshwater resources are grouped into two types, renewable and nonrenewable water resources, which are of interest to human beings. Groundwater and surface water are estimated as the long-term average flow of rivers on a yearly basis which are considered as renewable water resources, whereas deep aquifers, which do not have a significant replenishment rate on the human time scale, are deemed nonrenewable water resources (FAO 2003). The renewable water is further subdivided

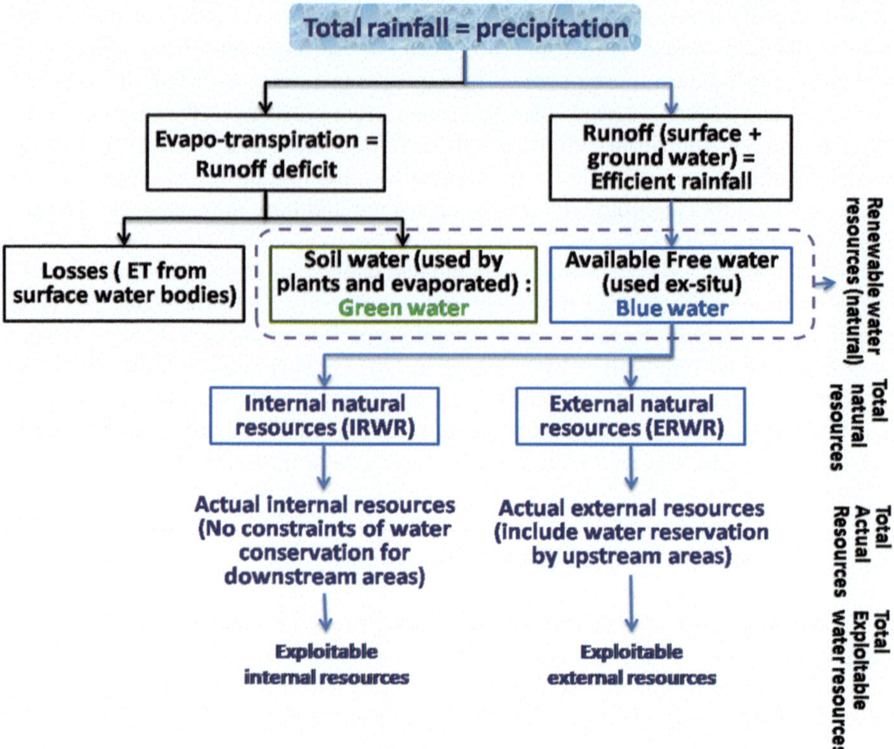

Fig. 8.2 Schematic representation of different water resource

into natural renewable and actual renewable. Natural renewable water resources are the total amount of a country's/downstream areas' water resources (internal and external resources), both surface water and groundwater, which is generated through the hydrological cycle. The amount is computed on a yearly basis. The actual renewable water resources are defined as the sum of internal renewable resources (IRWR) and external renewable resources (ERWR), taking into consideration the quantity of flow reserved to upstream and downstream countries/areas through formal or informal agreements or treaties and possible reduction of external flow due to upstream water abstraction. Unlike natural renewable water resources, actual renewable water resources vary with time and consumption patterns and, therefore, must be associated to a specific year (Mancosu et al. 2015).

8.2.2 Exploitable Water Resources

Both IRWR and ERWR together constitute the total natural fresh water. But not all natural freshwater, surface water, or groundwater is accessible for use/exploitation due to some constraints or natural barriers. Exploitable water resources, sometimes used synonymous with water development potential, consider factors such as economic and environmental feasibility of storing floodwater in dams, extracting groundwater; physical possibility of harvesting water which flows out to the sea; and the minimum flow requirements for navigation, maintaining environmental services and sustaining aquatic life.

Water resources can also be described as blue and green water based on its residential place and use but has no relevance to its colour. Blue water is the liquid water above and below the ground (rivers, lakes, groundwater), and green water is the soil water in the unsaturated zone derived from precipitation (Falkenmark 1986). The portion of water that is directly used and evaporated by plants (unsaturated soil water) is defined as green water. From the definition it is inferred that green water flow has two components: first, the productive part, or the transpiration by plants required for biomass production in terrestrial ecosystems and, second, the non-productive part or evaporation occurring from the land surface (Falkenmark and Rockström 2006). In general, blue and green water are described as renewable resources; however, practically blue water is freely assessed for use.

8.2.3 Coastal Aquifers

One of the important features of the coastal deposits is the occurrence of groundwater in the interbedded alluvial and fractured rocks. This available free water is also called blue water. An aquifer is a saturated geological formation that is permeable enough to yield significant quantities of water to wells and springs. Thus a porous and permeable water-bearing formation is called an aquifer. It can be confined or unconfined type depending on the overlying layers. This waster is exploited by different structures, viz. dug well, tube well/bore wells, and springs, which depend on the local conditions like geomorphology, water level, yield, and uses. This is most important water resource for the coastal and island region. The groundwater resources have two components, viz. static and dynamic. The static fresh groundwater reserves are the aquifer zones below the zone of groundwater table fluctuation. The dynamic component is replenished annually. According to the National Water Policy of many countries and as a sound practice, development of groundwater resources is to be limited to utilization of the dynamic component of groundwater (Mall et al. 2006).

In the coastal areas, sometimes, coarse sediments are found to occur along the coast where youthful rivers discharge. Generally due to the differential compaction and the nature of the bedrock topography, the coastal sediments attain a seaward dip. The distribution of freshwater aquifers is controlled by the dynamic equilibrium between hydrostatic heads in the fresh and saline water zones, influx of sea water into the streams and lagoons, and the relative mound of sea with respect to the land mass. Thus properly developed groundwater is the most important source of freshwater available for multiple uses in the coastal areas which are discussed in the following subsections.

8.3 Water Resources

8.3.1 Method of Assessment

For a global assessment of renewable water resources, a method was described by FAO (1996). It computes the total renewable water resources (TRWR) of a country and assesses the dependency ratio from neighboring countries. As described earlier, the TRWR is the sum of internal renewable water resources (IRWR) and external renewable water resources (ERWR). As the water resource data is available for a political boundary, it needs to be computed for natural boundary because surface water flow is influenced by the terrain as a part of hydrological cycle. In this context, FAO (2003) has defined IRWR as the volume of water (surface water and groundwater) generated from precipitation within a country or catchments. Surface water flows can contribute to groundwater replacement through seepage in the river bed or moves in through crack and crevices in the rock bed.

Conversely aquifers can discharge into rivers and contribute to their base flow which is the main source of river flow during dry periods. Sometimes, these two concepts do overlap, although surface water and groundwater are typically studied separately. ERWR are considered resources that enter from upstream countries through rivers or aquifers. ERWR is divided into two categories, viz. natural and actual ERWR. The natural ERWR is equal to the volume of the average annual flow of rivers and groundwater that enter into a country from neighboring countries, while the actual ERWR considers the incoming and outflows that derive from upstream and downstream countries through formal or informal treaties (Mancosu et al. 2015). Therefore, the actual resource is related to the amount of flow shared with neighboring countries which is viewed as geopolitical constraints for a particular country. The ratio between the ERWR and TRWR gives the dependency ratio of a country, an indicator that expresses the part of the water resources originated outside the country. The dependency ratio ranges between 0%, where there is no dependency on external water, and 100%, where a country totally depends on the water resources from neighboring countries (FAO 2003). All of these considerations and analyses are necessary to understand the dependency of countries on natural source (rain) and their neighboring countries.

8.3.2 Global Water Resources

There are several reports on global total freshwater resources which followed different methods to assess and so is the variation in the total estimated value. Further, over the years the accuracy of estimate has been drastically improving due to the use of new, efficient tools, and the coming together of countries in international forum as water crisis is a common challenge of humankind. However, water resources are estimated with reference to a particular year and may vary with the progress of time as it is depend on several dynamic components. The total freshwater resources in the world are estimated to be in the order of 43,750 km^3 year^{-1}, distributed

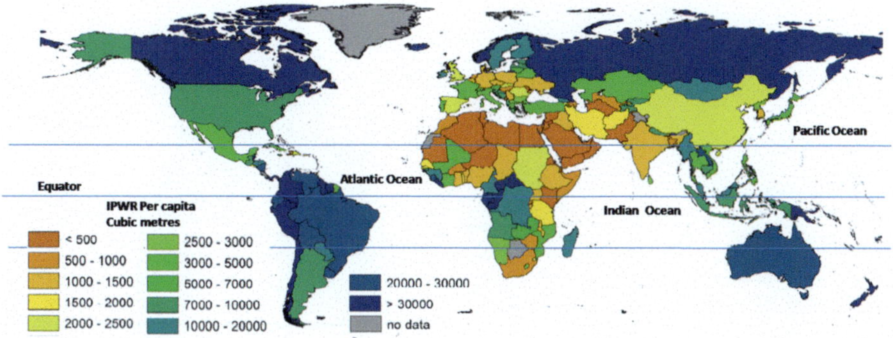

Fig. 8.3 Availability of water resources

throughout the world. At the continental level, America has the largest share of the world's total freshwater resources, with 45%, followed by Asia with 28%, Europe with 16%, and Africa with 9% (Shiklomanov 2000). In terms of resources per inhabitant in each continent, America has 24,000 m^3 $year^{-1}$, Europe 9300 m^3 $year^{-1}$, Africa 5000 m^3 $year^{-1}$, and Asia 3400 m^3 $year^{-1}$ (FAO 2003). At the regional level, tropical humid region has fairly good IPWR per capita due to good amount of annual rainfall and water resources (Fig. 8.3). However, at the country level, there is an extreme variability in terms of TRWR. Majority of the Gulf countries and countries located in the Northern Africa (Morocco, Algeria, Bahrain, Jordan, Kuwait, Libyan Arab Jamahiriya, Maldives, Malta, Qatar, Saudi Arabia, United Arab Emirates, and Yemen) are having very low TRWR of 500 m^3 per inhabitant. Using the water scarcity/threshold levels (Falkenmark 1986), it is estimated that Kuwait is the worst case with 10 m^3 per inhabitant. In terms of IRWR, the threshold of 1000 m^3 per inhabitant is considered the water stress level. By this approach, North Africa and the Middle East are the most critical cases, with values ranging from 0 to 1000 m^3 $year^{-1}$ per person,

8.3.3 Global Water Withdrawal

In an attempt to better capture the relation between supply and demand, the Millennium Development Goal Water Indicator purports to measure the level of human pressure on water resources, based on the ratio between water withdrawal by agriculture, municipalities, and industries over total renewable water resources. Arid and semi-arid regions in Asia and Africa have maximum withdrawal of more than 90% of RW as given in Fig. 8.4. In these areas surface flow is seasonal due to less rainfall. As a consequence this region exploits more groundwater resources than other region.

Agriculture accounts for roughly 70% of total freshwater withdrawals globally and for over 90% in the majority of least developed countries (LDCs) (FAO 2011). Without improved efficiency measures, agricultural water consumption is expected

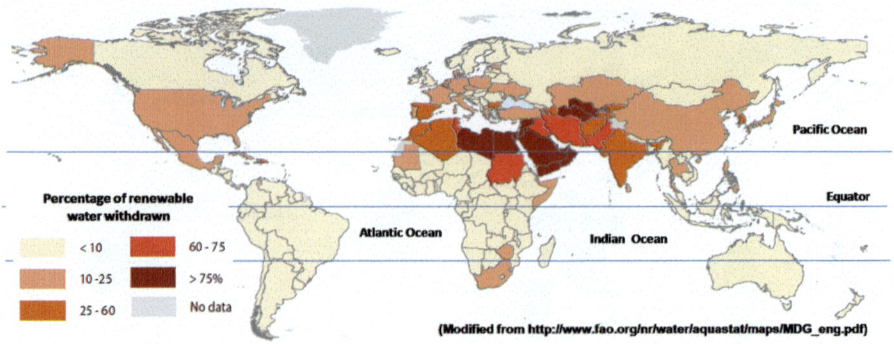

Fig. 8.4 Percentage of renewable water resources withdrawn

to increase by about 20% globally by 2050. Globally, some 38% of irrigated areas depend on groundwater (Siebert et al. 2013), which has contributed to a tenfold increase of groundwater abstraction for agricultural irrigation over the last 50 years. A t the same time, almost half of the world's population depends on groundwater for drinking.

8.4 Status of Freshwater Availability

8.4.1 Factors Affecting Surface Water Availability

In the last 50 years, the world has seen remarkable developments in water resources and agriculture. Substantial developments in water resource infrastructure and its distribution for various purposes have taken place. The world population grew from 2.5 billion in 1950 to 6.5 billion today, while the irrigated area doubled and water withdrawals tripled. Although agricultural productivity grew, freshwater availability has decreased. There are three major factors responsible for drop in freshwater availability. Diversion of water for food production and increase in population have primarily contributed to the decrease in freshwater availability. In hotspot industrial development region, the problem has different dimension. Due to unsustainable development and discharge of effluents and wastes, freshwater bodies are contaminated resulting in decrease of freshwater availability, while change in rainfall pattern (due to observed climate change) has contributed to the decrease in freshwater availability with significant regional differences.

- IWMI (2007) has estimated the world water scarcity which is given in Fig. 8.5. In the estimate little or no water scarcity indicates abundant water resources relative to use, with less than 25% of water from rivers withdrawn for human purposes.

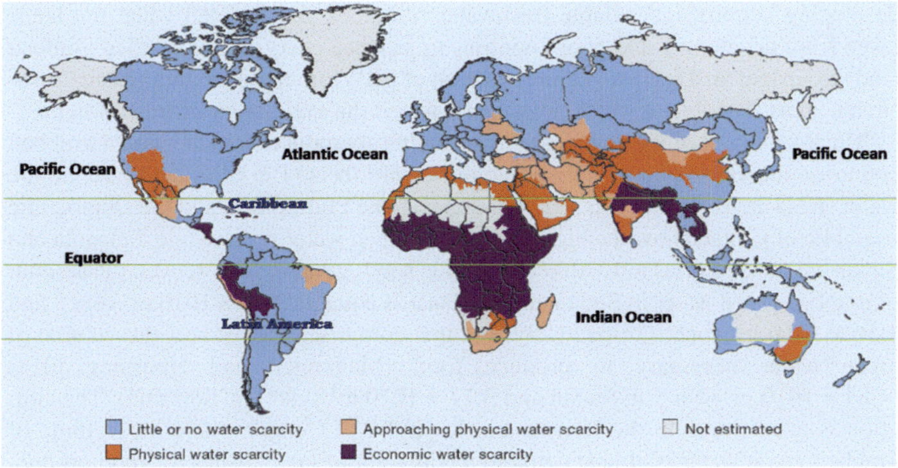

Fig. 8.5 Status of world water scarcity (IWMI 2007)

- Physical water scarcity indicates that water resources development is approaching or has exceeded sustainable limits. More than 75% of river flows are withdrawn for agriculture, industry, and domestic purposes (accounting for recycling of return flows). This definition implies that dry areas are not necessarily water scarce; proper development of water resources can improve the water availability.
- Approaching physical water scarcity where more than 60% of river flows are withdrawn. These basins experience physical water scarcity in the near future and will be greatly affected by climate change.
- Economic water scarcity occurs when human, institutional, and financial capital limit access to water even though water is available. Water resources are abundant relative to water use, with less than 25% of water from rivers withdrawn for human purposes, but malnutrition exists.

8.4.2 Agriculture and Water Scarcity

In recent times, greater pressure on the global land and freshwater resources resulted from the increased global demand for food and energy production. Since 2007–2008, the convergence of global crises in food, energy, finance, and the environment has driven a dramatic revaluation of land ownership (Borras et al. 2011). More specifically, the increased need for land is due to the increased global demand for biofuel (bioethanol) production. This is because of increase in global oil prices and green developmental policies adopted by many countries (Rulli et al. 2013). Such demand for land let to the investment in bioethanol production by taking land from developing countries. Available freshwater resources get depleted when the lands taken from developing and poor countries to produce energy plants (called grabbed land) are irrigated. This resulted in reduction of ERWR in the surrounding and downstream areas. This led to water stress and lowered the availability of fresh water.

The assessment of the water use for agricultural production in the top 24 grabbed countries shows that most grabbed countries are located in physical or economic water stress areas (Rulli et al. 2013). Indonesia, the Philippines, and the Democratic Republic of Congo show the highest grabbed green water rates, while Tanzania and Sudan are the countries most affected by blue water grabbing. Moreover, the amount of grabbed green water in Sudan and Tanzania is equal to 2.45 × 1010 m^3 $year^{-1}$ and 1.56 × 1010 m^3 $year^{-1}$, respectively, and this comprises almost half the volume of green water necessary to produce food (Mekonnen and Hoekstra 2011) (4.84 × 1010 m^3 $year^{-1}$ in Sudan and 3.77 × 1010 m^3 $year^{-1}$ in Tanzania). The situation is even worse in the Democratic Republic of Congo, where the volume of grabbed green water is almost equal to the amount of green water for food production (Mancosu et al. 2015).

By 2050, the available water per capita per year will drop below 1000 m^3, not only in the North African belt but also in Eastern and Southern Africa and the Middle East (Wallace 2000). In addition, some countries that are projected to experience physical water scarcity are already experiencing economic water scarcity. This occurs when investments on water resource infrastructure development (water supply pipe networks and reservoirs) needed to cope with the growing water demand are constrained by financial, human, or institutional capacity (IWMI 2007). Even though infrastructure might exist, high vulnerability to seasonal water fluctuations can lead to water scarcity for agriculture and domestic purposes. When infrastructure is inadequate, malnutrition can exist, even when water resources are abundant relative to water needs.

8.5 Coastal Aquifer

In most of the coastal areas, sea water lies at the base of the sand, and fresh groundwater lies on it as a lens which is the major source of freshwater availability. This is more pertinent to Small Island Developing States in the tropical region. The coastal sedimentary aquifers are among the most productive aquifers, and due to this the stress on them is also more. In general, groundwater is free from bacteriological pollution. When it is available in large quantities and undisturbed, it has an almost constant quality and temperature (Foster 1984). The equilibrium gets disturbed if large quantity of shallow groundwater is withdrawn. Caution needs to be exercised while developing these aquifers, as over-exploitation might result in various adverse environmental impacts including seawater intrusion and land subsidence. There are several factors which affect the coastal aquifers, and it requires concerted effort to manage the challenge.

8.5.1 Factors Affecting Coaster Aquifer

8.5.1.1 Land Subsidence

Large-scale withdrawal of groundwater (over-exploitation), especially from the artesian aquifers, can sometimes result in local land subsidence due to compression of the aquifers. Land subsidence poses serious problems to buildings and other structures and affects the equilibrium of freshwater-seawater interface region. Sometimes this causes inundation of low-lying areas, resulting in seawater ingress. The subsidence depends on the nature of subsurface formations, their extent, magnitude, and duration of the artesian pressure decline.

8.5.1.2 Seawater Intrusion

When groundwater is pumped out of coastal aquifers that are in hydraulic connection with the sea due to gradients saltwater from the sea may flow towards the well (Fig. 8.6). There is a dynamic equilibrium in the seawater-freshwater interface which gets disturbed due to over-exploitation of groundwater or reduced freshwater recharge. The resulting migration of saltwater into freshwater aquifers under the influence of groundwater development/over-exploitation is known as seawater intrusion. There is a tendency to indicate occurrence of any saline or brackish water along the coastal formations to seawater intrusion. The salinity can be due to several reasons, particularly results from leaching out of the salts from the aquifer material. In order to avoid mistaken diagnoses of seawater intrusion as evidenced by temporary increases of total dissolved salts, chloride-bicarbonate ratio is a criterion to evaluate the intrusion.

8.5.1.3 Upcoming of Saline Water

When an aquifer has an underlying layer of saline water and is pumped by a well penetrating only the upper freshwater portion of the aquifer, a local rise of the interface below the well occurs. This phenomenon is known as upconing. Generally the interface lies near horizontal at the start of pumping. The interface rises to progressively higher levels with continued pumping of water until eventually it reaches the well. This necessitates closing of the well because of the degrading influence of the

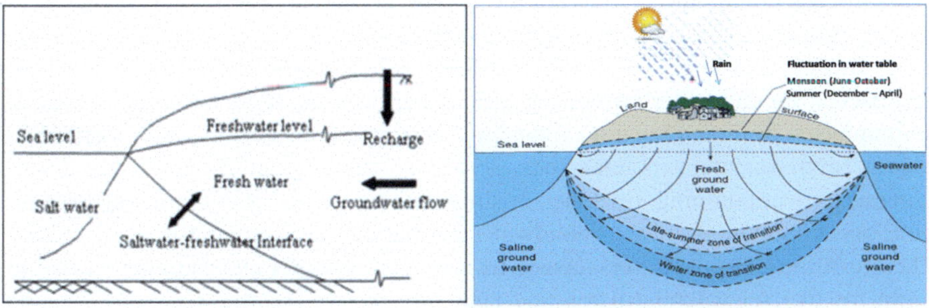

Coastal areas Island

Fig. 8.6 Freshwater-saltwater interface

saline water. When pumping is stopped, the denser saline water tends to settle downward and to return to its former position. Upconing of sea water is reported from the Lakshadweep and other small islands in the Indian Ocean. In these islands, the freshwater floats over saline water as a thin lens, and for every drop of one unit of the fresh water, the saline water rises by 40 units. Due to this, the islands do have very fragile groundwater system, and no pumping can be recommended here. The fresh water has to be skimmed to avoid upcoming.

8.5.1.4 Pollution of Surface Streams

Rivers and other streams flowing in the coastal areas carry sewerages and industrial effluents contributing to pollution of the coast and coastal aquifers. Further, discharge of effluents from industries located in the coastal region reaches the coastal wetlands and slowly percolates down to the groundwater.

8.5.1.5 Geogenic Salinity

This is the most common quality problem observed in the coastal aquifers. Here the salinity is due to leaching of salts in the aquifer material. In some cases, the formation water gets freshened year after year due to the leaching effect. This happens mostly for the water-soluble salts only.

8.5.1.6 Sea Level Rise

The observed and projected increase in mean sea level due to global warming poses a serious threat to the coastal aquifers, especially for the small island aquifers. The rise in the sea level will push the freshwater-seawater interface more inland along coastal aquifers and will submerge low-lying areas with saline sea water, thereby making the shallow aquifers saline. Water resources of the small islands located in the tropical region (Indian Ocean and Polynesian islands) will be significantly affected by the rise in sea level, and with the change in rainfall pattern, the negative effect will be even greater.

8.5.2 Management of Coastal Aquifers

Many coastal communities living in the tropical coastal areas do not have access to potable groundwater of adequate quality and quantity due to salinity and waterlogging. There is a growing consensus that the problem can be addressed by low-cost approaches for rainwater harvesting and recharge of aquifers which also involve communities in their planning, financing, and maintenance. It is also helpful to reduce environmental problems such as salinity intrusion and groundwater depletion due to over pumping, by understanding the importance of occurrence, distribution, and contamination of groundwater resources (Wickramaratne et al. 2006). Therefore, the coastal aquifers have to be managed carefully and cautiously to avoid problems like seawater intrusion and land subsidence. For these reasons detailed studies and regular monitoring are required.

- The aquifer geometry, distribution of the fresh water and saline water in the system, and hydrograph have to be studied in detail. It is essential to develop a clear understanding on the processes and systems, which can mitigate these problems.
- Constant monitoring of the pumping and movement of the freshwater saline water interface needs to be carried out to check the over-exploitation and prevent groundwater extraction in the grey areas.
- Tidal influence into the aquifer has to be studied in detail and is to be monitored periodically which will be helpful in designing coastal barriers, drainage channels, and slice gates.
- Season-wise safe yield of the aquifer has to be evaluated, and accordingly the extraction has to be restricted or regulated.
- Remedial measures have to be done wherever seawater intrusion has taken place and avoid as far as possible further disturbance to the water cycle in those areas.

8.6 Climate Change and Future Water Demand

8.6.1 Demand for Water

Water has become a scarce natural commodity due to its dwindling availability and increase in demand for various purposes. This has put enormous pressure on the available freshwater resources around the globe. The Intergovernmental Panel on Climate Change (IPCC 2007) report states that the magnitude of stress on water resources is expected to increase as a consequence of climate change, population growth, economic development, and land use change including urbanization. Thus, many studies focusing on the assessment of global water demand and its availability have been developed. In fact, water demand has reached critical levels in many areas of the world, especially in countries with limited water availability. The misuse of water resources, over-exploitation, lack of infrastructures to supply water, and also climate change are some of the reasons for water scarcity, despite the vast amount of water on the planet.

Globally, water consumption for all sectors amounts to 9% of total freshwater resources in the world with agriculture being the largest user, in turn accounting for approximately 70% of total water withdrawals which is equivalent to 2700 km^3 $year^{-1}$ (including losses) (FAO 2012). Agricultural sector receives up to two thirds of the total water withdrawals and accounts for almost 90% of the total water consumption in the world (Shiklomanov 2000). As more than 80% of global agricultural land is rainfed, water demand is met mostly from the green water resource (IWMI 2007). In Asia, Africa, and Central and South America, the values for specific water withdrawal range from 50% to 100% which experience great diversity in climatic conditions. Irrigation water withdrawals range between 96 km^3 in sub-Saharan Africa and 708 km^3 in East Asia; the highest values for specific water withdrawal are observed in South Asia, with 913 km^3 (Alexandratos and Bruinsma 2012).

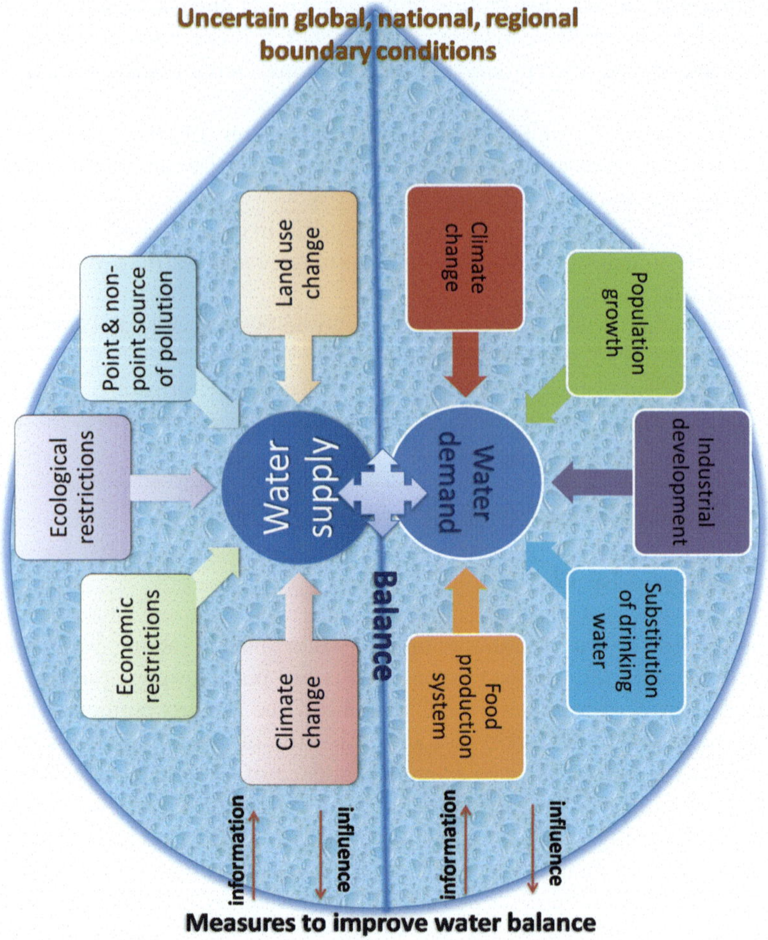

Fig. 8.7 Driving forces of future water supply and demand. (Modified from Hornbogen and Schultz 1998)

Various factors affecting water supply and demand are depicted in Fig. 8.7. In the future, the water demand will be influenced by population growth, industrial development, and food production besides climate change. At the same time, the water supply will be decided by land use change, ecological and economic restriction, and pollution besides climate change. The balance between these two will decide the freshwater availability (Hornbogen and Schultz 1998). The future global water situation and development until 2025 was analysed with different scenarios. Under the business as usual, the present contrast in water situation between industrialized and developing countries is likely to continue in the future. Withdrawal of water from the available resources in most of the industrialized countries is projected to decline due to technological and efficient water management. Thus the pressure on available freshwater resources will decline. At the same time withdrawal will continue to grow in developing countries due to urbanization and industrialization. This will increase the pressure on the available freshwater resources by increasing severe water stress area from 36.4 to 38.6 million km^2. The increase will be significant in Southern Africa, Western Africa, and South Asia which will be a limiting factor in the future for industrial and agricultural growth due to competition for water (Alcamo et al. 2000). On an average globally, 40% water deficit will be experienced by 2030 under a business as usual scenario.

8.6.2 Climate Change: Hydrological Cycle-Plant Relationship

According to the scenarios described in the IPCC Special Report on Emissions Scenarios, changes in precipitation and temperature may lead to changes in runoff and water availability, which in turn could affect crop productivity (Bates et al. 2008). The physical, chemical, and biological properties of freshwater lakes and rivers will be affected by the increase in temperature. This change is predicted to negatively affect many freshwater species and community and habitat composition. Moreover, as a consequence of sea level rise in low-lying coastal areas, the quality of freshwater aquifers is influenced by the intrusion of saline water.

The rising atmospheric CO_2 concentration is associated with higher temperatures and changes with precipitation patterns. Changes in precipitation patterns, the intensity and frequency of extreme events and soil moisture, runoff, and evapotranspiration fluxes have already been observed, and more important changes are expected in the future (Bates et al. 2008). The increase of precipitation at high latitudes and parts of the tropics and the decrease in some subtropical and lower mid-latitude regions are consistently projected by climate model simulations for the twenty-first century. Sillmann and Roeckner (2008) estimated a significant increase in extreme precipitation events in most regions of the world, especially in areas already relatively wet under present climate conditions. Analogously, dry spells are expected to increase, particularly in those regions that are characterized by dry conditions in the present-day climate. Altered frequencies of extreme events, associated with melting of winter snow and with consequent reduced storage of precipitation as snow, result in decreased water availability and significant impacts on crop yield. In addition, the combined effect of higher temperature and the reduction of water availability in regions affected by falling annual or seasonal precipitation leads to the increase of crop evaporative demands, with the consequent reduction in crop yield and agricultural productivity, where temperature constrains crop growth (FAO 2011).

Globally, at least 7130 km^3 of water are needed to satisfy crop evapotranspiration losses in agriculture, considering both blue and green water (irrigation delivery and on-farm system losses excluded); however, a considerable variation between regions is noticed (IWMI 2007). Irrigation is quite important in North Africa, the Middle East, the Near East, and Southern Asia, where more than 75% of food is produced by means of irrigation, and blue water crop evapotranspiration is about half of the total food crop evapotranspiration. Fischer et al. (2007) estimated increases in irrigation water requirements by over 50% from 2000 to 2080 in developing regions (Africa +300%) and by about 16% in developed regions. Furthermore, the most critical values of annual renewable freshwater resources are observed in the Middle East and Africa. In fact, the estimates made by Wallace (2000) highlighted that the populations in the Middle East and Southern Africa in the 1990s of the last century had between 1000 and 2000 m^3 of water per year. Whereas, during the same period, populations in the North African belt from Morocco to Egypt (including Sudan) had less than 1000 m^3 of water per person per year.

8.6.3 Research Gap and Future Prospects

Changes in the hydrological regime, precipitation and evaporation, and consequent change in soil moisture and the availability of freshwater resources are the most serious potential consequences of impending climate change in terms of its effect on man (Houghton et al. 1990). In spite of efforts by various stakeholders and global level organizations, lots of gap still persists in our understanding of the global water resources. Some of the gaps and required actions are listed below:

- Much improved quantitative assessments of different components of the global water cycle are essential to achieve accurate predictions of future water resources in a changed climate.
- It is essential to analyse recent experiences in climate variability and extreme events, and their impacts on regional water resources and groundwater availability.
- Study on changing pattern of rainfall, i.e. spatial and temporal variations and its impact on runoff and aquifer recharge pattern at regional and national level. Such efforts should clearly record the changes for its compilation and analysis and global level.
- Efforts should be made to study sea level rise due to increased runoff as projected due to glacial recession and increased rainfall through international collaborations.
- Seawater intrusions into costal aquifers should be assessed and mapped at subregional with site-specific analysis.
- Determine vulnerability of regional water resources to climate change, identify key risks, and prioritize adaptation responses.
- It is very essential to evaluate the efficacy of various adaptation strategies or coping mechanisms that may reduce vulnerability of the regional water resources and help to improve the water use efficiency.

The projected climate change, rainfall pattern, and water demand and supply clearly emphasize that in spite of variability and uncertainty, rainwater harvesting and its efficient use should be the centre of any adaptation strategy. The surface flow is seasonal in most of the coastal region and hence largely depends on subsurface resources. The subsurface storages have advantages of being free from adverse effects like inundation of large surface area, loss of cultivable land, displacement of local population, and substantial evaporation losses. The deeper groundwater levels in many parts of the coastal areas, either of natural occurrence or due to excessive groundwater development, may be substantially raised resulting in reduction on lifting costs and energy. The quality of natural groundwater would substantially improve in brackish and saline areas. The conduit function of aquifers can further help in natural subsurface transfer of water to various need centres, thereby reducing the cost-intensive surface-water conveyance system. Increased storage of rainwater in the form of groundwater will also push the saline water-coastal freshwater interface towards the seafront. The structures required for recharging groundwater reservoirs are of small dimensions and cost-effective such as check dams, percolation tanks, surface-spreading basins, pits, subsurface dykes, etc., and these can be constructed mostly with local knowledge and materials. This is a viable option for coastal and rainfed region as much of the future demand needs to be met from the uncertain rainwater through subsurface storage.

8.7 Water Resources and Climate Change: A Case Example of India

8.7.1 Climate Change Impact

There have been observed changes in surface temperature, rainfall, evaporation, and extreme events since the beginning of the twentieth century which are linked with the global climate change (IPCC 2007). Several studies around the globe show that these changes are likely to impact significantly upon freshwater resources availability and altered the hydrological cycle. In India, demand for fresh water has already increased manifold over the years due to urbanization, agriculture expansion, increasing population, rapid industrialization, and economic development. At present, changes in cropping pattern and land use pattern, over-exploitation of water storage, and changes in irrigation and drainage are modifying the hydrological cycle in many climate regions and river basins of India (Mall et al. 2006). Studies on interannual and long-term variability of monsoon and annual rainfall over India have indicated significant variations. Several climate models predicted an increase in precipitation by −24 to 15% over India, but regional changes may be different. Hence, observation and modeling at landscape level with input/linkage with global model might reveal finer information related to water resources.

8.7.2 Demand and Availability

The total catchment area of major rivers of India is 252.8 mha, and average annual potential in rivers is 1570.98 b.cu.m which is 85% of the total natural runoff. Similarly the total catchment area of minor rivers is 24.9 mha and the total available water resources of 1869 b.cu.m. Groundwater resources are divided into static and

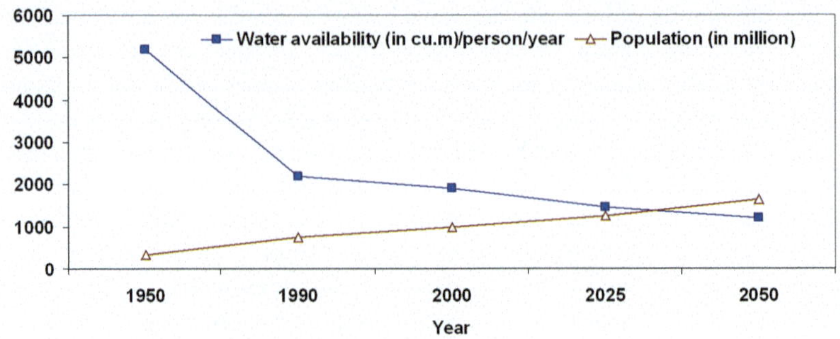

Fig. 8.8 Changes in per capita average annual freshwater availability and population growth

dynamic resources. The static fresh groundwater reserves (aquifer zones below the zone of groundwater table fluctuation) of the country have been estimated as 10,812 b.cu.m. The dynamic component is replenished annually, which has been assessed as 432 b.cu.m (Mall et al. 2006). According to the National Water Policy of India, development of groundwater resources is to be limited to utilization of the dynamic component of groundwater.

In this context the regional population pattern of India will greatly affect the per capita availability of fresh water. The observed and projected decline in per capita average annual freshwater availability and growth of population from 1951 to 2050 is depicted in Fig. 8.8 (MoWR 2003). It is inferred that after 2030, the freshwater availability will drastically come down exerting more pressure on the available resources and the water resource managers. This shows the two-sided effect on water resources—the rise in population will increase the demand for water leading to faster withdrawal of water, and this in turn would reduce the recharging time of the water tables (Mall et al. 2006).

It is projected that most irrigated areas in India including the coastal plains would require more water around 2025 to increase the crop and livestock productivity necessitated by the population explosion (Doll and Siebert 2001). The global net irrigation requirements would increase relative to the situation without climate change by 3.5–5% by 2025 and 6–8% by 2075. In India, roughly 52% of irrigation consumption across the country is extracted from groundwater as many of the places are still under rainfed. Therefore, it can be an alarming situation with decline in groundwater and increase in irrigation requirements particularly in coastal and island conditions due to climate change. At the same time efforts to provide supplemental irrigation from water storage in small and large dams/reservoirs during dry spells solves only part of the problem. Meanwhile maintaining the water quality is of equal importance to ensure adequate safe water supply not only for anthropogenic consumption, but also for the healthy survival of its habitats and aquatic life. Considering the interannual variability of rainfall in India, assessment of only volume may not be helpful until temporal and spatial variations of climate change and their impacts are assessed (Mall et al. 2006).

8.7.3 Long-Term Water Supply Prospects

At present, agriculture sector is the largest consumer of water with 83% of the available water which has increased progressively through the years as more and more areas were brought under irrigation (22.60 in 1950 to 82.5 mha in 2000). According to available estimates, due to judicious utilization, the demand on water in this sector is projected to decrease to about 68% by the year 2050, though agriculture will still remain the largest consumer (NCIWRD 1999).

In order to fulfil water demands in the future, harvesting of rainwater should contribute in meeting the water requirements in India. The rivers and rivulets of the Indian subcontinent are mainly monsoon-fed, with 80–90% runoff generated during the monsoon. The principle source for groundwater recharge is also monsoon precipitation. The country receives more than 75% monsoon rainfall from June to September, except in the eastern coast and total annual rainy days vary from 12 to 100. The interannual variability of the monsoon is expected to increase in the future due to possible climate change, making the monsoon less reliable as an assured source of water. Therefore, efforts are needed for more efficient groundwater recharge and harvesting of rainwater through identification, adoption, and adaptation of technological options. The subsurface reservoirs can store substantial quantity of water and are attractive and technically feasible alternatives for storing surplus monsoon runoff.

8.8 Conclusion

The concept of water resources is multidimensional and carries multistakeholder interest. It is not limited only to its physical measure, the 'flows and stocks', but encompasses qualitative, environmental, and socio-economic dimensions as well. Throughout the human history, water formed the centre of economic and social developments. Therefore reliable assessment of water resources, its equitable sharing, and proper utilization for sustainable development are imperative. Global freshwater resource is only 3% of total water on Earth out of which 70% is frozen in the polar region and above snow line in mountain ranges. Groundwater is the major source of freshwater availability followed by surface flow. The major demand for fresh water is from irrigation water requirement, industrial use, household consumption, and other purposes.

The only source of all these freshwater is rainfall. The available freshwater resources are defined as the average manual flow of rivers and recharge of aquifers generated from precipitation. The computation of the actual renewable water resources of a country/region takes account of possible reductions in flow resulting from the abstraction of water in upstream areas. The freshwater resources have two components, viz. renewable and nonrenewable, based on its nature. Natural renewable water resources are the total amount of a country's water resources (internal and external resources), both surface water and groundwater, which is generated through the hydrological cycle and computed on a yearly basis. Whereas deep aquifers without much significant replenishment rate on the human time scale are deemed nonrenewable water resources. Estimates show that with the current population growth rate and efficiency of water management practices, the world will face a 40% shortfall between forecast demand and available supply of water by 2030.

But these assessments and its quantification are based on the sound data collection. However, there remains much to do in order to obtain sound statistics on water resources, and particularly standardized data sets, at global level. The use of global data sets derived/compiled from country level statistics at greater scale with uniform method of data collection coupled with water-balance models can contribute to improving the assessment of water resources. Groundwater is being depleted for several purposes at a rate faster than it is being replenished either naturally or artificial means. At the same time, there is no scope to improve the surface water flow as it mostly depends on the precipitation. By 2025, about 1.8 billion people will be living in regions or countries with absolute water scarcity. The combined effects of growing populations, rising incomes, and expanding cities will see demand for water rising exponentially, while supply becomes more erratic and uncertain.

Meanwhile there are several disputes in utilizing and sharing this precious resource. Human activities have rendered water unusable at several places due to pollution, salinity, and over-exploitation. Therefore, water resource management has become a very important issue from several angles such as development of water resources for the future, conservation of freshwater bodies, protection of available water resources from pollution and over-exploitation, and minimization of disputes or conflicts at different scales. The problems are also complex in the coastal area due to seawater intrusion, salinity from the aquifer (in situ) material, pollution, sea level rise, and its impact on these aquifer systems. There should be proper regulations and monitoring which involve measures like precaution/ prevention; control/ restriction; and remedial/restoration measures. In summary, the available information suggests that water security remains a challenge for many tropical coastal countries today coping with complex water issues that cut across sectors.

References

Alcamo J, Henrichs T, Rosch T (2000) World water in 2025: global modeling and scenario analysis for the world commission on water for the 21st century, report A0002. University of Kassel, Kassel

Alexandratos N, Bruinsma J (2012) World agriculture towards 2030/2050: the 2012 revision, ESA Working Paper No. 12-03. FAO, Rome

Bates BC, Kundzewicz ZW, Wu S, Palutikof JP (2008) Climate change and water. IPCC Secretariat, Geneva

Borras SM, Hall R, Scoones I, White B, Wolford W (2011) Towards a better understanding of global land grabbing: an editorial introduction. J Peasant Stud 38:209–216

Collin ML, Melloul AJ (2012) Land-use planning guidelines for optimal coastal environmental management. J Environ Prot 3:485–501

Doll P, Siebert S (2001) Global modeling of irrigation water requirement. University of Kassel, Kassel
Falkenmark M (1986) Freshwater—time for a modified approach. Ambio 15:192–200
Falkenmark M, Rockstrom J (2006) The new blue and green water paradigm, breaking new ground for water resources planning and management. J Water Res Plan Manag 132:129–132
FAO (1996) Les Ressources en Eau, FAO/BRGM, Food and Agriculture Organization of the United Nations, Rome
FAO (2003) Review of world water resources by country, Water Reports 23. Food and Agriculture Organization of the United Nations, Rome. http://www.fao.org/docrep/005/y4473e/y4473e00.htm
FAO (2011) Climate change, water and food security, FAO Water Reports No. 36. FAO, Rome
FAO (2012) AQUASTAT Database, Food and Agriculture Organization of the United Nations. http://www.fao.org/nr/aquastat. Accessed 20 June 2017
Fischer G, Tubiello FN, van Velthuizen H, Wiberg DA (2007) Climate change impacts on irrigation water requirements: effects of mitigation, 1990–2080. Technol Forecast Soc 74:1083–1107
Foster SSD (1984) African groundwater development: the challenges for hydrologic science. In: Walling DE, Foster SSD, Wurzel P (eds) Challenges in African hydrology and water resources, IAHS Publications 144. IAHS Press, Wallingford, pp 3–12
Hornbogen M, Schultz GA (1998) Water: a looming crisis? UNESCO, Paris, pp 357–362
Houghton JT, Jenkins GJ, Ephraums JJ (1990) Climate change—IPCC Scientific Assessment, Inter Governmental Panel on Climate Change. University Press, Cambridge
IPCC (2007) Climate change : synthesis report contribution of working groups I, II and III to the fourth assessment report of the intergovernmental panel on climate change. Geneva, p 104
IWMI (2007) Water for food, water for life: a comprehensive assessment of water management in agriculture. Earthscan/International Water Management Institute, London/Colombo. www.earthscan.co.uk
Mall RK, Gupta A, Singh R, Singh RS, Rathore LS (2006) Water resources and climate change: an Indian perspective. Curr Sci 90(12):1610–1626
Mancosu N, Snyder RL, Kyriakakis G, Spano D (2015) Water scarcity and future challenges for food production. Water 7:975–992. https://doi.org/10.3390/w7030975
Mekonnen MM, Hoekstra AY (2011) National water footprint accounts: the green, blue and gray water footprint of production and consumption, Value of Water Research Report Series No. 50. UNESCO-IHE, Delft
MoWR (2003) Vision for integrated water resources development and management. Ministry of Water Resources, Govt of India, New Delhi, p 20
NCIWRD (1999) Report of the National Commission for Integrated Water Resources Development, Ministry of Water Resources, vol 1, p 542
Rulli MC, Saviori A, D'Odorico P (2013) Global land and water grabbing. Proc Natl Acad Sci U S A 110:892–897
Shiklomanov IA (2000) World water resources and water use: present assessment and outlook for 2005. In: Rijberman F (ed) World water scenarios analysis. World Water Vision
Siebert S, Henrich V, Frenken K, Burke, J (2013) Update of the global map of irrigation areas to version 5. University of Bonn/FAO, Bonn/Rome, p 178. https://doi.org/10.13140/2.1.2660.6728
Sillmann J, Roeckner E (2008) Indices for extreme events in projections of anthropogenic climate change. Clim Chang 86:83–104
Wallace JS (2000) Increasing agricultural water use efficiency to meet future food production. Agric Ecosyst Environ 82:105–119
Wickramaratne U, Weragoda SK, Dharmagunewardhane HA (2006) Importance of designing artificial groundwater recharging system by rainwater in coastal water supply schemes: a case study from the eastern coastal zone of Sri Lanka. national seminar on rainwater harvesting and water management, 11–12 November 2006, Nagpur, pp 215–225
World Resources Institute (2000) World resources 2000–2001. Washington, DC, 389 pp

Natural Disasters and Coastal Agro-ecosystems

9

Abstract

The coastal regions are prone to natural disasters, but the impact varies with the geographical location and adaptive capacity of the individual countries to cope with the post-disaster situations. Historical records show that cyclonic storm and flooding are the major disasters experienced, and these inflict heavy loss to agriculture and allied sectors. One of the main reasons for poor productivity and services of agro-ecosystem is due to uncertain weather and its associated natural disasters. On an average, the agriculture sector—including crops, livestock, fisheries, and forestry—absorbs approximately 22% of the economic impact caused by medium- and large-scale natural hazards and disasters. The high impact of natural hazards and disasters on agriculture calls for enhanced mainstreaming of disaster risk reduction and resilience building within the agricultural sectors. Adaptation options such as protective and production measures besides proper policy supports are discussed in this chapter.

Keywords

Cyclones · Flooding · Tropical coast · Agro-ecosystem · Disaster management

9.1 Introduction

Coastal region constitutes a transition zone where land and fresh water meet saline water and across which the effects of land on the ocean, and vice versa, are transferred and modified. The region is also enriched by land-based nutrients and nutrients that well up into the coastal waters from deeper levels of the ocean which makes it the most productive. Consequently different ecosystems have got evolved having great ecological and socio-economic significance. Coastal region also sustains economies and provides livelihoods through fisheries, agriculture, biodiversity based trade, ports, tourism, and other industries. For these reasons nearly half the

V. Ayyam et al., *Coastal Ecosystems of the Tropics - Adaptive Management*,
https://doi.org/10.1007/978-981-13-8926-9_9

world's major cities are located within 50 km of a coast, and coastal population densities are 2.6 times greater than those of inland areas (MEA 2006).

Although this region is endowed with good amount of rainfall, coastal alluvium, deltas and strong labour force, agricultural production is severely affected by geophysical and biological form of natural disasters due to its geographical settings. It is mainly attributed to the fact that disasters occurring in the coastal region affect the entire gamut of agriculture because biological component is strongly linked with physical component of the region making it a functional agro-ecosystem, and the disaster occurs in large extent of area. As a result it is often observed across the tropical coast that agricultural production is severely affected with high uncertainty except in certain irrigated crops and areas. In this context it is very important to understand the effect of natural disaster on agro-ecosystem rather than on single crop or livestock to plan for proper management strategy (Anaman 2003).

The agro-ecosystem productivity is also affected by the pattern of climate parameters which exaggerates the effect of other disasters occurring in this region. Observational data reveals that the average annual rainfall of the tropical coastal region is considerably high, and ambient temperature prevails for most part of the year. In spite of that, the productivity of agro-ecosystem is not at its best in several agroclimatic zones. The main reason for this is the uneven distribution of rainfall and extreme climatic events besides differences in agro-ecosystem management. Most importantly natural disasters like cyclones result in heavy rainfall thereby causing flood in the vicinity. It also accompanies with major storms and heavy wind in addition to flood which are highly detrimental to the function and ecosystem services of coastal agro-ecosystem. Though it is desirable, it is not possible to stop the cyclone and its related events; however, we can make ourselves well prepared to reduce the risk associated with such situations. The overwhelming effect of frequency and intensity of natural disaster masks the beneficial effect of other parameters linked to agro-ecosystem production.

On an average worldwide the present decade experienced 910 natural catastrophes in a year. Out of which 93% have been identified as weather-related disasters which had huge impact on the output from agro-ecosystem. An insight into the different categories of natural disasters showed that 45% were classified as meteorological in nature (storms), 36% were hydrological (floods), 12% were climatological (heat waves, droughts, etc.), and 7% were geophysical (earthquakes and volcanic eruptions) (FAO 2015a). Therefore it is pertinent to document the major disasters of the tropical coastal region and its impact on the agro-ecosystem. This chapter aims to assess the role of some of the climate-resilient technologies for coastal agriculture with a view to enhance the production and sustainable livelihood in the region. The situation also necessitates creating awareness among the stakeholders about the concept of natural disaster management in agro-ecosystem which in turn aids in development of integrated disaster management strategies.

9.2 The Concepts

9.2.1 Natural vs Agro-ecosystems

Ecosystem results from the integration of all living and nonliving factors of the environment and flow of energy and matter between different components. An agro-ecosystem can be viewed as a subset of a conventional ecosystem. But unlike natural system, human activity of agriculture lies at the core. At the same time, an agro-ecosystem is not restricted to the site or unit of agricultural activity (farm). It includes the region that is impacted by this activity, usually by changes to the complexity of species assemblages and energy flows, as well as to the net nutrient balance. In general, agro-ecosystem is a producer-consumer system in which humans attempt to direct the flow of energy and materials through the crops, livestock, soil flora and fauna, etc. At the same time the pattern of exchange and the rate of flow are limited by the innate growth patterns of crop plants and the nutritional and other maintenance requirements of the animals.

Conventionally, an agro-ecosystem, particularly one managed intensively, is characterized as having a simpler species composition and simpler energy and nutrient flows than 'natural' ecosystem. Similarly agro-ecosystems are often associated with elevated nutrient input and recycling to support higher productions which are taken out of the system than natural system. In this region, agro-ecosystems are under the influence of land-marine interface, livelihood, and other developmental activities. Further, coastal agro-ecosystems have to face greater challenges from weather-related phenomena and adverse edaphic conditions as well. All these conditions and pressure are reflected in the structure and functions of coastal agro-ecosystem.

9.2.2 Hazard and Disaster

Coastal ecosystems provide plenty of goods and services to the society, and at the same time natural disasters originating from the sea such as cyclone, super cyclone, tsunami, sea surges, hurricane, etc., are very common which affects its functions and the human society at large. In other words coastal agro-ecosystems are affected by these events to very large extent. Further coastal agro-ecosystem is also facing the challenges created by anthropogenic activities at many places. With the progress of time some of the activities/events may turn into disaster for agro-ecosystem due to hydrological or geophysical reasons. So it is highly advisable to be prepared with contingency measures for coastal ecosystems to face any kind of challenges at the best possible way. In this context two important terms are used to describe the situations, viz. natural hazard and disaster.

A natural hazard is a threat of a naturally occurring event having negative effect on humans and his wellbeing. This negative effect, when it happens, is called a natural disaster (Fig. 9.1). Hazard translates into disaster when there is vulnerable area sufficiently exposed to the hazard. The severity of disaster depends on the adaptive

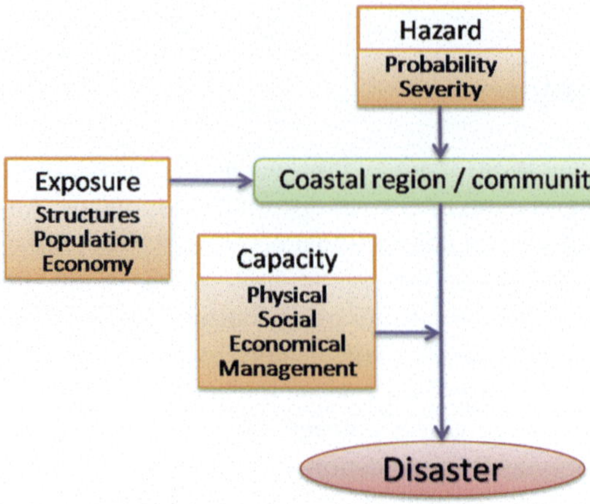

Fig. 9.1 Linkage between different elements of disaster

capacity of the society. In order to natural hazards are caused either by rapid or slow onset events which can be geophysical (earthquakes, landslides, tsunamis, and volcanic activity), hydrological (avalanches and floods) or climatological (extreme temperatures and drought), meteorological (cyclones), and others related to these two (wildfires, disease outbreak).

A natural disaster is a major adverse event resulting from natural processes of the Earth, for example, floods, hurricanes, tornadoes, volcanic eruptions, earthquakes, tsunamis, and other geologic processes. In other words, when the hazardous threat actually happens due to vulnerability and harms human wellbeing, we call the event a natural disaster. There are several attempts to define the events. A human centric approach defines disaster as an event if at least 10 people are killed and/or 100 or more are affected and/or an appeal for international assistance is made or a state of emergency declared. Anderson (1990) defined disaster based on the capacity of the society to manage. It is 'temporary events triggered by natural hazards that overwhelm local response capacity and seriously affect the social and economic development of a region'. United Nations defines disaster as a serious disruption of the functioning of society, causing widespread human, material, or environmental losses which exceed the capacity of the affected society to cope using only its own resources (UN 2015).

9.3 Vulnerability of Coastal Region

9.3.1 Disaster Imprints

In general, the coastal regions throughout the globe are highly vulnerable to several natural hazards, and by virtue of its proximity to sea, natural disasters are bound to happen. Broadly, natural disasters include hydrometeorological disasters and geophysical disasters (World Disaster Report 2003). In some areas biological hazard and

Table 9.1 Major natural disasters which affected the tropical region

Major natural disaster	Countries/islands affected	Major natural disaster	Countries/islands affected
2004 Indian Ocean Tsunami	Indonesia, Thailand, Sri Lanka, Southern India and Indian Ocean islands	Wind storm 2001	Cook Islands
Tsunami 2009	Samoa Islands	Cyclone Gene 2008	Fijian city capital of Suva
1933 Trinidad hurricane	Trinidad, Jamaica, Cuba	2010 Haiti Earthquake	Haiti
Typhoon Ophelia in January 1958	Jaluit Atoll (Marshall Islands)	Tropical cyclone Evan 2012	Samoa and Fiji
Hurricane INEZ 1966	Bahamas, Cuba, Dominican Rep, Haiti, Guadeloupe	2016, Hurricane Matthew	Bahamas, Haiti, and Jamaica
Cyclone Bebe in October 1972	Funafuti Atoll (Central South Pacific)	Fiji experienced 31 cyclones followed by sea surges between 1972 and 1995	Fiji
Hurricane Allen 1980	Haiti, Jamaica, St Lucia St Vincent	Flash floods, Apr 2014	Solomon Islands
1980 and 1987 Hurricane	Saint Vincent	Hurricane Maria, September 2017	Caribbean Islands
1989–1999, **six** major hurricanes	Antigua and Barbuda	Earthquake, September 2017	Mexico
Tropical Cyclone Pam	Vanuatu	Hurricane Irma, September 2017	Caribbean Islands
Tropical cyclone Ofa (February 1990)	Samoan Islands	Typhoon Damrey, November 2017	Vietnam, Philippines
1998, Hurricane Georges	Saint Kitts and Nevis		

Source: Nott (2003) and http://www.scienceve.com/10deadliestnaturaldisastersof twenty-firstcentury/

disasters arising out of it are included separately (disease epidemics and insect/animal plagues) as it is mostly manipulated by human activities. The hydrometeorological disasters include landslides/avalanches, droughts/famines, extreme temperatures and heat waves, floods, hurricanes, forest/scrub fires, windstorms, and others (insect infestation and waves/surges). On the other hand, the geophysical disasters include earthquakes and volcanic eruptions. Apart from this, human activity also result in some kind of serious challenges which turn into hazard such as climate change, unplanned urbanization, underdevelopment/poverty, as well as the threat of pandemics that will shape humanitarian assistance in the future. These aggravating factors will result in increased frequency, complexity, and severity of disasters.

The historical records of disasters pertaining to the tropical region show that no place in the tropical coastal region is free from disasters (Table 9.1). In addition to this deep depression, heavy wind and sea surges are also causing heavy damage to the coastal ecosystems. Among the different types of natural disasters recorded, cyclones, flood, and tsunami/sea surges are the major types affecting the region. The Caribbean region including the Gulf of Mexico mostly experiences cyclones (Charveriat 2000), whereas tsunami and sea surges are most prevalent in Pacific Ocean islands. Indian Ocean islands and surrounding coastal regions experience cyclonic storm, tsunami, and earthquake which inflict heavy damage almost annually. As the adaptive capacity of small island nations is low when compared to the coastal regions of larger countries and developed countries which are reflected in the disaster impact and management of after effects, here we are not elaborating the details and impact of each disaster rather this chapter focuses on the overall impact of disasters in this region.

9.3.2 Impact of Natural Disasters

The impacts of natural disaster can be direct or indirect. These impacts may be tangible or intangible; accordingly economic and social impacts of the disaster are described. And in many times, the impact is felt for long time depending on the adaptive capacity of the society after the disaster has occurred though the economic and infrastructural losses are visible immediately. It is inferred from the compilation of available records that over the years the impacts of natural disasters have increased either due to spread of human population into naturally hazardous areas or the number and intensity of natural hazard has increased due to anthropogenic activities.

In both cases, the impact is strongly felt on coastal agro-ecosystems. Such impact on agriculture is assessed either by valuing the food-livelihood loss or on absolute economic value of loss and gain. In economic impact assessment, only damage and losses to agriculture caused by the disaster is valued which is total but simple. But this method do not quantify the cascading and wider impact that disasters have on the food value chain, agroindustries, and sector growth or capture the implications for livelihoods and food security (FAO 2015b). Whereas in food-livelihood assessment, the impact of disasters on the agriculture sector as a whole and its potential consequences are grouped into different core categories, viz. direct physical damage, losses across the food value chain (backward-forward linkages), losses to manufacturing (agroindustries), impact on livelihoods, food security, and nutrition and sustainable development (Fig. 9.2). The physical damage caused by disasters has a direct impact on agricultural production with negative consequences along the food value chain, including backward linkages—disrupting the flow of agricultural inputs such as seeds and fertilizers—and forward linkages with processing and distribution, markets, and retailers. Disasters can destroy the infrastructure of input suppliers and postharvest facilities. They can interrupt food supply, market access, and trade. In medium- and large-scale disasters, high production losses can lead to increases in imports of food and agricultural commodities to compensate for lost production and meet domestic demand, increasing public expenditure. They can also reduce exports and revenues, with negative consequences for the balance of

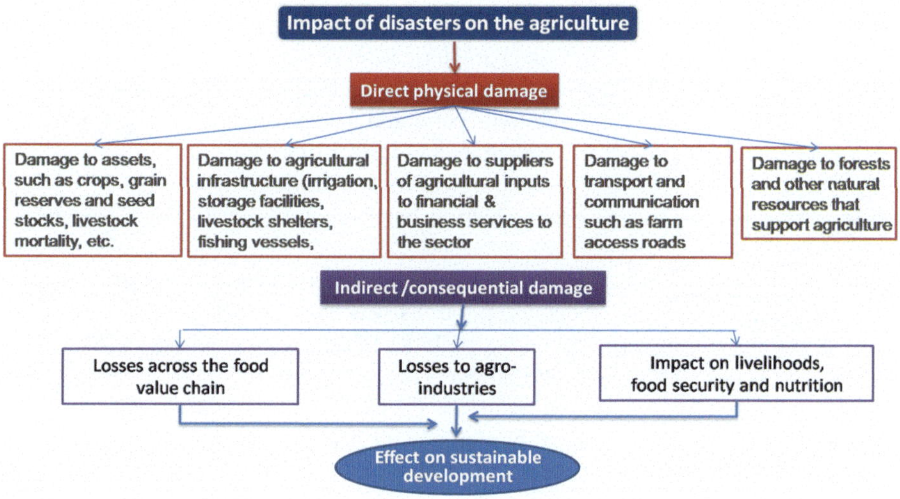

Fig. 9.2 The impact of disasters on agriculture sector and its wider consequences

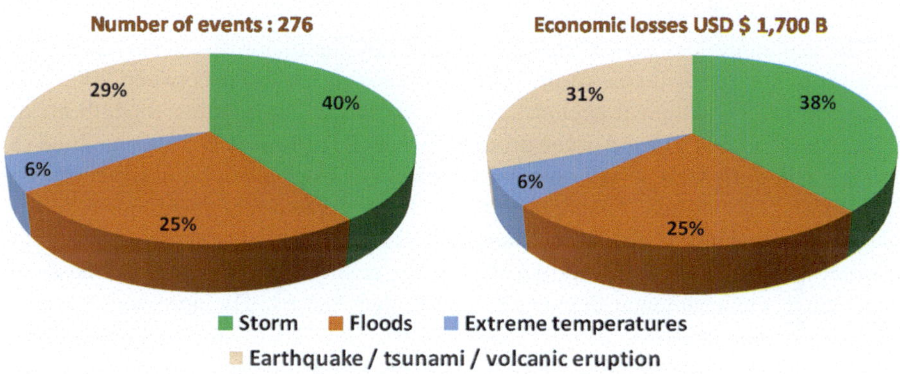

Fig. 9.3 Economic impact of natural disasters

payment (FAO 2015b). The overall effect of both direct and indirect impact will be felt on agricultural livelihoods, food security, and nutrition.

The assessment of economic impact on agriculture shows that economic losses from natural disasters have increased ninefold over the last 40 years (Anderson 1990). During the 1960s, for example, economic losses from natural disasters were estimated at $73 billion; the decade of the 1980s brought losses of $204 billion. From 1990 to 1999, total economic loss from natural disasters exceeded $630 billion (Fig. 9.3). The records show that storm is the major form of disaster (40%) which mostly strikes the coastal region wherein agriculture is the one of the major sector affected. In the 10 years since Hurricane Katrina, the world has seen an annual average of 260 major natural disasters, with average annual economic losses

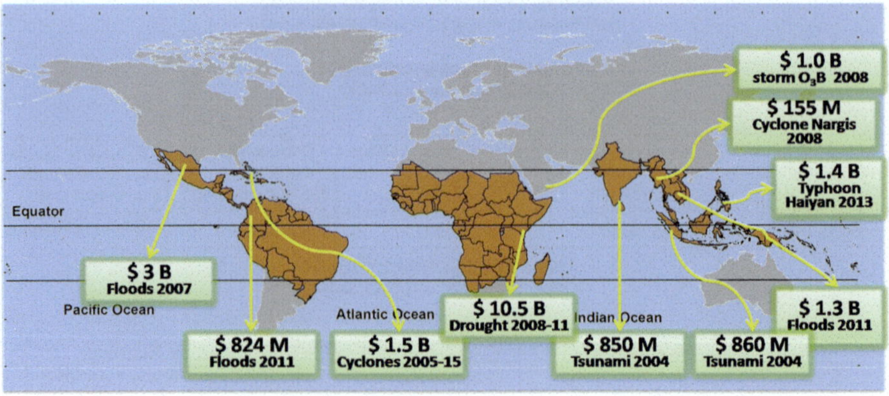

Fig. 9.4 Economic loss to agriculture sector due to major disasters of tropical region reviewed between 2003 and 2013. (Compiled from CIDI 2002)

of US$211 billion, insured losses of US$63 billion, and 76,000 lives lost (Aon Benfield 2017; UN 2015). Tropical storm accounts for 38% of economic loss estimated in USD followed by floods.

The economic loss across tropical countries by the major disasters causing the greatest damage and losses to the agriculture sector reviewed between 2003 and 2015 showed that tropical cyclones are the major type accounting for billions of dollar loss (Fig. 9.4). East African drought ($ 10.5 B) followed by Mexican flood ($ 3 B) recorded the largest economic loss in recent disaster history. Further it is seen that disasters are normal features in almost all the tropical region (islands not visible) and are extensive and larger magnitude (FAO 2015b). The indirect effects of disaster on agriculture appear progressively as a result of low incomes, decreases in production, environmental degradation, and other factors. The agro-ecosystem—including crops, livestock, fisheries, and forestry—besides its surrounding environment absorbs approximately 22% of the economic impact caused by medium- and large-scale natural hazards and disasters in developing countries. For climate-related disasters such as floods, droughts, and tropical storms, 25% of all damage and losses is on the agriculture sector. Agriculture is the single most affected sector by droughts, absorbing on average about 84% of all the economic impact.

Within agriculture, 42% of all damage and losses is on the crops subsector, followed by livestock with 36% (Fig. 9.5). The high impact of natural hazards and disasters on agriculture calls for enhanced mainstreaming of disaster risk reduction and resilience building within the agricultural sectors (FAO 2015b). At the same time, it is observed that there are major data gaps on the impact of natural hazards and disasters on the agriculture sectors in developing countries. This sector-specific data must be systematically collected and included in national and international disaster loss databases to better inform appropriate risk reduction policies and investments for and within the sector.

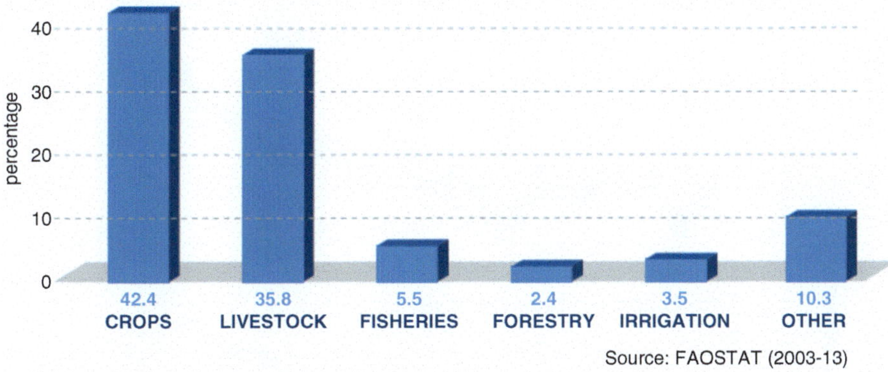

Fig. 9.5 Subsector-wise damage and losses to agriculture. (Source: FAOSTAT 2003–2013)

9.4 Climate Change and Natural Calamities

9.4.1 Observed Changes

The historical records of natural disasters and climatic parameters for the concurrent period highlight the link and vulnerability of coastal and island regions. The four most important parameters, viz. increase in temperature (air and ocean surface), sea level rise (SLR), change in rainfall pattern, and violent tropical cyclones, besides geographical features of the location largely decide the severity of natural disaster occurring in this region. Analysis of observed climatic data showed that, in general, global mean temperature increased by 0.6 °C during the twentieth century, while mean sea level rose by about 2 mm year^{-1}, although sea level trends are complicated by local tectonics and El Nino-Southern Oscillation (ENSO) events (Fig. 9.6). The third assessment report of the IPCC also found that much of the rainfall variability appeared to be closely related to ENSO events, combined with seasonal and decadal changes in the convergence zone. At the same time, recent studies show that annual and seasonal ocean surface air temperatures have increased by 0.6–1.0 °C since 1910 throughout a large part of the South Pacific, Southwest of the South Pacific Convergence Zone (SPCZ). The Caribbean, Indian Ocean and Mediterranean regions' warming ranged from 0 to 0.5 °C per decade for the 1971–2004 period (Trenberth et al. 2007).

But analysis of daily rainfall event for the period 1961 to 2003 showed that extreme rainfall trends are generally less spatially coherent than those of extreme temperatures (Griffiths et al. 2003). Even then, clear evidence exists that the number of storms reaching categories four and five globally has increased since 1970, along with increases in the Power Dissipation Index (Emanuel 2005) due to increases in their intensity and duration (Trenberth et al. 2007). Variations in tropical and extratropical cyclones, hurricanes, and typhoons in many tropical coastal and islands are dominated by ENSO and decadal variability which result in a redistribution of

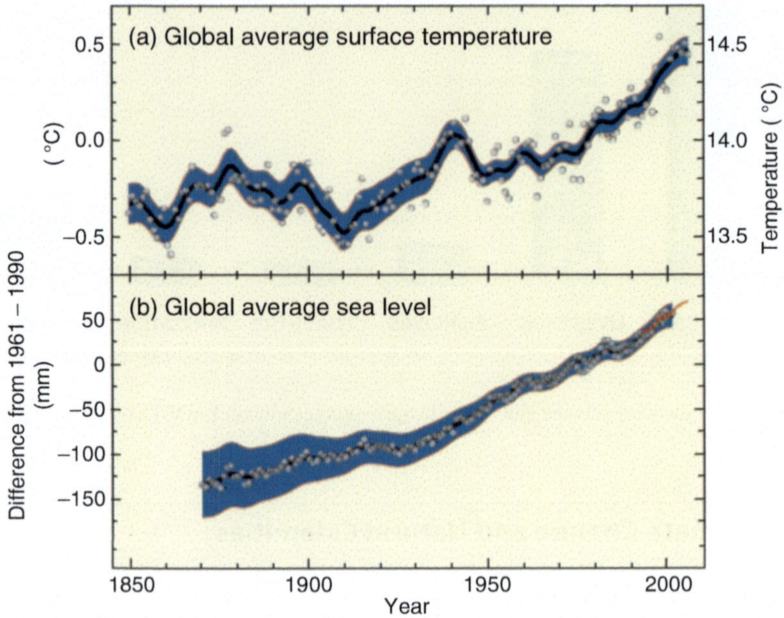

Fig. 9.6 Observed changes in (**a**) global average surface temperature, (**b**) global average sea level from tide gauge (blue) and satellite (red) data

tropical storms and their tracks, so that increases in one basin are often compensated by decreases in other basins.

Another important climate change event affecting the coastal region is sea level rise. Rising sea level has worldwide consequences because of its potential to alter ecosystems and habitability of coastal and island regions. In the analysis of 25 years of hourly data from 27 stations installed around the Pacific basin, the longest available sea level records show that the overall average mean relative sea level rise around the whole region is +0.77 mm/year (Mitchell et al. 2001). The Caribbean region experienced, on average, a mean relative sea level rise of 1 mm/year during the twentieth century. Considerable regional variations in sea level were observed in the records due to large-scale oceanographic phenomena such as El Niño coupled with volcanic and tectonic crustal motions of the Caribbean Basin rim, which affect the land levels on which the tide gauges are located. In the Indian Ocean, reconstructed sea levels based on tide gauge data and TOPEX/Poseidon altimeter records for the 1950–2001 period give rates of relative sea level rise of 1.5, 1.3, and 1.5 mm/year (with error estimates of about 0.5 mm/year) at Port Louis, Rodrigues, and Cocos Islands, respectively (Church et al. 2006). In the equatorial band, both the Male and Gan sea level sites in the Maldives show trends of about 4 mm/year (Khan et al. 2002), with the range from three tidal stations over the 1990s being from 3.2 to 6.5 mm/year (Woodworth et al. 2002).

Consequently the combination of SLR and tropical cyclones together could cause devastation to the coastal region. This is witnessed from the natural disaster records of the tropical region (Table 9.1). The table indicates the widespread occurrence of cyclones and sea surges in the tropical regions. For these reasons, waterlogging, flash floods, and coastal erosion have been experienced in this region significantly affecting the food production systems and the survival of coastal communities.

9.4.2 Tropical Cyclones and Extreme Events

Tropical cyclones are very important climatological feature affecting the tropical coastal region. Tropical cyclone form where there is a convergence of the necessary conditions, which include a large expanse of warm ocean water (generally, water temperatures must be at least 80 degrees Fahrenheit) and the relative absence of 'vertical shear', or winds that change appreciably in either magnitude or direction, thus 'shearing off' the cyclonic outflow at high altitudes. As the name suggests, tropical cyclones are most likely to form in the tropical regions of ocean basins, between about 5° and 20° latitude. The most active months are August and September in the Northern Hemisphere and January and February in the Southern Hemisphere (Das 2003). The rising warm, dense air near the ocean's surface creates an area of low pressure, technically known as a depression. When wind speeds reach 40 miles per hour, the depression reaches tropical storm status. At wind speeds of 74 miles per hour, the storm is designated as a hurricane, or typhoon.

Figure 9.7 indicates the average annual frequency of tropical storm formation in each of the world's major ocean basins. The number given on the left bar for each basin includes all cyclone formation of tropical storm strength or greater, whereas the number given on the right side bar indicates the ones that are of hurricane strength and above. It is observed that North Pacific both west and east experiences maximum number of hurricane and tropical storms. Basically it is due to the vast

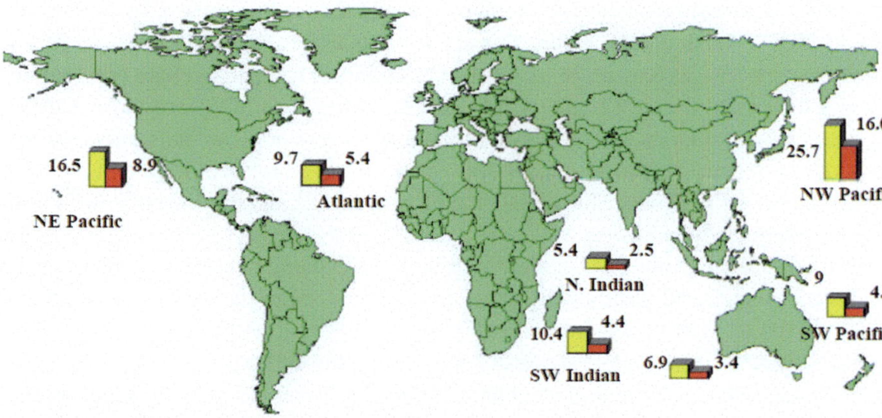

Fig. 9.7 Tropical cyclone formation and average number per year

expanse of the ocean. Though the Caribbean and Indian Ocean region experiences 9.7 and 5.4 average number of cyclones per year, the devastation is relatively very high because of dense coastal population, infrastructural status besides several vulnerable geographical features of the coastal areas.

9.5 Climate Change Projections and Its Impact

9.5.1 Projected Changes

Climate change (the rise in global temperatures: current 0.8 °C, future 2 °C by 2040, and 4 °C by 2100 from the preindustrial levels) is projected to increase the intensity and frequency of disasters (floods, increased precipitation) and the rise in temperature and sea level. The consequences of which will be felt on biodiversity, agriculture, fisheries, aquaculture, and livelihoods of people including wetland-dependent ecosystems. Global projection in seasonal surface air temperature and precipitation for the three 30-year periods (2010–2039, 2040–2069, and 2070–2099) relative to the baseline period 1961–1990 showed increase in surface air temperature and change in rainfall pattern for all regions and the islands (Ruosteenoja et al. 2003).

Sea level changes are of special significance, not only for the low-lying tropical islands but for many coastal areas where settlements, infrastructure, and facilities are concentrated. Projected globally averaged sea level rise at the end of the twenty-first century (2090–2099), relative to 1980–1999 for the six SRES scenarios, ranges from 0.19 to 0.58 m (Meehl et al. 2007). Climate models also indicate a geographical variation of sea level rise due to non-uniform distribution of temperature and salinity and changes in ocean circulation.

With reference to the tropical cycle for the future warmer climate, synthesis of the recent model results indicated increased peak wind speed and mean and peak precipitation intensities. The number of intense cyclones is likely to increase, although the total number may decrease on a global scale (Meehl et al. 2007). It is likely that maximum tropical cyclone wind intensities could increase, by 5–10% by around 2050 (Walsh 2004). Under this scenario, peak precipitation rates are likely to increase by 25% as a result of increases in maximum tropical cyclone wind intensities, which in turn cause higher storm surges. Other natural hazards of maritime origin are, viz. typhoons and storm surges, salt spray, erosion, and tsunamis. Storms and typhoons generate storm surges and waves, winds, and heavy rains mainly in regions at 10–40° latitude. This affects the most sensitive and biodiversity-rich coral reef and seagrass meadows as well.

In spite of uncertainty in magnitude, drivers, and spatial occurrence, the impacts of climate change and weather extremes are experienced in the tropical coastal and islands in the form of increasing monsoon period depressions, dry spell during post-monsoon period, intrusion of sea water into the freshwater aquifer, loss of agricultural land, coastal erosion, drastic changes in wetland habitats, etc. A large number of non-climate change stresses and disturbances, mainly driven by human activities, can also impact coral reefs in a significant way (Hughes et al. 2003). Schematic

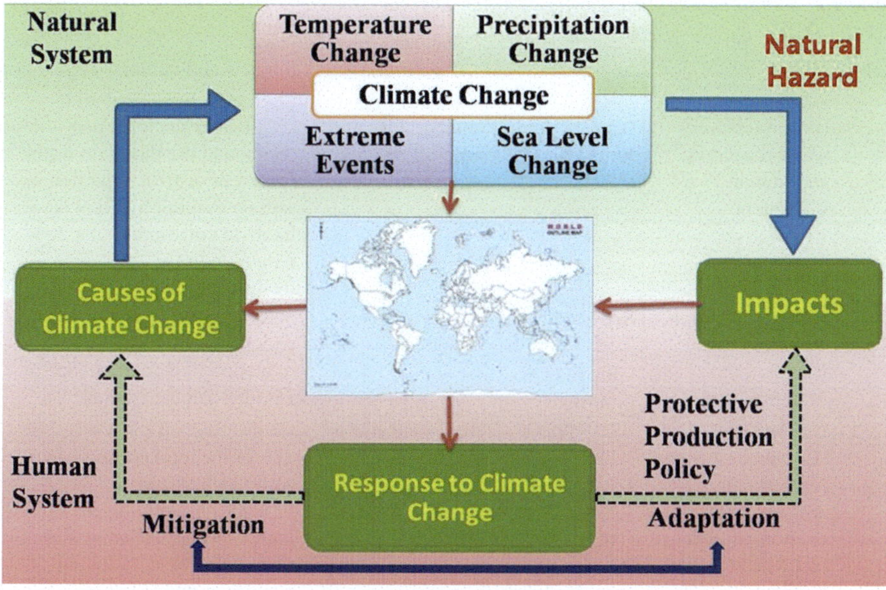

Fig. 9.8 Climate change and natural hazard impacts and adaptation measures

representation of climate change and natural hazard impacts on coastal ecosystem and its adaptation measures are given in Fig. 9.8.

9.5.2 Impact on Agro-ecosystem

A summary of predicted climate change under different scenarios and its likely impacts on different developing regions is given in Table 9.2 (Mimura et al. 2007). There is strong evidence that under most climate change scenarios, water resources in tropical islands are likely to be seriously compromised. Further, sea level rise is expected to exacerbate inundation, storm surge, erosion, and other coastal hazards, thus threatening vital infrastructure and other means of production that support the livelihood of coastal and island communities. This may pose serious threat in the future to the existence of not only human population in coastal and island region but also on other life forms with further ecological consequences.

The adverse effects of climate change on agricultural production are likely to be felt more in the lower latitude countries even though the magnitude of temperature change is projected to be less than in higher latitudes. Some of this higher vulnerability may be ascribed to environmental conditions where crops are grown close to their limits of heat tolerance or moisture availability (Parry et al. 1999). The impacts of changes in climate and climate variability on agricultural production will have substantial effects on small holder farmers in many parts of the tropics and subtropics, and the resulting reduced food security potentially will increase the risk of

Table 9.2 Future impacts and vulnerabilities of tropical countries

S. No.	Region and system at risk	Scenario and reference	Changed parameters	Impacts and vulnerability
1	Pacific islands: coastal erosion, water resources, and human settlement	SRES A2 and B2, FAO and World Bank (2001)	Changes in temperature and rainfall and sea level rise	Less rainfall coupled with accelerated sea level rise compound the threat on water resources; a 10% reduction in average rainfall by 2050 is likely to correspond to a 20% reduction in the size of the freshwater lens
2	American Samoa, 15 other Pacific islands: Mangroves	Sea level rise 0.88–2100 m Gilman et al. (2006)	Projected rise in sea level	50% loss of mangrove area in American Samoa, 12% reduction in mangrove area in 15 other Pacific islands
3	Caribbean (Bonaire, Netherlands Antilles): beach erosion and sea turtle nesting habitats	SRES A1, A1FI, B1, A2, B2 Fish et al. (2005)	Projected rise in sea level	On average, up to 38% (±24% SD) of the total current beach could be lost with a 0.5 m rise in sea level, with lower narrower beaches being the most vulnerable, reducing turtle nesting habitat by one-third
4	Caribbean (Bonaire, Barbados): tourism	None Uyarra et al. (2005)	Changes to marine wildlife, health, terrestrial features, and sea conditions	The beach-based tourism industry in Barbados and the marine diving-based ecotourism industry in Bonaire are both negatively affected by climate change through beach erosion in Barbados and coral bleaching in Bonaire
5	Caribbean and South Asian countries: corals, plantations	IPCC Third Assessment Report (Trenberth et al. 2007)	Change in temperature	Warming ranged from 0 to 0.5 °C per decade for the 1971–2004 period
6	South Asian Islands: agriculture	None Velmurugan et al. (2014)	Changes in temperature and rainfall	Accelerated coastal erosion, saline intrusion into freshwater lenses, and increased flooding from the sea cause large effects on human settlements

Modified from Mimura et al. (2007)

hunger and undernutrition (HLPE 2012). In this context it is vital to understand that local food production is vital to islands, even those with very limited land areas. In most of the islands, subsistence agriculture has existed for several hundred years which are significantly affected by the increase in natural calamities and climate change events.

It could become a major threat to world food security, as it has a strong impact on food production, access, and distribution. Developing and least developing islands states/nations are more vulnerable and will face additional challenges to attain the United Nations Millennium Development Goals by 2015. Averting this challenge requires strong and tangible actions from regional and national governments with a commitment for achieving international goals. Similarly farmers adapt by making changes in farming and land management decisions that reduce the negative consequences associated with changing climate (Jarvis et al. 2011). The adaptation options may also include increasing the resilience of existing farming systems, diversification, and risk management (Thornton and Herrero 2014).

9.6 Adaptation and Disaster Management in Agro-ecosystems

Tropical coastal areas including small islands are subject to floods due to its geographical location and proximity to sea and sometimes severe moisture stress caused by combination of natural, physical, and anthropogenic factors. In the tropical region at different season, one part of the tropical region is affected with flood, while other part happens to reeling under drought. Out of the total geographical area, about one-sixth area with 12% of the total tropical population is found to be susceptible to flood/storm and 8% of the total area suffers from severe moisture stress. This severely challenged the agro-ecosystem production and food security of the region and inland which depend on the region. Therefore, it is vital to identify the most severely flood- and drought-affected areas so that contingency measures can be taken at appropriate level. This requires recording and analysis of climatic data in spatio-temporal scale. Broadly, adaptation options include protective, production, and policy measures (Sivakumar 2006).

9.6.1 Protective Measures

In protective measure prominently there are two ways by which the coastal areas of continental countries and islands can be protected from the sea level rise and other natural hazards. They are artificial methods (hard solutions) and natural methods (soft solutions utilizing a natural buffer zone of coastal vegetation, sand dunes, lagoons, or coral reefs). Artificial methods are mainly kinds of seawalls (huge embankments, tsunami gates) that can be constructed on the coastal area for any predicted tsunami height. This requires considerable investment and suitable technologies. On the other hand, mangrove shield, protective coastal vegetation, adjustment/modification of coastal economic activity, and relocating the habitation away from the vulnerable coastal areas are some of the soft options available to cope with the threat of climate change, sea level rise, and natural calamities (Roy et al. 2002).

Among these two approaches, coastal vegetation provides a range of protection options against sea level rise and other natural calamities besides economic revenue sources for coastal communities. The potential benefits of vegetation are mainly the capacity to mitigate the adverse impacts of natural disasters like tsunamis, storms, and cyclones, to control coastal erosion, and to enhance fishery potentials of the adjacent coastal waters by providing organic and inorganic nutrients (Tanaka et al. 2009). However, a vegetation barrier cannot completely stop a tsunami or storm surge, and its effectiveness depends on the magnitude of the storm surge as well as the structure of the vegetation. The effectiveness of vegetation also changes with the age and structure of the forest. In order to realize the potential benefits of coastal vegetation, a holistic coastal resource management approach incorporating suitable coastal vegetation has to be implemented with due consideration to the socio-economic conditions of the coastal communities.

Mangroves are the most suitable species and dominant component of coastal vegetation providing protection to the coastal areas besides enhancing biodiversity. When a natural disaster, such as tsunami, strikes, mangroves act as green belts and protect the coastline, forming a natural barrier. Since mangroves play a significant economic and ecological role in the islands and coastal areas, their loss or damage would result in across-the-board effects like reduction in the income of fishermen, loss of commercial fisheries related to mangrove ecosystem, and loss of habitat for a number of fishes, molluscs, insects, crustaceans, reptiles, birds, and monkeys. Mangrove destruction will also affect coastal development and human safety as they serve as the natural bio-shield against many natural calamities. The mangrove ecosystems are relatively less prone to the vagaries of climate change as they are more resilient to gradual changes.

The changes in temperature and levels of carbon dioxide in the atmosphere will affect the growth rate and spread of the mangroves but are unlikely to destroy the ecosystem. Several anthropogenic causes including forest use, diversion of freshwater, land reclamation, agriculture, and coastal development have influenced mangrove habitats. Climate change, particularly its associated sea level rise of higher magnitude, is a major threat to mangrove coastal areas, and it is essential to develop ways to reduce vulnerability through strategic management planning. Alternatively, construction of dykes along the most vulnerable portion of the coast is also used in certain islands. Protection by dikes needs to consider not just the extent of average sea level rise but also the effect of more frequent and intense storm surges (Macintosh and Ashton 2002). However, protection from sea level rise using engineering solutions is not a viable option due to cost and ecological sustainability.

9.6.2 Production Measures

9.6.2.1 Scope for Resilient Agriculture

The natural disasters cannot be stopped, but at the same time, we can improve the level of preparedness to face such disasters. There are some practices which can provide better resilience to cyclones and floods with good recovery after the incidence has occurred. Therefore, it has the potential to reduce the extent of crop damage. In this context, there exists a tremendous scope for cyclone and flood-resilient agricultural systems. The resilience capacity of the community plays positive role in reducing the vulnerability of the community. The role of both mitigation and adaptation of climate change that impacts in minimizing the yield losses is significant. Resilient agricultural system can be defined as a combination of different agricultural practices which aim at providing better resistance and adoptability to the crops to various natural disasters such as floods and drought.

9.6.2.2 Role of Agro-advisories to Manage the Disaster

Under such adverse situation, weather-based agro-advisories have vital role to play both during pre- and post-incidence occurrence for better management of situation in agro-ecosystem and minimize the losses. Synoptic methods, statistical methods, and numerical weather prediction models developed based on observational network can be used for generation of rainfall, temperature, thunderstorm, dust storm, clouds, cyclonic storms (their courses and stages), heavy rainfall warnings, frost warning, squall warning, heat waves, etc. Digital and image information at 10 min interval from Doppler weather radar is very useful in addition to many other observed data in predicting thunderstorm activity. Specialized products derived from satellites such as NDVI, surface insolation, LST, albedo, reference evapotranspiration, soil moisture, aridity index, satellite-based rainfall, etc. also play very significant role in disaster management in agriculture. Such generated agromet products are useful if they were promptly delivered to the end users to reduce disaster management risk for farmer community. Information and communications technology (ICT) should be effectively used for dissemination of such information. Due to this information, farmers can tackle the situations under different extreme weather events and ultimately reduce the impact of adverse weather (Chattopadhyay and Chandras 2018).

9.6.2.3 Use of Deep Water and Flood-Tolerant Rice Varieties

Flash floods are frequently witnessed during the monsoon season due to heavy rains within a short period resulting in huge crop loss. Further flooding may occur due to heavy downpour in the catchment areas and drainage congestion at the coastal front. Under such situations, flood-tolerant rice varieties should be used by the farmers to reduce the yield loss (Johnson 2003). For the East Coast of India, waterlogging tolerant rice varieties such as Durga, Hangseswari, and Varshadhan is suggested which has resulted in additional rice yield and economic returns of the farmers. Further, in cyclone and flood-prone areas, the level of submergence at the time of transplanting is higher than the seedling height which makes the seedling establishment very difficult. To overcome this constraint, overaged seedlings of 60 days old are recommended which can be produced in community nurseries. This method provides an yield advantage of about 32% over the normal seedlings (30 days old), and most importantly this practice helps to tide over the waterlogging situation (Roy Chowdhury et al. 2016).

9.6.2.4 Mat Nursery and System of Rice Intensification

The mat nursery was suitable to prepare seedlings in a short period of time using plastic mats under field conditions where time is not available to prepare wet nursery for transplanting. In this method a plastic mat is spread out in the field on which think soil is placed over which the seeds are sown. After 15 days the plastic mats are rolled back to take the rice seedlings to the field for transplanting. This technique has been under use in several parts of the tropics. Various paddy varieties (Taichung-sen-Yu and Quing Livan No. 1 in normal soil; BTS 24, SR 26B, and CST 7-1 in problem soil) can be raised in a flat field as a field demonstration for the benefit of the farmers. The System of Rice Intensification (SRI) is used along with the MAT nursery. The critical stage is the first 20 days after transplanting, when seedlings are vulnerable to damage by continuous rainfall. The mat nursery technique yield savings in field preparation and seed costs compared to conventional nursery preparation. The SRI method produced significantly higher yields (3.95 t/ha) than conventional planting methods (2.00 t/ha).

9.6.2.5 Aquatic Crops with Economic Importance

In areas of perennial waterlogging and cyclone-prone areas where even rice cultivation is difficult, cultivation of aquatic crops with economic importance is recommended. Some of the suitable crops, water chestnut, *Typha* sp., and *Colocasia* sp., either as sole crop or with fish, have potential to provide additional income to farmers. In the waterlogged areas of Bangladesh, crops are cultivated on the floats prepared using aquatic plants.

9.6.2.6 Land Modification

In medium and lowlands under high rainfall region and cyclone and flood-prone areas, land modification techniques such as raised bed and furrow, three tier forming, and farm pond with broader dykes are recommended to manage waterlogging and salinity. Land modification also known as land shaping is nothing but the articulation of land arrangement so as to overcome hydrologic problems in agricultural area for potential crop cultivation (Ambast et al. 2011). The same design could also be used for rainwater harvesting and fish culture. This would be highly effective in utilization of the available water for higher crop productivity, cropping intensity, and economic net returns. This technique also helps to manage salinity besides waterlogging and crop diversification resulting in higher economic net returns to the farmers. Further details are provided in Chap. 20 on innovative land management.

9.6.2.7 Bio-drainage

In flood-prone and waterlogged coastal areas, the practice of bio-drainage using *Casuarina* and *Eucalyptus* plantations would act as a viable flood resilient system as they improve soil drainage and operates better microclimate. Bio-drainage is nothing but removal of soil excess water by standing trees through fast transpiration for which the trees have certain special characters. Pumping of excess soil water/free water happens when deep-rooted plants use their bioenergy. With intensity of solar radiation, the transpiration from leaf surface also increases. This allows intercrop cultivation and helps in advanced planting of winter/summer crop resulting in higher water and land productivity. The farmers will get higher economic net returns due to fuel wood, intercrop, and second crop produce. In addition, water and soil conservation measures in flood catchments in cyclone-prone areas are needed to moderate peak flows and water suspended sediments and siltation of waterways (Kumar et al. 2014). The practice of de-sanding, de-siltation, and land shaping will help in bringing desirable shift in soil physical condition under post-flood situation.

9.6.2.8 Brackish Water-Based Integrated Farming System

Land affected with waterlogging/salinity in the coastal region can be converted to brackish water aquaculture ponds using spillways to regulate the entry of sea water. Along the dyke crops, spinach, amaranth, okra, bitter gourd, bottle gourd, and pumpkin can be cultivated. Sweet potatoes can be planted along the slopes of the dykes. Plantation crops such as coconut, banana, and morinda can withstand slight waterlogging for a brief period and can be planted during the rainy season. Fodder slips of hybrid napier and para grass planted along the sides of the ponds are another viable options to rejuvenate dairy unit after any calamities. Apart from the income from fish, the crops on the embankments provide additional income from fruits and vegetables in 1 year. This intervention can produce significant increase in income from fields that were almost completely destroyed by the tsunami/seawater ingress. Some of the technological packages for flood and moisture stress management for coastal and tropical island conditions are given in Fig. 9.9.

9.6.3 Policy Support and Other Long-Term Management Plan

Apart from the protective and production measures, proper policy and long-term management plan is highly warranted for effective disaster management particularly in coastal agriculture (Petak and Atkisson 1982). First, the local and national government should establish a policy framework for disaster management beginning from defining the natural disaster. With reference to the agro-ecosystem, adequate support for the farmers of the region who have incurred losses during and after the disaster is highly desirable. Comprehensive policy to support crop insurance, land development, marketing facility, and input on credit till the reasonable recovery of agriculture to its original state of production is very much essential. In this context, international cooperation is highly valuable for the recovery of the coastal region in case of large disaster such as December 2004 Indian Ocean tsunami as disasters do not recognize geographical boundaries. Major disasters may often simultaneously affect several countries which require international assistance. Therefore, it will be the national endeavour to develop close cooperation and coordination with international organizations working in the area of disaster management.

Another important aspect of disaster management in agriculture is disaster contingency plan because contingency planning helps in providing better resilience in

Management of moisture stress through IFS and land management

Effect of broad beds and furrow system during flooding

Growing of drought tolerant horse gram during dry season

Awareness campaign for drought management (Sorghum)

Fig. 9.9 Management of flood and moisture stress

post-cyclone and flood period resulting in lesser extent of damage to components of agriculture. Contingency plan can be defined as a plan aimed and executed for an outcome other than in the usual or expected plan. In other words, it is frequently used for risk management when an exceptional risk in future. In general, the change in sowing or planting time of crops, change in seed rate, change in schedule of fertilizer use, and use of short duration varieties and improved crop genotypes form the core component of contingency crop planning. In case of cyclone and its associative flood events, we should recommend the use of waterlogging-tolerant rice varieties. Another example is the practice of zero tillage as crop management intervention helps in better crop establishment and productivity of sunflower, okra, and bitter gourd under post-flood situation. Hence, it should be the policy of the government to get well prepared for sustaining agricultural productivity, and this necessitates the concept of contingency planning and promotion of climate-resilient agriculture in the vulnerable areas.

The changing climate is a major concern for agricultural productivity in general and food security in particular (Kumar et al. 2014). This has also resulted in higher frequency of natural disasters seriously affecting the coastal region. For these reasons every government now has or should have a long-term management plan in line with the national priorities and international efforts. As a result there has been a paradigm shift in the government focus from the earlier rescue and relief-centric approach to a holistic approach covering all aspects of disaster/cyclone management. The following are the long-term plan required for extenuating the threat of cyclone.

- Improvement of onshore warning and alert system along with construction of cyclone shelter at suitable places. The installation of warning systems in several Caribbean countries has reduced the loss of human life due to hurricanes. Prohibition of permanent settlement in floodplains, enforced by selective insurance coverage, has significantly reduced flood damage in many vulnerable areas (Anonymous 1990).
- Construction/renovation of canals and embankments for improved drainage.
- Shelter belt plantation to reduce the effect of wind.
- Construction of missing road links to facilitate the rehabilitation and movement of goods.
- Conservation and regeneration of mangrove forests.
- Construction of saline embankments and one-way sluice gates.
- Dredging of river mouths.
- Institutional capacity building and hazard reduction studies.
- However, concerted efforts are required to evolve sound national strategies for the management of disasters by emphasizing research and development needs in the following aspects.
  - Preparation of disaster management plans at different levels
  - Strengthening of cyclone tracking and monitoring system
  - Early warming dissemination
  - Storm surge modeling
- Awareness generation among the local communities and stakeholder for disaster risk mitigation.

9.7 Conclusion

The tropical coastal areas including island nations across the globe are highly vulnerable to natural hazards especially cyclones, sea surges, earthquakes, floods, drought, and landslides. Several disasters have happened in this region more often than any other region on Earth. The frequency and severity of disasters triggered by natural hazards have continued to rise over the last three decades. Besides other negative effects on the coastal ecosystems, disasters jeopardize agricultural production and development and often have cascading negative effects across national economies.

Agro-ecosystem in the coastal area supports millions of population but highly vulnerable to natural disasters and projected to be negatively affected by climate change in the future. At the same time, the system has to feed the evergrowing human population to ensure food security. This necessitates developing a multi-pronged and integrated long-term disaster management strategy for coastal agro-ecosystem. This includes preventive, mitigation, and preparedness measures as response and rehabilitation measures. Climate-resilient production technologies should be developed for the coastal region to face the challenges of climate change. It is also very pertinent to note that disasters do not recognize geographical boundaries. Major disasters may often simultaneously affect several countries. Therefore, vulnerable nations should develop close cooperation and coordination with international organizations working in the area of disaster management.

In the context of cyclone management in agriculture, several options for land, crop, and livestock management have been evaluated at different places. One such viable option is the development of community nursery and seed bank to supply seedlings of rice in case of cyclone damage. At least seed bank should be developed for 10% of the area under each block which can be utilized for transplanting in post-cyclone period. This necessitates evolving a collaborative strategy to face the cyclone events in an effective manner. In summary, a well-prepared and scientific-based multipronged approach is essential in which various stakeholders working in the line of disaster management play significant role to effectively address pre- and post-disaster situations.

References

Ambast SK, Ravisankar N, Velmurugan A (2011) Land shaping for crop diversification and enhancing agricultural productivity in degraded lands of Andaman and Nicobar Islands. J Soil Salinity Water Qual 3(2):83–87

Anaman KA (2003) Assessing the economic and social impacts of extreme events in agriculture and use of meteorological information to reduce adverse impacts. In: Agrometeorology related to Extreme Events. WMO No. 943, World Meteorological Organization, Geneva

Anderson M (1990) Analyzing the costs and benefits of natural disaster responses in the context of development, Environment working paper 29. World Bank, Washington, DC

Anonymous (1990) Disaster, planning and development: managing natural hazards to reduce loss. Department of Regional Development and Environment Executive Secretariat for Economic and Social Affairs Organization of American States, Washington, DC

Benfield A (2017) Weather, climate & catastrophe insight. Annual Report 2017. www.aon.com/annual reports 2017. Accessed 22 Oct 2018

Charveriat C (2000) Natural disasters in Latin America and the Caribbean: an overview of risk, Research Department working paper no. 434. Inter-American Development Bank, Washington, DC

Chattopadhyay N, Chandras S (2018) Agro-meteorological advisory services for informed decision making in India. In: Chandrakasan Sivaperuman C, Velmurugan A, Singh AK, Jaisankar I (eds) Biodiversity and climate change adaptation in tropical islands. Academic, USA, pp 763–783

Church JA, White N, Hunter J (2006) Sea level rise at tropical Pacific and Indian Ocean islands. Glob Planet Chang 53:155–168

CIDI (2002) International disaster situation reports. Center for International Disaster Information. http://www.cidi.org/disaster. Accessed on 18 Aug 2018

Das HP (2003) Incidence, prediction, monitoring and mitigation measures of tropical cyclones and storm surges. In: Agrometeorology related to extreme events. WMO No. 943, World Meteorological Organization, Geneva

Emanuel K (2005) Increasing destructiveness of tropical cyclones over the past 30 years. Nature 436:686–688
FAO (2015a) The impact of natural hazards and disasters on agriculture and food security and nutrition a call for action to build resilient livelihoods. Food and Agriculture Organization, Rome
FAO (2015b) The impact of disasters on agriculture and food security. Food and Agriculture Organization, Rome. www.fao.org/resilience
FAO and World Bank (2001) Farming systems and poverty- improving farmers' livelihoods in a changing world. FAO and World Bank, Rome
FAOSTAT (2014) Country statistics. Food and Agriculture Organization of the United Nations. http://faostat.fao.org/site/567/default.aspx#ancor. Accessed 18 July 2018
Fish MR, Cote IM, Gill JA, Jones AP, Renshoff S, Walkinson A (2005) Predicting the impact of sea level rise on Caribbean sea turtle nesting habitat. Conserv Biol 19:482–491
Gilman E, Van Lavieren H, Ellison J, Jungblut V, Wilson L et al (2006) Pacific Island mangroves in a changing climate and rising sea, UNEP regional sea reports and studies 179. United Nations Environment Programme, Regional Sea Programme, Nairobi, p 58
Griffiths GM, Salinger MJ, Leleu I (2003) Trends in extreme daily rainfall across the South Pacific and relationship to the South Pacific Convergence Zone. J Climatol 23:847–869
HLPE (2012) Food security and climate change. A report by the high level panel of experts on food security and nutrition of the Committee on World Food Security, Rome, Italy
Hughes T, Baird A, Bellwood D, Card M, Connolly S et al (2003) Climate change, human impacts, and the resilience of coral reefs. Science 301:929–933
Jarvis A, Lau C, Cook S, Wollenberg E, Hansen J, Bonilla O, Challinor A (2011) An integrated adaptation and mitigation framework for developing agricultural research: synergies and trade-offs. Exp Agric 47(02):185–203
Johnson G (2003) Assessing the impact of extreme weather and climate events on agriculture with particular reference to flooding and rainfall. In: Agrometeorology related to extreme events. WMO No. 943, World Meteorological Organization, Geneva
Khan TMA, Quadir DA, Murty TS, Kabir A, Aktar F, Sarker MA (2002) Relative sea level changes in Maldives and vulnerability of land due to abnormal coastal inundation. Mar Geod 25:133–143
Kumar A, Brahmanand PS, Nayak AK (2014) Management of cyclone disaster in agriculture sector in coastal areas. ICAR-Indian Institute of Water Management, Bhubaneswar, p 108
Macintosh DJ, Ashton EC (2002) A review of mangrove biodiversity and conservation. Centre for Tropical Ecosystem Research. University of Aarhus, Denmark
MEA (2006) Millennium ecosystem assessment, marine and coastal ecosystems and human well-being: synthesis. United Nations Environment Programme, Nairobi, p 40
Meehl GA, Stocker TF, Collins W, Friedlingstein P, Gaye AT, Gregory J, Kitoh A, Knutti R, Murphy J, Noda A, Raper S, Watterson LG, Weaver A, Zhao ZC (2007) Global climate projections. Climate change 2007: the physical science basis. In: Solomon S, Qin D, Manning M, Chen Z, Marquis M, Averyt KB, Tignor M, Miller HL (eds) Contribution of working group I to the fourth assessment report of the intergovernmental panel on climate change. Cambridge University Press, Cambridge, pp 747–846
Mimura N, Nurse L, McLean RF, Agard J, Briguglio L, Lefale P, Payet R, Sem G (2007) Small islands – climate change 2007: impacts, adaptation and vulnerability. In: Parry ML, Canziani OF, Palutikof JP, van der Linden PJ, Hanson CE (eds) Contribution of working group II to the fourth assessment report of the intergovernmental panel on climate change. Cambridge University Press, Cambridge, pp 687–716
Mitchell W, Chittleborough J, Ronai B, Lennon GW (2001) Sea level rise in Australia and the Pacific. Proceedings science component: linking science and policy, Pacific Islands Conference on Climate Change, Climate Variability and Sea Level Rise, Rarotonga, Cook Islands. National Tidal Facility, The Flinders University of South Australia, Adelaide, pp 47–58
Nott JF (2003) Intensity of prehistoric tropical cyclones. J Geophys Res 108(D7):4212. https://doi.org/10.1029/2002JD002726

Parry M, Cynthia R, Ana I, Gunther F, Mathew L (1999) Climate change and world food security: a new assessment. Glob Environ Chang 9:551–567
Petak WJ, Atkisson AA (1982) Natural hazard risk assessment and public policy: anticipating the unexpected. Springer-Verlag, New York
Roy BC, Mruthyunjaya, Selvarajan S (2002) Vulnerability of climate induced natural disasters with special emphasis on coping strategies of the rural poor in Coastal Orissa, India. Paper presented at the UNFCC COP 8 Conference organized by the Government of India, UNEP and FICCI, 23 October to 1 November 2002, New Delhi, India
Roy Chowdhury S, Rautary SK, Panda RK, Das M, Srivastava RC, Ambast SK (2016) Technological options for agricultural water management in Eastern regions of India. ICAR-IIWM, Bhubaneswar, p 39
Ruosteenoja K, Carter TR, Jylha K, Tuomenvirta H (2003) Future climate in world regions: an intercomparison of model-based projections for the new IPCC emissions scenarios, The Finnish Environment 644. Finnish Environment Institute, Helsinki. 83 pp
Sivakumar MVK (2006) Climate prediction and agriculture: current status and future challenges. Clim Res 33(1):3–17
Tanaka N, Nandasena N, Jinadasa AK, Sasaki KBSN, Tanimoto Y, Mowjood MIM (2009) Developing effective vegetation bioshield for tsunami protection. J Civil Environ Eng Syst 26:163–180
Thornton PK, Herrero M (2014) Climate change adaptation in mixed crop-livestock systems in developing countries. Glob Food Sec 3(2):99–107
Trenberth KE, Jones PD, Ambenje PG, Bojariu R, Easterling DR, Klein Tank AMG, Parker DE, Renwick JA, Rahimzadeh F, Rusticucci MM, Soden BJ, Zhai PM (2007) Observations: surface and atmospheric climate change. In: Solomon S, Qin D, Manning M, Chen Z, Marquis M, Averyt KB, Tignor M, Miller HL (eds) Climate change 2007: the physical science basis, Contribution of Working Group I to the Fourth Assessment Report of the Intergovernmental Panel on Climate Change. Cambridge University Press, Cambridge, pp 235–336
UN (2015) Agriculture and disaster risk. United Nations. http://www.wcdrr.org/documents/wcdrr/prepcom1/ UN/ATTFONWO.pdf. Accessed 20 Nov 2017
Uyarra MC, Cote IM, Gill JA, Tinch RRT, Viner D, Watkinson AR (2005) Island-specific preferences of tourists for environmental features: implications of climate change for tourism-dependent states. Environ Conserv 32:11–19
Velmurugan A, Swarnam TP, Ambast SK, Ravisankar N, Subramani T (2014) Land degradation and its spatio-temporal changes induced by natural events in Andaman Islands. J Andaman Sci Assoc 19(1):65–74
Walsh K (2004) Tropical cyclones and climate change: unresolved issues. Clim Res 27:77–83
Woodworth PL, Le Provost C, Richards LJ, Mitchum GT, Merrifield M (2002) A review of sea level research from tide gauges during the World Ocean Current Experiment. Oceanogr Mar Biol 40:1–35
World Disaster Report (2003) Focus on ethics in aid, International Federation of Red Cross and Red Crescent Societies. www.ifrc.org/en/publications-and-reports/world-disasters-report/wdr2003/ Accessed on 16 Sept 2018

Part III

Threats and Strategies for Coastal Ecosystem Management

Sand Mining and Strategies for Its Management

10

Abstract

Sand extraction from beaches and inland dunes and dredging from ocean and river beds have significantly increased in recent decades. Sand is mainly used in construction and often in manufacturing as an abrasive or in concrete. Sand dunes present on shorelines are formed by a combination of wind and waves and stabilized with native dune grass. Dunes protect the coastal areas from the destructive forces of damaging waves and wind. Similarly sand deposits in the river stream and in the estuaries help to maintain the water balance and protect the riparian areas. However, mining of sand from coastal dunes and river bed creates negative consequences in the coastal ecosystems. Most importantly instream sand mining lowers streambed elevation, accelerates erosion, and affects the adjoining groundwater system. Sand mining involving both onshore and offshore sites seriously affects the coastal ecosystem and threatens the survival of several flora and fauna. On the other hand, there is a mounting pressure on the available sand deposits for various purposes. There are three logical approaches to meet the challenge and protect the coastal ecosystem, viz. regulation of use of existing deposits, reclamation of mined sites, and alternative sources of sand. These aspects are discussed in this chapter in detail.

Keywords

Mining · Coastal dunes · River bed · Regulation and reclamation

10.1 Introduction

The three major drivers of environmental change—population growth, economic development, and climate change—are interlinked and exerting a range of pressures on natural resources with a consequential negative impact on the coastal environment. More specifically urbanization and coastal development involving transport,

V. Ayyam et al., *Coastal Ecosystems of the Tropics - Adaptive Management*,
https://doi.org/10.1007/978-981-13-8926-9_10

services, farming, and industry are major pressures on terrestrial and marine biodiversity, environmental quality, water resources, air quality, and cultural and natural heritage. In recent decades concern about changes in the size and composition of coastal populations in the tropical countries has also been growing.

In the same context, certain human activities have profound effect on the coastal environment. For example, sand mining and dredging is one such activity grown in high demand which is directly linked to coastal development, human need, and environmental degradation (Anonymous 2018a). Sand mining is the extraction of sand, mainly through an open pit but sometimes mined from beaches and inland dunes or dredged from ocean and river beds. It is mainly used in construction and often in manufacturing as an abrasive or in concrete (Mensah 1997). Mining activity targets different sources as some uses require higher purity and specific size of sand than others. For example, sand used in concrete must be free of seashell fragments, and in construction it is used directly.

In general, sand mining in the coastal areas includes two mutually opposing activities. In other words sand mining results in two contrasting effect. Primarily sand mining presents great opportunity for the coastal countries to extract industrially valuable minerals such as rutile, ilmenite, and zircon besides direct use in construction. These minerals normally occur combined with sand and then are separated in water by virtue of their different densities. On the other hand, sand mining is a direct cause of erosion and land degradation and affects the local wildlife both offshore and onshore. Many of the wild animals in the coastal region depend on sandy beaches for nesting clutches. For these reasons, sand mining in these locations has led to the near extinction of gharials (crocodile), threatened several species of sea turtles, and birds (Pitchaiah 2017). Several studies have also shown that disturbance of underwater and coastal sand by mining activities causes turbidity which is greatly harmful for corals and reef environment that need sunlight (Supriharyono 2004). Further, removal of physical coastal barriers, such as natural sand dunes, leads to flooding of beachside communities and the destruction of picturesque beaches which will seriously affect the tourism sector and livelihood of thousands of people who depend on it (CBD 2014).

Review of mining activities has shown that mining of beach and dune sand is a global phenomenon not restricted to a particular region or country. In recognition of the damage caused by sand mining and the need to preserve beaches for future generations, mining is regulated by law in many coastal countries (Bush et al. 1995). Yet, it remains an important global problem seriously affecting the coastal environment and biodiversity because of illegal mining and over-exploitation affecting the coastal ecosystem in several ways. With this background, this chapter highlights the sand mining particularly in the coastal areas and its consequential effect on the coastal environment, biodiversity, and other livelihood support. Also alternative options are discussed aimed at meeting the developmental needs and sustainable use of such resource for providing livelihood to the coastal population.

10.2 Occurrence of Sand Deposits and Its Significance

In this chapter we discuss sand deposits of two kinds, viz. sand deposited by riverine system in the river bed and by the combined action of waves and wind in the coastal areas. Deposition of water-flown materials begins when there is not much gradient for the water to flow further. In other words the streams originating in the land deposit the sand in the coastal and estuaries when they can no longer carry them to the sea due to loss of gradient, obstruction to water flow, or reduction in volume of water. Beach formation begins with the eroded continental materials such as sand, gravel, and cobble fragments. These materials are transported to sea by river systems and along the coast by the longshore current. Sand dunes are present on shorelines where fine sand is transported landwards by a combination of wind and waves and stabilized with vegetation (Carter and Woodroffe 1994). Dunes can help protect coastal property from the destructive forces of storm surges and tsunamis. Formation of these two types of sand deposits is discussed in the following subsections.

10.2.1 Formation of Coastal Sand Dunes

Most of the open-ocean shorelines have sand deposits that exist in equilibrium with sea wave, ocean currents, and wind processes. This type of sand deposits exists in several connected coastal environments. In general sand dunes are present on shorelines where fine sediment is transported landwards by a combination of wind and waves and stabilized with vegetation (dune grasses). In the onshore there are beaches with sand deposited by waves and currents and sand dunes deposited by the wind. Off-shore, there is sand stored on the shoreface reaching from the beach out to a depth of 10 m or so and in deltas by the action of tide and river (Carter and Woodroffe 1994). At many places sand and fine materials are eroded from nearby cliffs by waves and transported along the shoreline by longshore drift. Similarly rivers and coastal streams transport sand to the coast from inland deposits and river bank. Waves wash this sand up on the beach and also remove it during storms. Once sand has been deposited on a beach, it is transported by the wind by the process of saltation and surface creep (Fig. 10.1). When the prevailing wind direction is onshore, the sand gradually migrates landwards from the water's edge (Arnott 2010).

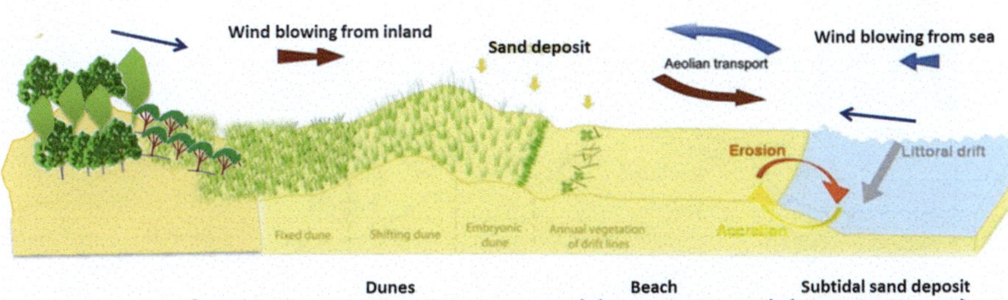

Fig. 10.1 Formation of sand dunes as a function of wind and waves

However, to a certain extent, the shape and size of dunes depend on the shape of the beach and the wind direction. On a shallow-sloped beach, the energy of waves is dissipated, and the suspended grains of sand in the sea water are deposited, resulting in greater dune formation potential. Steeper beaches do not dissipate this energy, and consequently the sand is more frequently washed back to the sea (Anonymous 2018b). For these reasons the dunes on steeper beaches are smaller; in contrast it is wide and large in a shallow-sloped beach.

10.2.2 Types of Sand Dunes

In general a series of dunes are formed parallel to the coast depending on the wind direction, wave action, and obstruction to pile up the sand. The dune at the back of a natural beach or end of the dry beach is called the foredune (Fig. 10.2). In other words this is a ridge of sand forming with its axis or crest parallel to the shoreline. Such a dune or sand ridge in association with the natural beach is called a *beach ridge*. In these locations the vegetation serves as an anchor for wind-blown sediment and generates continuous dune growth. It is often seen that removal of grass or dune vegetation by storms, fire, trampling, or overgrazing, the dunes get destabilized and begin to migrate landward (Arnott 2010).

It is also observed that the size of a dune is mainly a function of sand supply among other factors. The larger the sand supplies from the beach, the higher the dunes. Prevailing wind directions, beach width, and time available to build a dune are other factors. At the same time, it is not clear why more sand comes ashore in some locations and builds big dunes while little far away locations dunes are small and low. Most foredunes have occasional gaps between them where the storm waves penetrate landwards. These openings are referred to as *overwash passes* that are very important for the islands' evolution especially in a time of rising sea level. Overwash passes allow accumulation of storm-derived sand which elevates the islands. Some overwash passes are the result of artificial cuts by human activities through dunes at the ends of roads and streets. In general, overwash deposits on the back sides of dunes are easily distinguishable from dune sands because of their high shell content (Anonymous 2018b). At some locations larger subtidal beach also

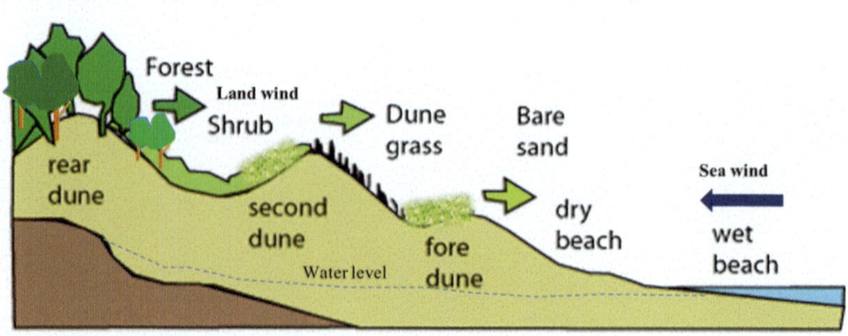

Fig. 10.2 Types of dunes and scope for sand mining

exist in the offshore area formed by the littoral drift and eroded materials from the beach. In summary, all these dunes, deposits, and processes are part of coastal sandy ecosystem, where in physical and biological forces are in dynamic equilibrium with each other.

10.2.3 Sand Budget in a Stream

Unlike the coastal sand dunes and sandy beaches, sand deposits in a riverine system and estuaries largely derive their sandy material from inland catchments and erosion of stream banks. The nature of catchment, rainfall, gradient, and water flow in the river largely decides the type and amount of sand materials carried by the river, while wind has limited role. Any attempt to utilize these materials (termed as disturbances) by sand-and-gravel mining either in-channel or near-channel changes the sediment budget and results in consequential changes in the channel hydraulics. These interventions can have variable effects on aquatic habitat, depending on the magnitude and frequency of the disturbance, mining methods, particle size characteristics of the sediment, the characteristics of riparian vegetation, and the magnitude and frequency of hydrologic events following the disturbance.

Therefore temporal and spatial responses of alluvial river systems are a function of geomorphic thresholds, feedbacks, lags, upstream or downstream transmission of disturbances, and geologic/physiographic controls. Sometimes the streams get dried up due to storing of water in the upstream areas (human intervention), but it is reiterated that maintaining these streams is essential to face any unforeseen or extreme events. Determining the sand budget for a particular stream reach requires site-specific topographic, hydrologic, and hydraulic information. This information is vital to determine the amount of sand that can be removed from the area without causing undue erosion or degradation, either at the site or at a nearby location, upstream, or downstream.

10.2.4 Sandy Coastal Ecosystem

Coastal sand dunes provide a range of habitats for plants and animals though it is very harsh. The habitats nearest to the beach (towards water front) are salty, devoid of nutrients, and harsh. Thus the plants (grasses) colonizing this dunes should able to tolerate abiotic stresses such as lack of moisture and nutrients, exposure to salt spray, wind abrasion, and ground instability (Fig. 10.3). With increasing distance inland, the number of species and the amount of ground cover also increases, reflecting the time available for soil development and the increased availability of moisture and nutrients. Shrubs and woodland tend to establish only on the oldest, stable dunes farther inland. At sites with limited human disturbance, this woodland may approach the climax vegetation for sandy soils in an area. The sequence of plants that occupy the different habitats within dune systems is called succession. Similarly

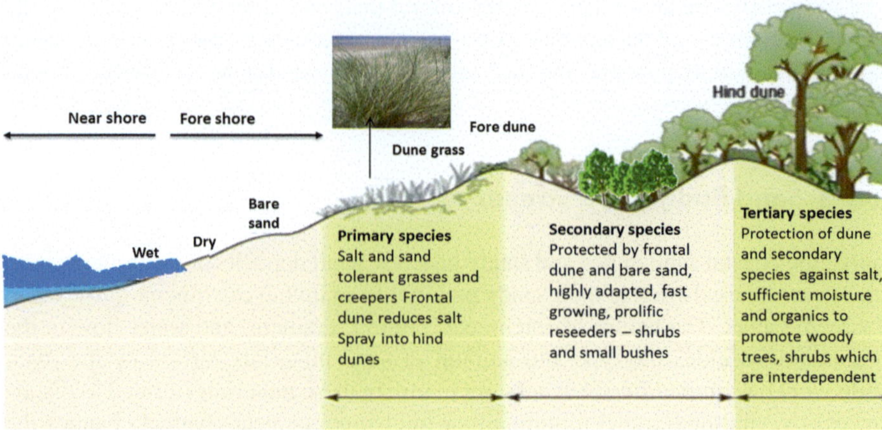

Fig. 10.3 Vegetation succession in a sand dune ecosystem

the fauna inhabiting or depending on these vegetations also varies with adaptation for salty and harsh conditions in the frontal dune to woodland in the farther dunes.

As discussed earlier in the frontal dunes, only well-adapted grasses tend to grow. These dune grasses stabilize the sand in which it is growing, and the way the grass spreads will inevitably affect the shape of the dune. Thus it is vital to conserve the local grasses which are well adapted to the local harsh conditions. Sometimes grasses can be introduced from outside to stabilize the dunes. At the same time, it is observed that due to the clustering nature of some grasses, dunes may have gaps or overwash passes. Such gaps, as discussed earlier, facilitate the penetration of storm waves onto the island. In contrast, dunes with fewer gaps prevent or reduce storm overwash except in the larger storms. It is seen that penetration of the storm waves through the dunes is important because it brings overwash sand to the island, raising its elevation. Vertical growth through the accumulated overwash and dune sand is seen as progressive development as it is very important for a barrier island to evolve with a rising sea level (Anonymous 2018b). Any disturbance to these active sand deposits will have serious consequences in the natural evolution of sandy ecosystem.

10.2.5 Ecological and Economic Significance

Sand deposition and formation of sand dune are the collective function of wind, waves, and transport of materials by the water. Once the sand gets accumulated in the beach by the wave action, it is picked up by the wind blowing landwards. The movement of particle is the function of size of particles and strength of the prevailing wind. But the debris present in the beach slows down the wind or sometimes large enough to block it, causing sand to accumulate in the wind shadow portion of the object. This is followed by the growth of the dune vegetations, that is, dune grass seeds find their way to the new piles of sand, germinate, sprout, and trap more sand. The rotting vegetation seaweed or salt marsh grasses that caused the sand accumulation to begin with now furnish nutrients to help the new seedlings survive. This means beginning of organic recycling. As the plant grows, it continues to cause more and more sand to accumulate resulting in new dune formation with the progress of time (Carter and Woodroffe 1994). If the condition is favourable for the formation of multiple dune ridges caused primarily by variation in wind, the inter ridge area will be populated by shrubs, vines, and small trees, and ultimately in the succession of plants, a maritime forest may grow. Such an established system is known as sandy ecosystem.

Thus under natural conditions, the types and density of vegetation fairly indicate the age and length of stability of dunes. Grasses may be established within a season, but shrubs take 10–20 years to become established, but even centuries are required for the natural establishment of maritime forest (Fig. 10.4). In other words as these embryonic dune grow, the vegetation keeps moving up and out, holding much of the sand in place. Along with the vegetation, several faunal population tend to find their habitat in it. Together with flora and fauna, sand deposits become a unique coastal habitat. These deposits are linked to one another such that addition or removal of sand from one area affects all of the other environments. Some coasts have sand bodies that are somewhat disconnected from the modern coastal sand deposits. This sand may occur well below or above present sea level representing deposits formed in the past when the level of the ocean was higher or lower than today.

Fig. 10.4 Stabilization of sand dunes and wet beach

Casuarina shelter belt on the frontal dune Grazing of dune vegetation as green fodder

Fig. 10.5 Role of vegetation on the coastal dunes

Dunes can help protect coastal areas and infrastructures from the destructive forces of damaging waves and wind, but human activities have been severely altering the ecological role of the sand dunes (Fig. 10.5). These sand deposits are valuable source of rare minerals. Beaches can provide a cheap source of sand for use in concrete and for a variety of other construction purposes. Beach sand is inexpensive because it is already unconsolidated and is easy to remove with front-end loaders or even by hand loading of trucks (Anonymous 2018a). When mature forest is removed for development or for sand mining, the vegetative equilibrium is disrupted. Further, overgrazing by animals in the coastal areas also removes certain plant species, damages salt marshes, and prevents much under-canopy growth in maritime forests.

10.3 Mining and Structural Damage

Sand-and-gravel mining in stream channels and mining of sand in the active sand dunes or beach can damage structures built in the coastal areas and river bank by counter effect. Several studies have documented the bed degradation caused by the two general forms of instream mining, viz. pit excavation and bar skimming. Bed degradation, also known as channel incision, occurs through headcutting (Fig. 10.6). In headcutting, excavation of a mining pit in the active channel lowers the streambed, creating a nick point that locally steepens channel slope and increases flow energy (Pitchaiah 2017). During high flows, a nick point becomes a location of bed erosion that gradually moves upstream.

Headcutting mobilizes substantial quantities of sediments from the streambed, transports downstream, and deposits in the excavated area and locations further downstream. In gravel-rich streams, effect on downstream of mining sites may be short-lived when mining ends due to reestablishment of balance between sediment input and transport, whereas the effects in gravel-poor streams may develop rapidly and persist for many years after mining has stopped. However, regardless of downstream effects, headcutting in both gravel-rich and gravel-poor streams remains a

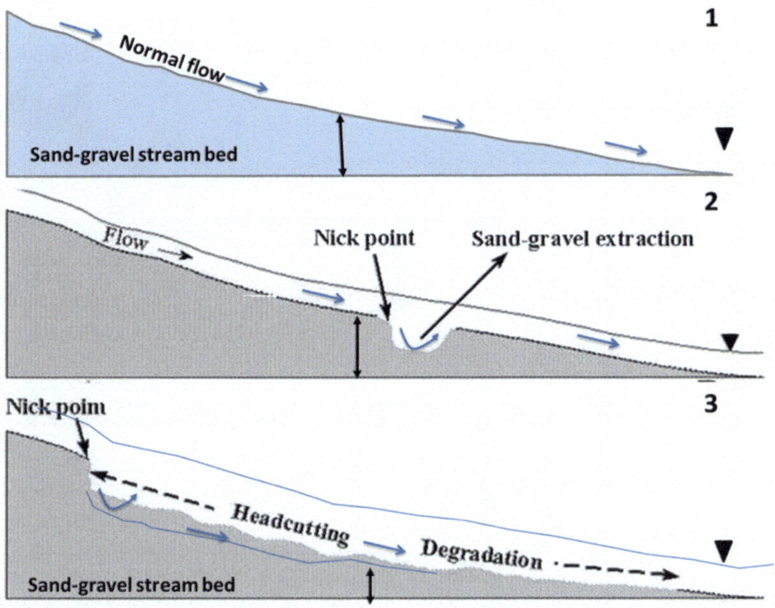

Fig. 10.6 Stream cut erosion

major ecological concern. Headcuts often move long distances upstream and into tributaries and in some watersheds moving as far as the headwaters until resisted by geological or man-made structures.

A second form of bed degradation occurs when mineral extraction increases the flow capacity of the channel. A pit excavation locally increases flow depth (Fig. 10.7), and a bar skimming operation increases flow width. Both conditions produce slower stream flow velocities and lower flow energies, causing sediments arriving from upstream to deposit at the mining site. As stream flow moves beyond the site and flow energies increase the amount of transported sediment leaving the site become less than the sediment carrying capacity of the flow. This sediment-deficient flow tends to pick up more sediment from the stream reach below the mining site, furthering the bed degradation process. This condition continues until the equilibrium between input and output of sediments at the site is reestablished.

Similar effect is observed below dams which trap sediment and release water downstream. Instream mineral excavation at downstream of dams compounds this problem, although other factors such as levees, bank protection, and altered flow regimes also promote channel incision. Sand/mineral extraction rates in many streams are often in excess of sediment supply from the watershed, suggesting that extraction is largely responsible for observed channel changes. Channel incision not only causes vertical instability in the channel bed but also causes lateral instability in the form of accelerated stream bank erosion and channel widening. Incision increases stream bank heights causing bank failure when the mechanical properties of the bank material cannot sustain the material weight. Channel widening causes

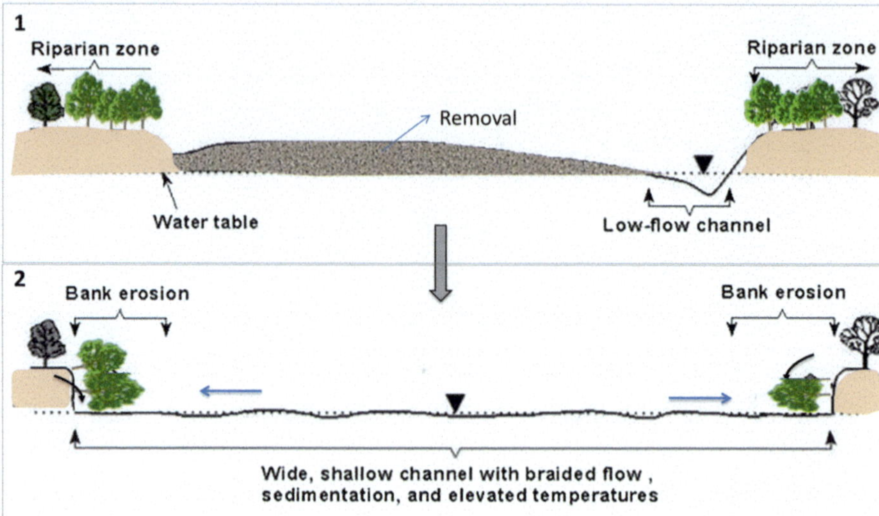

Fig. 10.7 Channel cross sections (1) with a typical sand-gravel bar in relation to the low-flow channel, riparian zone, and water table and (2) the wide shallow channel as a result of mining characterized by bank erosion, braided flow, and sedimentation

shallowing of the streambed (Fig. 10.7) as deep pools fill with gravel and other sediments. Therefore, shallowing and widening of the channel increase channel instability and transport of sediments downstream.

Instream mining lowers the stream bottom, which may lead to bank erosion. Similarly depletion of sand in the streambed and along coastal areas causes the deepening of rivers and estuaries and the enlargement of river mouths and coastal inlets. It may also lead to saline water intrusion from the nearby sea. The effect of mining is compounded by the effect of sea level rise.

10.4 Environmental Impacts of Sand Mining

Sand and gravel had been used in the construction of buildings, roads, and public utilities since the beginning of settled human civilizations. But, extraction of these materials was much below the natural replenishments so that natural processes and biodiversity were not affected. In contrast, since the beginning of the twentieth century, demand for sand and gravel has been on the rise. In order to meet the demand, sand mining has increased in amount and extent. This has seriously affected the environmental quality and many natural processes.

In general the environmental impact of mining includes erosion, formation of sinkholes, loss of biodiversity, and contamination of soil, groundwater, and surface water by chemicals from mining (Bush et al. 1995). Excessive instream sand-and-gravel mining causes the degradation of rivers. Sand mining generates extra vehicle traffic which also negatively impairs the environment. Besides creating environmental damage, the contamination resulting from leakage of chemicals also affects the health of the local population. For example, Moroccan coast has a very large coastal sand body resupplied by regular erosion of cliffs and episodic injections from rivers in flood. Sand mining completely damaged the coastal environment (Pitchaiah 2017). In the following subsections, some of the most important impacts are discussed in details.

10.4.1 Habitat Loss

Every species requires specific habitat conditions besides suitable climate for their long-term survival. Native species from the coastal areas including estuaries and streams are uniquely adapted to the habitat conditions that existed before humans began large-scale alterations. With the progress of time, such adaptation resulted in endemism. However, sand mining and other human activities in many of the coastal countries have significantly affected the habitability of these areas which include both aquatic and terrestrial habitats. In this context, habitat destruction is defined as the process by which natural habitat is rendered incapable of supporting its native species. In this process, the organisms that previously used the site are displaced or destroyed, reducing biodiversity (Pitchaiah 2017). Sometimes such modified habitat may favour the establishment of invasive species affecting the composition and survival of native species.

Habitat disruptions have favoured some species over others and caused overall declines in biological diversity and productivity in some of the tropical coastal countries where sand mining is rampant. In most of the streams and rivers, habitat quality is strongly linked to the stability of channel bed and banks. As discussed earlier, the stability of sand-bed and gravel-bed streams depends on a delicate balance between stream flow, sediment supplied from the watershed, and channel form. But the changes in sediment supply and channel form triggered by sand mining disrupt habitat development processes. In other words bed degradation and sedimentation cause substantial negative effects on aquatic life. Furthermore, movement of unstable substrates results in downstream sedimentation of habitats which depends on the intensity of mining, particle size, stream flow, status of surrounding vegetation, and channel morphology. Channel reaches become more uniformly shallow as deep pools fill with gravel and other sediments, reducing habitat complexity, riffle-pool structure, and numbers of large predatory fishes. In contrast, channel widening results in shallowing of the streambed hindering movement of fishes and other aquatic organisms between the pools.

Other than sand mining, clearing of natural habitats for agricultural purpose also results in habitat destruction. The complete removal of vegetation and destruction of the soil profile or mining destroys terrestrial habitat both above and below the ground with a consequent reduction in faunal population and diversity. Other important causes of habitat destruction include logging, trawling, over-exploitation, and urban sprawl. In summary, habitat loss is viewed as a process of natural environmental change caused by habitat fragmentation, geological processes, and climate change, all of them linked to human activities. In continuation, habitat destruction is viewed as the primary cause of species extinction which is discussed in the following subsection.

10.4.2 Biodiversity

Without much doubt it has been observed that destruction of the habitats has been the major cause of biodiversity losses while direct poisoning caused by mine-extracted material and indirect poisoning through food and water also significantly affected flora, fauna, and microorganisms in several coastal countries. Habitat modification reflected in change in soil depth, pH, and temperature also disturbs communities in the area. Endemic species are more sensitive to these changes since they need very specific environmental conditions which are prevalent only in that location. Destruction or modification of the habitat condition puts them at the risk of extinction.

As discussed earlier, sand mining at large scale created serious environmental hazard affecting beaches, creeks, and rivers all over the world. Rampant sand mining destroys environment and biodiversity besides creating water insecurity and land erosion (Bush et al. 1995). Developing countries in the tropics, in general, consider sand as a free and abundant natural resource which is practically inexhaustible as they surge towards development. Sand is necessary for construction and drives the construction boom on which rapid development is dependent, but most often mafias form around the illegal sand mining trade. Sand is commonly mined from sand dunes, river beds and river banks, in creeks, beaches, and on the sea bed. These locations are particularly sensitive to environmental damage and support highly biodiverse environment which are linked to wetlands, sand dunes, coral reefs, and beaches. Both marine and terrestrial flora and fauna are affected by mining, and it also affects recreational, fishing, and agricultural activities. Further, nesting areas of birds are also affected by sand mining, destroying the natural habitat on which migratory and other birds are dependent (CBD 2014). Some of the very specific effects on different biological components are discussed below:

- Soil texture and water quality are significantly affected in the mined sites with the consequential impact on the plant communities. Most of the plants have a low concentration tolerance for metals in the soil, but sensitivity differs among species. In the mined areas, grass diversity is less affected by high contaminant concentration than shrubs and herbs. In this context, some species are more resistant and will survive these levels, and some non-native species that can tolerate these concentrations in the soil will migrate in the mine surrounding lands to occupy the ecological niche. Similarly most of the cultivated crops can grow on weakly contaminated sites, but yield is generally lower than it would have been in regular growing conditions. Plants also tend to accumulate toxic materials (heavy metals) in their aerial organs, and sometimes it enters into human food chain through fruits and vegetables (Chae and Iain 1996).

- Mining and other related activities alter/change pH, CO_2 level, salinity, and temperature which significantly affect the coral reef environment. Sometimes coastal sand mining destroys the ecological status/structure of coral reefs (Supriharyono 2004).
- The mining industry can impact aquatic biodiversity through direct poisoning due to increased bioavailability of some of the toxicants (Daniel et al. 2002; Georg et al. 2009). Contaminants can also affect aquatic organisms with high concentrations of suspended sediment and reduced sunlight penetration. Metal oxide deposition can limit biomass by coating algae or their substrate, thereby preventing colonization. Further, it will cause biomagnification through food chain.
- Algae communities are less diverse in acidic water containing high zinc concentration while mine drainage stress decreases the algal primary production. Diatoms and microbial community are greatly modified by any chemical change (Daniel et al. 2002).
- In many mined areas, it has been observed that water insect and crustacean communities are modified around mined sites resulting in a low trophic completeness and community being dominated by predators. Similarly fishes are also affected by high pH, temperature variations, and chemical concentrations resulting from mining close to the water bodies (Gerhardt et al. 2004).

10.4.3 Effect on Groundwater

It is often seen that uncontrolled sand mining transforms the river beds into large pits. Consequently the groundwater table drops leaving the drinking water wells dry located on the embankments of mined rivers. We have already discussed about how instream mining lowers the elevation of stream flow and the floodplain water table. As a consequence, this can severely affect water table-dependent woody vegetation in riparian areas and may decrease wet periods in riparian wetlands. Instream mining lowers the stream bottom that results in accelerated bank erosion. Further, mining of sand close to the sea favours intrusion of saline water into coastal freshwater body causing contamination. In other words, depletion of sand in the streambed and along coastal areas causes deepening of rivers and estuaries and the enlargement of river mouths and coastal inlets. Consequently it may lead to saline water intrusion from the nearby sea affecting the groundwater. The effect of mining is also compounded by the effect of sea level rise in the coastal areas and most of the tropical islands. This suggested that mining should be viewed as ecologically linked activity not in isolation.

10.4.4 Changes in Water Quality

Coastal and instream sand mining activities will have an impact upon water quality of both surface and subsurface water. The impact depends on the intensity of mining, topography, and closeness to the water bodies among other factors. The impacts include increased turbidity at the mining site due to suspension of sediment, sedimentation, seawater intrusion, oil spills, or leakage from excavation machinery and transportation vehicles. Increased river bed and bank erosion increases suspended solids in the water at the excavation site and further downstream. Suspended solids may adversely affect water users and aquatic ecosystems in the coastal areas. The impact is particularly significant if water users downstream of the site are abstracting water for domestic use. Suspended solids can significantly increase water treatment costs as well. Wastewater from the mine containing dissolved heavy metals such as lead, arsenic, cadmium, etc. may get leaked into the surroundings and result in groundwater contamination. It has been observed from several large mining sites across the tropics that water produced from mine drainage, mine cooling, aqueous extraction, and other mining processes increased the potential for these chemicals to contaminate groundwater and surface water (Chae and Iain 1996).

10.5 Management of Sand Mining

From the above discussion, it is clear that sand mining involving both onshore and offshore sites is seriously affecting the coastal ecosystem and threatens the survival of several flora and fauna. At the same time, pressure on the available sand deposits is increasing as every country strives for urbanization and coastal development (Anonymous 2018a, b). Therefore, somewhere a logical line should be drawn considering the human need and ecological consequences which also encompass future alternative use of sand deposits and the coastal environment. There are three logical approaches to this problem, viz. regulation of use of existing deposits, reclamation of mined sites, and alternative sources of sand.

10.5.1 Regulation

- As far as possible, avoid mining of relatively recent coastal sand bodies. In those areas, immediate restoration/reclamation activities should be initiated to avoid further ecological consequences such as flooding, salinization, and habitat loss.
- Gravel-rich streams are less susceptible than gravel-poor streams to disturbance from bar skimming. This is because replenishment by excess gravel from upstream sources would partially mitigate channel disruption. Mining of bars in gravel-rich streams should be preferred over mining in gravel-poor streams. Reaches that receive high rates of deposits may be better locations for mining than reaches that do not.
- Excavation of sand or gravel deposits should be limited to unconsolidated areas containing primarily smaller material (at least 85% of material is less than 3 in in diameter) that is loosely packed. The extraction sites containing any woody perennial vegetation greater than 1-in diameter should be avoided.

- Providing buffer zone in the high sand demand areas is a very essential aspect of regulation strategy. Buffers should separate the excavation area (extraction site) from the low-flow channel and the adjacent channel bank. These buffers lessen the risk of mining-induced headcuts. For example, a 20-ft-wide undisturbed buffer should be maintained between the removal area and the water line at the time of excavation and between the removal area and adjacent bank.
- Similarly in the riparian zone, a 25-ft-wide streamside riparian corridor should be maintained in an undisturbed condition landwards of the high bank for the length of the sand-and-gravel removal site; disturbed areas in this riparian zone shall be limited to access roads.
- Sand or gravel should not be excavated below the elevation of the water table at the time of removal. If the stream is dry, the excavation must not proceed beyond the lowest undisturbed elevation of the stream bottom, which is a function of local hydraulics, hydrology, and geomorphology.
- Mining activities should avoid relocation, straightening, and modification of established water conveyance channels.

10.5.2 Alternative Choices

- Coasts with sand deposits which are having low inland slopes and deposited in the geologic past when the level of the ocean was higher or lower than today offer an alternate choice as mining sites. As these older sand bodies are often no longer directly connected to modern coastal ecosystems, sand mining would be less damaging than the mining of the current coastal sand bodies that is in equilibrium with the sea.
- There are alternative sources of sand available, most of which will be at least slightly more expensive than the presently used coastal sand. Crushing of rocks is one of the best alternatives that many Caribbean Island nations have turned to.
- Explore the offshore sand deposits; identify the potential sites with very less impact on the flora and fauna for future use. However, such sites around coral and smaller islands should be avoided.
- In well-regulated mines, careful measurements of water and soil are undertaken to exclude any type of water contamination that could be caused by the mine's operations. Adequate enforcement should be there for restricting mine operators to meet standards for protecting surface and groundwater from contamination. This is best done through the use of nontoxic extraction processes as bioleaching.

10.5.3 Reclamation

- If mining continues on some limited basis, reclamation (reshaping) of the impacted landscapes must be carried out. Similarly reclamation of abandoned mining sites should be considered to restore the flora and fauna through long-term planning and support. Further details of reclamation of mined sites are provided in a separate chapter (Phytoremediation).

10.6 Conclusion

Coastal sand dunes are formed when wind erodes and transports sediments up to a beach and deposits them on an obstacle, whereas sand particles carried by the streams from the catchment areas are deposited in the river bed and estuaries. These sand dunes and river bed sand deposits play very important ecological role particularly in the coastal ecosystem. Sand dunes protect the coastal areas from the sea waves and provide habitat for large number of coastal flora and fauna including several endemic species. Several studies have proved that sand sustains the rivers and facilitates percolation of water to groundwater.

However due to steady rise in demand for sand, these ecologically valuable materials are being mined. Sand mining is rampant in the tropical coastal countries as they surge ahead in the developmental race. At several places in these countries, geologically recent sand deposits and river beds are targeted for sand extraction by employing heavy machineries. This results in negative environmental consequences often overweighing its intended use. The environmental impact of mining includes soil erosion, formation of sinkholes, loss of habitat and biodiversity, and contamination of soil, groundwater, and surface water by chemicals from mining processes. Besides creating environmental damage, the contamination resulting from leakage of chemicals also affects the health of the local population. Instream sand mining results in the destruction of aquatic and riparian habitat through large changes in the channel morphology. Degraded stream habitats result in loss of fisheries productivity, biodiversity, and recreational potential.

Therefore, without considering the importance of sandy ecosystem, commercial exploitation for short-term gains by indiscriminately mining sand deposits from the rivers, dunes, and beaches is destroying the whole environment and reducing the ecological services to the local people. In order to maintain the balance between the environment and sand mining, measures such as regulation of use of existing deposits, reclamation of mined sites, and exploring alternative sources of sand should be followed. This should be backed by appropriate policy and adequate legal mechanisms to conserve the sand deposits. At the same time, mining operators, in conjunction with cognizant resource agencies, must work to ensure that sand mining is conducted in a responsible manner.

References

Anonymous (2018a) Mining of coastal sand: a critical environmental and economic problem for Morocco. http://coastalcare.org/wp-content/pdf/morocco_mining.pdf. Accessed on 14 Aug 2018

Anonymous (2018b) Sand dunes. http://coastalcare.org/educate/sand-dunes/#how-do-sand-dunes-form. Accessed on 20 Aug 2018

Arnott DR (2010) Introduction to coastal processes and geomorphology. Cambridge University Press, Cambridge, p 442

Bush DM, Richard MT, Liboy JG, Lisbeth Hyman L, Neal WJ (1995) Living with the Puerto Rico shore. Duke University Press, Durham, p 216

Carter RWG, Woodroffe CD (1994) Coastal evolution: late quaternary shoreline morphodynamics. Cambridge University Press, Cambridge, 517 pp

CBD (2014) Effect of sand mining on biodiversity. Twelfth meeting of the conference of the parties to the convention on biological diversity, 6–17 October 2014 – Pyeongchang, Republic of Korea. https://www.cbd.int/cop2014/default.shtml. Accessed 20 July 2018

Chae JM, Iain T (1996) Heavy metals contamination of soils and plants in the vicinity of a lead-zinc mine, Korea. Appl Geochem 11:53–59

Daniel LM, Peter DS, Jeffrey SB (2002) Soil microbiological properties 20 years after surface mine reclamation: spatial analysis of reclaimed and undisturbed sites. Soil Biol Chem 34:1717–1725

Georg S, Wolfram A, Thomas L, Marianne P, Marieluise W, Irene KL, Max B (2009) Metalloid contaminated microhabitats and their biodiversity at a former antimony mining site in Schlaining, Austria. Open Environ Sci 3:26–41

Gerhardt A, Janssens de Bisthoven L, Soares AMVM (2004) Macroinvertebrate response to acid mine drainage: community metrics and on-line behavioural toxicity bioassay. Environ Pollut 130:263–274. https://doi.org/10.1016/j.envpol.2003.11.016

Mensah JV (1997) Causes and effect of coastal sand Mining in Ghana. Singap J Trop Geogr 18:69–88

Pitchaiah PS (2017) Impacts of sand mining on environment – a review. SSRG Int J Geo Inform Geol Sci 4(1):1–6. http://www.ijettjournal.org

Supriharyono (2004) Effects of sand mining on coral reefs in Riau islands. J Coast Dev 7(2):89–100

Climate Change and Its Impact on the Coastal Region

11

Abstract

Coastal zones are highly populated and among the world's most diverse and productive environments. Coastal areas include complex ecosystems such as coral reefs, mangrove, salt marshes, seagrasses, etc. Global climate change accelerated by human activities affects the physical, biological, and biogeochemical characteristics of the coastal regions. Consequently the ecological structure, their functions, and the goods and services of the coastal regions are being modified. The ecosystem resilience will be greatly reduced through human impacts as well as rising sea levels, increasing sea temperatures, and other climate ocean-related changes, including prevailing wave activity and storm waves and surges. Sea level rise and increased seawater temperatures are projected to accelerate beach erosion and cause degradation of natural coastal defences such as mangroves and coral reefs resulting in negative effect on the socio-economic aspects of coastal population. Therefore, integrated approaches are essential at various levels to manage the climate change impact on coastal region.

Keywords

Global warming · Sea level rise · Ocean acidification · Coral bleaching · Flooding

11.1 Introduction

In recent times, 'climate change' has become the most referred term in scientific reports which largely focuses on the effect of greenhouse gas emissions resulting from anthropogenic activities. In common parlance it is the observed changes in surface temperature, rainfall, evaporation, extreme events, and greenhouse gas emission since the beginning of the twentieth century (Velmurugan et al. 2015). Climate change is defined as significant and lasting change in the statistical distribution of weather parameters over periods ranging from decades to millions of years.

V. Ayyam et al., *Coastal Ecosystems of the Tropics - Adaptive Management*,
https://doi.org/10.1007/978-981-13-8926-9_11

In other words it may be a change in average weather conditions or the distribution of events around that average. Scientific studies have revealed that climate change may be limited to a specific region or occur across the whole Earth. This is not a new phenomenon because climate of the Earth was never static; it had been subjected to changes throughout the geological times. Studies on climatic history of the Earth revealed that the influence of natural forces on climate and the associated events were greater. Therefore, climate change is defined as change of climate which is attributed directly or indirectly to human activities that alter the composition of the global atmosphere often referred as global warming which is in addition to natural climate variability observed over comparable time periods (UNFCC 1992).

Although climate change is a global phenomenon, global warming is not spatially uniform across the globe. At the same time, there is almost no region in the world that has not experienced some rise in average temperature or change in rainfall pattern. Analysis of global precipitation pattern showed variations and some notable trends in recent decades. The Third Assessment Report of IPCC (2001) also found much of the rainfall variability appeared to be closely related to ENSO events, combined with seasonal and decadal changes in the convergence zones. Such increase in mean temperature and sea level rise is of great concern for the coastal and small island region. In order to manage or minimize the negative impacts, proper adaptation measures are required which should be based on the observational record of climate parameters and projections relevant to the region. Focus should also be paid to the mitigation measures as it is linked with the climate change and socio-economic development of this region (Fig. 11.1). In other words the adaptation measures are regional in nature but climate change is a global phenomenon.

Analysis of observational data showed a global mean temperature increase of around 0.6 °C during the twentieth century. However, there are regional differences as the rate of increase in the Pacific and Caribbean exceeded the global average. Besides temperature, mean sea level rose by about 2 mm/year which is of great

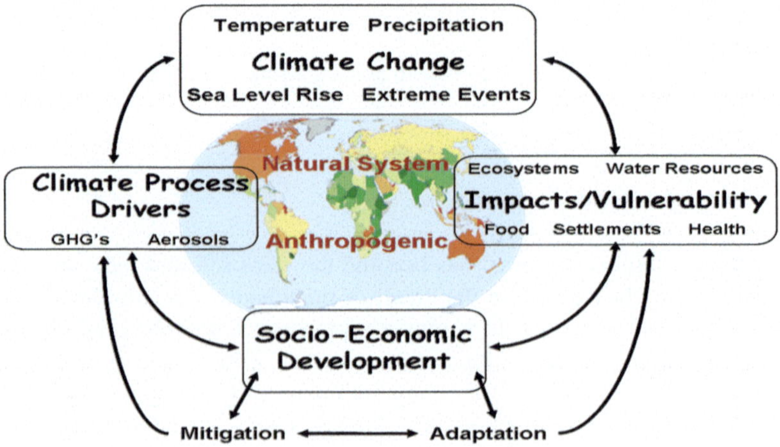

Fig. 11.1 Interlink between climate change drivers, responses, and development

concern because it triggers other consequential events and affects the human well-being. The reanalysis of temperatures averaged over land and ocean surfaces showed consistent warming trends in all small island and coastal regions over the 1901–2004 period (Trenberth et al. 2007) which invited the attention of climatologists world over. Owing to their high vulnerability and low adaptive capacity, small islands and the coastal region have legitimate concerns about their future, based on observational records, experience with current patterns and consequences of climate variability, and climate model projections (Mimura et al. 2007).

In this chapter we discuss the climate change and its effect on the tropical coastal region. The coastal region, where land meets ocean, is one of the most dynamic natural systems. Here, the three main components of our planet—the hydrosphere, the lithosphere, and the atmosphere—meet and interact, forming interconnected systems. In this chapter the definition of coastal system (region) is adopted from IPCC Fourth Assessment Report (2007a). As per the report, coastal systems are considered as the interacting low-lying areas and shallow coastal water including their human components. It is evident that coasts are of great ecological and socio-economic importance. They sustain economies and provide livelihoods through fisheries, ports, tourism, and other industries. They also provide ecosystem services such as regulating atmospheric composition, cycling of nutrients and water, and waste removal (Curran 2002).

Climate is one of the two most important physical factors the other being topography determining the survival and nature of all living beings, from individuals and communities to populations and entire species, by heavily influencing the natural systems on which they depend. Therefore normal functioning of the ecosystem, here we discuss of coastal ecosystem, mainly depends on the climate; any significant variations will have consequences. There may be species loss if flora and fauna cannot adapt to climate change or migrate to suitable habitats fast enough. The species composition of communities and ecosystem services will be affected, which in turn risks severely reducing productive functions and diminishing ecosystem resilience. In addition, the various effects on marine biotic populations are likely to modify the web of interactions at the community level (Gilman et al. 2010). In other words the response of some species to climate change may constitute an indirect impact on the species that depend on them. In addition, for many species, the primary impact of climate change may be mediated through effects on synchrony with species food and habitat requirements.

At the same time, there could be variations/deviations from the projected climate change at the regional level. Regional climate depends on the temperature, precipitation, and winds experienced over the long term at that location. These characteristics are determined by other factors, including the latitude and altitude of the region, its topography, large-scale atmospheric and oceanic circulation patterns, and the region's proximity to large bodies of water. It is also worth mentioning here that sea level rise and associated impacts are complicated by local tectonics and El Niño-Southern Oscillation (ENSO) events which originate in the sea. The following subsections discuss all these aspects in details.

11.2 Climate Change and the Global Scenario

11.2.1 Climate Change: The Dichotomy

In the last two decades, scientific community across the globe has been engaged in debate on global climate change, while other stakeholders are concerned about its possible effects. In this context, two of the most important effects of climate change on coastal and ocean ecosystems are increasing temperatures and sea level rise as a result of increasing greenhouse gas emissions (Freely et al. 2004). The dichotomy is about the source of these gases. One school of thought put the blame on human activities, and the other articulate that it is a natural process linked with the evolution of Earth and still continuing. Nonetheless, it is observed that the oceans act as a significant temperature sink (Levitus et al. 2005), absorbing an estimated 80% of the additional heat generated by global warming (IPCC 2007). This has led to an increase of 0.1 °C in the average temperature of the upper 700 m of the oceans over the last 40 years. Further, the rate of sea temperature rise is predicted to increase over the next decades, with the estimated rise in the mean surface temperature for the end of the twenty-first century lying between 1.1 and 6.4 °C (IPCC 2007).

According to the conclusion at the Conference of the Parties to the UNFCCC (1/CP.10) and the Intergovernmental Panel on Climate Change (IPCC 2007), the impact of the climate change in the coming decades will affect billions of human beings, especially in developing countries, potable water, and food (UNFCCC 2007). The effect of temperature rise based on the year 2000 data suggested that Earth may warm to 3 °C in the next three decades, i.e. in 2030. As a result people from the developing world particularly in coastal and island nations will face increase of hunger, climatic sensitive diseases, and 20–30% of risk of extinction of flora and fauna in the sensitive areas, and some or portion of the small island states may get submerged due to sea level rise. Over and above, the present scenario suggested that as the climate change and its impact are inevitable which forced the think tankers all over the world and from the vulnerable areas to have elaborate planning and assess its future financial implications. Concerned with these climate change burden, developing nations put the onus on the developed world to provide aid for conserving the environment, while the western countries are suggesting for green technology-driven adaptation measures.

On the other hand, looking through the history of the Earth, it is suggested that climate change may be a cyclic effect happening in the environment. Anthropogenic greenhouse emission is not the only factor responsible for global climate change. The contribution of GHG by anthropogenic activities is insignificant when compared with the natural emissions. The time frame of observation was not aware of the individual human lifetime. Further, in the earlier time frames from the Earth history, it is suggested that the so-called intellectual humans may witness this effect for the first time. Therefore, it is highly essential to record these events in each and every stage to see how the climatic change affected in micro level and what kind of impact are produced in the environment. Attention and efforts should focus on adaptive measures to face the emerging challenges. If this information and measures have not evolved, humans may lose in-depth knowledge about the genetic variations, environment impact of genetic, biodiversity, extinction of fauna and flora, and evolution of fauna and flora forever.

11.2.2 Global Observations

There is broad consensus that contemporary global climate change is reality and that much of the ongoing change is a direct result of human activity (IPCC 2007; Hansen et al. 2007). In particular, burning fossil fuels and changing land use have driven atmospheric carbon dioxide concentrations (CO_2[atm]) up from a pre-industrial value of about 280 parts per million (ppm) to 385 ppm in 2008 (Meure et al. 2006). Annual increases are now exceeding 2 ppm emission trend that is excess of the worst-case scenario of the Intergovernmental Panel on Climate Change (IPCC 2007b). Direct link between global temperature and CO_2 [atm] was established by several studies because of this CO_2 released into the atmosphere has become a matter of serious concern (IPCC 2007). The increased heating in the lower atmosphere/ Earth's surface (radiative forcing) resulting from the 'greenhouse' effect caused by increasing atmospheric CO_2, methane, and other gasses (at a value of about 3 W m^{-2}) is unprecedented in at least the last 22,000 years (Joos and Spahni 2008). It has already had direct physical consequences for the marine and coastal environment and organisms living there. These include increase in mean global sea surface temperature, by 0.13 °C per decade since 1979, and ocean interior temperature, by >0.1 °C since 1961; increasing wind velocity and storm frequency; changes in ocean circulation, vertical structure, and nutrient loads (IPCC 2007b); as well as rising sea level—by more than 15 cm in the last century (Rahmstorf 2007) and presently by a mean of about 3.3 mm per year (Fig. 11.2).

It is well established that the oceanic and atmospheric gas concentrations tend towards equilibrium, and hence increase in CO_2 [atm] drives more CO_2 into the ocean, where it dissolves (water) forming carbonic acid (H_2CO_3). This in turn increases ocean acidity (pH). Observations indicated that in the last 200 years, on an average ocean pH has dropped by 0.1 (30% increase in hydrogen+ ion concentration) (The Royal Society 2005). It has been suggested that a CO_2 [atm] of 450 ppm is a critical threshold beyond which catastrophic and irreversible change might occur; this would bring a global mean temperature rise of 2 °C above pre-industrial values. Similarly the associated thermal expansion and sea level rise could severely erode or submerge large coastal tracts and small islands in the tropics. At present rates, this threshold will be passed by 2040, but climate-related systems are notoriously nonlinear (Lenton et al. 2008). By 2040, some particularly sensitive marine ecosystems such as coral reefs and its associated fauna could get destroyed or may be lost, and other unexpected consequences may arise (Hester et al. 2008). However, long-term global climatic data observation networking and its analysis are still lacking which needs international collaborations. At the same time, consistency in observations and data collection should improve and should compare with other regional level observations.

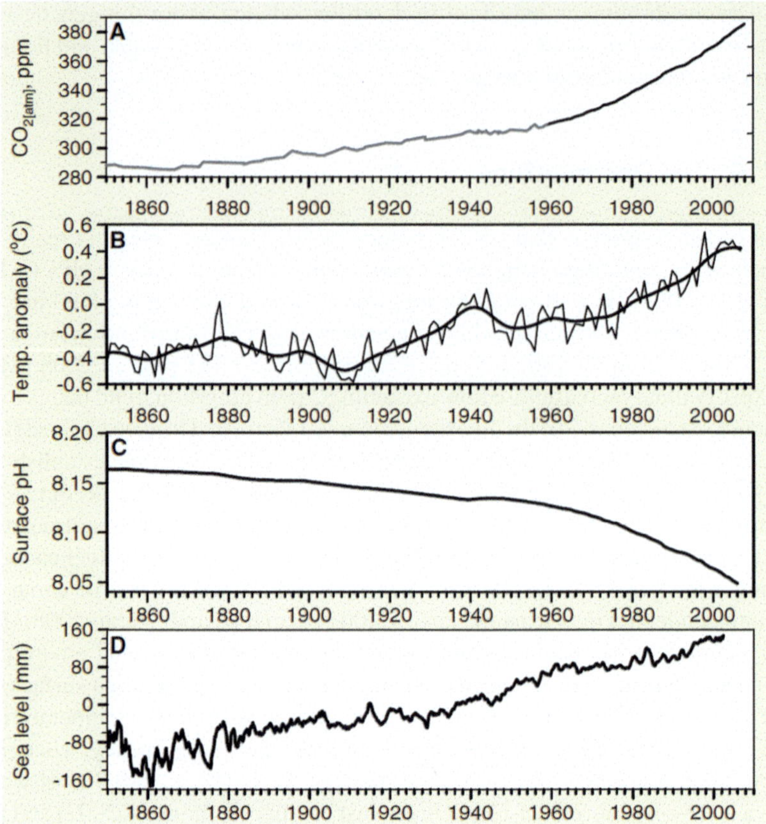

Fig. 11.2 Indices of physical climate change relevant to coastal/marine systems. (**a**) CO_2 [atm] from Law Dome ice cores (grey line) and Mauna Loa direct observations (black line [www.esrl.noaa.gov/gmd/ccgg/ trends/]). (**b**) Annual and smoothed combined global land and marine surface temperature anomaly (http://www.cru.uea.ac.uk/cru/info/ warming/). (**c**) Ocean surface pH calculated from CO_2 [atm]. (**d**) Relative global sea level. (Jevrejeva et al. 2006)

11.2.3 Future Climate Projections

Global climate models also known as general circulation models (GCMs) are the most complex of climate models, since they attempt to represent the main components of the climate system in three dimensions. GCM uses hundreds of mathematical equations to describe processes that happen on our planet, viz. wind, temperature anomaly, ocean currents, and plant growth. GCM is a good tool to perform climate change experiments from which climate change scenarios can be constructed (Fig. 11.3), for example, how wind patterns affect the transport of storms from ocean to coastal and across regions, relations between ocean currents and the amount of heat in the atmosphere, plant growth, and the amount of carbon dioxide in the atmosphere.

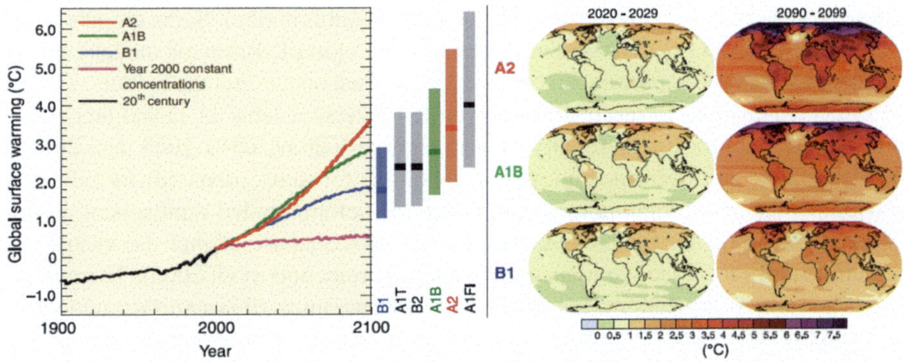

Fig. 11.3 Predicted changes in surface warming. **Left panel**: Solid lines are multi-model global averages of surface warming (relative to 1980–1999) for the SRES scenarios A2, A1B, and B1, shown as continuations of the twentieth-century simulations. The orange line is for the experiment where concentrations were held constant at year 2000 values. The bars in the middle of the figure indicate the best estimate (solid line within each bar) and the likely range assessed for the six SRES marker scenarios at 2090–2099 relative to 1980–1999. The assessment of the best estimate and likely ranges in the bars includes the atmosphere-ocean general circulation models (AOGCMs) in the left part of the figure, as well as results from a hierarchy of independent models and observational constraints. **Right panels**: Projected surface temperature changes for the early and late twenty-first century relative to the period 1980–1999. The panels show the multi-AOGCM average projections for the A2 (top), A1B (middle), and B1 (bottom) SRES scenarios averaged over decades 2020–2029 (left) and 2090–2099 (right). {WGI 10.4, 10.8, Figures 10.28, 10.29, SPM}. (Ref: IPCC 2007)

The performance of individual GCMs may differ for individual climate variables as well as for different regions of the world. Typically, GCMs are validated for their ability to reproduce spatial patterns (Huth 1997) of selected variables and their annual cycles (Nemesova et al. 1998). A prerequisite for the inclusion of a model in climate projections is that it adequately simulates present-day climate conditions. Therefore, statistical methods were used to test the fidelity with which individual models simulate observed climate conditions over the Asia/Pacific region. Two statistical metrics, pattern correlation and root mean square error, were used to measure model performance for the AR4 report. A pattern correlation coefficient of 1.0 indicates a perfect match.

Some of the climatic models are very large and contain so many equations and are so complex that they need to be run on supercomputers. Usually this is done to make climate predictions for one or more centuries into the future; sometimes models are run backwards to find out how climate may have changed in the past. Global climate models represent how natural processes of our planet work using an imaginary three-dimensional grid. The grid isn't really there but it is part of the model. It covers the surface of the modelled Earth and extends upwards in layers through the modelled atmosphere. At every intersection in the model's grid, the model makes its calculations. Small processes that happen between a model's grid points cannot be 'seen' in the model results, but they can be described in other ways. Some models have wide spacing in the grid. This means that there are fewer points where the model calculates.

These models run faster and are useful when less detail is needed. Some models have very closely spaced grid points. These are much more detailed climatic models, and it can take a long time for the supercomputer, even a fast one, to run the model.

Projected changes in seasonal surface air temperature (Table 11.1) and precipitation (Table 11.2) for the three 30-year periods (2010–2039, 2040–2069, and 2070–2099) relative to the baseline period 1961–1990 were prepared for all the subcontinental scale regions of the world using seven coupled atmosphere-ocean general circulation models (AOGCMs). All the models projected increase in surface air temperature for all the regions under study (Ruosteenoja et al. 2003). The model prediction-based projections suggested a gradual warming of sea surface temperature (SSTs) and a general warming trend in surface air temperature in all the regions and seasons (Lal et al. 2002). Regarding precipitation, the models simulated only a marginal increase or decrease (10%) in annual rainfall over most of the small islands in the South Pacific. During summer, more rainfall is projected, while an increase in daily rainfall intensity, causing more frequent heavier rainfall events, is also likely (Lal 2004). Most importantly the projected globally averaged sea level rise at the end of the twenty-first century (2090–2099), relative to 1980–1999 for the six SRES scenarios, ranges from 0.19 to 0.58 m (Meehl et al. 2007). In all SRES scenarios, the average rate of sea level rise during the twenty-first century exceeds the 1961–2003 average rate (1.8 ± 0.5 mm/year). Climate models also indicated a geographical variation of sea level rise due to non-uniform distribution of temperature and salinity and changes in ocean circulation. Further, it is likely that maximum tropical cyclone wind intensities could increase, by 5–10% by around 2050 (Walsh 2004). Under this scenario, peak precipitation rates are likely to increase by 25% as a result of increases in maximum tropical cyclone wind intensities, which in turn cause higher storm

Table 11.1 Projected increases in air temperature (°C) by region, relative to the 1961–1990 period

Region	2010–2039	2040–2069	2040–2069
Mediterranean	0.60–2.19	0.81–3.85	1.20–7.07
Caribbean	0.48–1.06	0.79–2.45	0.94–4.18
Indian Ocean	0.51–0.98	0.84–2.10	1.05–3.77
Northern Pacific	0.49–1.13	0.81–2.48	1.00–4.17
Southern Pacific	0.45–0.82	0.80–1.79	0.99–3.11

Table 11.2 Projected changes in precipitation (%) by region, relative to the 1961–1990 period

Region	2010–2039	2040–2069	2040–2069
Mediterranean	−35.6 to +55.1	−52.6 to +38.3	−61.0 to +6.2
Caribbean	−14.2 to +13.7	−36.3 to +34.2	−49.3 to +28.9
Indian Ocean	−5.4 to +6.0	−6.9 to+12.4	−9.8 to +14.7
Northern Pacific	−6.3 to +9.1	−19.2 to +21.3	−2.7 to +25.8
Southern Pacific	−3. 9 to +3.4	−8.23 to +6.7	−14.0 to +14.6

surges. However, changes in tropical cyclone tracks are closely associated with ENSO and other local climate conditions. All these projections together will have profound consequences on the coastal and marine ecosystems in the tropics.

11.3 Climate Change Impacts and Physical Environment

11.3.1 An Overview

Coastal habitats are very unique, and its ecosystems are some of the richest areas for biodiversity. Coastal ecosystems provide important environmental, social, and economic service functions besides livelihood to millions of people all over the world. Even small changes associated with climate change can have consequential implications in some of the physical component/compartments of the coastal region posing greater challenges to the normal functioning of coastal ecosystems particularly if these changes take place very rapidly. A summary of climate change events and its potential impacts on physical environment, ecosystem services, and biological system as individual and community of coastal region is summarized in Table 11.3. However, owing to the great diversity of coastal environments; regional and local differences in projected relative sea level and climate changes; and differences in

Table 11.3 Summary of climate change impacts on coastal region

S.No.	Climate change events	Impact on		
		Physical environment	Ecosystem functions and services	Individuals/ communities
	Primary events	Coastal erosion	Provisioning	Coral reefs
1	Air temperature	Submersion/flooding	Regulating	Reef biodiversity
2	Sea surface temperature	Changes in coastal geomorphology	Sociocultural	Seagrass beds
3	Sea level rise	Nutrient loading	Biogeochemical	Mangroves and associates
4	Rainfall pattern	Salinization of land and groundwater	Agro-ecosystem production	Coastal wetlands
5	Ocean acidification	Deposition/ sedimentation		Littoral flora
		Sediment transport		Rock shore fauna
	Associated/ linked events	Increase in invasive species		Coastal fauna
6	Wind pattern			Agricultural crops
7	Ocean currents			Domestic animals
8	Humidity			Human wellbeing
9	Salinity distribution			
10	Upwelling			
11	Extreme events			

the resilience and adaptive capacity of ecosystems, sectors, and countries, the impacts summarized here will be highly variable in time and space.

The main climatic factors linked with the global warming which are very much relevant to the coastal ecosystems are temperature and rainfall pattern. Similarly the consequential parameters are sea level rise and ocean acidification. Apart from this ocean currents, wind pattern, humidity, salinity distribution, upwelling in the sea, and occurrence of extreme events affect the physical environment of coastal region. All these factors together affect the normal functioning of coastal ecosystems; besides these factors affect fundamental physiological processes and the survival of living systems.

There may be exceptions and deviations at regional level from these global level observations. At regional scale organisms and the environmental factors that determine their survival may be distributed as gradients that blend into one another at the edges of the space they occupy. Despite the common types of stresses exerted on coastal regions globally, the specific aspects of particular ecosystems should not be ignored, even as microclimates are exceptions to the effects of climate systems over very large scales. Further, scientific communities around the world are not fully aware of the interactions of ecosystems with their surroundings, especially in coastal and marine environments. Because often it was experienced that many indirect effects of global change on ecosystem functioning will likely reveal themselves as gradual impairments rather than as readily apparent losses of ecosystem integrity.

11.3.2 Physical Environment

As the physical components of climate change respond as linked system, while analysing the climate change effects on coastal environment, many other measures have to be used to study and monitor the consequences of climate change in oceans and coastal areas. This include sea levels, sea surface temperatures, intrusion of saltwater into fresh water, rates of evapotranspiration from oceans and other water bodies, snow and ice cover, and changes to polar ice sheets and glaciers. As the ecosystem is composed of both living and nonliving systems, it often exhibit nonlinear responses brought on by crossing thresholds that alter their composition and key processes—changes that in turn affect ecosystem stability, resilience, and functions. Once some critical threshold is passed, even relatively small stresses may trigger rapid ecosystem degradation and loss of integrity. Climate change may very likely be the factor pushing some ecosystems over the limit.

The IPCC Fourth Assessment Report states that coasts are highly vulnerable to cyclones, extreme waves, storm surges, altered rainfall and runoff patterns, and ocean acidification (Nicholls et al. 1999). Climate change contributes to accelerated sea level rise by the thermal expansion of near-surface ocean waters and the increased melting of glaciers and ice sheets. Sea level rise in turn has numerous impacts. Flooding of coasts, estuaries, and river deltas can alter the physical structure of habitats and lower habitat availability and suitability, compromising the biota on which organisms higher up on food webs depend. Increased coastal erosion can reduce or remove beach areas and protective barrier islands and interfere with nearshore currents and their physical transport patterns. Subsequent changes in drainage and irrigation patterns and modifications of fluvial flows can reroute sediment transport nutrient runoff into coastal waters. Saltwater intrusion into coastal wetlands, especially estuaries, can negatively affect these ecosystems and contaminate groundwater and other inland freshwater sources. This is a major concern for human wellbeing and agro-ecosystem productivity. Higher sea levels can also intensify storm surges. Global warming will also increase sea surface temperatures as surface waters absorb heat from higher air temperatures. Such altered temperature regimes can significantly affect the reproduction and survival of species unless they can adapt quickly enough (Nicholls et al. 2008).

From the available empirical data strong regional and local variations in the impacts of climate change on coasts has also been observed as a result of non-climate factors which are linked to global warming. Without any doubt increase in sea level rise is consistent with global warming. However, the projected inundation and impacts would not be uniform throughout the coastal areas in the tropics, because local sea level changes depend on both changes in sea surface height and land surface height. Tectonic uplift and land uplift due to postglacial rebound reduce local sea level rise, while local subsidence will accelerate it. In combination with weather extremes, changes in LSL could modify the pattern of coastal inundation with severe impacts on sensitive coastal ecosystems and habitats.

Increase in LSL would also increase the destructive force of coastal storms and wave action leading to increased coastal erosion and an acceleration of ecosystem losses. As a consequence, the total shoreline retreat could be much larger than the model projection based on the present-day topography. For example, 1 m rise in LSR would inundate 17% of the coastal areas in Bangladesh, 20% farmland in Bangladesh, more than half of Maldives, and 15–20% of area of Polynesian islands. Coastal inundation pattern leads to both the conversion of dry land to wetland and the conversion of wetlands to open water, thus destroying coastal habitats or forcing the ecosystems to migrate inland. However, human interference and managed ecosystems often inhibit migration resulting in loss of valuable ecosystems. Further, such alteration of biological system may result in change in pest equilibrium which may result in build-up of base population which affects crops, human, and livestock when the situation turns highly favourable.

11.4 Impacts on Coastal Ecosystems

11.4.1 General Nature

Climate change events such as increase in surface temperature, sea level rise, changes in rainfall pattern, etc. interact with and exacerbate the impacts of already existing environmental and human drivers on the coastal ecosystems. The consequences are felt at the region (coastal), ecosystem, and individual/community level which are positioned in a hierarchical manner (Fig. 11.4).

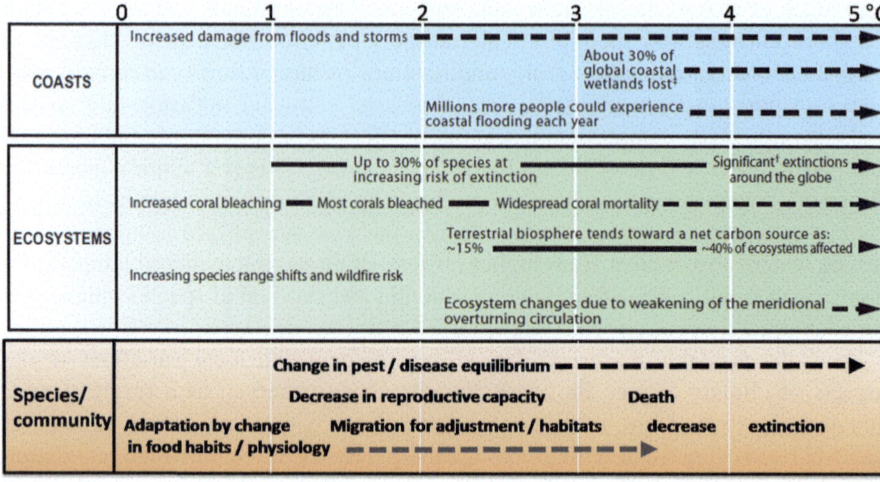

Fig. 11.4 Examples of impacts associated with global average temperature change (relative to 1980–1999 °C)

- Temperature increase, change in rainfall patterns, drought, sea level rise, coastal flooding, damage from cyclones, and change in salinity can affect habitat suitability for an individual species and communities. There will be species loss if flora and fauna cannot adapt to climate change or migrate to suitable habitats fast enough. The species composition of communities and ecosystem services will be affected, which in turn risks severely reducing productive functions and diminishing ecosystem resilience.
- With reference to the impacts on the biological system, it is felt at all levels of biodiversity, viz. genes, species, habitats, and ecosystems across the globe.
- The implications for humans include negative impacts on food security, spread of diseases and vectors into new regions, and loss of productive agricultural land and water bodies. Further this will also cause changes in seasonal cycles, plant and animal life cycle variations, and changes in essential environmental services.

Oceans absorb more CO_2 from the atmosphere when the concentration of CO_2 is more than the equilibrium state which results in ocean acidification. Approximately a quarter of all the CO_2 released into the atmosphere by human activities is absorbed by the oceans (Canadell 2007). This process creates carbonic acid in the sea water as shown in Fig. 11.5. Ocean acidification has a corrosive effect on the corals and shelled organisms by dissolving calcium from their body shells that form the basis of marine and coastal food webs. In a glimpse of the potential future implications of ocean acidification, coastal and marine locations with naturally occurring high CO_2 levels showed a high number of invasive species and low biodiversity (Orr 2009).

At the same time, it can be argued that populations and species can respond in different ways to large-scale environmental changes. It is unlikely that the response

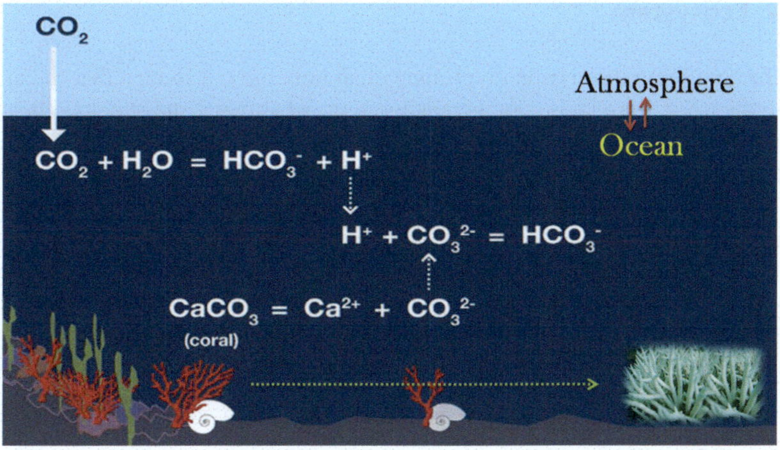

Fig. 11.5 Ocean acidification and its impacts on corals

of biological system would be always negative. Species may benefit, remain unchanged, decline in numbers, or become extinct due to climate change.

- It may not face migration, but they may continue in the original distribution range, adapt, and disperse to new regions.
- Similarly changes in the physiology or survivorship and reproduction of a species may affect any other species—including humans—that depend on it. As species respond individually and collectively, chain reactions will take place in the food webs of coastal and marine areas.
- In the same context, pressures from human activities can bring about dramatic changes on several types of ecosystems, including coral reefs, kelp forests, and wetlands which go along with the climate changes (Scheffer et al. 2001).

Alternatively there may be some benefits from climate change as well. These include increased precipitation over some areas that are currently water scarce and increased production from some agro-ecosystems which also benefit from carbon fertilization. Some species that are currently restricted by temperature regimes may expand their range. Growing and breeding seasons for species of the higher latitudes may be longer. However, a climate variable that favours one species or ecological community may be detrimental to another at that time or later. These facts suggest that most of the regions will likely experience the impact of climate change, which means both positive and negative effects will likely take place in the same large areas depending on the biogeographical and physical conditions besides other factors. The magnitude of any positive effects will depend on the ways species are able to adapt to the changes, whereas the possibility of unforeseen negative consequences are derived from limited observations or model predictions, but largely the effects will unfold only with the progression of time.

11.4.2 Ecosystem

From the foregoing discussion on climate change impact, it is seen that the coastal areas in the tropics are facing an increasing range of stresses and shocks. The insufficient mitigation measures, unfolding variations in the climatic parameters, and non-climate related threats to the ecosystem and the magnitude of them now pose a threat to the resilience of ecosystem including humans which are likely to be exacerbated by climate change in the future. IPCC (1996) have identified the areas of greatest sensitivity to accelerated sea level rise and storms. Summary of predicted/observed impacts of climate change on different coastal ecosystems are given in Table 11.4. These areas are low-elevation coral atolls and reef islands, as well as low-lying deltaic, coastal plain, and barrier coasts, including sandy beaches, coastal wetlands, estuaries, and lagoons.

11.4.2.1 Coral Reefs

Coral reefs form unique ecosystem which occur in a variety of fringing, barrier, and atoll settings throughout the tropical and subtropical world. Coral reefs support nearly 25% of all marine species, and they are among the world's most diverse and productive ecosystems (Mulhall 2009). Yet, they are among the most threatened and high-profile ecosystem victims of climate change in the oceans. Coral reefs occur mainly in relatively nutrient-poor tropical waters, but because of very efficient nutrient cycling, complex biological interactions, and energy transfer through the system, their ecological productivity is high. Reefs provide food to humans, habitat to plants and animals, and protection of coasts from storm and wave damage by acting as natural breakwaters. They are also important attraction for tourism, particularly in developing countries.

The major climatic change impact on this unique ecosystem is coral bleaching (a process attributed to increasing ocean temperature and ocean acidification), while corals are already threatened by over-exploitation and coastal development. Coral bleaching occurs due to the death of zooxanthellae that live within the coral in a symbiotic arrangement which is sensitive to the rise in sea surface temperature. Over 34% of the vast and diverse coral reefs of Asia are reported to have been lost in 1998, largely as a result of coral bleaching induced by the 1997–1998 El Niño event. In the same context, it is projected that increase in sea temperature is predicted to cause widespread coral collapse and degradation of coral ecosystem.

The other climate change-related blow to the survival of corals and its diversity is in the form of ocean acidification that is corrosive to coral growth. The process of calcium removal (corrosion) is discussed in the previous section. Recent data on ocean acidification suggest that by the middle of the next century, the process of coral calcification will decrease to the level where erosion (removal of calcium from coral due to dissolution in the acidic environment) will be greater than new growth. The coral reef ecosystem is also severely affected by high rate of sedimentation and increase in sea level. All these factors combined will affect the reach of sunlight to the coral base and cause bleaching. The 2008 report of the Global Coral Reef Monitoring Network sums up the status, threats, and trends concerning the coral

Table 11.4 Effect of climate change on coastal ecosystem

Sl. No.	Coastal ecosystem	Climate change impact	Country/region documented or studied	References
1	Deltaic coasts	Seawater intrusion into freshwater aquifers due to rising sea level	Thailand, arid Israel, Vietnam, South China, Indian Ocean atolls	Moore (1999), Melloul and Goldberg (1997), Chen and Stanley (1998), Singh and Gupta (1999), and Sherif and Singh (1999)
			Nile delta	
		Land loss and submersion due to wave erosion	Nile (Egypt)	Stanley and Warne (1998)
			Ganges (India, Bangladesh)	Umitsu (1997)
		Flooding due to SLR and increased precipitation	Coastal areas of tropical Asia	IPCC (2007)
2	Beaches and barrier coasts	Retreat of gravel barrier beaches. Also related to frequency and severity storm	Coastal areas with more sediment load	Forbes et al (1997)
		Higher rainfall causing more runoff and sediment load forms gravel beach ridge	South American coast and Peru	Sandweiss et al. (1998)
		Cliff erosion increased with El Nino incidence	Pacific coasts of South America	Kaminsky et al. (1998)
3	Coastal wetlands	Mangroves will be affected due to sea level rise	Jamaica	Alleng (1998)
		Accretion of tidal marsh tracks sea level rise	Coastal region but depend on organic sediment supply	Van de Plassche et al. (1998), Nuttle et al. (1997), and Stumpf and Haines (1998)
		Variable sea level rise	Gulf of Mexico	
4	Coral reef	Coral bleaching due to global climate change and ocean acidification	Throughout the tropical region	Brown (1997), Brown et al. (2000), Hoegh-Guldberg (1999), Wilkinson (1999), and Kleypas et al. (1999)
		Increase in SST	Andaman waters	Krishnan et al. (2010)
		Reduced species diversity	Caribbean region	Edinger et al. (1998)
		Susceptible to diseases	Caribbean region	Aronson et al. (2000)
5	Seagrass beds	Increased coastal water temperatures, sea level rise, and extreme events can contribute to increased stresses	Tropical coast	Michel and Pandya (2010)
6	Rock shore	Habitat destruction and desiccation	Tropical coast and islands	Pimm (2008) and Subba Rao and Dey (2000)

(continued)

Table 11.4 (continued)

Sl. No.	Coastal ecosystem	Climate change impact	Country/region documented or studied	References
7	Littoral/ Mangroves	SLR and storm; Plate movement (associated factor) results in reduced cover, species diversity and composition	Tropical coast and islands	IPCC (2007), Pallewatta (2010), and Alongi (2008)
8	Estuaries	Multiple stress, invasive species	Global level	IPCC (2007)
9	Coastal agriculture	Salinity and waterlogging due to seawater intrusion	Tropical islands	Velmurugan et al. (2014) and IPCC (2007)
		Erosion of coastal land and salinization of groundwater	Global level	

reefs of the Indian Ocean (Wilkinson 2008). The status indicated that coral reefs located in the Indian Ocean are either highly degraded or degraded to a medium degree. Loss of habitat, sedimentation, overfishing, and pollution remain non-climatic threats for coral reefs. Reefs in South Asia were subjected to the massive bleaching event in 1998 resulting in significant reduction of coral cover but are recovering. Reefs of the western atoll chain of the Maldives and the Bar Reef of Sri Lanka showed relatively good recovery, but there is little or no recovery in other reefs near Sri Lanka and the eastern atoll chain of the Maldives. A few reefs that are highly protected as nature preserves of French territory islands in the Polynesian group which are part of coral triangle remain relatively intact.

11.4.2.2 Seagrass Beds

Seagrass beds are very unique system in the shallow waters. It is formed by unique groups of species of flowering plants that grow completely submerged in shallow coastal waters where sunlight and nutrients are enough to grow. These beds are very productive and provide a host of ecosystem services, ranging from filtering and trapping sediments to nutrient cycling. They are an important nursery habitat for innumerable fish and invertebrate species and a source of food for many coastal organisms in tropical areas. In tropical waters, seagrasses often occur in proximity to coral reefs or mangroves, suggesting interactions/link that benefits the organisms inhabiting both types of habitat. Seagrasses are considered as ‘coastal canaries’. Because they need high levels of light, they reflect changes in turbidity and water depth far more rapidly than comparable species in coastal waters (Michel and Pandya 2010). In all regions observations showed that sediment or nutrient inflows resulting from accelerated coastal erosion are the greatest stressors leading to seagrass loss and degradation. Climate change factors such as increased coastal water temperatures, sea level rise, sediment load due to accelerated erosion in the catchment area, and extreme events such as storms and storm surges can contribute to increasing these stresses on seagrass meadows.

11.4.2.3 Tidal Wetlands: Mangroves and Tidal Marshes

An estimate by Nicholls et al. (1999) suggested that by the 2080s, sea level rise could cause the loss of as much as 22% of the world's coastal wetlands. In the tropical coastal region, wetland loss is recorded primarily as a result of direct human action reinforced by climate change. Coastal wetlands are likely to respond to climate change and sea level rise through (MEA 2003) any of the three ways:

- Accretion
- Migration inland
- Habitat loss

There are two main types of tidal wetland, viz. mangroves and salt marsh, which are discussed in this chapter although serious impacts on other coastal vegetation types, including subtidal seagrasses, can be expected (Short and Neckles 1999).

Mangroves are one of the unique ecosystems occupying the seafront portion mostly along the coastal and estuaries. It provides protection to the land behind and supports wide biodiversity. Reconstructions of ancient coastal geomorphological changes indicate that mangroves can respond to shifts in sea level. Vertical accretion by sediment accumulation may allow mangroves to keep up with mounting sea levels. But where sea level rise exceeds the rate of accretion, habitat loss is inevitable. The response of tidal marshes to sea level rise is similarly affected by organic and inorganic sediment supply and the nature of the backshore environment (Brinson et al. 1995; Nuttle et al. 1997). In general, tidal marsh accretion tracks sea level rise and fluctuations in the rate of sea level rise (van de Plassche et al. 1999; Varekamp and Thomas 1998). At the same time in some areas, the current rate of marsh elevation gain is insufficient to offset relative sea level rise.

Alternatively wetland types can also migrate or shift inland. Where migration is not possible or is prevented by human interventions such as converting coastal zones to other types of land use, habitat loss will be the result. Many mangrove localities are experiencing subsidence and therefore higher relative rates of sea level rise. Other climate change drivers, such as shifts in rainfall patterns, increased CO_2 levels, and changes in storm and wave actions, will affect mangroves in different ways, sometimes positively. It is also possible that mangroves may not respond as a single entity, but that constituent species will adapt in different ways, leading to changes in distribution as well as composition of forest community and structure.

But mangroves are subjected to the destructive forces of human beings and climate change. About one third of wetlands, such as intertidal zones and floodplains, and vegetated ecosystems, such as mangroves, can be critical to the buffering of flooding and erosion arising from climate change. The world has witnessed the large-scale destruction to this ecosystem around the Indian Ocean region by the December 2004 tsunami, and such events are likely to occur in the future as well. Increase in droughts/dry days and declining precipitation in most delta regions of

Bangladesh, India, parts of China are being witnessed in recent decades. Reduction of freshwater flows and saltwater intrusion in the Indus Delta and Bangladesh have seriously compromised these regions' mangrove forests. In the Southeast Asia, these wetlands are being converted into rice paddies due to the pressure from the ever-growing populations. In the African continent, land use change from wetland to human-managed agro-ecosystem has been witnessed in large tracts.

11.4.2.4 Deltaic Coast

Deltas develop where rivers deposit more sediment at the shore than can be carried away by waves. River deltas are among the most valuable since ancient times and heavily populated and support large biodiversity. At the same time, it is highly vulnerable coastal systems because of natural processes and partly because of human-induced stresses. In addition to increasing erosion, many of the world's low-lying deltas in the coastal regions will be exposed to potential inundation due to sea level rise and subsidence of deltas. Deltas that are deteriorating as a result of sediment starvation, subsidence, and other stresses are particularly susceptible to accelerated inundation, shoreline recession, wetland deterioration, and interior land loss (Day et al. 1997).

Non-climate-related human activities such as draining for agricultural development; levee building to prevent flooding; and channelization, damming, and diking of rivers to impede sediment transfers have made deltas more vulnerable to sea level rise. Some of the sediment starvation areas include the Rhone and Ebro deltas (Jimenez and Sanchez-Arcilla 1997) and polder projects in the Ganges-Brahmaputra (Jelgersma 1996). Where local rates of subsidence and relative sea level rise are not balanced by sediment accumulation, flooding and marine processes will dominate. Sediment transport by the Nile, Indus, and Ebro Rivers has been reduced by 95% and in the Mississippi by half in the past 200 years, mostly since 1950 (Day et al. 1997). Further stress was caused by subsurface fluid withdrawals and draining of wetland soils.

11.4.2.5 Beaches and Barriers

Sea waves and tidal processes are the major forces shaping and maintaining sandy coasts which occupy about 20% of the global coastline (Bird 1993), in which a smaller proportion consists of gravel and cobble-boulder beaches and related land forms, occurring in tectonically active and high-relief regions. It is often observed that coral rubble beaches and islands are common in low-latitude reef areas. In some locations sand and gravel mix together or occur as separate barriers along the coast. They provide habitat niche to many of the flora and fauna in the coastal regions and are economic value to the human beings. Therefore, analysis of climate change impacts on the coastal zone should also include beaches and barriers of sand and gravel.

Coastal analysis showed that over the past more than ten decades, about 70% of the world's sandy shorelines have been retreating, about 20–30% have been stable, and less than 10% have been advancing. Nicholls (1998) has indicated that accelerated sea level rise and LSL in coming decades make general erosion of sandy shores more likely. Bird (1993) postulated that with global warming and its consequent sea level rise, there will be tendencies for currently eroding shorelines to erode further, stable shorelines to begin to erode, and accreting shorelines to wane or stabilize. At the same time, the local changes in coastal conditions and particularly in sediment supply may modify the trend in future. Therefore while assessing coastal response to sea level rise, coastal orientation in relation to dominant storm wind and wave approach direction forms important aspects of the analysis (Héquette et al. 1995). The sediment redistribution may lead to varying rates and/or directions of shoreline migration between zones of sediment erosion and deposition (Short et al. 2000).

11.4.3 Biogeochemical Response of Tropical Coastal Systems

Biogeochemical cycles are very important component of global circulations. It is vital for the normal functioning of ecosystems and derives ecosystem services. Global climate change could adversely affect the biogeochemistry and ecology of aquatic systems, and in combination with anthropogenic factors, the effects will be devastating. The major factors affecting biogeochemical cycles are increase in temperature, rainfall pattern which could result in higher weather rate, nutrient/sediment load, and erosion. Further the tropical coasts receive the majority of river inputs into the ocean, linked with other systems through food chain/web including human beings, and they harbour a variety of diverse ecosystems. Therefore, the biogeochemical response of coastal systems to environmental change and the interplay of natural and anthropogenic control factors in the recent geological past with an emphasis on tropical regions need to be documented.

In general, weathering of minerals is faster in low-latitude than in higher-latitude regions with a maximum in the Southeast Asia and Western Pacific region. Estimates revealed that nitrogen and phosphorus inputs into the ocean have trebled between the 1970s and 1990s due to human activities. As a consequence of increased nutrient inputs and a change in the nutrient mix, excessive algal blooms and changes in the phytoplankton community composition towards non-biomineralizing species have been observed in many regions. This has its consequential effect in food webs and biogeochemical cycles of coastal seas including the release of greenhouse gases (Jennerjahn 2012). As opposed to the trends in high-latitude regions, tropical coastal seas display increasing organic matter inputs and reduced calcification rates which have important implications for calcifying organisms and the carbon source or sink function of the coastal areas and nearshores. Particularly coral reefs which are thriving in warm tropical waters are suffering from ocean acidification. Nevertheless, they are not affected uniformly, and the sensitivity to ocean acidification may vary among coral reefs and shell species. Therefore, the prediction of future scenarios requires an improved understanding of present and past responses to environmental change with particular emphasis put on tropical regions.

11.4.4 Socio-economic Impacts

Due to various reasons, about 40% of the world's population lives within 100 km of the coast. With the increase in population density and economic activity in the coastal region, pressure on coastal ecosystems also increases resulting in shoreline alteration (Fig. 11.6). At the same time, climate change will have direct effect on the socio-economic aspects of the coastal population and indirectly through impact on coastal ecosystem services. Recent extreme events and non-climate-related hazards have highlighted many of the challenges inherent to coastal/island regions for adapting to climate change. For example, the reality of island states vulnerability is powerfully demonstrated by the devastation experienced on the Caribbean island of Grenada when Hurricane Ivan made landfall in September 2004. Similarly Indian Ocean tsunami of December 2004 which struck the coastal areas/islands situated in Indian Ocean region powerfully demonstrated the destructive power of tsunami waves.

There are lessons learnt from the previous hazards of larger magnitude, but still one constraint on successful management of climate-related risks to coastal systems is the limited ability to characterize in appropriate detail how these systems, and their constituent parts, will respond to climate change drivers and to adaptation initiatives (Finkl 2002), particularly understanding the extent to which natural coastal systems can adapt and continue to provide essential life-supporting services to the human society. In this context the socio-economic profile of each region/subregion of the coastal areas needs to be characterized to understand its vulnerability and ability to adapt. Although 40% of the people live within 100 km of coast, density, social profile, and economic status vary with countries and subregions which largely decide the adaptive capacity. It means that for the same intensity/hazard of any climatic parameters, the impact on the population is likely to vary. For

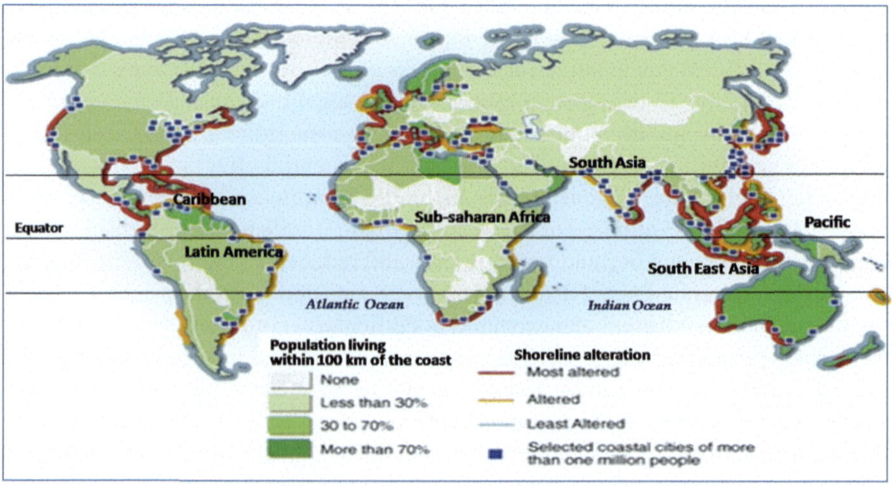

Fig. 11.6 Population living in tropical coastal areas and shoreline alteration

example, December 2004 tsunami affected settled population and their economic activity including infrastructure more than tribals of Indian Ocean Islands. At the same time, tribal faced more problems on account of food shortage and health hazards. In the same context, the impact of sea level rise and other climate change events on ecosystem services and the population benefiting it are likely to vary with the local conditions. For example, harmful algal blooms associated with increased SST and the influx of nutrients into an estuary can result in economic harm through shellfish closures, impacts on tourism, reduction of estuarine primary productivity, deterioration of fishery habitat (e.g. seagrass beds), and mortality of fish and shellfish. This manifested effect will have different socio-economic implications depending on the dependency and technical capacity of the region affected. Consequently it will impart regional character to the socio-economic impact of climate change.

As per the IPCC Common Methodology, several authors have categorized the potential socio-economic impacts of coastal regions (Bijlsma et al. 1996; Klein and Nicholls 1999) which are presented in Table 11.5. In general,

- There will be negative impacts on several sectors, including tourism, freshwater quality and supply, fisheries and aquaculture, agriculture, human settlements, financial services, and human health due to climate change.
- The number of people potentially affected by storm surge flooding is expected to double (or triple) in the next century if proper and effective adaptation measures are not taken.

Table 11.5 Climate change impact on socio-economic aspects and constraints for management

Socio-economic impacts	Constraints
Increased loss of property and coastal habitats due to sea level rise	Lack of dynamic predictions of land form migration
Increased flood risk and potential loss of life due to inundation and wave action	Insufficient or inappropriate shoreline protection measures
Damage to coastal protection works and other infrastructure	Data integration hampered by divergent information management systems and agencies
Increased disease risk due to changed equilibrium	Lack of definition of key indicators and thresholds relevant to coastal management
Loss of renewable and subsistence resources	Inadequate knowledge of coastal conditions, tribal customs, and appropriate management measures
Impacts on agriculture and aquaculture through decline in soil and water quality loss of tourism, recreation, and transportation services	Lack of long-term data for key coastal descriptors
Loss of nonmonetary cultural resources and values of native population	Fragmented and ineffective institutional arrangements and weak governance
Large-scale catastrophe and inundation may create climate refugees particularly from the small island nations	Societal resistance to change due to different customs and traditions

- Protection of low-lying island states (e.g. Maldives, Lakshadweep) and nations with large deltaic areas is likely to be very costly.
- As a result of the above adaptation measures to sea level rise and climate change, there will be important tradeoffs, which include environmental, economic, social, and cultural values.

Towards the end of the twenty-first century, projected sea level rise will affect low-lying coastal areas with large populations. The cost of adaptation could amount to at least 5–10% of GDP of these countries. Coastal areas, especially heavily populated mega delta regions in South, East, and Southeast Asia, will be at greatest risk due to increased flooding from the sea and, in some mega deltas, flooding from the rivers. In South America there is a risk of significant biodiversity loss through species extinction in many areas of tropical Latin America coastal communities, and habitats will be increasingly stressed by climate change impacts interacting with development and pollution. By mid-century, climate change is expected to reduce water resources in many small islands, e.g. in the Caribbean and Pacific, to the point where they become insufficient to meet even human demand during low-rainfall periods.

Climate change impacts on natural systems can have profound effects on socio-economic systems (Harvey et al. 1999) as well because there is a dynamic interaction between natural and socio-economic systems. Instead of being considered as two independent systems, they are increasingly regarded as developing in a coevolutionary way, as shown by the feedback loop from the socio-economic system to the natural system. In some coastal societies and islands, the significance of cultural values is equal to or even greater than that of economic values. At the same time, patterns of human development and social organization in a community are important determinants of the vulnerability of people and social institutions to sea level rise and other coastal hazards. Consequently, in spite of several measures initiated by different national governments and international organizations, still there are several constraints. For example, technological capacity is a component of social and economic resilience, although adaptation strategies may involve more than engineering measures. Technological options can be implemented efficiently only in an appropriate economic, institutional, legal, and sociocultural environment. In this context, indigenous (traditional) technical knowledge should also be considered as an option to increase resilience, and to be effective any technology must fit in with traditional social structures (Veitayaki 1998).

11.5 Conclusion

Several changes that might be expected to accelerate with global warming in the future are already observed in some regions and ecosystems. The major climate change-associated events are global sea level rise, increases in SST, ocean acidification, and changing rainfall pattern besides regional decreases in sea-ice cover. Impacts associated with these changes have included shoreline erosion, coastal geomorphological changes, wetland loss, seawater intrusion into freshwater lenses, and impacts on coral reefs and associated biodiversity.

Analysis of compiled data on climate parameters and global projections revealed that global climate change will affect the biogeophysical characteristics of the oceans and coasts, modifying their ecological structure and affecting their ability to sustain coastal residents and communities. Owing to these challenges, the biodiversity in the coastal region tend to get adapt, migrate, or become extinct seriously hampering the ecosystem services it provide. Climate change impacts in the coastal region are likely to reflect local geological, physiographical, ecological, and socio-economic conditions within a broader regional or global context. Coastal communities—particularly on low-lying deltas, atolls, and reef islands—face threats of inundation, increased flooding, and saltwater intrusion, with impacts on health and safety, water supply, artisanal fisheries, agriculture, aquaculture, and other infrastructure. Agricultural production is highly dependent on weather, climate, and water availability and is adversely affected by weather- and climate-related disasters.

At the same time, climate change tends to disproportionately affect the more vulnerable segments of society, such as the poor and the marginalized, as they are already living in highly vulnerable areas and having poor adaptive capacity. The key challenge in addressing the climate threat to coastal zones is timely adaptation to global warming. Other effects of climate changes on natural and human environments are emerging, although many are difficult to discern due to adaptation and non-climatic drivers. Nevertheless, serious socio-economic, technical, political, and ecological problems hamper the planning and implementation of adaptive strategies.

References

Alleng GP (1998) Historical development of the port royal mangrove wetland, Jamaica. J Coast Res 14:951–959

Alongi DM (2008) Mangrove forests: resilience, protection from tsunamis, and responses to global climate change. Estuar Coast Shelf Sci 76(1):1–13

Aronson RB, Precht WF, Macintyre IG, Murdoch T (2000) Coral bleach-out in Belize. Nature 405:36

Bijlsma L, Ehler CN, Klein RJT, Kulshrestha SM, McLean RF, Mimura N, Nicholls RJ, Nurse LA, Pérez Nieto H, Stakhiv EZ, Turner RK, Warrick RA (1996) Coastal zones and small islands. In: Watson RT, Zinyowera MC, Moss RH (eds) Climate change: impacts, adaptations, and mitigation of climate change. Scientific-technical analyses. Contribution of Working Group II to the second assessment report of the Intergovernmental Panel on Climate Change. Cambridge University Press, Cambridge, pp 289–324

Bird ECF (1993) Submerging coasts: the effects of a rising sea level on coastal environments. Wiley, Chichester, p 184

Brinson MM, Christian RR, Blum LK (1995) Multiple states in the sea-level induced transition from terrestrial forest to estuary. Estuaries 18:648–659

Brown BE (1997) Coral bleaching: causes and consequences. Coral Reefs 16:129–138

Brown BE, Dunne RP, Goodson MS, Douglas AE (2000) Bleaching patterns in reef corals. Nature 404:142–143

Canadell JG (2007) Contributions to accelerating atmospheric CO_2 growth from economic activity, carbon intensity, and efficiency of natural sinks. Proc Natl Acad Sci 104(47):18866–18870
Chen X, Stanley DJ (1998) Sea-level rise on eastern China's Yangtze Delta. J Coast Res 14:360–366
Curran S (2002) Interactions between coastal and marine ecosystems and human population systems: perspectives on how consumption mediates this interaction. Ambio 31(4):264–268
Day JW, Martin JF, Cordoch L, Templet PH (1997) System functioning as a basis for sustainable management of deltaic ecosystems. Coast Manag 25:115–153
Edinger EN, Jompa J, Limmon GV, Widjatmoro W, Risk MJ (1998) Reef degradation and coral biodiversity in Indonesia: effects of land based pollution, destructive fishing practices and changes over time. Mar Pollut Bull 36:617–630
Finkl CW (2002) Long-term analysis of trends in shore protection based on papers appearing in the journal of coastal research, 1984-2000. J Coastal Res 18:211–224
Forbes DL, Orford JD, Taylor RB, Shaw J (1997) Proceedings of the Canadian Coastal Conference'97, Guelph, Ontario: Inter decadal variation in shoreline recession on the Atlantic coast of Nova Scotia. In: Canadian Coastal Science and Engineering Association, Ottawa, ON, Canada, pp 360–374
Freely R, Sabine CL, Lee K, Berelson W, Kleypas J, Fabry VJ, Millero FJ (2004) Impact of anthropogenic CO_2 on the $CaCO_3$ systems in the oceans. Science 305:362–366
Gilman SE, Urban MC, Tewksbury J, Gilchrist GW, Holt RD (2010) A framework for community interactions under climate change. Trends EcolEvol 25:325–331
Hansen J, Sato M, Ruedy R, Kharecha P, Lacis A, Miller R, Nazarenko L, Lo K, Schmidt GA, Russell G (2007) Dangerous human-made interference with climate: a GISS model E study. Atmos Chem Phys 7:2287–2312
Harvey N, Clouston B, Carvalho P (1999) Improving coastal vulnerability assessment methodologies for integrated coastal zone management. Aust Geogr Stud 37(1):50–69
Héquette A, Desrosiers M, Forbes DL (1995) Proceedings of the international conference on coastal change: The role of shoreline configuration and coastal geomorphology in nearshore sediment transport under storm combined flows, Canadian Beaufort Sea. In: Workshop Report 105, Inter governmental Oceanographic Commission, Paris, France, pp 563–570
Hester KC, Peltzer ET, Kirkwood WJ, Brewer PG (2008) Unanticipated consequences of ocean acidification: a noisier ocean at lower pH. Geophys Res Lett 35:19601
Hoegh-Guldberg O (1999) Climate change, coral bleaching and the future of the world's coral reefs. Mar Freshw Res 50:839–866
Huth R (1997) Potential of continental-scale circulation for the determination of local daily surface variables. Theor Appl Climatol 56:165–186
IPCC (1996) Climate change impacts, adaptations and mitigation of climate change: scientific-technical analyses. Contribution of Working Group II to the second assessment report of the Intergovernmental Panel on Climate Change. Cambridge University Press, Cambridge, p 880
IPCC (2001) Contribution of Working Group I to the third assessment report of the Intergovernmental Panel on Climate Change. In: Houghton JT, Ding Y, Griggs DJ, Noguer M, van der Linden PJ, Dai X, Maskelland K, Johnson CA (eds) Climate change : the scientific basis. Cambridge University Press, Cambridge, p 881
IPCC (2007a) Climate change: synthesis report contribution of Working Groups I, II and III to the fourth assessment report of the Intergovernmental Panel on Climate Change. Geneva, Switzerland, p 104
IPCC (2007b) Contribution of Working Group I to the fourth assessment report of the Intergovernmental Panel on Climate Change. In: Solomon S, Qin D, Manning M, Chen Z, Marquis M, Averyt KB, Tignor M, Miller HL (eds) Climate change: the physical science basis. Cambridge University Press, Cambridge, p 996
Jelgersma S (1996) Land subsidence in coastal lowlands. In: Milliman JD, Haq BU (eds) Sea-level rise and coastal subsidence. Kluwer Academic Publishers, Dordrecht, pp 47–62
Jennerjahn TC (2012) Biogeochemical response of tropical coastal systems to present and past environmental change. Earth Sci Rev 114(1–2):19–41

Jevrejeva S, Grinsted A, Moore JC, Holgate S (2006) Nonlinear trends and multiyear cycles in sea level records. J Geophys ResOceans 111:C09012

Jimenez JA, Sanchez-Arcilla A (1997) Physical impacts of climatic change on deltaic coastal systems (II): driving terms. Clim Chang 35:95–118

Joos F, Spahni R (2008) Rates of change in natural and anthropogenic radiative forcing over the past 20,000 years. Proc Natl Acad Sci, USA 105:1425–1430

Kaminsky GM, Ruggiero P, Gelfenbaum G (1998) Monitoring coastal change in Southwest Washington and Northwest Oregon during the 1997/98 El Niño. Shore & Beach 66(3):42–51

Klein RJT, Nicholls RJ (1999) Assessment of coastal vulnerability to climate change. Ambio 28(2):182–187

Kleypas JA, Buddemeier RW, Archer D, Gattuso J, Landon C, Opdyke BN (1999) Geochemical consequences of increased carbon dioxide in coral reefs. Science 284:118–120

Krishnan P, Dam-Roy S, George G, Anand A, Murugesan S, Kaliyamoorthy M, Vikas SR (2010) Elevated Sea Surface Temperature (SST) induces mass bleaching of corals in Andaman. Curr Sci 100:1800–1804

Lal M (2004) Climate change and small island developing countries of the South Pacific. Fijian Studies 2:15–31

Lal M, Harasawa H, Takahashi K (2002) Future climate change and its impacts over small island states. Clim Res 19:179–192

Lenton TM, Held H, Kriegler E, Hall JW, Lucht W, Rahmstorf S, Schellnhuber HJ (2008) Tipping elements in the Earth's climate system. Proc Natl Acad Sci USA 105:1786–1793

Levitus S, Antonov J, Boyer T (2005) Warming of the world ocean 1955–2003. Geophys Res Lett 32:L02604. https://doi.org/10.1029/2004GL021592

Meehl GA, Stocker TF, Collins W, Friedlingstein P, Gaye AT, Gregory J, Kitoh A, Knutti R, Murphy J, Noda A, Raper S, Watterson IG, Weaver A, Zhao ZC (2007) Global climate projections. Contribution of Working Group I to the fourth assessment report of the Intergovernmental Panel on Climate Change. In: Solomon S, Qin D, Manning M, Chen Z, Marquis M, Averyt KB, Tignor M, Miller HL (eds) Climate change: the physical science basis. Cambridge University Press, Cambridge, pp 747–846

Melloul AJ, Goldberg LC (1997) Monitoring of seawater intrusion in coastal aquifers: basis and local concerns. J Environ Manag 51:73–86

Meure CM, Etheridge D, Trudinger C, Steele P, Langenfelds R, van Ommen T, Smith A, Elkins J (2006) Law Dome CO_2, CH_4 and N_2O ice core records extended to 2000 years BP. Geophys Res Lett 33:L14810

Michel D, Pandya A (2010) Coastal zones and climate change the Henry L. Stimson Center, Washington, DC

Millennium Ecosystem Assessment (2003) Ecosystems and human well-being: millennium ecosystem assessment. Island Press, Washington, DC, p 245

Mimura N, Nurse L, McLean RF, Agard J, Briguglio L, Lefale P, Payet R, Sem G (2007) Small islands. Climate change (2007): impacts, adaptation and vulnerability. Contribution of Working Group II to the fourth assessment report of the Intergovernmental Panel on Climate Change. Cambridge University Press, Cambridge, pp 687–716

Moore WS (1999) The subterranean estuary: a reaction zone of ground water and sea water. Mar Chem 65:111–125

Mulhall M (2009) Saving the rainforests of the sea: an analysis of international efforts to conserve coral reefs. Duke Environ Law Policy Forum, p 19-2

Nemesova I, Kalvova J, Klimperova N, Buchtele J (1998) Comparison of GCM-simulated and observed climates for assessing hydrological impacts of climate change. J Hydrol Hydromech 46:237–263

Nicholls RJ (1998) Assessing erosion of sandy beaches due to sea-level rise. In: Maund JG, Eddleston M (eds) Geohazard in engineering geology, vol 15. Engineering Geology Special Publications, Geological Society, London, pp 71–76

Nicholls RJ, Hoozemans FMJ, Marchand M (1999) Increasing flood risk and wetland losses due to sea-level rise: regional and global analyses. Glob Environ Chang 9:S69–S87

Nicholls RJ, Wong PP, Burkett VR, Woodroffe CD, Hay JE (2008) Climate change and coastal vulnerability assessment: scenarios for integrated assessment. Sustain Sci 3(1):89–102
Nuttle W, Brinson M, Cahoon D (1997) Processes that maintain coastal wetlands in spite of rising sea level. EOS Trans Am Geophys Union 78(25):257–261
Orr JC (2009) Research priorities for ocean acidification. Report from the second symposium on the Ocean in a High-CO_2 World-2009. http://ioc3.unesco.org/oanet/HighCO2World.html
Pallewatta N (2010) Impacts of climate change on coastal ecosystems in the Indian Ocean region. In: Michel D, Pandya A (eds) Coastal zones and climate change. The Henry L. Stimson Center, Washington, DC, pp 3–16
Pimm SL (2008) Biodiversity: climate change or habitat loss-which will kill more species? Curr Biol 18:117–119
Rahmstorf S (2007) A semi-empirical approach to projecting future sea level rise. Science 315:368–370
Ruosteenoja K, Carter TR, Jylha K, Tuomenvirta H (2003) Future climate in world regions: an intercomparison of model-based projections for the new IPCC emissions scenarios. The Finnish Environment 644. Finnish Environment Institute, Helsinki, p 83
Sandweiss DH, Maasch KA, Belknap DF, Richardson JB III, Rollins HB (1998) Discussion of Lisa E. Wells (1996) The Santa beach ridge complex. J Coast Res 14:367–373
Scheffer M, Carpenter S, Foley JA, Folke C, Walker B (2001) Catastrophic shifts in ecosystems. Nature 413(6856):591–596
Sherif MM, Singh VP (1999) Effect of climate change on sea water intrusion in coastal aquifers. Hydrol Process 13:1277–1287
Short FT, Neckles HA (1999) The effects of global climate change on seagrasses. Aquat Bot 63:169–196
Short AD, Trembanis A, Turner I (2000) Beach oscillation, rotation and the Southern Oscillation: Narrabeen Beach, Australia. Abstracts of the 27th International Conference on Coastal Engineering, July 16–22, 2000, Sydney, Australia, vol 1, Paper 4. Australian Institution of Engineers, Australia, Barton, ACT, Australia
Singh VS, Gupta CP (1999) Groundwater in a coral island. Environ Geol 37:72–77
Stanley DJ, Warne AG (1998) Nile Delta in its destruction phase. J Coast Res 14:794–825
Stumpf RP, Haines JW (1998) Variations in tidal level in the Gulf of Mexico and implications for tidal wetlands. Estuar Coast Shelf Sci 46:165–173
Subba Rao NV, Dey A (2000) Catalogue of marine mollusks of Andaman and Nicobar Islands. Zoological Survey of India, Kolkata, pp 1–323
The Royal Society (2005) The impacts of ocean acidification on other marine organisms and ecosystems. Policy document 12/05. London, UK, p 60. https://royalsociety.org/topics-policy/publications/2005/ocean-acidification
Trenberth KE, Jones PD, Ambenje PG, Bojariu R, Easterling DR, Klein Tank AMG, Parker DE, Renwick JA, Rahimzadeh F, Rusticucci MM, Soden BJ, Zhai PM (2007) Observations: surface and atmospheric climate change. Contribution of Working Group I to the fourth assessment report of the Intergovernmental Panel on Climate Change. In: Solomon S, Qin D, Manning M, Chen Z, Marquis M, Averyt KB, Tignor M, Miller HL (eds) Climate change 2007: the physical science basis. Cambridge University Press, Cambridge, pp 235–336
Umitsu M (1997) Landforms and floods in the Ganges Delta and coastal lowland of Bangladesh. Mar Geod 20:77–87
UNFCC (1992) United Nations framework convention on climate change, Bonn, FCCC/INFORMAL/84, p 24
UNFCCC (2007) United Nations framework convention on climate change, climate change: impacts, vulnerabilities and adaptation, Bonn, p 60
Van de Plassche O, Van der Borg K, de Jong AFM (1998) Sea level climate correlation during the past 1400 years. Geology 26:319–322
Van de Plassche O, van der Borg K, de Jong AFM (1999) Sea level climate correlation during the past 1400 years: reply. Geology 27:190

Varekamp JC, Thomas E (1998) Sea level rise and climate change over the last 1000 years. EOS Trans Am Geophys Union 79:69–75

Veitayaki J (1998) Traditional and community-based marine resources management system in Fiji: an evolving integrated process. Coast Manag 26:47–60

Velmurugan A, Swarnam TP, Ambast SK, Ravisankar N, Subramani T (2014) Land degradation and its spatio-temporal changes induced by natural events in Andaman islands. J Andaman Sci Assoc 19(1):65–74

Velmurugan A, Dam Roy S, Krishnan P, Swarnam TP, Jaisankar I, Singh AK, Biswas TK (2015) Climate change and Nicobar Islands: impacts and adaptation strategies. J Andaman Sci Assoc 20(1):7–18

Walsh K (2004) Tropical cyclones and climate change: unresolved issues. Clim Res 27:77–83

Wilkinson CR (1999) Global and local threats to coral reef functioning and existence: review and predictions. Mar Freshw Res 50:867–878

Wilkinson CR (2008) Status of coral reefs of the world. Global Reef Monitoring Network, Reef and Rainforest Research Center, Townsville

12 Adaptation and Mitigation Strategies for Coastal Areas

Abstract

Climate-related changes represent potential additional stresses on the coastal ecosystems that are already under pressure from degradation and non-climatic pressure. The potential impact of these pressures on the coastal areas and livelihood of coastal communities necessitate the need for researchers, policymakers, residents, and other key stakeholders to work together to establish a framework for adaptation that is integrated within current coastal management processes and practices and takes a broader view of the subject. There are three important coastal adaptation strategies, viz. protection, accommodation, and retreat. All these measures involve technologies, often available or known to the scientific communities. The major challenge to adaptation is access to the adaptation measures, participation by all the stakeholders, and coordination at different levels. At the same time, mitigation methods are centred on adjustment or modification in transportation, shelter, food production, energy, and carbon storage and bioengineering. This chapter describes various options available for adaptation and mitigation strategies for natural as well as human-managed systems.

Keywords

GHG emissions · Resilience · Vulnerability · Planned retreat · Autonomous adaptation

12.1 Introduction

Climate change is one of the major challenges to the human society in this century. Increased concentrations of greenhouse gases in the atmosphere from anthropogenic activities besides natural releases have led to increased global mean surface air temperatures, climate and weather variability, changes in precipitation patterns, and variations in seasonal cycles such as monsoons. The main climate variables of

V. Ayyam et al., *Coastal Ecosystems of the Tropics - Adaptive Management*,
https://doi.org/10.1007/978-981-13-8926-9_12

interest to coastal region and ecosystems are mean surface air temperature, rainfall, humidity, wind speeds, and the frequency and intensity of extreme climate/weather events such as storms and droughts (Michel and Pandya 2010). There are other consequential events due to global warming which affect the coastal and island region, viz., sea level rise, increase in sea surface temperature, intrusion of saltwater into fresh water, and changes in rates of evapotranspiration from oceans and other water bodies (Nicholls and Nicholls 2008). These changes are interconnected and complex which affect fundamental physiological processes and the survival of living systems globally.

Climate change is not limited to a particular region or continent; it is a global change driver that interacts with and exacerbates the impacts of already existing environmental and human drivers on the natural world. But living systems exhibit some resilience; hence, every system has its own critical limit below which it tries to adjust or adapt itself to external changes without affecting its function. Above the critical limits, ecosystems often exhibit nonlinear responses to crossing threshold levels (beyond its resilience) that alter their composition and key processes. Once some critical threshold is passed, even relatively small stresses may trigger rapid ecosystem degradation and loss of integrity. Consequently that affects ecosystem stability, resilience, and functions. At the same time, populations and species can respond in several ways to large-scale environmental change. They may benefit or remain unchanged; they may continue in the original distribution range, adapt, disperse to new regions, decline in numbers, or become extinct. As species respond individually and collectively, chain reactions will also take place in the food webs at the ecosystem level (IPCC 2007).

The effects will not, however, be uniform in all the regions. Certain coastal environments will be at greater risk, such as tidal deltas and low-lying coastal plains, for example, sandy beaches and barrier islands, coastal wetlands, estuaries and lagoons, and coral reefs and atolls. On this basis, increased coastal flooding is expected to be most severe in South and South-East Asia, Africa, the southern Mediterranean coasts, the Caribbean, and in most islands in the Indian and Pacific Oceans (UNFCCC 2006).

Impact studies show that due to external pressure on coastal ecosystems, the species composition of communities and ecosystem services are affected. As the coastal communities are benefited and dependant on the provisioning service, the changes will bring negative impacts on food security. In agro-ecosystem the negative effect could be through spread of diseases and vectors into new regions, loss of productive agricultural lands and water bodies, changes in seasonal cycles, and changes in ecosystem functions manifested in the form of yield loss and degradation of environmental services. To cope with such emerging situations, societies have several adaptation options which help to reduce the impacts. Simultaneously by reducing GHG emissions (mitigation), the rate and magnitude of change can also be reduced (IPCC 2007). With these understandings, this chapter focuses on adaptation and mitigation options for the coastal areas including the agro-ecosystems to achieve sustainable development and conservation priorities.

12.2 The Conceptual Framework

12.2.1 Mitigation and Adaptation

The observed variations and projected changes in the climate of earth will have consequential changes in the ecosystem functions of the coastal areas which will undoubtedly affect the provision of ecosystem services for coastal communities. Therefore maintaining or increasing ecosystem services into the future will require integrating adaptation and mitigation strategies (Fig. 12.1). Mitigation strategies are those that reduce the magnitude of climate change through reduction in GHG emission or by carbon sequestration. On the other hand, adaptation is the alteration to human or natural systems that are designed to reduce the negative impacts of climate change (IPCC 2007a). Adaptation strategies should be much more varied and needs to be broader than mitigation efforts, particularly in the coastal and island regions. Further adaptation options are planned/developed often for a specific system or location.

Mitigation and adaptation strategies designed to address changes in climate may have additional positive or negative effects on ecosystem functioning, services, and human wellbeing. For example, replacing fossil fuels with green energy sources will reduce GHG emissions while simultaneously reducing air pollution, acid rain, and smog which benefit the human health. Therefore, this chapter overviews some of the most effective and important mitigation and adaptation strategies for addressing changes to the climate system caused by increased GHG emission.

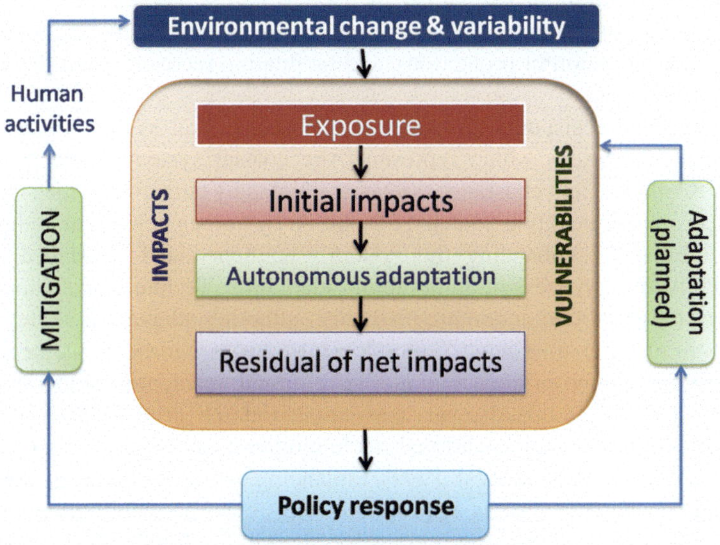

Fig. 12.1 Conceptual representation of adaptation, mitigation, and vulnerability associated with climate change. (Modified from IPCC 2001)

12.2.2 Resilience and Vulnerability

In the context of climate change and coastal management, vulnerability of a particular system or region is most important factor increasing the impact of climate impact. Climate vulnerability is the degree to which a system is susceptible to or unable to cope with adverse effects of climate change, including climate variability and extremes. Climate vulnerability depends on exposure, sensitivity, and adaptive capacity (adapted from IPCC 2012). Climate exposure is the extent and magnitude of a climate and weather event. Sensitivity is the degree to which the area of concern/system is susceptible to a climate impact. Adaptive capacity is the ability of the area of concern/system to adjust or respond to the changing conditions.

On the other hand, the concept of coastal resilience has become more important in planning for adaptation due to the need for developing climate-resilient technologies. Coastal resilience has ecological, morphological, and socioeconomic components, each of which represents another aspect of the coastal system's adaptive capacity to external disturbances. There are several natural features that contribute to resilience of the coastal areas by providing ecological buffers, including coral reefs, salt marsh, and mangrove forest and morphological protection in the form of sand and gravel beaches, barriers, and coastal dunes.

As the human component is inseparable from coastal region and the ecosystem, it is imperative to assess socio-economic resilience as well. It is the capability of a society to prevent or cope with the impacts of climate change and sea level rise, including technical, institutional, economic, and cultural ability. Enhancing this resilience should be seen as or equivalent to reducing the risk of the impacts on society. It can be strengthened mainly by decreasing the probability of occurrence of natural hazard (managed retreat or protection), avoiding or reducing its potential effects (accommodation or protection), and facilitating recovery from the damages when impacts occur.

Resilience and resistance are functions of the natural system's capacity for autonomous adaptation, which represents the coastal system's natural adaptive response. Resilience and resistance often are affected by human activities which can be either positive or negative. Planned adaptation by human beings can reduce natural vulnerability by enhancing the system's resilience and resistance, thereby improving the effectiveness of autonomous adaptation. Technological capacity is a component of social and economic resilience, although adaptation strategies may involve several other measures. Technological options can be implemented efficiently only in an appropriate economic, institutional, legal, and sociocultural context (Klein et al. 2000). It is also very convenient and may provide new possibilities by considering indigenous (traditional) technologies as an option to increase resilience (Veitayaki 1998; Nunn et al. 1999). Enhancing coastal resilience in these ways increasingly is regarded as an appropriate way to prepare for uncertain future changes while maintaining opportunities for coastal development. Therefore these understandings and possibilities support the view that enhancing resilience is a potentially powerful adaptive measure.

12.3 Mitigation

As discussed earlier mitigation measures are basically aimed at reducing the GHG emissions, and with reference to agro-ecosystems, it provides for carbon sequestration along with production practices. In general the mitigation strategies can be grouped into five major categories, viz. transportation, shelter, food production, energy, and carbon storage and bioengineering (Table 12.1). Some of these measures also improve the ecosystem functioning and human health. These are briefly described below.

GHG emissions and road transportation represented 74% of the total carbon dioxide emissions from transportation (IPCC 2007b). Transport network in several instances fragments the natural habitats. Further vehicular pollution also affects the ecosystem function particularly the health of plants and animals including human population. There are three basic approaches to these problems, viz. increase the efficiency of transportation, shift in the mode of transportation, and reduce the amount of transportation needed.

Shelters including other commercial infrastructures are regionally variable aspects of human culture. The design of these structures is dependent on local climate, land access, available materials, and regional culture/style. In recent times significant amount of GHG emission comes from the way we design, construct, and inhabit the built environment. On an average 8% of annual carbon emissions worldwide can be attributed to building material extraction, production, transportation,

Table 12.1 Mitigation strategies for major GHG emission activities

Sl. No.	Activity	Mitigation strategy	Challenges
1	Transportation	Change in mode of transport, vehicle type and source of fuel, increased vehicle efficiency	Lack of political will, access to technology, poorly designed urban centres/settlement pattern, limited public transport, and poor quality of service and cultural values
2	Shelter/ housing	Alternative building materials and building design, use of nonconventional sources, etc.	Cost of alternative materials, design control by local authorities, lack of incentives
3	Energy	Reduced consumption, use of renewable sources, energy efficiency	Lack of political will, cost and access to green technology
4	Industries	Reduction of pollutant, treatment of effluent and waste, use of green technology, follow emission audit system	Lack of political will, cost of green technology, competition to improve profit etc.
5	Food production	Alternative production methods, conservation, change in diet/food	Access to green technologies, cost of alternative methods, poverty, government policies, cultural values
6	C sequestration	Use of organic materials, reforestation, soil storage	Lack of incentives and motivation, lack of government support, cost of technology, etc.

and installation (Mazria and Kershner 2008). Therefore, it is advisable to go for regenerative buildings, integrated landscapes, and carbon conscious materials.

Energy consumption pattern and method of production influence GHG emission which in turn affect ecosystem functioning and services besides human health. Significant reduction in GHC emission is possible by reducing energy consumption accomplished through a combination of technological advances and changes in human behaviour (IPCC 2007b). Further, shift to renewable energy sources has great potential to reduce GHG emissions and reduce impact on ecosystems and services.

Agro-ecosystem or food production system releases considerable amounts of GHG into the atmosphere. In general it accounted for 10–12% of the total global anthropogenic emission (Rosenzweig and Tubiello 2007; Smith et al. 2008). Land use change, N fertilizer use, livestock production, rice farming, transportation and land preparation machineries, and biomass burning are among the major direct sources of GHG emission (Burney et al. 2010). There are ample ways for mitigating climate change impact through modification in agricultural practices and food consumption. This includes soil carbon-enhancing practices and altered production methods such as improved production efficiencies, land use changes, crop and rangeland management, and livestock and manure management (Rosenzweig and Tubiello 2007; Smith et al. 2008).

Alternatively some of the atmospheric carbon can be stored in soil and live biomass for considerable period of time, by which the GHG in the atmosphere can be brought down. At best it can compensate for the release. Agriculture and degraded soils can act as large carbon sinks with the potential to sequester 55–78 Gt of carbon globally. Soil organic carbon is the major component of this large soil C pool. C sequestration can be facilitated by adding biomass to the soil and minimizing soil disturbances, cover crop, mulches, crop rotation, agroforestry, etc. These practices additionally improve the soil fertility as well (Lal 2004).

12.4 Adaptation

In the context of climate change and coastal ecosystems, adaptation is the adjustment in natural or human systems in response to actual or expected climatic stimuli or their effects, which moderates harm or exploits beneficial opportunities. Various types of adaptation are proposed, viz. anticipatory and reactive adaptation and autonomous and planned adaptation (IPCC 2001). Many of these adaptation strategies are either general concepts or strategies that apply to most ecosystems and regions.

12.4.1 Adaptation Strategy

The process of coastal adaptation can be conceptualized to address the fact that climate change and variability, together with other stresses on the coastal environment, produce the impacts. The mitigation efforts should try to remove the cause of the impacts, while adaptation options should modify or adjust to the impacts.

Table 12.2 Options and technologies for adaptation in coastal areas

Protect	Retreat	Accommodate
Hard structures—such as dykes, seawalls, detached breakwaters, tidal barriers	Establishing setback zones	Establish early warning and evacuation systems
Soft structures—dunes or wetland restoration/creation, beach nourishment	Relocating threatened buildings	New agricultural practices such as land shaping, using salt-resistant crops
Indigenous options—walls of wood, stone or coconut leaf, afforestation with local tree species	Phasing out development and other works in exposed areas	Hazard insurance in the zone
	Creating upland buffers	New building codes for safety
	Rolling easements	Improved drainage to avoid flooding
		Desalination systems

Modified from UNFCCC (2006)

Climate change generally will exacerbate existing problems such as coastal flooding, erosion, saltwater intrusion, and degradation of ecosystems (Bijlsma et al. 1996). At the same time, non-climate stresses can be an important cause of increasing coastal vulnerability to climate change and variability. Under such situations, adaptation options should incorporate with policies in other areas, such as disaster mitigation plans, land use plans, and watershed resource plans. In other words, adaptation options are best addressed when they are incorporated in the integrated coastal management and sustainable development plans.

It is very much essential to evaluate any such plan or strategy which is regarded as a crucial component of the vulnerability assessment using common methodology. However, in general, the methodology has been less effective in assessing the wide range of technical, institutional, economic, and cultural elements in different localities (Klein and Nicholls 1999). Therefore, successful coastal adaptation should embrace more than just selecting one of the technical options to respond to sea level rise; it is a more complex and iterative process, with a series of policy cycles (Klein et al. 2000). Four steps can be distinguished in the process of coastal adaptation. They are information collection and awareness raising, planning and design, implementation and monitoring, and evaluation.

At the same time, it has been observed that often adaptive responses are undertaken reactively rather in a stepwise, planned, and anticipatory fashion. In order to have effective adaptive mechanism to address the climate change and other challenges, coastal communities have three possible adaptation options, viz. protect, accommodate, and accommodate (Biljsma et al. 1996). Some of the most important measures under each option are given in Table 12.2.

- Protect—This option aims to guard the land from the sea so that existing land uses can continue, by constructing hard structures (e.g. seawalls) as well as using soft measures (e.g., beach nourishment). But practical experiences suggest to

consider soft options that involve restoring dunes or creating or restoring coastal wetlands or continuing with indigenous approaches such as afforestation.
- Retreat—This involves no attempt to protect the land from the sea in the declared zone. In other words the simplest approach might be to establish a setback zone requiring development to be at a specified distance from the water's edge. And there are also intermediate options in the form of 'easements'—legal agreements that restrict the size or density of structures within areas at risk and specify permitted types of shoreline stabilization. The area to which these apply can also be designed to automatically move or 'roll' landward as the sea advances.
- Accommodate—This involves continuous occupation of the land but makes some adjustments. For example, elevating buildings on piles, growing flood- or salt-tolerant crops, and establishment of early warning systems.

Policy criteria and coastal development objectives of a region or country significantly influence the process of adaptation. Other critical influences include values, awareness, and factors such as historical legacies, institutions, and laws. Collaborative efforts of this kind can support a process of shared learning and joint problem-solving, thereby enabling better understanding, anticipation of, and response to climate change.

12.4.2 Coastal Adaptation Options

12.4.2.1 General Features

Planning for strategies for future shifts in climate should essentially require expanding the temporal and spatial scales of plans, policies, and management (Lawler 2009). In that case, future planning will need to address both short-term and long-term changes in systems. Planning will need to occur at multiple, integrated scales and will need to involve local, regional, and sometimes national participants (Kareiva and Enquist 2008). At the same time, due consideration should be given to the climate justice. This means poor communities tend to bear the burden of climate change more than their human counterparts.

Due to changes in spatial scale and uncertainties, it is necessary to have cooperation and coordination among the different stakeholders from different jurisdictions and management units (Kareiva and Enquist 2008). In addition, as emission from one location can affect the climate of other location, addressing the global nature of climate change will involve cooperation at an international level (Hannah 2010).

Another important aspect of adaptation is the adaptive management which is a flexible approach. It is an iterative process in which the system is monitored following the initial management action, and new actions are designed and implemented in response to the results of the initial actions. It helps to manage uncertainty in the systems, and thus it has great potential for managing systems in a changing climate (Arvai et al. 2006).

More diverse systems tend to be more stable to environmental changes than less diverse systems. Thus restoring or protecting the biological diversity of a system will likely to increase resilience of that system to climate change (Glick et al. 2009).

12.4.2.2 Specific Options for Ecosystem Adaptation

Human activities have already impacted freshwater systems. Under current climate change projections, almost all freshwater ecosystems will face significant impacts before the end of this century. Most of these impacts will be negative and will have consequential effect on the human livelihoods and communities that depend upon them (Poff and Zimmerman 2010). Therefore adaptation measures should focus on change in precipitation, rising water and air temperature, increasing evapotranspiration, and changing regimes of ground- and surface water. Broadly there are four approaches to the freshwater problems, viz. ecological water management, connectivity, protected areas, and restoration strategies.

The interface between the land, sea, and air found in coastal areas is a high energy system subject to storms, winds, and tides. There are two main approaches that will likely prove most useful in coastal systems, viz. building resilience through protection and restoration and facilitating the transition to new ecosystems. Protection and restoration of natural buffers such as barrier islands, mangroves, and coral reefs not only require little financial construction and maintenance but also maintain the wealth of ecosystem services associated with these systems. This includes nursery and rearing habitat for many commercial seafood stocks. On the other hand, in many coastal regions, mangrove area has come down drastically for favour of aquaculture. This loss destroys a wide range of ecosystem services provided by these forests, which range from carbon sequestration to flood protection, spawning, and rearing grounds for many fish and other marine species. Restoration of mangroves facilitates the increased adaptation to storm surge. Similarly protecting corals from stresses such as excess sediment and nutrient load, pollutant inputs from terrestrial water, unregulated diving, etc. can increase its ability to adapt to the projected changes.

Identification and conservation of particular species or zooxanthellae and algal combination (corals means two organisms, viz. coral polyps and zooxanthellae which live in symbiotic relationship) are more resilient to warming. The purpose of adaptation is to reduce the net effect of climate change and sea-level rise. Only recently adaptation within natural ecosystems has been considered a possibility which is based on coastal resilience. An example is provided by coral reefs. Applying the two types of adaptation, Pittock and Jones (1999) suggests that 'autonomous adaptation' is what reefs would do by themselves, whereas 'planned adaptation' involves conscious human interference to assist in the persistence of some desirable characteristics of the coral reef system. The first type of adaptation may involve more rapid growth of coral, changes in species composition, or evolution of particular species in response to changed temperatures or other conditions. Planned adaptation might involve 'seeding' of particular reefs with species adapted to higher temperatures or attempts to limit increased sediment, pollutant, or freshwater flow onto reefs.

12.4.2.3 Agricultural Systems

Minor climatic variations can have a major impact on agricultural output even in a single growing season, so long-term agricultural productivity and food security will certainly be affected by ongoing climate change—a matter of increasing concern since over the next 30 years, global food production will need to double to feed the planet's growing population (UNFCCC 2006). Several studies have concluded with evidence that increased temperatures and changes in precipitation regimes are likely to have negative impacts on global food production systems and food security by aggravating the existing regional disparities between developing and developed countries (Easterling et al. 2007). Slight increase in temperature could reduce the crop performances that are already growing at supraoptimal temperatures in tropical countries (Easterling and Apps 2005). It is understood now that even with significant mitigation measures, adaptation strategies are necessary to address the impacts of climate change on agriculture.

Plants can cope with climatic extremes on their own but to certain extent. Some have a natural adaptive capacity, such as certain rice cultivars that flower early in the morning, enabling them to avoid the damaging effects of higher temperatures later in the day. Human efforts need to identify such mechanism available in nature for developing crops with suitable adaptation mechanisms. Therefore, most of the adaptation for crop production is likely to rely on human intervention. At the same time, farming communities have considerable experience of coping with adverse climatic events, such as droughts and floods, and with salinity. For example, local communities have discovered or introduced new forms of irrigation or diversified to varieties that are higher yielding or have greater tolerance for drought or salty conditions prevailing in their locality (UNFCCC 2006). They have also changed land topographies by using 'grass waterways'—areas where grass is left to grow permanently to drain run-off water and avoid erosion. Some of the important options are listed in Table 12.3.

Adaptation strategies for more resilient agricultural systems may include altering the timing and location of cropping, switching crop species and cultivars, selecting livestock breeds, crop rotations, and efficiently using and managing water resources. These are called autonomous strategies. However, for many agricultural systems, the magnitude of climatic changes will necessitate more intensive planned adaptation strategies that foster significant transformation of systems or infrastructures. Such intensive strategies are very essential for sub-Saharan Africa and South Asia where climate will have major impacts on agriculture (Nelson et al. 2009). These planned adaptation options include developing new infrastructure for irrigation, transport, water use, and reuse technologies; shifts in the distribution of agriculture; building the capacity to make continuing adjustments; and investing in new technical and management options. Among all improving crop performance and water conservation are very important and essential adaptation options across tropical coastal areas.

Maddrell (1996) found that over time scales of 35 and 100 years, managed coastal retreat is the most cost-effective adaptation option in reducing flood risks and protection costs for coastal facilities. Pope (1997) has suggested several ways

Table 12.3 Some important adaptation options for agriculture

Response strategy	Purpose	Some adaptation options
Use different crops	To grow suitable crops adapted to the challenge conditions	Introduction of suitable crop and varieties
		Carry out research on new varieties
Change land topography	In order to improve water uptake and reduce wind erosion	Subdivide large fields
		Maintain grass waterways
		Roughen the land surface
		Build windbreaks
Increase water storage and use efficiency	Improve water use to increase water productivity, availability, and control erosion	Line canals with plastic films
		Where possible, use brackish water
		Concentrate irrigation in periods of peak growth
		Use drip irrigation
Land management	To adjust to the waterlogged and saline environment, avoid entry of sea water during high tide/sea surges	Land shaping
		Dykes and one way sluice gate
		Drainage improvement in low-lying areas
		Floating agriculture (local technology for sea water intruded areas)
Change farming practices	To conserve soil moisture and nutrients, reduce run-off and control soil erosion	Mulch stubble and straw
		Rotate crops
		A void monocropping
		Use lower planting densities
Change the timing of farm operations	To adjust to the change in climatic pattern or variations in weather condition	Advance sowing dates to offset moisture stress during warm period
		Prepare crop-weather calendar
Weather management	Informed decision-making	Weather-based agromet advisories

of responding to coastal erosion and flooding that have relevance in the context of climate change. Documented changes in tidal characteristics as a result of the construction of sea dikes and seawalls also have implications for shore protection in the face of rising sea level. Several alternatives to seawalls have been suggested as adaptation measures to reduce coastal erosion and saltwater intrusion from rising sea levels including improving drainage quality and channel capacity, increasing pumping facilities to reduce the water table, constructing a barrier across the mouth of the river, and developing new crops that are tolerant of a higher groundwater table (Chen and Zong 1999).

12.5 Conclusion

Coastal areas of continental countries and islands are facing the challenges emanating from the sea and other anthropogenic impact originated in the land. The situation is further getting aggravated by the global climate change by affecting the physical, biological, and biogeochemical characteristics of the coastal regions, modifying their ecological structure, their functions, and the goods and services they provide. Consequently planning for adaptation to the emerging situations besides the adjustment to the present conditions has become imperative. Simultaneously it would be responsible to take up mitigation measures to reduce the GHG emission from the adjustment made to the present activities and planned changes.

The vulnerability of a coastal community to the climatic and non-climatic natural hazards would depend on the adaptive capacity. This can be improved by at different levels and in various aspects of the socio-economic systems. At the same time, it is important to realise that adaptive capacities of coastal ecosystems including communities vary among species, sectors, and geographical regions. The general method of adaptation involve protect, accommodate, and retreat. All of them involve some form of adjustments and technology which in the broadest sense includes not just materials or equipment but also diverse forms of knowledge. The greatest challenge is to ensure access to knowledge and technology by the poor communities so as to ensure the climate justice. At the same time, it is highly essential that methods of adaptation ultimately need to be boosted by political and social action.

Much of the increase in yields in recent decades resulted from improved genetic material, management practices, irrigation, and increased use of fertilizers or other inputs. Climate change and its impacts on the coastal ecosystem demand that one of the main focuses of international and national research agenda for adaptation to climate change should be the search for a new generation of crop varieties adapted to the changing environment. Equally progress in bringing climate resilient to coastal agriculture system is possible largely by technology transfer. Because crop varieties are sensitive to local conditions, much of this has therefore been between regions with similar agroclimatic conditions. Integrated assessment and management of coastal ecosystems and a better understanding of their interaction with human development will be important components of successful adaptation to climate change. It is also highly desirable the integration of traditional practices into assessments of vulnerability and adaptation for tropical coastal region.

References

Arvai J, Bridge G, Dolsak N, Franzese R, Koontz T, Luginbuh A et al (2006) Adaptive management of the global climate problem: bridging the gap between climate research and climate policy. Clim Chang 78(1):217–225. https://doi.org/10.1007/s10584-006-9094-6

Bijlsma L, Ehler CN, Klein RJT, Kulshrestha SM, McLean RF, Mimura N, Nicholls RJ, Nurse LA et al (1996) In: Watson RT, Zinyowera MC, Moss RH (eds) Coastal zones and small islands. Climate change 1995: impacts, adaptations and mitigation of climate change: scientific-technical analyses. Contribution of working group II to the second assessment report

of the intergovernmental panel on climate change. Cambridge University Press, Cambridge, pp 289–324
Burney JA, Steven J, Davisc SJ, Lobella DB (2010) Green house gas mitigation by agricultural intensification. Proc Natl Acad Sci U S A 107(26):12052–12057
Chen X, Zong Y (1999) Major impacts of sea-level rise on agriculture in the Yangtze Delta area around Shanghai. Appl Geogr 19:69–84
Easterling W, Apps M (2005) Assessing the consequences of climate change for food and forest resources: a view from the IPCC. In: Salinger J, Sivakumar M, Motha RP (eds) Increasing climate variability and change. Springer, Dordrecht
Easterling WE, Aggarwal PK, Batima P, Brander KM, Erda L, Howden SM et al (2007) Food, fibre and forest products. In: Parry ML, Canziani OF, Palutikof JP, van der Linden PJ, Hanson CE (eds) Climate change 2007: Impacts, adaptation and vulnerability, Contribution of Working Group II to the Fourth Assessment Report of the Intergovernmental Panel on Climate Change. Cambridge University Press, Cambridge., 2007, pp 273–313
Glick P, Staudt A, Stein B (2009) A new era for conservation: review of climate change adaptation literature. National Wildlife Federation, Washington, DC, p 69
Hannah L (2010) Climate change biology. Academic, Burlington, p 470
IPCC (2001) Contribution of Working Group I to the Third Assessment Report of the Intergovernmental Panel on Climate Change. In: Houghton JT, Ding Y, Griggs DJ, Noguer M, van der Linden PJ, Dai X, Maskelland K, Johnson CA (eds) Climate change: the scientific basis. Cambridge University Press, Cambridge, p 881
IPCC (2007) Assessing key vulnerabilities and the risk from climate change. In: Parry ML, Canziani OF, Palutikof JP, van der Linden PJ, Hanson CE (eds) Climate change 2007: impacts, adaptation and vulnerability, Contribution of Working Group II to the Fourth Assessment Report of the Intergovernmental Panel on Climate Change. Cambridge University Press, Cambridge, pp 779–810
IPCC (2007a) An assessment of the Intergovernmental Panel on Climate Change 2007: synthesis report, p 73
IPCC (2007b) Climate change 2007. In: Solomon S, Qin D, Manning M, Chen Z, Marquis M, Averyt KB, Tignor M, Miller HL (eds) The physical science basis. Contribution of Working Group I to the Fourth Assessment Report of the Intergovernmental Panel on Climate Change Cambridge, Cambridge University Press, p 996
IPCC (2012) In: Field CB, Barros V, Stocker TF, Dahe Q, Dokken DJ et al (eds) Managing the risks of extreme events and disasters to advance climate change adaptation: special report of the Intergovernmental Panel on Climate Change. Cambridge University Press, Cambridge/New York. https://doi.org/10.1017/CBO9781139177245.333333
Kareiva P, Enquist C (2008) Synthesis and conclusions—preliminary review of adaptation options for climate-sensitive ecosystems and resources. In: Julius SH, West JM (eds) A report by the U.S. climate change science program and the subcommittee on global change research. U.S. Environmental Protection Agency, Washington, DC, pp 9–66
Klein RJT, Nicholls RJ (1999) Assessment of coastal vulnerability to climate change. Ambio 28(2):182–187
Klein RJT, Buckley EN, Nicholls RJ, Ragoonaden S, Aston J, Capobianco M, Mizutani M, Nunn PD (2000) Coastal adaptation. In: Metz B, Davidson OR, Martens JW, van Rooijen SN, Van Wie McGrory L (eds) Methodologies and technological issues in technology transfer, A Special Report of IPCC Working Group III. Cambridge University Press, Cambridge/New York, pp 349–372
Lal M (2004) Climate change and small island developing countries of the South Pacific. Fijian Stud 2:15–31
Lawler JJ (2009) Climate change adaptation strategies for resource management and conservation planning. Ann N Y Acad Sci 1162(1):79–98. https://doi.org/10.1111/j.1749-6632.2009.04147.x19432646
Maddrell RJ (1996) Managed coastal retreat, reducing flood risks and protection costs, Dungeness Nuclear Power Station, UK. Coast Eng 28:1–15

Mazria E, Kershner K (2008) The 2030 blueprint: solving climate change saves billions. Architecture 2030, Santa Fe

Michel D, Pandya A (2010) Coastal zones and climate change. The Henry L. Stimson Center, Washington, DC

Nelson GC, Rosegrant MW, Koo J, Robertson R, Sulser T, Zhu T, Ringler C, Msangi S, Palazzo A, Batka M, Magalhaes M, Santos RV, Ewing M, Lee D (2009) Climate change- impact on agriculture and costs of adaptation. International Food Policy Research Institute, Washington, DC, p 30

Nicholls H, Nicholls RJ (2008) Global Sea-level rise and coastal vulnerability. Sustain Sci 3(1):5–7

Nunn PD, Veitayaki J, Ram-Bidesi V, Venisea A (1999) Coastal issues for oceanic islands: implications for human futures. Nat Res Forum 23:195–207

Pittock AB, Jones RN (1999) Adaptation to what, and why? Environ Monit Assess 61:9–35

Poff NLR, Zimmerman JKH (2010) Ecological responses to altered flow regimes: a literature review to inform the science and management of environmental flows. Freshw Biol 55(1):194–205

Pope J (1997) Responding to coastal erosion and flooding damages. J Coast Res 13(3):704–710

Rosenzweig C, Tubiello FN (2007) Adaptation and mitigation strategies in agriculture: an analysis of potential synergies. Mitig Adapt Strateg Glob Chang 12(5):855–873. https://doi.org/10.1007/s11027-007-9103-8

Smith P, Martino D, Cai Z, Gwary D, Janzen H et al (2008) Greenhouse gas mitigation in agriculture. Philos Trans R Soc B 363:789–813

UNFCCC (2006) Technologies for adaptation to climate change. United Nations Framework Convention on Climate Change, Bonn, p 40

Veitayaki J (1998) Traditional and community-based marine resources management system in Fiji: an evolving integrated process. Coast Manag 26:47–60

Strategies and Collaborations for Management of Coastal Areas

13

Abstract

The coastal area is the dynamic interface between the land and the sea and may extend inland and seaward to a variable extent, depending upon the different deciding factors. These areas are being modified by human activities, and coastal ecosystems are under severe pressure from human exploitation. The condition is also exaggerated by climate change. Different coastal countries have devised and implemented various programmes in the past to deal with these problems with varying success. Therefore a comprehensive, integrated framework for policy planning and coastal management, called integrated coastal zone management (ICAM), has become imperative. The overall goal of ICAM programme is to ensure sustainable use of coastal natural resources and conservation of coastal ecosystems. This chapter deals with various steps involved in the planning process of ICAM, legal and institutional framework, collaboration at different levels, and underlying principles. The analysis of different situations and requirements of coastal countries supports the view that participatory coastal area development provides the best option, but involvement of local government and community accountability in coastal management will help to make efforts more effective and sustainable.

Keywords

Coastal planning · Framework · Integrated coastal area management · Community participation

13.1 Introduction

The coastal area is the interface between the land and the sea and may extend inland and seaward to a variable extent, depending upon the objectives and needs of the particular programme in any country. Tropical coastal areas because of its unique

V. Ayyam et al., *Coastal Ecosystems of the Tropics - Adaptive Management*,
https://doi.org/10.1007/978-981-13-8926-9_13

environment contain diverse and most complex ecosystems on Earth. Unlike human-defined coastal area boundaries for a particular purpose, natural boundaries are variable which includes inshore waters, intertidal areas, and extensive tracts of contiguous land (Chua 1986; ICLARM 1986; Clark 1992a, b; Boelaert-Suominen and Cullinan 1994). Coastal areas are not only economically valuable for agriculture, fisheries, forest-based products, commerce, industry, navigation, and recreation but also include ecologically important environments which provide valuable ecosystem services.

In spite of their importance and intimate interaction with human society, coastal areas are being modified, and ecosystems are severely pressurized by human activities. It is a matter of serious concern that these economic, social, and environmental benefits are at great risk and ecosystems are pushed towards degradation and species extinction. This is further aggravated by increased economic activity within the coastal zones resulting in costly damages to property, society, economy, and environment.

On the other hand, climate change events interact with and exacerbate the impacts of already existing environmental and human drivers on the natural world (IPCC 2007). Changes in climate will affect biodiversity at all levels: genes, species, habitats, and ecosystems. Temperature increase, change in rainfall patterns, drought, sea level rise, coastal flooding, damage from cyclones, and change in salinity can affect habitat suitability (FAO 1992). Populations and species can respond in several ways to large-scale environmental changes. As species respond individually and collectively, chain reactions will take place in the food webs of coastal and marine areas that will affect other species—including humans—that depend on it.

There were several attempts in the past at different planning levels across different countries to manage these coastal problems and conserve the ecosystems with varying success. The experiences proved that coastal area management is too complex to be handled by traditional sectoral planning and management at one level (NOAA/CZM 1980; Hayes 1985; Snedaker and Getter 1985). This realization led to the recognition that utmost it is not possible to conserve any one particular resource in the absence of a comprehensive, integrated framework for policy, planning, and management such as integrated coastal zone/area management. Further planning for integrated coastal area management (ICAM) must be well coordinated and structured between sectoral implementing agencies at regional level with strong support derived from national and international cooperation. A balanced management perspective is vital for integrated coastal management in which inter-sectoral relationships are fully understood, trade-offs recognized and anticipated, benefits and alternatives critically assessed, appropriate management interventions identified and implemented, and necessary institutional and organizational arrangements worked out (FAO 1992). An overview of processes, framework, and principles involved in integrated coastal area management is given in Fig. 13.1. The details are discussed in the subsequent sections.

The framework clearly indicates that resource conservation and economic development concerns should be balanced and addressed adequately specific to coastal ecosystems and must be combined and integrated at different levels. Well-planned,

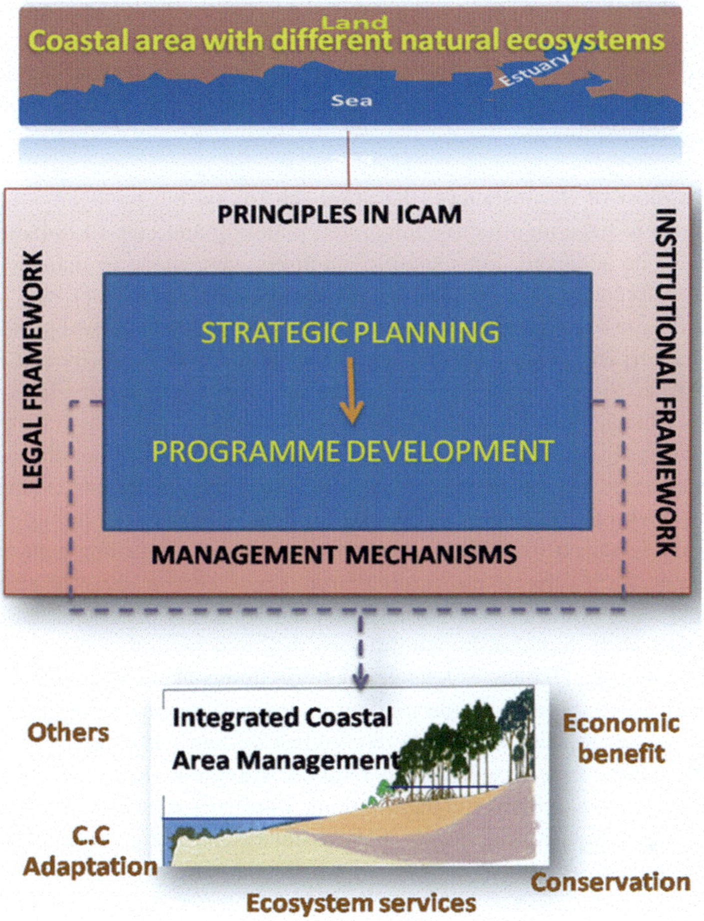

Fig. 13.1 An overview of processes of integrated coastal area management

conservation-oriented development planning of coastal zone will provide more opportunity to the economic and social prosperity of coastal communities in long term. In contrast environmentally destructive development will sooner or later have a negative socio-economic impact. This chapter provides an overview of need, scope, and process of coastal zone management for resource conservation and ecosystems functioning in tropical coastal countries to get sustainable economic benefit.

13.2 Problems Associated with Coastal Management

The experiences of tropical coastal countries in coastal management revealed that failure to fully harness the potential for sustainable coastal resource use and conservation is not because of lack of technology and information to fix the problems but failure to utilize this knowledge and technology to good effect. Here 'unsuccessful' on the part of tropical coastal countries means failure/partial success in achieving the desired results by utilizing the available resources to accomplish the planned target. There are four prominent reasons for unsuccessful in coastal management plan among other things.

- First, in most of the instances, the approach to coastal management has been inadequate to have an effective integrated planning and lack of holistic view of what is to be achieved. For example, attempts were made to manage fisheries without reference to the impacts of pollution on the ecosystem which provide habitat for the fishes. In sub-Saharan Africa, the newly developed ports and harbours lie perpendicular, or nearly so, to the littoral zone, thereby causing acute downdrift erosion. This problem has been documented in Benin, Togo, Nigeria, Liberia, Ghana, Cote d'Ivoire, South Africa, Tanzania, and Somalia, among others. In most of these cases, attempts at solving the harbour-induced erosion have further exacerbated the problem (Ibe 1986). Similarly we manage water quality, without regard to the impacts of overfishing and habitat destruction. Analysing this failure showed that the tendency to tackle specific rather than the whole problem is due to the administrative structure of having separate agencies for fisheries, agriculture, conservation, ports, harbours, etc. It should be addressed holistically with shared responsibilities by different sectors.
- Secondly in general, there is a tendency to do things at small a scale in space or time. But it is to be understood that environmental management is really about managing our impacts so that the ecological system can operate successfully. Managing large coastal ecosystem requires sufficient time and collaboration with agencies in neighbouring jurisdictions. Implicitly it means that the results of a reduction of negative impacts on the coastal systems may not be observable in a short period of time or even longer and the efforts have not been institutionalized.
- Third important reason for ineffective management was due to flaw in developing inclusive planning. This means that the planning process fails to recognize or place human at the centre of the planning and management. And it should also be realized that humans have cultural, societal, religious, and traditional belief systems that strongly influence their responsiveness to new ideas. Success in management could be achieved only by working with people, over long periods of time to build their understanding of the need, and their receptivity to the actions to be implemented.
- Lastly the identified leadership at local or regional level lacks the tools and resources to effectively implement it. In some places the leadership lacks knowledge/understanding of the problems of coastal areas.

13.3 The Need and Guidelines for ICAM

In most of the tropical coastal countries, it is becoming more and more difficult to manage any one selected natural resource or promote one economic sector in isolation, in opposition to comprehensive, integrated framework for policy planning and coastal management. As the natural resources and problems of coastal areas are so different from their terrestrial counterparts, it requires a different approach for its effective management. Threats to the productivity of these unique resource systems arise from development activities and their side-effects, such as reef and beach mining, shoreline filling, marine construction, lagoon pollution, sedimentation, and other activities that are distinct from those on land. Therefore, a special and distinct management methodology is required, such as integrated coastal management programme.

Experiences of some of the coastal countries in coordinated and integrated programme proved to be an effective framework for dealing with conflicts arising from interactions of the various uses of coastal areas. The integrated coastal management approach helps to incorporate modern principles of planning and resources management, intensive information bases, and interdisciplinary processes. Fishing, agriculture, mining, shipping, defence, public health, and recreation are complex activities requiring cooperative management and inter-sectoral coordination. It requires the full involvement of all the stakeholders through a comprehensive and integrated programme.

For many countries, there is urgency about adopting a systematic programme to conserve coastal resources from further degradation as the pace of degradation and extent has become high. This is more severe in West African coast (Portmann et al. 1989) and Indonesia (Siry 2006) needing integrated coastal management. Similarly the world population growth and migrations, six out of ten people now live within 60 km of the coast, and the population of the coastal zone is projected to double within the next 20–30 years. In addition, global environmental change (e.g. sea level rise) may have its biggest impact on coastal areas (IUCN 1990). For these reasons it is highly essential to have integrated developmental programme in place.

Such programme will help to coordinate and integrate the interests of all appropriate economic sectors and stakeholders (Caddy 1990). It is particularly useful in solving problems that exist between various sectors, for example, in resolving conflicts among fisheries, tourism, oil and gas development, and public works where these sectors are all attempting to use the coastal zone simultaneously (Sorensen and McCreary 1990). Further it provides for developing national support and international cooperation in dealing with the shared interest of human beings.

The overall spirit of the ICAM or its relevant policy or legal mechanism in any coastal country could derive its guiding principles from the 1992 United Nations Conference on Environment and Development (UNCED 1993). This provides significant support for integrated management of coastal areas. Chapter 17 of Agenda 21 outlines a commitment by coastal nations to 'integrated management and sustainable development of coastal areas and the marine environment under their national jurisdiction'. Paragraph 17.5 of Programme A of Chapter 17 sets out the guidelines for integrated coastal area management as being to:

- Provide for an integrated policy and decision-making process to promote compatibility and a balance of uses
- Identify existing and projected uses of coastal areas and their interactions
- Concentrate on well-defined issues
- Apply preventive and precautionary approaches in planning and implementation
- Promote the application and development of methods that reflect changes in value resulting from uses of marine and coastal areas, including pollution, marine erosion, loss of resources, and habitat destruction
- As far as possible, provide access, for concerned individuals, groups, and organizations to relevant information and opportunities for consultation and participation in planning and decision-making at appropriate levels

13.4 Objectives of ICAM

In common parlance integrated coastal area management refers to a special type of programme for coastal countries developed for the purpose of conserving coastal resources or environments (Sorensen and McCreary 1990) through proper management of developmental activities. Use of the term coastal area management implies that the governmental unit administering the programme has distinguished a special coastal area which is known as 'zone' having a geographic extend and combines both ocean and terrestrial domains. Hence it is emphasized that the extent of zone may vary with counties and purpose. At the same time, if the government is not notified of the zone, then it is known as area having geographical extent in the coastal region. Therefore, in the planning process, it is preferable to use the term ICAM. Such programme also includes the interests of all parties and different economic sectors that are stakeholders in coastal development. This latter aspect 'the involvement of all parties of interest' is what distinguishes 'integrated coastal zone management'. To be successful, ICAM has to be a distinct process focusing on distinct issues with clear and unambiguous goals and objectives.

The overall goal of the ICAM programme should be to ensure optimum sustainable use of coastal natural resources, uncompromised maintenance of high levels of biodiversity, and proper conservation of critical habitats. In this context, the broad and tangible objectives of ICAM include enhancing agricultural production, supporting fisheries, protecting the community from storm ravages, handling tourists, promoting public health, sustaining services from mangrove forests, preserving coral reefs, and finally providing livelihood support to coastal communities. All these activities require coordinated community action for their accomplishment.

At ground level the major objective of ICAM is to coordinate the initiatives of the various coastal economic sectors (e.g. shipping, agriculture, fisheries) towards long-term optimal socio-economic outcomes, including resolution of conflicts between sectors and arranging beneficial trade-offs. This means that a multi-sector approach could jointly guide the activities of the key economic sectors under an effective coastal planning and management system centred on coastal ecosystem conservation (FAO 1992).

As an example of the above, both coastal tourism and fisheries depend to a large extent on a high level of environmental quality, including coastal water quality. Both sectors are impacted by pollution, loss of wildlife habitat, and aesthetic degradation from uncontrolled oil and gas development. Therefore, tourism and fisheries sectors should collaborate and work through ICZM to extract safeguards from the oil industry.

At the same time, it should be understood that resource management and environmental conservation aimed at providing livelihood are the motivations for ICAM planning and management programmes which should be compatible with economic growth. In fact, enhanced long-term economic development may be the overall driving force which advocates the integrated coastal management to reduce the negative impacts in the coastal areas. In addition, a comprehensive ICAM programme needs to include the following components for any meaningful impact, viz.:

- Arrangements of policy, goals, legal authorization, and an enforcement mechanism
- Coordination with several sectors through a coordination office
- A project review/permit mechanism and environmental impact assessment procedure call upon scientist to review these things
- Promote cooperation and coordination at different levels and protection of common interest of humankind

13.5 Basic Frameworks of ICAM

As stated earlier in this chapter, maintaining the productivity of coastal natural resources alongside the conservation of coastal ecosystems requires a holistic and comprehensive approach. A first step towards this certain basic legal and institutional framework based on common and sound principles should be developed to ensure its successful implementation and continuation (Fig. 13.1). In the same context, it is also necessary to define a broad management zone—one extending from the coastal hinterlands and lowlands (terrestrial or the dry side) to the coastal waters and the deep sea (aquatic or the wet side). Followed by this a multi-sector management programme must be devised so that all stakeholders and all concerned government agencies are involved (FAO 1992). The framework for management should also seek broad public support for its successful implementation and follow-up.

13.5.1 The Legal Framework

13.5.1.1 Importance of Law

For successful implementation of ICAM, proper legal framework (law) is essential. In the absence of clear understanding of the legal and institutional arrangements governing coastal management and appropriate legal mechanisms to implement it, the long-term prospects of an ICAM initiative will be seriously jeopardized. Implementing ICAM involves changes, modifications, and new legal establishments to ensure the implementation. Accordingly the boundaries of a managed coastal area may change as different problems are addressed at different scale/management units (Fig. 13.2). Every coastal country aiming at ICAM might attempt at changing the way existing institutions operate, creating new institutions, changing the rights of users of coastal resources, and introducing new mechanisms to regulate human activities within, or that may affect, coastal areas (FAO 1992). This will invariably require repealing or amending existing legislation and when ever need arises enacting entirely new legislation. It can be expected that the implementation of these changes may also give rise to new conflicts between implementing agencies, between implementing agencies and other stakeholders in the coastal area, and between stakeholders. In addition, legislation can be used to provide tools and mechanisms to enable and facilitate the integrated management of coastal areas and to establish mechanisms for resolving disputes (Sorensen and McCreary 1990).

13.5.1.2 Rights to Own and Use Coastal Resources

Another important policy decision required in ICAM is whether to change the regime governing property or user rights in order to attain the objectives. This strongly requires ground-level understanding of the functioning of the social system and resource use. There is wide difference and several possibilities as seen from the existing pattern of resource use. For example, in many of the tribal areas, tribal

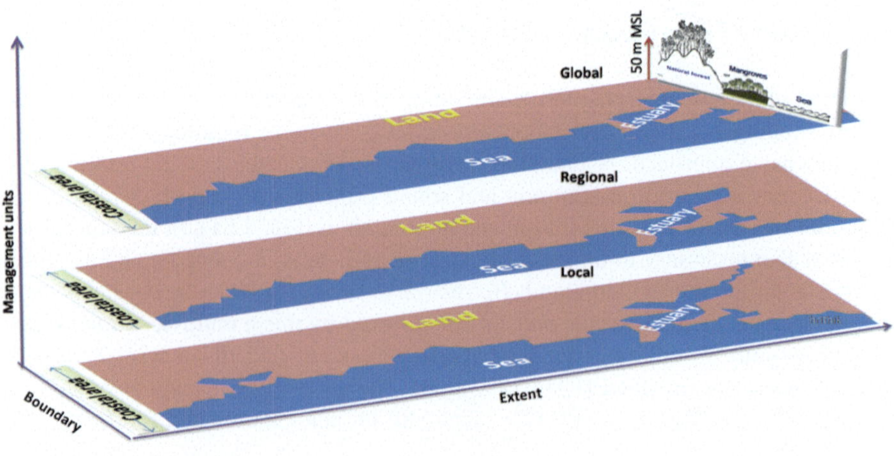

Fig. 13.2 Changing boundaries of managed coastal areas at different management units

society collectively owns the resources, and in highly commercialized areas, individual or group owns the right to harness the resources. This requires understanding the function of different property regimes and governance arrangements; without this, serious policy errors can be made (Mathiews 1995). Legally the concept of ownership has been described as a 'container' of a number of rights in relation to the property which means the right to possess it, to use it, to enjoy its fruits, and to dispose of it. The owner may transfer particular rights to other parties for a period on agreed terms (rent a tract of land for grazing for 3 years), but the container of ownership remains with the owner. In many of the tribal societies in tropical island nations, the ownership remains with the society as a whole, but user rights are granted to the members of the tribal group based on their social norms. The criteria for categorization may vary between legal systems, but many legal systems incorporate distinctions. When property rights are discussed from an economic perspective, distinctions are often made between state property, private property, common property, and open access regimes (FAO 1992).

13.5.1.3 Evaluating Regulatory Frameworks

It is highly appreciable to make initial assessment of existing regulatory mechanism of ICAM aimed at evaluating the extent to which existing laws, property and use rights, and institutional arrangements are promoting or hampering the attainment of the objectives of the coastal management programme. Based on the outcome, required modifications can be planned. Most of the existing laws are applicable mainly to a particular sector or human activity and are administered by the line government agency responsible for that sector. However, since ICAM is concerned with area management, it will have to necessarily review a wide range of laws, administrative regulations, institutional practices, and customary laws. The fundamental reason is to effectively address the needs of the ICAM. Therefore ICAM should include the following:

- ICAM should clearly define the rights and duties of owners and users of coastal resources in line with the policy decision of a particular coastal country, for example, the constitution and laws governing land tenure and water use rights.
- It is essential to appropriately govern the management and exploitation of natural resources such as laws regulating land tenure, human settlements, mining, oil and gas exploitation, industrial development, pollution control, tourism, and other primary activities. Any gap in these aspects if identified should be covered with suitable regulations.
- Provide a structure within which human and other resources can be organized and coordinated for the purpose of attaining long-term strategic objectives of economic development.
- Promote the conservation and protection of the environment, wildlife, and biological diversity.
- Establish mechanisms such as conciliation, mediation, and arbitration for averting or resolving disputes concerning the use of coastal resources.

Sometimes the obligations, principles, standards, recommendations, etc. contained in international agreements and documents may provide a useful measure against which to evaluate national legal frameworks for ICAM. In addition, it would be highly appreciable for drafters of national laws to take into account the provisions of binding international treaties, such as the United Nations Convention on the Law of the Sea, 1982, and other conventions such as Convention on Biological Diversity (Rio de Janeiro, 5 June 1992); the Convention for the Protection of the World Cultural and Natural Heritage (Paris, 16 November 1972); and the Ramsar Convention on Wetlands (Ramsar, 2 February 1971).

13.5.1.4 Legal Principles

There should be some fundamental legal principles based on which every coastal country attempts to devise its own framework for ICAM in consistent with the existing international and national laws. In the process it is inappropriate to transplant a law from one legal system to another where the social context and legal traditions are very different. In this context, in order to promote ICAM, some of the principles such as the precautionary principle, the principle of preventive action, the polluter pays principle, rational and equitable use of natural resources, and responsibility to avoid transboundary environmental damage could be incorporated into national laws in order to promote ICAM. Most of them are derived from international environmental laws. At the same time, it is emphasized that these principles are likely to provide practical guidance to the drafters of national legislation rather than on attempting to discuss all relevant principles (Boelaert-Suominen and Cullinan 1994).

13.5.2 The Institutional Framework

During the policy formulation process for ICAM, goals and objectives are determined based on the national priorities and international obligations; and the means to achieve them are formulated as strategies and plans. It is expected that the process may vary considerably from one country to another which is reflected in the institutional arrangements supported by legal framework. In general, however, the setting of area goals and the strategy to achieve them is undertaken by a coordinating agency in conjunction with the concerned line departments, other government agencies, and representatives of other stakeholders. The plans through which the strategy is implemented are formulated by the responsible executing agencies, usually the line ministries and subnational and local governments, often in conjunction with non-governmental organizations. Consultation with local people is highly desirable but rarely undertaken. As a result when ICAM plans are prepared as stand-alone plans—distinct from national or subnational development plans and from sectoral development and renewable resource management plans—it is often difficult to find the human and financial resources to successfully implement them. In contrast, when ICAM is an integral part of sectoral development and renewable resource management plans, this problem may be less pronounced.

13.5.2.1 Defining the Boundaries of ICAM

As discussed earlier, the coastal area is the interface between the land and the sea and may extend inland and seaward to a variable extent, depending upon the objectives and needs of the particular programme in any country. While adopting the principle of ICAM, governments have to firmly decide whether to apply ICAM at the area, subnational or national level. Coastal countries, especially small island countries, need to adopt a national ICAM policy considering their geographical extent, but other countries can also have specific ICAM plans only for certain threatened parts of the coast rather than for the entire coastal area of the country. However, the absence of an ICAM plan does not necessarily imply that coastal resources are not managed. In sectoral plans, plans for environmental management or local community management may fulfil coastal management requirements successfully. In most coastal states, there are differences in climatic, biophysical, economic, social, political, institutional, and cultural variety among different parts of the coast. Therefore, different resource allocation decisions and, perhaps, different routes to implementing them will be required. Moreover, with the exception of small island states, the complexity of national coastal areas makes it impractical to manage at the national level. In most countries, therefore (other than for small states), coastal management is usually undertaken at the area level (FAO 1998).

A mandate to manage a narrow strip of transitional area, comprised of mostly tidally influenced habitats, may win approval more easily than a broader zone including deeper parts of the sea and higher and drier parts of the shore lands. The zone should include, at least, the coastal lowlands, intertidal areas, salt marshes, wetlands and beaches, and offshore features such as coral reefs and island habitats. It is often the coastal common property resources of the 'wet side' that are emphasized in ICZM rather than the private property of the 'dry side'. These resources can be adequately managed by a programme that gives priority to the edge zone. The relationship between national, sectoral, and integrated coastal area planning is illustrated in Fig. 13.3. It shows that ICAM should correspond with the sectoral and national planning for greater acceptance and adoption (FAO 1992).

There are many factors influencing a boundary such as biophysical, economic, social, jurisdictional, and/or organizational in character. For these reasons the boundaries of a managed coastal area may change as different problems are addressed. Table 13.1 shows coastal zone boundaries of different coastal countries. Experiences of successful integrated coastal management stress the need to ensure that the area under management encompasses the source of the environmental changes that have an adverse effect in the coastal area. Consequently some areas defined for integrated coastal management may be relatively small, and others may be larger than some countries (FAO 1992). Although ICAM is usually area-based, it is desirable to have a national ICAM outline plan to ensure a proper legal framework and integration with national development planning and meet international obligations.

It is also advisable to limit initial implementation of ICAM to small selected pilot sites, which provide an opportunity to explore the broader institutional requirements and demonstrations prior to the expansion of ICAM to larger and more

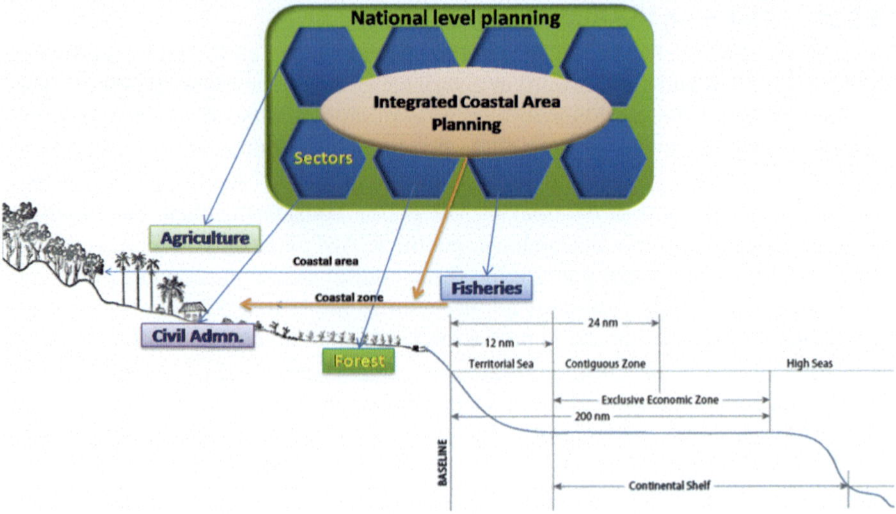

Fig. 13.3 Relationship between national, sectoral, and integrated coastal area planning. (Modified from FAO 1992)

Table 13.1 Coastal zone boundaries of different countries

Country	Inland boundary	Ocean boundary
Brazil	2 km from MHW	12 km from MHW
China	10 km from MHW	15 m isobath (depth)
Costa Rica	200 m from MHW	MLW
India	500 m from HTL	LTL
Israel	1–2 km variable	500 MLW
Madagascar	300 m from MHW	Variable
Philippines	Flexible	Continental shelf
Queensland	400 m from MHW	3 NM from the CB
South Australia	100 m from MHW	3 NM from the CB
Spain	500 m from highest storm or tide line	12 NM (limit of territorial sea)
Sri Lanka	300 m from MHW	2 km from MLW
Sweden	100–300 m	2 NM
United Kingdom	Flexible	12 NM
USA	Flexible	12 NM
Zimbabwe	300 m from MHW	1–2 km variable

complex sites. Accordingly with the progress of time, the boundaries of coastal management will also vary.

13.5.2.2 Governance and Institutional Arrangements

Effective coastal management is dependent on the existence of appropriate institutions and organizations. Changes can be gradually introduced where these arrangements are found inadequate or constraining. At the beginning programme activities will have to be adapted to existing institutions and organizations; later adjustments are phased over longer term. Analysing the present coastal area management situation points to the fact that only few countries have actually start with comprehensive, enabling legislation for coastal area management on a nationwide basis. Many countries take more expedient, piecemeal approaches in which existing laws and policies are amended or coordinated by means of inter-agency guidelines, or new, complementary laws are enacted (FAO 1998). At the same time, it is observed that only few countries have the capacity to deal with nationwide management of coastal areas followed by its application to local situations. If organized correctly, pilot site plans can be used as a testing ground for domestic capabilities in technical, institutional, and organizational approaches for managing coastal areas. The successful model can also be upscaled to other similar areas.

Almost in all the coastal countries, there could be a range of possible institutional arrangements, but coastal governance is a long-term undertaking that requires stepwise changes in behaviour and institutional reform to overcome prevailing unsustainable path dependencies (Olsen et al. 2009). In general this can be grouped into multi-sectoral integration and systematic integration. In multi-sectoral integration, common goals are translated into independent sectoral planning by coordinating the various agencies responsible for coastal management. This is done on the basis of a common policy and bringing together all the stakeholders to work in cooperation towards achieving the common goals by following mutually agreed strategies. Whereas in systematic integration, an entirely new, integrated institutional structure is created by placing management, development, and policy initiatives under a single institution. But this essentially requires cooperative goal definition, planning, and dynamic action. In practice a well-planned programme with proper coordination will help to harness the strength of various government departments, local organizations, and other stakeholders to achieve the common goal. Further, line ministries or existing government agencies are typically highly protective of their core responsibilities, as these relate directly to their power base and funding. The establishment of an organization with broad administrative responsibilities overlapping the traditional jurisdictions is likely to meet with resistance rather than cooperation (FAO 1992). Therefore, entrusting with agreed responsibilities to achieve common goal through integration and coordination should be preferred.

As there are different players and issues involved in the development of coastal areas, integration is an important mechanism for successful implementation of ICAM. Three distinct types of integration are important for ICAM, viz. systems integration, functional integration, and policy integration:

- Systems integration relates to the need to ensure that all relevant interactions and issues concerning coastal development, regardless of their origins, are considered.
- Functional integration aims to ensure that management interventions are consistent with coastal area management goals and objectives. This brings more focus on the management.

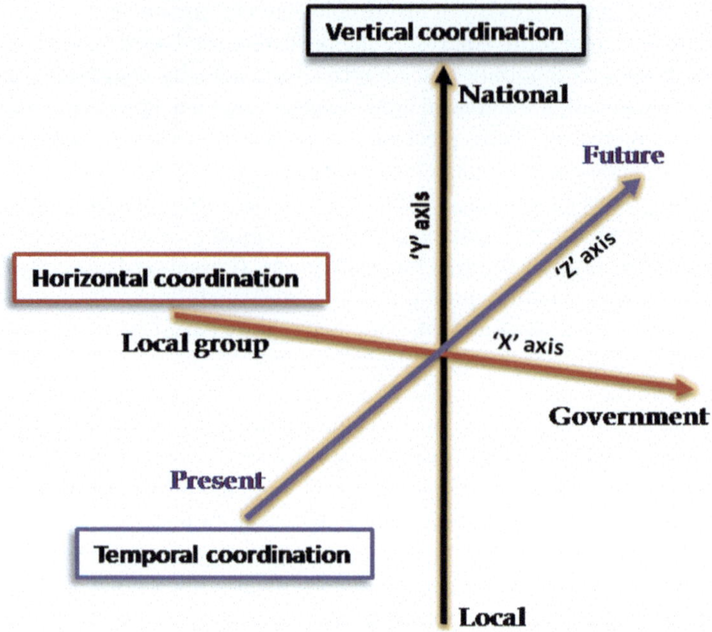

Fig. 13.4 Schematic representation of coordination at various levels in ICAM

- Policy integration concerns the need to incorporate management policies, strategies, and plans with development policies, strategies, and plans.

Similar to integration, coordination is an important mechanism in coastal area management which brings together disparate resource users, community groups, non-governmental organizations, research institutions, and government agencies at various levels, including central, regional, and local. The hierarchical relationship between national, sectoral, and integrated coastal area planning, and a different coordination is illustrated in Fig. 13.4 (modified from FAO 1998). A well-structured coordination provides a forum for negotiation and conflict resolution by minimizing overlaps and gaps in responsibilities. It is very important that coordination should happen at several levels, viz. vertical, horizontal, and temporal.

- Vertical coordination refers to cooperation among various hierarchical levels of government, viz. central, regional, and local. This will also help in sharing responsibilities and greater understanding with reference to ICAM.
- Horizontal coordination focuses on gaining cooperation within a specific level of hierarchy, such as at the local level among local government, sectors, and various stakeholders. This facilitates for addressing a local problem more effectively.
- Temporal coordination refers to achieving optimal phasing of management actions which helps in continuity and target accomplishments.

From the foregoing discussions, it is evident that ICAM plan should be incorporated into national economic and sector development plans to ensure appropriate institutional and budgetary support. It should also ensure that ICAM goals are compatible with the national and sectoral development plans.

13.5.2.3 Prerequisites for ICAM

In general, whatever the exact nature of planning and implementation process, the success of ICAM will revolve around active public involvement at decision-making and due consideration for cultural system of the locals besides an effective coordinating mechanism. Therefore ICAM process should bring together a variety of social actors, ranging from resource users and communities to legal, recognized, or customary institutions, from various hierarchical levels as a prerequisite before its implementation. It should also:

- Identify various levels or types of public involvement
- Share information in the form of dissemination to groups affected or interested in a decision
- Do consultation where feedback is sought from affected groups and interested parties prior to the making of decisions
- Ensure participation by multiple stakeholders in the decision-making process itself

The second important prerequisites for successful ICAM are working towards a higher human dimension, where project activities are integrated into the culture of people. The participatory process can develop an innovative mix of traditional and 'imported' expertise, according to each specific context experienced in different locations in a coastal country.

13.5.2.4 Institutional Coordination and Enabling Environment

The selection of the most appropriate institution is a critical part of the planning and implementation process as a whole. The criteria that should be considered in this selection are:

- Institutional experience in integrated planning and ability to survive changes and adapt
- Compatibility between the proposed role as coordinating agency and the institution's mandate
- Strong constituency base and ability to perform and lever link with required government and other stakeholders
- A strong broad-based institutional capacity
- A holistic approach to policy issues

In addition to the institutional coordination, there should be some favourable condition to work effectively. This includes:

- Broad agreement on respective roles, jurisdictional boundaries, provision of resources, and responsibilities at different levels of governance
- Adoption of a holistic approach and provision for effective negotiation (bargaining) between different levels of governance and among the stakeholders
- Information sharing, joint assessment, and acceptance of costs and benefits of environmental changes and proposed policy interventions, when appropriate
- Participation by all stakeholders and due consideration for livelihood of local communities and provision for alternative solutions

13.5.2.5 Building Capacities for ICAM

Because coastal problems involve a range of actors and a multiplicity of complex interrelated issues, an institutional capacity to resolve conflicts and institutional supervision will be required to manage the participation and relationships of interested parties and the different interests represented, as well as to identify suitable options and agree on objectives and means to attain them.

- The institutions contributing to ICAM should have the collective capacity to allocate resources (man, machinery, and finance) according to the goals and strategic approach adopted. Not all individual participating organizations should be able to satisfy this requirement, but it is important that the system as a whole is able to deliver the following services (Olsen 1993; Osborne 1993): Thus, forming a team that is proficient in strategy and methods requires in-service training.
- During implementation, local staff should be exposed to a variety of learning experiences relevant to the coastal management. In addition, the participatory monitoring and evaluation process should have a continuing educational and formative orientation by identifying lessons learned and further learning needs.
- A mix of hiring ad hoc project staff and mobilizing professionals from relevant line agencies and institutions would allow higher efficiency and wider dissemination of the approach among local institutions.
- Depending on local circumstances, consideration should also be given to establishing funding mechanisms for conflict resolution and providing training and education for third parties to mediate, facilitate, or arbitrate in disputes.
- Administrative procedures and dispute resolution offices or commissions might also be established to anticipate and resolve conflicts arising within and between institutions.

13.5.2.6 Sustainability Indicators

Several indicators are increasingly used to provide a convenient description of the current state of available resources, weigh the performance of any programme, and predict the system responses. However, these indicators need to be developed according to their perceived applications which require reliable statistics and raw data. Indicators could be used to monitor and evaluate the changes, the processes, and the sustainability of beneficial changes. Different types of indicators are developed depending on the scale (global, national, and local) and the level of aggregation and purpose (ecological and economical).

In the context of coastal area management, indicators of ecological sustainability (land suitability, pollution impacts) can be combined with economic data to construct scenarios based on available information. By including in the cost of production the cost of the deterioration in the natural resource base, policy choices can be made aimed at mitigating the negative impacts on natural resources, and this will help to allocate resources more efficiently. Placing a reasonable value on the costs and benefits of preserving natural resources is essential in order to make a workable estimate of externalities or indirect costs. The criteria guiding the selection of sustainability indicators for ICAM are given below:

- Policy relevance: Ensure that the indicators should address environmental changes of primary concern in the country or area where they are being applied.
- Predictability: The indicators should provide information that anticipates environmental change and thus allows decisions to be taken before change becomes irreversible.
- Performance: It should ensure human and institutional capacities are able to manage the policy and planning process effectively through participatory and transparent approaches.
- Measurability: The indicators should provide information in quantitative or qualitative terms. So that it can be either convertible to a monetary value or used for informed decision-making.
- Interdependency: The selected indicators should provide information that clearly links cause and effect and reflects the dynamism between the ecological, social, and economic environments. This may help to better understand the interdependency and the driving forces that need to be measured.

Some of the common indicators are human development indicators, environmental indicators, performance indicators, interdependence indicators, and pressure-state-response indicators (PSR). Among all indicators, PSR is the most widely used framework for sustainability measurement which addresses the chain of events that lead to environmental impacts.

13.6 The Policy and Planning Process

The main ICAM process is centred on the successive completion of process cycles. In turn each cycle addresses management issues, formulates and implements strategies and plans, and monitors and evaluates the performance. The basic assumption is that an appropriate strategic basis and plan will lead to sound implementation. The process is divided into steps, for which individual definition varies slightly between authorities but that, in principle, are inception, performance review, data collection and research, analysis, strategy formulation, plan formulation/absorption into sectoral plans, and plan implementation. The ICAM process should be viewed as long-term, continuous, and iterative, as shown in Fig. 13.5.

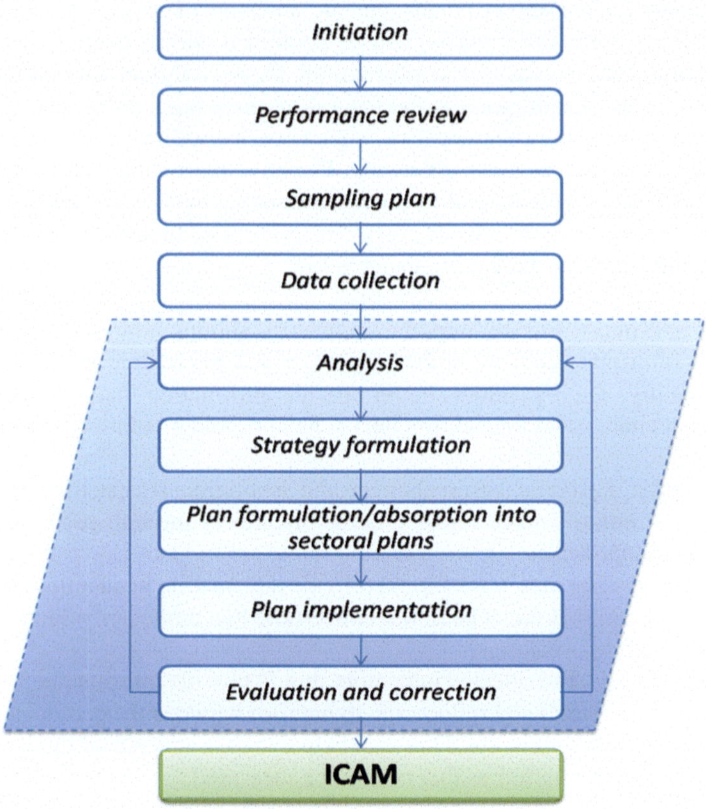

Fig. 13.5 Policy and strategic planning in ICAM

The first cycle usually begins with the most urgent issue and has a limited geographical and institutional coverage. Through successive cycles, the scope and scale of the strategy and plan are increased to incorporate new and more complex problems and stakeholders. This is followed by performance review including monitoring and evaluation which is an integral element throughout the process. Performance review not only suggests regular revisions and adjustments during the implementation phase but represents the motor that moves ICAM plans into successive cycles. Thus, as achievements are consolidated and confidence is gained, there will be an evolution from a small ICAM pilot project to an expanded application resulting in complete, integrated, and effective national-level coastal management programme.

The end products of the ICAM planning process are the formulation and implementation of a viable management strategy. This is more pertinent to developing adaptation response to island nations which are economical and provide livelihood to local communities (Ram-Bidesi et al. 2011). As stated earlier in this discussion, there is no fixed method/way to organize ICAM, rather there are different options for coastal countries to begin the process such as:

- Start with a local demonstration project
- Coordinate 'expanded' sectoral plans containing an ICAM part
- Create an entirely new integrated plan. This is essential for small island nations

The most important underlying aspect is that the process leading to the formulation of the coastal strategy involves all stakeholders and sectors, so that irrespective of any implementation path, the agreed goals and objectives are included. For example, in Indonesia, decentralized planning process was used. It attempted to manage coastal zone through a cooperative process, building trust and common understanding with strong political will, strengthening local government capacity, and enhancing upward and downward accountability mechanisms in handling multiple users, interests, and stakeholders in coastal zone (Siry 2006).

Based on the above discussions, it can be concluded that throughout the ICAM process, negotiation will constitute a base for sound participatory decision-making, and conflict-solving should be based on the dominant cultural values. Negotiation involving a variety of actors will be necessary to represent different views. This mechanism also paves the way for hearing from different interest groups present in coastal communities with various expectations for solutions who face a range of different institutional, operational, and financial constraints. This also takes into account the needs felt by local actors and the normative needs as identified by the technical staff of collaborating institutions (FAO 1998).

13.7 Legally Enforceable Management Mechanisms

The coastal zone is used extensively for a large number of activities. These multiple uses are not always compatible with the natural process resulting in wide array of problems for resource users and decision-makers. Coastal land is used for human settlement, agriculture, trade, industry, and amenity. The coastal sea presents problems related to transport, fishing, dumping, mining, sedimentation, pollution, etc., stemming from an intensification and diversification of ocean uses. There is also over-exploitation of coastal biological resources and an array of conflicts due to the interaction among user groups in congested marine areas (Clark 1991). All these contrasting interests require suitable conflict-solving mechanism and other legal arrangements. Therefore, wide varieties of legal instruments have been used to promote ICAM objectives. However, identifying appropriate, practical legal mechanisms to give effect to the concepts and principles such as sustainable development that underlie ICAM remains one of the major challenges facing legislative drafters. This necessitates the development or inclusion of certain legal mechanisms for the successful development and implementation of coastal area management programme (FAO 1992). Some of the most important mechanisms based on the identified/experienced problems are pollution control, creating protected areas, critical areas, biodiversity, natural hazards, restoration, environmental and socio-economic assessment, and conflict resolution.

13.7.1 Pollution Control

Pollution control is most important for the conservation and health of coastal areas and all biological form. In most of the coastal countries, pollution control authority is operating, and such authority will retain continuing responsibility for pollution in coastal area as well. So, the ICZM plan or any development project should confirm with it. In addition, ICZM programmes should focus on special coastal pollution sources not presently addressed, perhaps those caused by watershed disturbance and runoff and, especially, those that affect critical habitats. Of particular concern is sediment runoff from construction sites, farmlands, and forest-cutting and land-clearing operations. Also, pollutants flushed by storm runoff (sewage, toxic industrial waste) into coastal waters can create toxicity (biocides, oil wastes, etc.) and bring excessive nutrients (fertilizer, animal wastes). Special attention should be paid to controlling industrial wastes and improving sewage treatment to reduce risks to public health from consuming seafood and bathing in contaminated water (IUCN/UNEP/WWF 1991).

13.7.2 Protected Areas

The most certain way to conserve biodiversity and protect critical areas is to acquire ownership or just assert ownership when the areas are already owned by government. While the primary ICZM programme utilizes the regulatory powers of government to achieve resource conservation, by controlling private development activities and resolving potentially divisive conflicts among competing users of the coastal zone, this accomplishes only a part of the job. Additional and more focused natural area conservation can be accomplished by exercising the proprietary rights (rights by virtue of ownership) of government through declaration of protected areas—resource reserves, natural areas, and/or national parks. In this case the legal authority (government agency/department) has the right to determine the way their land or water areas are used, including the right to deny access. Clearly the system of protected areas will be more effective if it is addressed by the ICZM programme. Conversely, protected area planning is essential to the full development of a country's ICZM programme. In other words, coastal and marine nature reserves should form a complementary and integral part of ICZM programmes. Such reserves are a necessary component of coastal resource stewardship.

Protected areas complement and help to accomplish other objectives of ICZM by conserving nursery areas for fisheries production, enhancing tourism revenues and recreational benefits, preserving wilderness values, promoting baseline scientific and management studies, and so forth. Together PA and ICZM ensure the maintenance of a healthy resource base upon which to build sustainable coastal development. It is also possible for the resource users and private property owners to assist with coastal resource conservation through voluntary actions and through provision of incentives.

13.7.3 Critical Areas

The coastal zone is essential to marine life and supports a large part of the world's living marine resources, certainly more than the open sea. While the whole ecosystem is the focus of management, each of its structural parts and essential processes must be conserved—with the emphasis placed on critical areas. In the development of ICZM programme, priority should be given to identify the special coastal habitats that should be given a high degree of protection as 'critical areas'. Detailed information on them should be available to all interests, particularly to resource users like fishermen and the public and the private development community. Although ICZM programmes are created primarily to manage physical development, they have to be based upon detailed knowledge of resource vulnerabilities and a candid examination of conservation issues. In tropical coastal countries, ICZM conservation programmes may be centred on mangrove forests, coral reefs, submerged seagrass meadows, beach-dune systems, and lagoons/estuaries (including embayment) known to be especially valuable. While it is useful and practical to focus on individual habitat types, one must not forget that they exist only as components of wider coastal systems; therefore, they should be managed as parts of a greater system. Habitats identified as critical areas may provide sanctuary to:

- Valuable species during vulnerable phases of their lives (shrimp, seabirds, fishes)
- Endangered species (whales and turtles)
- Migratory species (waterfowl and wading birds)

These habitats are used for feeding and spawning, as nurseries, and for shelter. Safeguarding these critical marine habitats can help conserve species, maintain fisheries, and support tourism, thereby ensuring sustainable development.

13.7.4 Biodiversity

The very purpose and motivation for designating critical areas are to conserve biodiversity particularly for those species and the habitats that have been designated as especially valuable or in danger of extinction, viz. endangered birds, turtles, and marine mammals. These designations may be done in response to an international treaty or by independent national action. In addition to habitat management, there may be other appropriate actions to be taken under the ICZM programme such as banning exploitation of the species. Similarly mangrove protection is an example for critical habitat protection which is essential for fishes, shellfishes, and other marine life. ICZM attempts to prevent destruction of mangrove forests by constraining excessive mangrove clearing and other destructive developmental activities which are unreasonable. Protection of species and its habitats would help to avoid long-term net-negative effect on biodiversity, the coastal economy, and the welfare of coastal people. For example, the Sundarbans delta (UNESCO World Heritage site located in India and Bangladesh) is the world's largest continuous stretch of mangroves. Along with the coral reefs of the Maldives, India, and Sri Lanka and the dryland mangroves of Pakistan, it supports thousands of floral and faunal species which provide livelihood to millions of people (Paul and Nishida 2007). More details are on biodiversity given in a separate chapter on 'Biodiversity Conservation and Restorative Measures' in this book.

13.7.5 Natural Hazards

In most of the densely populated tropical coastal nations, the risks of natural disasters to inhabitants of the coastal areas are rising because of population increases, climate change events, and poorly planned development projects. Coastal people become more vulnerable to natural hazards such as floods, typhoons, or tsunamis when land reclamation projects encourage settlement in low-lying areas or when land clearing and construction removes protective vegetation, reefs, or sand dunes (Sorensen and McCreary 1990).

Reducing the impacts of hazards begins with preservation of coastal landforms and protective cover that provide natural resistance to wave attack, flooding, and erosion from hurricanes and storms (COS 2009). These landforms differ significantly around the world. Human activities that remove or degrade protective landforms, for instance, by removing beach sand, weakening coral reefs, bulldozing dunes, or destroying mangrove swamps, diminish the degree of natural protection the coast affords. For example, if dunes are removed by sand mining or because of obstruction to ocean views, the risk to coastal development behind the former dunes is greatly increased. Similarly, mangroves serve to dissipate wave energy and to stabilize the land areas behind them from the erosive forces of storms (Clark et al. 1980). In several Indian Ocean rim countries, damage to the coastal areas by the Indian Ocean tsunami of December 2004 was comparatively less behind the protective cover provided by the mangrove stand. A particularly disastrous example is Bangladesh where hundreds of thousands of people were victims in major cyclonic sea storms. Therefore, in the ICZM context, the following two management measures are the most important for natural hazard management:

- Conserve protective features: There is a need to safeguard to the maximum extent possible all the natural resources and features that protect the coast from storm surge and waves in high hazard areas (prohibit sand removal, avoid mangrove clearing, protect coral reefs, etc.).
- Establish a setback line: There is a need to delineate a coastal construction setback line landward of the 'high hazard zone' and prohibit construction seaward of it.

The conservation of protective features for hazard protection would be handled by the ICZM programme in the same manner as conservation of natural habitats.

13.7.6 Restoration

Land use change, pollution, habitat destruction, accelerated erosion, and interference with water circulation, among other effects, have seriously reduced the economic and environmental benefits of most coastal natural systems. While it is an important function of ICZM to constrain further ecological losses, it is also important to plan for the repair of natural systems damaged in the past by way of restoration. For instance, damaged or destroyed coral ecosystem may not recover fully or not at all. But mangroves and wetlands can be restored to large extent by concerted efforts. If a wetland has been diked for rice culture or aquaculture, it would be relatively easy to reconvert it to a nearly natural wetland condition. Similarly degraded dunes and beaches can be rebuilt. Mangrove saplings can be replanted in the natural habitat and in degraded areas. Polluted estuaries can be cleaned up. The technology is well developed for most such restoration projects, following the 'nature synchronous' approach. In summary, the ICZM programme should include an ecosystem rehabilitation work element as well. This will require an investigation of the condition of coastal resources, an economic evaluation of losses and benefits of rehabilitation, and recommended priorities for rehabilitation projects.

13.7.7 Environmental and Socio-economic Assessment

It is vital to have impact assessment of any developmental activity enshrined in any programme. This provides vital information to conserve and allow judicious utilization of natural resources. At the centre of the ICZM strategy and control process is the environmental impact assessment (EIA). The information and analysis generated in the EIA provide the basis for the decision on whether a coastal project is approved as is, or approved with conditions attached, or not approved (Burbridge and Maragos 1985; Beanlands 1985; Jernelov and Marinov 1990). EIA is used to predict the environmental effects of a project or a programme, providing a systematic way to review coastal project impacts, to modify them to reduce negative impacts on resources, and to reject the worst of them.

Lately it has become common to include economic and social considerations as part of the assessment. Burbridge (1987) recommends that ecological and economic impacts be jointly evaluated. With increased perception of ecological functions, it is possible to improve the economic expression of their value to society. For example, mangroves, once considered worthless unless harvested (or made into shrimp ponds), are now seen as extremely valuable ecological resources showing the importance of balancing conservation and utilization. In recent times a human ecology perspective takes into account traditional uses, rights, and special needs of tribal population residing in the coastal areas. Regional development is premised on the principle of increased social equity.

13.7.8 Conflict Resolution

As the ICZM process operates at the interface between land and water, there may be intense conflict between private (or quasi-private) property-based operations in shorelands and public (common) property-based activities in the tidelands and coastal waters. There is also possibility for sectoral conflicts or conflicts between different hierarchical levels with the government systems. Thus, the ICZM process should have an important mediating role by any legal entity or any related institutions, between conflicting wetside and dryside interests. This essential role should not be played out sector against sector; rather, the mediating/coordinating entity must look at all sectors with legitimate interests to find the most broadly compatible solutions. The objective of ICZM process is to prepare a comprehensive plan in which the various development sectors are assessed for their effects on the various resources in a given geographic area (coastal area).

13.8 Conclusion

Coastal areas are dynamic interface or transition areas between land and sea, including large inland lakes. Coastal areas are ecologically very important as they provide a number of environmental goods and services besides numerous economic benefits to the coastal communities. However, ecosystem and natural resource damage caused by anthropogenic activities in many of the coastal areas have seriously affected the normal functioning of the ecosystems and reduced its environmental goods and services.

At the same time, there are ranges of biophysical constraints that will make achieving ICAM sustainability goals challenging. During the twenty-first century, these biophysical and social constraints include energy scarcity, climate change, the loss of ecosystem services, the limitations of neoclassical economics, and human settlement patterns (Giosan et al. 2014). The multifaceted character of many of the issues and the urgent need to have a holistic approach to management and adoption of integrated coastal zone management with national support and international collaboration have become imperative. In some parts of the tropical world, ICZM may be initiated in response to a planning mandate but more often because of a crisis such as a use conflict, a severe decline in a resource or ecosystem services, or a devastating experience with natural hazard.

The boundary of coastal areas as envisaged in the coastal management plan may extend inland and seaward to a variable extent, depending upon the objectives and needs of the particular programme undertaken in any coastal country. However, small island countries need to adopt an island-specific national coastal management policy considering their geographical extent. It can be started as a strategy for the entire coastal country or can focus initially on a special management areas based on priority.

ICZM incorporates modern principles of planning and resources management and intensive information bases on interdisciplinary processes. It should be planned with the involvement of all stakeholders with a required legal and institutional framework and due consideration for national priorities and international obligations to ensure its implementation and sustainability. During its implementation there should be interaction between the stakeholders at different levels with the active involvement of coastal communities. The overall objective of ICZM is to provide for sustainable use of coastal natural resources and to conserve the ecosystems while facilitating the degraded system to recover. But experiences indicate that there is no single correct way to organize and implement ICAM strategies and plans. Each country will develop its own customized plan, as evidenced by the diversity found in existing worldwide programmes. Nevertheless, there are some basic features that are common among them. A prudent ICAM will ensure increased fisheries productivity, tourism revenues, sustained mangrove forestry, and security from natural hazard devastation. Consequently the potential long-term benefits of coastal management would be evident in environmental quality and natural area protection with sustainable economic benefit to the coastal communities.

References

Beanlands GE (1985) Basic approaches to EIA. In: Proceedings of the Caribbean Seminar on Environmental Impact Assessment. Caribbean Conservation Association, pp 64–81

Boelaert-Suominen S, Cullinan C (1994) Legal and institutional aspects of integrated coastal area management in national legislation. FAO, Rome, p 118

Burbridge PR (1987) Economic considerations in management of coastal resources. In: CAMP Network Newsletter, May 1987, pp 2–3

Burbridge PR, Maragos JE (1985) Analysis of environmental assessment and coastal resources management needs (Indonesia). Report of the International Institute for Environment and Development, Washington, DC

Caddy JF (1990) Enclosed and semi-enclosed seas: principal issues and future actions. UN/FAO, Fishery Resources and Environment Division, Informal paper, 5p

Chua TE (1986) Managing ASEAN coastal resources. Tropical coastal area management, ICLARM, Manila. 1(1):8–10

Clark JR (1991) The status of integrated coastal zone management: a global assessment. CAMPNET, University of Miami/RSMAS, Miami, 118p

Clark JR (1992a) Carrying capacity and tourism in coastal and marine areas. Parks Magazine 2(3):13–17

Clark JR (1992b) Integrated management of coastal zones. FAO Fisheries Technical Paper. No. 327. Rome, Italy, 167p. http://www.fao.org/docrep/003/T0708E/T0708E02.htm ch2

Clark JR, Banta JS Zinn JM (1980) Coastal environmental management. Fed. Ins. Admin. FLA-4, 161 p

COS (2009) Pacific ocean synthesis: scientific limited literature review of coastal and ocean threats, impacts. Center for Ocean Solutions. The Woods Center for Environment, Stanford University

FAO (1992) Integrated management of coastal zones. FAO Fisheries technical paper 327. Food and Agriculture Organization of the United Nations, Rome. http://www.fao.org/docrep/003/T0708E/T0708E01.htm#ch1

FAO (1998) Integrated coastal area management *and* agriculture, forestry and fisheries. FAO guidelines, Food and Agriculture Organization of the United Nations, Rome. http://www.fao.org/docrep/W8440e/W8440e00.htm#TopOfPage

Giosan L, Syvitski JPM, Constantinescu S, Day J (2014) Protect the world's deltas. Nature 516:31–33

Hayes MO (1985) Beach erosion. In: Clark J (ed) Coastal resources management: development case studies, Coastal Management Pub. No. 3, RPI and National Park Service, pp 67–200

Ibe (1986) Harbour development related erosion at Victoria Island, Lagos. In: Proceedings of the first international conference on geomorphology, Manchester, UK: 15–21 September 1985, International Geomorphology, Part 1, pp 457–465

ICLARM (1986) Tropical coastal area management newsletter, 1, 1. International Center for Living Aquatic Resources Management, Manila

IPCC (2007) Assessing key vulnerabilities and the risk from climate change. In: Parry ML, Canziani OF, Palutikof JP, van der Linden PJ, Hanson CE (eds) Climate change 2007: impacts, adaptation and vulnerability. Contribution of Working Group II to the fourth assessment report of the Intergovernmental Panel on Climate Change. Cambridge University Press, Cambridge, pp 779–810

IUCN (1990) Caring for the world. A strategy for sustainability. IUCN, Gland. Second Draft, 135 p

IUCN/UNEP/WWF (1991) Caring for the earth: a strategy for sustainable living. IUCN/UNEP/WWF, Gland, 228 pp

Jernelov A, Marinov U (1990) An approach to environmental impact assessment for projects affecting the marine environment. UNEP, Regional Seas Repts and Studies, No. 122, p 37

Mathiews DR (1995) Commons versus open access: the collapse of Canada's east coast fishery. Ecologist 25:213

NOAA/CZM (1980) Coastal management program for the Commonwealth of Puerto Rico. NOAA, Office of CZM, Washington, DC, p 16

Olsen SB (1993) Will integrated coastal management programs be sustainable?: the constituency problem. Ocean Coast Manag 21:201–225

Olsen SB, Page GG, Ochoa E (2009) The analysis of governance responses to ecosystem change: a handbook for assembling a baseline. LOICZ Reports and Studies no. 34. GKSS Research Center, Geesthacht. http://www.loicz.org/imperia/md/content/loicz/print/rsreports/

Paul T, Nishida J (2007) South Asia region, annual review – 2007–08. South Asia Sustainable Development Sector Department. World Bank, www.worldbank.org/sar

Portmann JE, Biney C, Ibe AC, Zabi S (1989) State of the Marine Environment in the West and Central African Region. UNEP Regional Seas Reports and Studies no. 108

Ram-Bidesi V, Lal PN, Conner N (2011) Economics of coastal zone management in the pacific. IUCN and Suva, Gland, Fiji: IUCN, xiv + 88

Siry HY (2006) Decentralized coastal zone management in Malaysia and Indonesia: a comparative perspective. Coast Manag 34:267–285

Snedaker SC, Getter CD (1985) Coastal resources management guidelines. Coastal Management Publication No. 2, NPS/AID Series. Research Planning Institute, Columbia, 205 p

Sorensen JC, McCreary ST (1990) Institutional arrangements for management of coastal resources. Coastal Management Publication No. 1, (Rev. ed.) U.S.Nat. Pk. Serv./USAID Series, 194 p

UNCED (1993) United Nations conference on environment and development-1992. Agenda 21. Rio de Janeiro, Brazil, p 300. https://www.unsystem.org/content/united-nations-conference-environment-and-development-unced-1992

Part IV

Conservation of Coastal Ecosystem

Biodiversity Conservation and Restorative Measures

14

Abstract

Coastal ecosystems of the tropics include innumerable flora and fauna of both terrestrial and aquatic habitats. Coastal ecosystems are increasingly threatened, degraded, and destroyed by marine- and land-based human activities. Major adverse impacts to ecosystems stem from climate change, unsustainable extraction of marine resources, physical alteration of coastal landscapes, destruction of marine and coastal habitats, and pollution of land and water which act cumulatively. Such a situation and the emerging climate change impacts are seriously hampering the livelihood of coastal communities and drive some of the endemic species to extinction. Therefore conservation of biodiversity assumes greater significance. Several attempts have been made over a period of time by regional and national governments, international organizations, and nature conservationist against species extinction which was the driving force for the emergence of global conservation regime. This chapter provides a broad overview of the current state of biodiversity conservation and management in the world with an emphasis on coastal areas of tropical countries.

Keywords

Biodiversity · Mangroves · Coral reef · Conservation · Ecosystem restoration

14.1 Introduction

Biodiversity has received worldwide attention due to the recognition of its important role in supporting human wellbeing and ecosystem services. At the same time, concern over the loss of biodiversity has resulted in global consensus on the conservation of biodiversity, the sustainable use of its components, and the fair and equitable sharing of benefits arising out of the utilization of genetic resources.

V. Ayyam et al., *Coastal Ecosystems of the Tropics - Adaptive Management*,
https://doi.org/10.1007/978-981-13-8926-9_14

Biological diversity or biodiversity is defined as the variety of the planet's living organisms and their interactions. In other words, biodiversity is the occurrence of different types of ecosystems, different species of organisms with the whole range of their variants and genes adapted to different climates, and environments along with their interactions and processes. Biodiversity includes the genetic and diversity of life forms such as plants, animal microbes, etc. living in a wide range of ecosystems. The term biodiversity encompasses all of life's variation, expressed in genes, individuals, populations, species, communities, and ecosystems (SCBC 2010).

As the term defines biodiversity as life's variations, it is essentially a dynamic entity and has changed throughout the history of life on Earth. This change encompasses two aspects: first it is an evolutionary process, and progressive another is loss of biodiversity which has negative consequences. The mechanisms responsible for biodiversity change are evolutionary processes of speciation and extinction, along with ecological processes. This has been altered by the anthropogenic activities which threaten biological diversity on a worldwide scale but over an extremely short geological time period. Such anthropogenic centric threat to biodiversity includes taxonomically specific exploitation, introduced species, and genetic or behavioural degradation, all of which can interact and ultimately result in extinction. These threats combine with the community and ecosystem level alteration of habitat degradation, fragmentation, destruction, pollution, and global climate change resulting in disruption and alteration of community and ecosystem structure and function.

However the biological diversity is not uniformly distributed throughout the globe conditioned by the natural forces till the influence of human activity significantly altered it. There are certain regions in the world which naturally favoured the proliferation of biodiversity than others. The tropical region particularly coastal areas has uniquely naturally variable ecosystems, including tropical rainforests, freshwater lakes and streams, salt marshes and mudflats, mangrove and coastal littoral forests, seagrass, fringing, and offshore coral reefs (SPREP 2012). The tropical region ecosystems have high species turnover and an unusual richness of endemic terrestrial and freshwater species (Kinch et al. 2010). The Pacific islands, within the Coral Triangle region, are broadly considered the centre of highest marine biodiversity on the planet (Veron et al. 2009). Marine species richness tapers off towards the eastern islands of Polynesia, with proportionally increasing endemism in some taxa. Though biodiversity is not distributed uniformly throughout the globe, its conservation is a shared responsibility of human beings in the world. This is because the biodiversity is linked with the human wellbeing and part of ecosystems which are of human concern.

14.2 Threat to Biodiversity and Ecosystem Services

Ecosystems provide supporting, regulating, provisioning, and cultural services (MEA 2005) to which the biodiversity is a part. Anthropogenic activities directly or indirectly use ecosystem services and derive benefits from the biodiversity, but the subsequent negative impact changes ecosystem functions and ecosystem service flows. Although majority of the human population in tropical region depend on the rich biodiversity directly or indirectly to derive their livelihood, the loss of biodiversity and land conversion for other purposes has not stopped. Several human activities directly affect the biodiversity of coastal region resulting in biodiversity loss and serious ecological consequences as well. On the other hand, climate change has emerged as a prominent cause for the loss of biodiversity. Consequently, habitat loss and degradation, invasive species, over-exploitation, pollution, disease, and human-forced climate change have profoundly impacted biodiversity (Kingsford et al. 2009).

The human settlement or expansion history into the tropical region indicates that tropical forests cleared for agriculture and plantations and increased species intensity dwindled the number of plant and animal species and their genetic diversity. At the same time, growing pollution degrades ecosystems, and climate change further threatens biodiversity (species, ecosystems, and genes). Consequently several species have gone extinct, and more are threatened. Invasive alien species, spread by expanding international trade, further threatens native biodiversity.

Many of the tropical countries particularly the Pacific region islands were impacted by human settlements during the historical times. For example, about 50% of indigenous birds (40 species) were eliminated from the Hawaiian Islands when Polynesians settled in the island (McNeill 1994). The Caribbean region experienced introduction of new species for commercial exploitation and food from several parts of the tropical world. Thus, it is clear that human communities relied on ecosystem services delivered through subsistence agriculture, fishing, and hunting, within strongly linked socio-ecological systems where remaining biodiversity was maintained through small-scale disturbance and cultivation (Thaman 2014). Apart from direct human interference, climate change can exacerbate anthropogenic threats such as extensive deforestation and over-exploitation of marine resources on biodiversity of tropical countries besides excessive withdrawal of ground water in the coastal areas. There is a direct link between climate change, ground water withdrawal, and coastal salinization, but it will have ecological consequences that will seriously affect the coastal flora and fauna. The IPCC 2002 Technical Report (V) also very explicitly recognized the importance of human influence on biodiversity loss and the additional level of threat imposed by climate change that has already begun to affect biodiversity (IPCC 2007). Short- and long-term climate change impacts will interact with prevailing threats to species and ecosystems synergistically and will complicate and put additional burden on biodiversity conservation efforts. Therefore it is imperative for to conserve the biodiversity for continued flow of ecosystem services and reduce the environmental degradation.

Because of these reasons, it can be argued that biodiversity conservation should balance mix of species to assure a flow of benefits over a range of conditions rather than the optimum rate of exploitation of a single species. Biodiversity conservation contributes to producing and delivering ecosystem services and ecosystem resilience, and ecosystem services are the primary motivation for biodiversity conservation. Therefore, there is a critical need to design and implement effective ways of conserving biodiversity and also an urgent need to understand the condition of our biodiversity and ecosystem services through regular state of biodiversity reporting (Keppel et al. 2014). Although conservation has not yet succeeded in bringing about many favourable changes in the environment, it has had a beneficial influence in government and other institutions. The state of knowledge, access to expertise, awareness of the problems, and ways to deal with them has been greatly enhanced over the past few decades. In the same context, we are in a better position to make real strides towards conserving biodiversity than ever before.

14.3 Biodiversity Types

Although the term biodiversity is defined comprehensively, little uncertainty or gap arises basically because of different perception. Because the term denotes tiny microbes to a large life which are linked with its surroundings and hierarchically linked through food chain/web. But without doubt, all them denote or explain part of the same but in different manifestations of a particular ecosystem. In general biodiversity refers to the variety of all forms of life on Earth that are different plants, animals, microorganisms, the genes they contain, and the ecosystem they form. The manifestation of biodiversity occurs through biological resources, viz. genes, species, community, and ecosystems, and ecological processes of which they are part. In the context of tropical coastal regions, this is more pertinent as these islands are known to possess very unique ecosystem and the species are highly endemic. The genetic variability available in the tropical region has been continuously created in the evolutionary process which combines with the physical environment of the coastal region and forms very unique ecosystems such as corals, rock shore, benthic, mangrove, etc. Therefore logically biodiversity should be considered at three different levels.

Understanding these diversity levels are vital to characterize and record the biodiversity which will ultimately help in biodiversity conservation. Although genetic diversity is the basis for creating other types of diversity, the environmental influence cannot be underestimated. For that reason, biodiversity includes genetic variability (for which different varieties of spices have appeared in the course of evolution) and diversity of life forms such as plants and animal microbes living in a wide range of ecosystems of which these species are part. It is also observed that the diverse living forms of the ecosystem get modulated with the global environmental changes as well. Biodiversity at different levels are discussed below:

14.3.1 Genetic Diversity

It describes the variation at the micro level which is carrier of genetic information. This is in the form of number and types of genes as well as chromosomes present in different species. In general, the magnitude of variations in genes of a particular species increases with increase in size and environmental parameters of the habitat. The genetic variation arises through sexual reproduction or by gene and chromosome mutation which is spread in the population by recombination of genetic materials during cell division after sexual reproduction. Genetic diversity in a population is very important because,

- It helps in speciation or evolution of new species as in the case of Coral Triangle and tropical evergreen forest where thousands of species got evolved.
- Presence of genetic variation is useful in adaptation to changes in environmental conditions. For example, mangroves and littoral vegetations developed adaptation depending on the salinity, marshy condition, and seawater inundation. This change in growing environment, evolution of genetic variations, and their interactions resulted in special kind of adaptation mechanism and speciation resulting in endemism. In addition, this helps to enhance the biodiversity of the region.
- Genetic diversity is very important for agricultural productivity and development. This is seen in the context of novel or useful genes present in the wild relatives of cultivated species and subsequent development of different agro-ecosystems including farming system diversity.

14.3.2 Species Diversity

This is next level of diversity which occurs at species level. It describes the variety in the number and richness of the species with in a region. The species richness may be defined as the number of species per unit area. The richness of a species tells about the extent of biodiversity of a site and provides a means for comparing different sites. The species richness depends largely on climatic conditions. The number of individuals of different species with in a region represents species evenness or species equitability. On the other hand, when a species is confined entirely to a particular area, it is termed as endemic species. Endemism is a unique feature of islands and other isolated places biodiversity where the organisms are confined in isolation and confronted with different living environment.

14.3.3 Ecosystem Diversity

It describes the assemblage and interaction of spices living together and the physical environment in a given area. It relates varieties of habitats, biotic communities, and ecological processes in biosphere. It also tells about the diversity within the ecosystem. It is also sometimes referred as land escape diversity because it includes placement and size of various ecosystems. For example, the landscapes like grasslands, deserts, mountains, rocky, sand dunes, etc. show ecosystem diversity. The ecosystem diversity is due to diversity of niches, trophic levels, and ecological processes like nutrient cycling, food webs, energy flow, role of dominant species, and various related biotic interactions. The major advantage of such type of diversity can generate more productive and stable ecosystems or communities capable of tolerating various types of stresses.

14.4 Rationale for Conservation

With loss of species, we lose the ultimate source of our crops, livestock, wildlife and the genes we use to improve agricultural resilience, the inspiration for manufactured products, and the basis of the structure and function of the ecosystems that support humans and all life on Earth (McNeely et al. 2009). Above and beyond material welfare and livelihoods, biodiversity contributes to security, resiliency, and freedom of choices and actions (MEA 2005). In this context, conservation and protection of entire ecosystem are very important because conservation of biodiversity is a common concern and shared responsibility of human beings all over the world. Quantitative measures of biodiversity focus on a taxonomic unit, typically the species, although aspects of ecological diversity in a region can also be measured. The most threatened areas inclusive of habitat with high species diversity are labelled as 'biodiversity hotspot' which mostly includes tropical rainforests, coastal areas, and islands. For these reasons these areas should receive high priority on conservation programme.

Therefore, presence of wide biodiversity or the number of species alone is an inadequate indication of conservation priority because several areas can share the same species. In contrast, areas with high levels of endemism are irreplaceable; hence, it is imperative to conserve these places because the unique species they contain cannot be saved elsewhere. In many cases areas with significant endemic species are also facing greater risk of disappearing owing to human impact. As species become threatened and vanish, the consequential effect will be felt on the broader ecosystems and myriad benefits to human wellbeing that depend upon biodiversity. Bringing an end to global biodiversity loss requires that limited available resources be guided to those regions that need it most. The biodiversity hotspots do this based on the conservation planning principles of irreplaceability and vulnerability.

14.5 Status and Distribution of Biodiversity

Biodiversity provides the basis for sustenance of life on Earth and is the key for safeguarding the wealth of the world for future generations. Management of biodiversity requires measurement and quantification because the term encompasses such a wide range of biological phenomena. In general, biodiversity is assessed with reference to genes, species, and ecosystems and to make comparisons based on quantitative values ascribed to one or more of these levels. On the other hand, the distribution of biodiversity is not uniform across the globe; it varies with region and climatic zones. This is because of difference in favourable climate, edaphic, and other life-supporting conditions.

The tropical region is very rich in biodiversity and gifted with a wide variety of flora and fauna which are unique than other regions of the world. Studies have indicated that biodiversity of the tropical region constitutes more than half of the world's species (Wilson 1988). This probably exceeds more than that as far as marine biodiversity of the tropical waters is concerned. Nevertheless the species diversity is not equally spread over the total tropical area but is more or less clumped in centres of high diversity, and these are sometimes characterized by high levels of endemism (Myers 1990). In order to be able to conserve tropical biodiversity, it is valuable to know species distribution and its abundance.

On the other hand, the tropical region particularly island species appear to be patchy in its spatial distribution often limited by the geographical isolation of the islands. The vast expanse of ocean around the islands limits its spread, and the natural selection continued for the island conditions resulting in high level of endemism as observed today. Furthermore, the ecology of the majority of tropical coastal and island species is poorly known, and this is especially true of the largest taxonomic group. It is widely accepted that the biodiversity present in tropical forests, mangroves, and coral ecosystem represents an enormous potential resource of species and genetic variation which may benefit humanity (Balick and Mendelsohn 1991; ICBP 1992).

The floral and faunal diversity (endangered/IUCN Red Listed) of major tropical nations/islands located in the four tropical oceans mainly of coastal region are given in Table 14.1. It is observed that the tropical region is rich in vascular plants, corals, fishes, and birds with high endemism due to the reasons discussed earlier. For example, the Indian Ocean region has highest number of vascular plants with endemism; therefore the Western Ghats (India)-Sri Lanka region has been recognized as one of the global biodiversity hotspots. Another important biodiversity hotspot region involving tropical island is the Caribbean region. The data shows that Haiti, Cuba, and Jamaica together contribute more than half of the regional biodiversity. At the same time Jamaica and Cuba are having largest number of endangered/threatened species. The Pacific region islands called Coral Triangle are known for its richest marine biodiversity in the world particularly the coral ecosystem. It is regarded as the global epicentre of marine biodiversity (Allen 2007) and a global priority for conservation. At the same time, this region has highest number of endangered or threatened species which are very unique to the island ecosystem. Nevertheless, majority of the island population is dependent on it, but over-exploitation, aggressive fishing practices, and pollution besides other anthropogenic activities are responsible for its current threatened status. In essence due to more number of vital ecosystem services and livelihood security derived from the biodiversity, it should receive top priority for protection. A word of caution should be borne in mind while interpreting the data as it is not exhaustive but indicative only.

In the tropical coastal and island region due to copious amount of rainfall and plenty of sunlight, tropical evergreen and moist deciduous forest are commonly seen with large biodiversity. Such island nations are located in the Pacific region, but there are island nation with reasonable resources located in the Indian Ocean (the Andaman, Nicobar, Comoro, Mascarene, and Seychelles islands), a few in the Atlantic (Principe, São Tomé, and Annobon), and many in the Caribbean. The total area of the rainforests in tropical island is very small, and the forests on each island are much less species rich than their continental counterparts. But the high rate of endemism means that together they support a significant proportion of all rainforest species. At the same time, their biotas are also highly endangered: around 75% of

Table 14.1 Endangered/IUCN Red Listed species of some tropical countries

Country name	Flora	Mammals	Birds	References
Indian Ocean region				
India	247	116	136	BSI (2011) and ZSI (2015)
Andaman and Nicobar Islands	43	05	13	BSI (2011) and Sivaperuman and Raghunathan (2012)
Bangladesh	486	138	566	Anonymous (2011)
Sri Lanka	280	15	11	BDS/MoE and DNBG (2012)
Thailand	90	81	130	http://*earthsendangered.com*/search-regions3.asp
Myanmar	37	60	110	http://*earthsendangered.com*/search-regions3.asp
Madagascar	284	92	52	http://*earthsendangered.com*/search-regions3.asp
Mauritius	86	03	11	NRCBD (2015)
Maldives				Anderson and Baldock (2001)
Seychelles	44	02	20	Government of Seychelles (2014)
Pacific region				
Philippines	217	55	129	http://*earthsendangered.com*/search-regions3.asp
Papua New Guinea	149	52	86	http://*earthsendangered.com*/search-regions3.asp
Solomon	13	19	38	GoS (2014)
Fiji	63	05	13	https://chmfiji.wordpress.com/about/overview-of-fijis- biodiversity
Palau	05	02	05	Kitalong (2010)
Caribbean region				
Antigua and Barbuda	04	–	–	Daltry (2007) www.earthendangered.com
Bahamas	07	06	13	BMMRO (2009)
Belize	22	05	08	Meerman (2005)
Cuba	42	16	18	www.cubanaturaleza.org and www.earthendangered.com
Grenada	02	–	09	GoG (2014)
Guyana	18	05	07	FAO (2005)
Haiti	22	05	14	Posner et al. (2010)
Jamaica	186	05	15	GoJ (2010)
Atlantic region				
Guinea-Bissau	05	15	14	Catarino (2004)
São Tomé and Principe	02	–	–	NBSAP II (2015)

terrestrial vertebrate extinctions in the last 400 years have been on islands, although by no means all in rainforest (Primack 2006). Currently island rainforests suffer from the additional problem of an apparently much greater susceptibility to invasion by exotic plant and animal species (Teo et al. 2003).

Table 14.2 Reef area and diversity of tropical countries

Sl. no	Region/island	Reef area (km²)	Coral diversity	Sl. no	Region/ island	Reef area (km²)	Coral diversity
1.	Comoros	430	NA/314	15	Fiji	10,020	177/398
2.	Seychelles	1690	206/310	16	Guam	220	140/220
3.	Mauritius	870	161/294	17	Palau	1150	154/384
4.	India	5790	208/345	18	Marshall Islands	6110	222/340
5.	Sri Lanka	680	100/318	19	Kiribati	2940	110/365
6.	Maldives	8920	212/244	20	Wallis and Futuna	940	NA/363
7.	Malaysia	3600	281/568	21	Tuvalu	710	NA/366
8.	Singapore	<100	176/186	22	Tokelau	<50	NA/210
9.	Indonesia	51,020	443/602	23	Samoa	490	NA/211
10.	Philippines	25,060	421/577	24	American Samoa	220	150/212
11.	Papua New Guinea	13,840	378/517	25	Tonga	1500	NA/218
12.	Solomon Islands	5750	101/398	26	Niue	170	NA/189
13.	New Caledonia	5980	151/359	27	Cook Islands	1120	51/172
14.	Vanuatu	4110	296/379	28	French Polynesia	6000	174/168

Compiled from Spalding et al. (2001), FAO country statistics, IUCN

Within this range they are far from evenly distributed, with large areas confined to remote island regions and offshore areas far from major land masses. Among the tropical island nations, Indonesia and the Philippines constitute 76,000 km^2 reef areas which are almost half of the reef area in the tropical region. Among the smaller countries, French Polynesia has the largest area under reef followed by Marshall Islands, Maldives, Fiji, and Papua New Guinea (Table 14.2). Coral ecosystem is highly diverse and one of the best in the marine biodiversity realm. The islands and island nations located in the biodiversity triangle exhibit high diversity and endemism.

14.6 Biodiversity Conservation

14.6.1 Importance

The environmental crisis accelerated by industrial activities is multidimensional, but biodiversity crisis is more fundamental and urgent. The loss of biodiversity will not only eliminate the species from the Earth, but it will have serious consequences in the food chain and livelihood of people besides contribution to the ecosystem services. Biodiversity is being depleted by the loss of habitat, fragmentation of habitat, over-exploitation of resources, human-sponsored ecosystems, climatic changes, pollution invasive exotic spices, diseases, shifting cultivation, poaching of wildlife, etc. Since human beings are enjoying all the benefits from biodiversity, they should take proper care for the preservation of biodiversity in all its form and good health for the future generation. It is also vital and responsibility of human beings to prevent the degradation and destruction of the habitats which helps to maintain the biodiversity at its optimum level.

Similarly, if global warming continues as predicted over the next 50 years, up to one quarter of all plant and animal species currently existing on Earth would become extinct, because they have nowhere to escape from their current habitat to new habitat which is suitable for them. Further, most of the habitats are under some kind of human influence. Therefore, all strategies for conservation of biodiversity involve relationships between local residents who live in the critical habitat of threatened and endangered species, government agencies, international organizations, and scientists, especially conservation biologists.

Conservation of biodiversity is protection, upliftment, and scientific management of biodiversity so as to maintain it at its threshold level and derive sustainable benefits for the present and future generation. In other words, conservation of biodiversity is the proper management of the biosphere by human beings in such a way that it gives maximum benefits for the present generation and also develops its potential so as to meet the needs of the future generations. Mainly the conservation of biodiversity has three basic objectives, viz. to maintain essential ecological processes and life-supporting systems, to preserve the diversity of species, and to make sustainable utilization of species and ecosystems.

In conservation planning efforts around the globe, to systematically list number of species in threat of extinction was initiated in 1963 known as IUCN Red List system (Fig. 14.1). Over time, hundreds of scientists have worked to create listing criteria with nine categories that have been carefully defined to be maximally useful as a diagnostic tool to help establish extinction risk over all taxa. Species are assigned to one of nine categories, which indicate their threat status or their status in the review process (ICBP 1992).

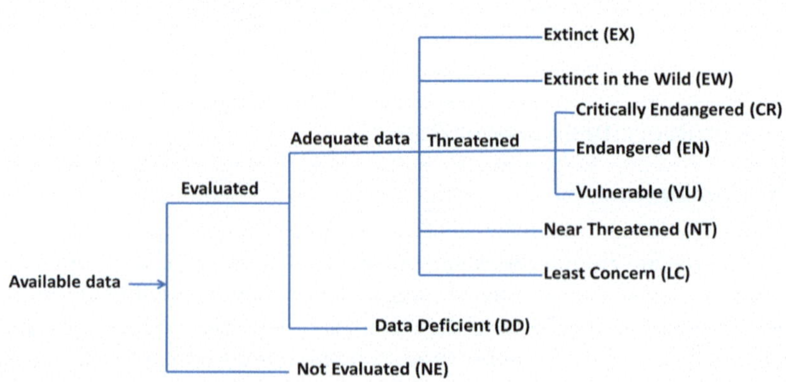

Fig. 14.1 IUCN Red List categories

14.6.2 Strategies for Biodiversity Conservation

The basic feature of biodiversity conservation strategy should protect and improve the ecological integrity and long-term viability of the more intact (core) landscapes of the region rather than protecting patches. Otherwise, these diverse patches should be integrated to have larger areas which are viable. Within these intact areas, priority actions should be taken to reverse the impacts/restore the original, remove threats, and reinstate ecological processes. Reverse declines are to reinstate ecosystems that have been differentially lost in locations where this will meaningfully contribute to stemming species declines and reinstating critical ecological processes so that the ecosystem is normal. Within these areas, the priority actions are to reinstate open/tree systems and improve the habitat value of shrubby systems. Efforts should be made to recover threatened species and an ecological community; the intent of this strategy is to ensure the long-term persistence of species and ecosystems at immediate risk of extinction in the wild. The measures may be specific to individual species and ecosystems, but primarily it focus on increasing distribution and abundance of species at best halting or reversing declining trends. Focus should also be given to assess and address the emerging threats before their impacts on biodiversity are fully realized. All these actions will succeed only if adequate amount of representative data are collected but are guided by the current amount of knowledge. This calls for upgradation and updation of the information and alongside on the strategy. Some of the important strategies for the conservation of tropical biodiversity are as follows:

- The foremost activity should be conservation of useful animals, plants, and their wild relatives both in their natural habitat (in situ) and in zoological or botanical gardens (ex situ). In the same context, all the varieties of food, forage and timber plants, livestock, agriculture animals, and microbes should be conserved in different places as many of the coastal regions and tropical islands face the threat of sea level rise.
- At the same time, all the economically important organisms, plants, and marine life in protected areas should be identified, catalogued, and conserved.
- It is also important to identify and safeguard the critical habitats for each species with adequate legal support. Care should be taken for the development of reserves and protected areas.
- Promote bilateral and multilateral agreement to protect the habitats of migratory birds by involving local, national and international institutions, and people.
- Efforts should be made for setting up of national parks and wildlife sanctuaries to safeguard the genetic diversity and their continuing evolution. Concurrently environmental laws should be strictly imposed.
- There should be prioritization to preserve unique ecosystems. In wildlife conservation programme, priority should be given to endangered species over vulnerable species and to vulnerable species over rare species.

- The conservation strategy should aim at discouraging international trade in wildlife, promote prevention of illegal poaching and hunting of wildlife, and reduce the over-exploitation of useful products as far as practicable.
- Efforts should be made to create public awareness regarding biodiversity and its importance for the living organisms.

14.7 Biodiversity Conservation Methods

Several regional and international efforts by scientists, different government, and international organizations against species extinction have been the main driving force for the emergence of the global conservation regime. There are two different streams to achieve the conservation goals (Adams 2004). One stream of governance models that became the cornerstone of global biodiversity conservation since the beginning of the twentieth century has focused predominantly on protected areas (Adams and Hutton 2007). However, over the past few decades, another important stream of governance models has concentrated on community-based conservation. Although this stream has received worldwide attention and recognition, it predates the establishment of the formal protected area network. In parallel with the development of community-based conservation approaches, there has been a growing recognition of the need for protected areas to also involve and promote the livelihoods of communities living in or around them.

Basically there are two types of conservation methods, viz. in situ and ex situ conservations. In situ method conserves biodiversity in its natural state without affecting the ecosystem in which the species or community occurs. Different conservation methods are depicted in Fig. 14.2, and the details are discussed in the following subsections.

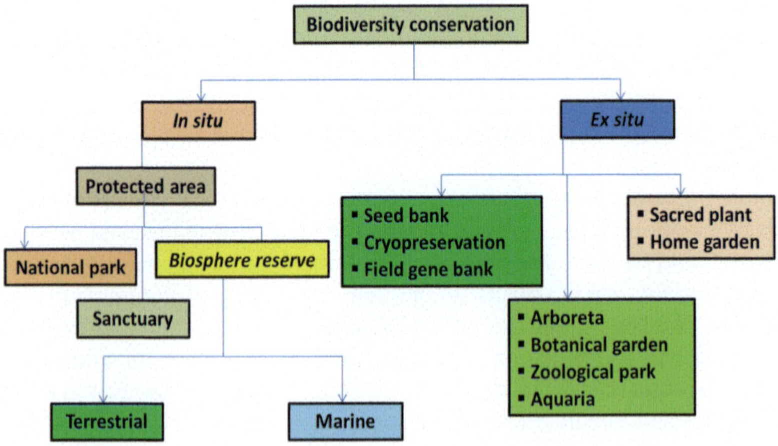

Fig. 14.2 Different biodiversity conservation methods

14.7.1 In Situ Conservation

Any species tend to grow or live better in its natural state and in community which provides some specific advantage. This is essential for the sustenance of that species. In general conservation of species in their natural habitat or natural ecosystem is known as in situ conservation. In the process, the natural surrounding or ecosystem is protected and maintained so that all the constituent species are conserved and benefited. Concurrently the factors which are detrimental to the existence of species concerned should be eliminated or reduced to insignificant level by suitable mechanism. The protection and management of biodiversity through in situ conservation involve certain specific areas known as protected areas which include national parks, sanctuaries, and biosphere reserves. The advantages of in situ conservation are:

- It is a cheap and convenient way of conserving biological diversity particularly for tropical islands which are isolated and surrounded by sea.
- It offers a way to preserve a large number of organisms simultaneously, known or unknown to us.
- The existence in natural ecosystem provides opportunity to the living organisms to adjust to differed environmental conditions and evolve into a better life form.

The protected areas are biogeographical areas where biological diversity along with natural and cultural resources are protected, maintained, and managed through legal and administrative measures. The demarcation of biodiversity in each area is determined on the basis of climatic and physiological conditions. In these areas, hunting, firewood collection, timber harvesting, etc. are prohibited so that the wild plants and animals can grow and multiply freely without any hindrance. Some protected areas in India are cold desert (Ladakh and Spiti), hot desert (Thar), saline swampy area (Sundarbans and Rann of Kutch), tropical moist deciduous forest (Western Ghats and North East), marine park (Mahatma Gandhi MNP, Port Blair), etc.

Protected areas include national parks, sanctuaries, and biosphere reserves. Remote sensing-based assessment of areal distribution of different protected areas of Madagascar is given in Fig. 14.3. This helps to conserve some of the unique and endemic species found in Madagascar. In the Caribbean region, Cuba has the largest number of marine protected areas followed by the Dominican Republic (Table 14.3). This region has higher number of marine protected areas than the terrestrial. There are 37,000 protected areas throughout the world. As per World Conservation Monitoring Centre, India has 581 protected areas, national parks, and sanctuaries.

14.7.1.1 National Parks

These are small reserves maintained by the government meant for the protection of wildlife and their natural habitats. The area of national parks ranges from 1 km^2 to few thousand km^2. The boundaries are well demarcated and circumscribed. The activities like grazing forestry, cultivation, and habitat manipulation are not

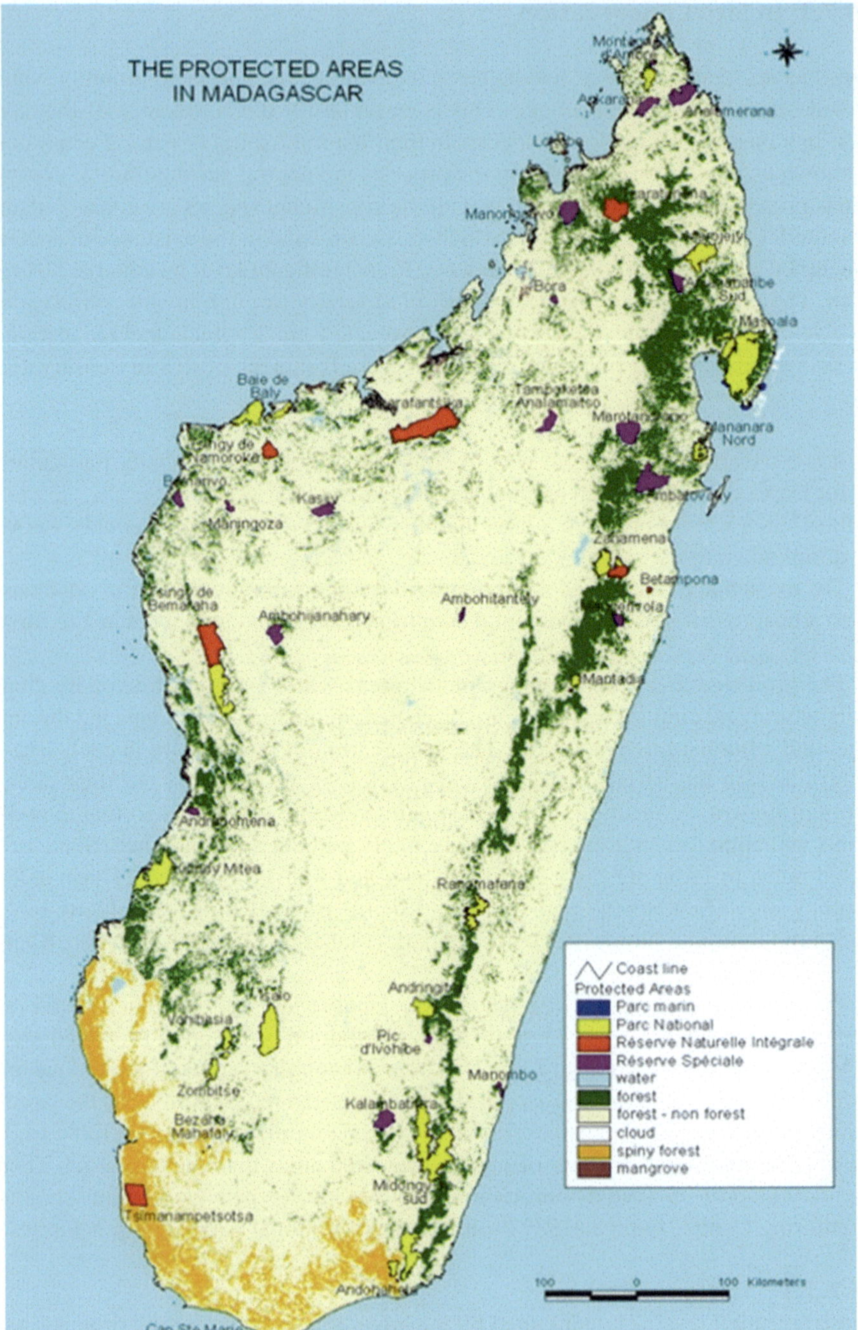

Fig. 14.3 Protected areas in Madagascar (based on remote sensing data)

Table 14.3 Marine protected areas in the Caribbean

Country/island	Marine protected areas	Others	Total
Cuba	15	15	30
Dominican Republic	7	4	11
Jamaica	4	0	4
Haiti	0	0	0
St. Lucia	2	1	3
Trinidad and Tobago	1	7	8
Barbados	1	1	2
Belize	2	0	1
Dominica	1	0	9
Bahamas	9	0	1
St. Vincent and Grenadines	1	0	1

Source: GBRMPA, WB, WCU (1995)

permitted in these areas. The earliest national parks are the Yellowstone National Park in the USA and the Royal National Park near Sydney, Australia. These parks were chosen because of their scenic beauty and recreational values. At present many such protected areas throughout the world protect rare species and their habitat. World Conservation Monitoring Centre (WCMC) has recognized more than 37,000 protected areas around the world.

14.7.1.2 Sanctuaries

These are the protected areas where only wild animals (fauna) are present. The activities like harvesting of timbers, collection of forest products, cultivation of lands, etc. are permitted as long as these do not interfere with the objective of the project. In other words controlled biotic interference is permitted in sanctuaries, which allows visiting of tourists for recreation as well.

14.7.1.3 Biosphere Reserves

The concept of biosphere reserves was launched in 1975 as a part of UNESCO's 'Man and Biosphere Programme', dealing with the conservation of ecosystems and the genetic resources contained therein. Biosphere reserves or natural reserves are multipurpose protected areas with boundaries circumscribed by legislation. The main aim of biosphere reserve is to protect the entire ecosystem. It helps to preserve genetic diversity in representative ecosystems by protecting wild animals, traditional lifestyle of inhabitant and domesticated plant/animal genetic resources. These are scientifically managed allowing only the tourists and carryout scientific studies. In some countries tourist are also not allowed if it also includes tribal population.

A terrestrial biosphere reserve consists of different zones serving different purposes, viz. core, buffer, and transition zones (Fig. 14.4). The natural or core zone comprises an undisturbed and legally protected ecosystem nurturing the nature as such. The buffer zone surrounds the core area and is managed to accommodate a greater variety of resource use strategies and research and educational activities. The transition zone is the outermost part of the biosphere reserve. This is an area of

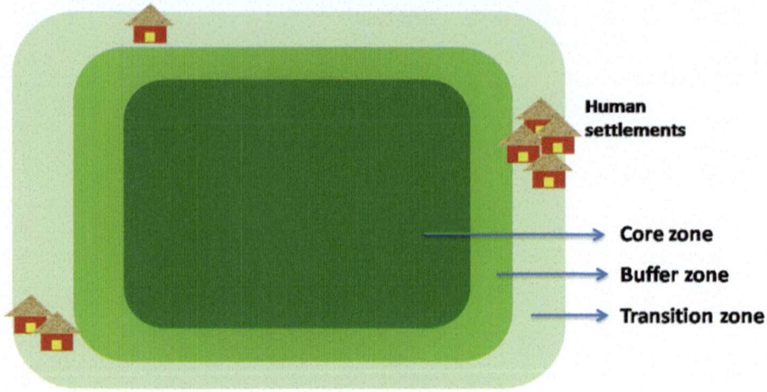

Fig. 14.4 Zonations in a terrestrial biosphere reserve

active cooperation between reserve management and the local people, wherein activities like settlements, cropping, forestry, recreation, and other economic uses continue in harmony with people and conservation goals. Some of the significant services of biosphere reserves are:

- These help in the restoration of degraded ecosystem.
- The main role of these reserves is to preserve genetic resources, species, ecosystems, and habitats without disturbing the habitants.
- These maintain cultural, social, and ecological sustainable economic developments.
- These support education and research in various ecological aspects

Based on the success of establishment of biosphere reserves all over the world, the World Network of Biosphere Reserves (WNBR) composed of 669 biosphere reserves in 120 countries, including 20 transboundary sites, were established for achieving the 3 interconnected functions of conservation, development, and logistic support (Fig. 14.5). The WNBR of the MAB Programme promotes North-South and South-South collaboration and represents a unique tool for international cooperation through sharing knowledge, exchanging experiences, building capacity, and promoting best practices. It fosters the harmonious integration of people and nature for sustainable development through participatory dialogue, knowledge sharing, poverty reduction and human wellbeing improvements, respect for cultural values, and society's ability to cope with change, thus contributing to the Millennium Development Goals.

14.7.1.4 Sacred Forests and Lakes

In India and some other tropical islands, a traditional strategy for the protection of biodiversity has been in practice in the form of sacred forests or groves. These forest patches of varying dimensions sometimes with water bodies are protected by local

Fig. 14.5 Location of biosphere reserves in Asia-Pacific region

people due to their religious sanctity. In general, they are most undisturbed forests without any human which protect aquatic flora and fauna (Fig. 14.6a, b).

14.7.1.5 International Conventions

Conservation of biological diversity and sustainable use of its components came into the limelight in 1972 (United Nations Conference on Human Environment, Stockholm). In 1973, the UNEP identified conservation of biodiversity as a priority area; hence there was need to get the legal mandate for conservation of world resources. There were negotiations for a legally binding instrument to address biological diversity and its loss to enhance fairness and equity in sharing of the benefits of biodiversity. This led to the opening of the Convention on Biological Diversity in 1992 at Rio de Janeiro under the United Nations Conference on Environment and Development (UNCED) popularly called the Earth Summit.

Most of the countries and islands are party to the Convention on Biodiversity (CBD) and most other international conventions, including the Convention on Wetlands of International Importance, as well as such international initiatives as the UNESCO's Man and Biosphere programme and World Heritage. Subscribing to these international treaties ensures that the native government should make efforts to conserve the entire ecosystem. Other international biodiversity conventions and conservation organizations related to biodiversity conservation are:

- The Ramsar Convention on Wetlands of International Importance (Fig. 14.6a)
- International Union for Conservation of Nature (World Conservation Union)
- Convention on International Trade in Endangered Species (CITES)
- International Convention for the Protection of Birds
- International Board for Plant Genetic Resources
- African Convention on the Conservation of Nature and Natural Resources

a

b

Fig. 14.6 Role of Ramsar wetland and national park in conservation. (**a**) Chilika Lake, India, is the largest salted water lake of Asia. Chilika Lake Bird Sanctuary hosts a wide range of flora, fauna, and aquatic fauna and attracts large number of aquatic birds including migratory birds. (**b**) Bhitarkanika National Park, India, is hotspot of biodiversity, surrounded by the wildlife sanctuary, Gahirmatha Beach, and marine sanctuary. The mangrove sanctuary is home to white saltwater crocodile, olive ridley sea turtles, Indian python, and king cobra

- UNESCO programme on Man and Biosphere
- International Convention for the Regulation of Whaling
- Convention on the Conservation of Migratory Species of Wild Animals
- World Resources Institute
- World Wide Fund for Nature

Further, the US Agency for International Development (USAID) is committed to helping developing countries maintain biologically diverse habitats and environmental services while supporting sustainable development and economic growth. For the past three decades, the agency has helped countries maintain and manage the variety of species, genetic resources, and ecosystems in natural settings such as forests, grasslands, wetlands, and coastal, marine, and freshwater habitats. USAID currently implements many types of conservation programmes including eco-regional and protected area management, community-based natural resources management, enterprise-based conservation initiatives, public-private conservation alliances, innovative long-term conservation financing, environmental education, and policy development and reform.

14.7.2 Ex Situ Conservation

Ex situ conservation involves maintenance and breeding of endangered plants and animals under partially or wholly controlled conditions in specific areas like zoo, gardens, nurseries, etc. That is, the conservation of selected plants and animals in selected areas outside their natural habitat is known as ex situ conservation. The stresses on living organisms due to competition for food, water, space, etc. can be avoided by ex situ conservation, thereby providing conditions necessary for a secure life and breeding. For ex situ conservations, the species to be conserved should be identified followed by adoption of different ex situ methods such as long-term captive breeding and propagation for the species which have lost their habitats permanently, short-term propagation and release of the animals in their natural habitat, animal translocation and reintroduction, and advanced technology in the service of endangered species. The different advantages of ex situ conservation are:

- It gives longer life time and breeding activity to animals.
- Genetic techniques can be utilized in the process.
- Captivity breed species can again be reintroduced in the wild.

The ex situ conservation strategies include botanical gardens, zoological gardens, conservation stands, and gene, pollen, seed, seedling, tissue culture and DNA banks. Seed gene banks make the easiest way to store germplasm of wild and cultivated plants at low temperature. While in field gene banks, preservation of genetic resources is being done under normal growing conditions.

14.7.2.1 Cryopreservation

This type of in vitro conservation is done in liquid nitrogen at a temperature of −196 °C. This is particularly useful for conserving vegetatively propagated crops, e.g. potato. Cryopreservation is the storage of material at ultralow temperature (−196 ° C) either by very rapid cooling, as used for storing seeds, or by gradual cooling and simultaneous dehydration, as being done in tissue culture. In cryo-preservation, representative materials of existing biodiversity can be stored for a considerable long period of time in compact low-maintenance refrigeration units.

14.7.2.2 Botanical Gardens

This is a method of ex situ conservation of threatened and endangered species in their respective regions. This is maintained by governments at different levels, by educational institutions, and with the international assistance as well. There are more than 1500 botanic gardens and arboreta in the world containing more than 80,000 species. Many of these botanic gardens now have seed banks, tissue culture facilities, and other ex situ technologies.

14.7.2.3 Zoological Gardens (Zoos)

There are more than 800 professionally managed zoos around the world with about 3000 species of mammals, birds, reptiles, and amphibians. Many of these zoos have well-developed captive breeding programmes. The conservation of wild relatives of crop plants, animals, or cultures of microorganisms provides breeders and genetic engineers with a ready source of genetic material. Many of the tropical islands are rich in the number of endemic faunal species it possesses, while its record in agro-biodiversity is very impressive as well.

14.8 Conservation of Marine Environment (Coral Reef)

The marine environment up to 6-m depth of water in the coastal/island ecosystem is dominated by unique ecosystem called coral reefs. Mangrove dominates where sedimentation is occurring in the seafront and estuaries. Towards the open sea and around islands in the tropics, coral biodiversity is the common feature. Coral reefs and their associated habitats have an almost unimaginable and hugely underappreciated value to mankind, providing billions of people with a source of food and income. Their biological richness comes close to that of tropical forests. This is why coral reefs have been dubbed the ‘rainforests of the sea’.

Changes to the clarity, temperature, or salinity of the water can cause corals to die or bleach. Coral bleaching refers to the evacuation of zooxanthellae, the colour-producing organism in coral reefs, when conditions become less than ideal. When zooxanthellae evacuate, coral polyps die, which leaves the white calcium carbonate exoskeleton behind called ‘coral bleaching’. Sedimentation due to soil erosion and disturbance to the coral bed algal bloom due to excessive fertilization of coastal land result in coral bleaching. In several developing regions in the coastal areas/tropical islands, the coral bed is being removed/destroyed to facilitate construction of ports and other infrastructure resulting in large-scale damage to the coral ecosystem. Poaching or extraction of corals directly for various purposes besides aquarium particularly some of the critical species is also causing irreparable damage to the coral ecosystem. Therefore it is utmost essential to conserve the entire coral ecosystem so as to conserve the innumerable marine biodiversity. Some of the most important measures to conserve corals reef biodiversity are as follows.

14.8.1 Marine Protected Areas

One of the key mechanisms of protecting coral reefs and its environment is the establishment of marine protected areas (MPAs). MPAs have become a mainstream tool for conserving marine biodiversity and are advanced under several global and regional instruments, including the CBD. MPAs are zoned to allow for complete restriction of harvest in some areas, restricted use in others, and managed use in yet others (Agardy 1994). This also provides for sustainable use of coral reefs by the local people. In other places coral species should be protected under general species protection laws across the region. Some MPAs are established and managed specifically for a purpose. For example, it is established for preservation of a historical site as a refuge for a particular species or breed. Multiple-use MPAs are zoned to allow for complete restriction of harvest in some areas, restricted use in others, and managed use in yet others (Agardy 1994). There are now 14,688 MPAs covering almost 15 million km^2 in the world. While MPA coverage has grown significantly over the last decade, the geographical distribution of MPAs is very uneven with a small number of countries making up the majority of the area covered by MPAs.

14.8.2 Monitoring of Coral Reefs

It is highly essential to develop a list of protected flora and fauna for coral reef sensitive species. This should be followed by monitoring the trends and patterns of use and the health of reefs. This will help to develop effective management strategies. Worldwide, there are several organizations that monitor the status of coral reefs. The Global Coral Reef Monitoring Network (GCRMN) coordinates efforts to improve the management of coral reefs through knowledge sharing and capacity building and works closely with Reef Check and Reef base. The latter is a global database of coral reef-related information. After the 1998 coral bleaching event, and with the ongoing threat of coral degradation as a consequence of other human activities, the Coastal Oceans Research and Development in the Indian Ocean (CORDIO) was commenced in 1999.

14.8.3 Legislation

Prevention of overharvesting and protection through legislation is vital activity to protect and regulate the exploitation of coral ecosystem. Several countries have enacted coastal zone management/regulation under which marine areas are protected. Most of this protection is afforded to marine vertebrates, but some countries—such as India and Sri Lanka—have laws protecting several species of coral, molluscs, and echinoderms.

- In Sri Lanka, all stony corals are protected by law (Flora and Fauna Protection Act, 1993).
- In India, all stony corals, all black corals, all fire corals, and all sea fans are protected by law (Wildlife Protection Act, 1972).

14.8.4 International Collaboration

- The trade of 2000 species of stony corals, black coral (*Leiopathes* sp.), giant clams (*Tridacna* sp.), queen conch (*Strombus gigas*), seahorses, and sea turtles is prohibited or restricted internationally under the Convention on International Trade in Endangered Species of Wild Fauna and Flora (CITES).
- All hard (stony) coral species are listed on CITES Appendix II—which means trade is restricted (http://www.cites.org/).
- Saving reef habitat alone will not stop this decline, because our threats are not just local but global. However, novel ex situ conservation techniques, such as genetic banks using frozen samples, hold strong promise to help offset these threats.

14.8.5 Provision for Alternate Livelihood

The connection between poverty and coral reef ecosystems is significant. Nearly two thirds of all countries with reef areas are developing countries, and a quarter of these are least developed countries (UNDP 2002). It is now well recognized that such communities need to be offered alternatives for their livelihoods in order to ensure that coral reefs are not further damaged, as well as to alleviate poverty in coastal/island areas. Therefore, coastal managers should focus towards more integrated and participatory approaches to reef management and conservation. Such approaches involve identifying and supporting alternative livelihoods to reduce dependence on coral reefs and ecotourism, as well as enhancing current livelihood activities to make them more cost and resource efficient. In certain countries/coastal areas, limited and controlled local use of coral reefs which are non-destructive is now advocated instead of blanket restrictions on use. This is applicable not only to reef but also other reef biodiversity. Rights to reef access and resolution of conflicts over resource use, community involvement, and collaborative management should be part of reef management (Whittingham et al. 2003). The selective tourism can be opted to the level of carrying capacity of the coral reef environment available in each location or island. This is more pertinent to the small island nation in the tropics. In the case of selective tourism, protection of flora and fauna through notification, and monitoring of environment, a particular island can be considered as a model island for sustainable protection of coral reef environment.

14.8.6 Reef Resilience

Adapting to climate change is, perhaps, the biggest challenge that coastal managers are facing today in respect to coral reef conservation and management. This is further aggravated by the low adaptive capacity of many of the tropical countries. Understanding why some reefs do not succumb to bleaching while others nearby do (*resistant*) and why some 'bounce' back quickly while others do not (*resilient*) has become extremely important. The nature conservancy and its partners have developed an R2-Reef Resilience Toolkit that is designed to help managers prepare for and respond to coral bleaching events (Wilkinson 2004). Such a technological support to understand reef resilience requires not only financial resources, but also international collaboration is required as the event is not restricted to particular country or region. This may happen to a considerable area in the tropics.

14.8.7 Building Awareness

Nevertheless, equal emphasis need to be given for awareness among the coastal communities and other stakeholders of the need for coral biodiversity conservation. Building awareness about coral reefs, their diversity, and the services they provide helps greatly in mitigating the threats to these fragile ecosystems across the globe. Awareness at the community level is most effective as it can help to encourage users of coral reefs to change their behaviour to sustainable use of these ecosystems. At the same time, awareness at national level through the media and conservation education is essential to ensure that policymakers integrate coral reef conservation into all stages of development. In response to the growing threats to coral reefs around the world, 1997 and 2008 were designated the International Year of the Reef (IYOR).

14.9 Restoration

14.9.1 Rationale

Many of the world's ecosystems including many of marine have undergone significant degradation with negative impacts on biological diversity and peoples' livelihoods due to anthropogenic activities. In the case of the oceans, the anthropogenic signature is strongest in the coastal zone (Vitousek et al. 1997) particularly coral reefs, seagrass beds, and mangrove forests. Perhaps one of the most important human disturbances on coral reefs is overfishing (Hodgson 1999) followed by destruction of coral beds. Similarly many of the coastal lagoons and lakes natural or man-made (wetlands) are under severe anthropogenic influence and at varying state of degradation process. These negative impacts need to be reversed with adequate and appropriate measures which are termed as restoration in the context of conserving the existing biodiversity and its habitat. In the current state of biodiversity and several habitats, restoration is one of the most important measures of biodiversity conservation.

14.9.2 Attributes of Restoration

Ecosystem restoration is the 'process of assisting the recovery of an ecosystem that has been degraded, damaged, or destroyed'. The ultimate goal of restoration is to create a self-supporting ecosystem that is resilient to perturbation without further assistance. It is very important measures in the coastal areas, wetlands, and other biodiversity areas wherever possible. After restoration, the ecosystem should have the following attributes (SER 2004):

- Similar diversity and community structure at least in progression in comparison with reference sites
- Presence of indigenous/endemic species
- Presence of functional groups necessary for long-term stability
- Capacity of the physical environment to sustain reproducing populations
- Normal functioning/processes
- Integration with the landscape
- Elimination of potential threats
- Resilience to natural disturbances
- Self-sustainability with restored ecosystem services

14.9.3 Mangroves and Corals

Restoration, using principles from ecology and techniques from engineering, is feasible for all three major coastal ecosystems in the tropics, namely, coral reefs, seagrass beds, and mangrove forests. The level of difficulty, however, varies with the different types of ecosystems. When viewed in these terms, mangrove forests are probably the easiest habitats to restore, followed by seagrass beds and then coral reefs. Mangrove silviculture has been practised in some Asian countries since the nineteenth century (Kaly and Jones 1998). This gained momentum in the South and Southeast Asian nations and small islands located in the Indian Ocean after the December 2004 tsunami. Mangrove ecosystem acts as a natural bio-shield in protecting the coastal areas against the sea surges and reducing the impact of tsunami waves. There are few examples of restoration of reef as well. An example of success

Fig. 14.7 Corals and its associated species under conservation

in restoration on a relatively large scale is the work of Guzman (1991) on reefs on the Pacific coast of Costa Rica which have been devastated by a combination of natural and human disturbances. A total of 110 live fragments of colonies from nearby reefs were transferred to the affected sites and attached with wire to 30-cm-long steel stakes. After 3 years, there was 79–83% survivorship and a 41–115% increase in new colonies due to fragmentation. Some of the coral reef-associated species flourishing under protected regime are given in Fig. 14.7.

14.9.4 Coastal Wetlands

Owing to an evergrowing human population and the subsequent need for land wetlands are bound to get adversely impacted. The outcome of adverse impacts may vary from wetland degradation to wetland loss which needs to be reversed. One of the proven effective strategies of wetland mitigation is wetland creation that restores wetlands to a landscape that suffered wetland loss. While wetland restorative measures mostly concentrates on removal of pollutants, reducing or removing human impact on the wetlands and allowing the nature to revert back the wetland.

Wetland creation efforts lead to construction of 'replacement wetlands' also called 'mitigation wetlands'. The total area in an affected landscape has to be similar to or larger than the original wetland area impacted in that landscape. The replacement wetland is constructed mainly for ecological wellbeing of the landscape. Wetland creation to replace lost wetlands in a landscape may require reestablishing original vegetation, hydrology, or other parameters to restore original or closer to original functions of wetlands. Further, this should also able to support and restore the original fauna that were existing there. There are several examples of successful restoration of wetland ecosystem including the migratory birds. For constructing a replacement wetland, an area having diameter of at least 25 m can be selected as wetland construction site in a flat barren land. Creation of a replacement wetland may be carried out using manual tools or heavy equipment depending on type of site selected, size of the replacement wetland required, and complexity level of the structure of the replacement wetland.

Similarly freshwater lakes and larger man-made pond can be restored to its productivity level by preventing nutrient loads through engineering methods like sediment dredging and removal, treatment of catchment areas, etc. It is also witnessed that increase in zooplankton populations can result in lower algal population. This is achieved by reducing or removing temporarily resident fish population. It is essential to have top-down trophic control of algal populations and prevention of vegetation growth towards lakes. Efforts should be made to prevent pollution and raise the pH by lime treatment so that the ecosystem normal functions are restored.

To compensate for wetland loss in a rural or suburban landscape, restoration of the landscape by creating ponds may be beneficial in several ways besides the primary purpose of ecological restoration. However, seasonality is an important characteristic of a large number of natural and human-made village ponds, countryside ponds, and suburban ponds that have been supporting man's socio-economic needs particularly in small island nations in the tropics. Fortunately, reestablishment of seasonal ponds for ecological restoration purpose is also possible (Biebighauser 2008). Such wetlands may contribute to the socio-economic wellbeing of the villages in the landscape by helping in recharging/restoring groundwater levels besides supporting the endemic species which are well adapted to the seasonality of the wetlands.

14.10 Conclusion

As marine and coastal ecosystems provide important economic, sociocultural, and environmental benefits, their sustainable management, protection, conservation, and restoration are crucial. Compilation of data and direct evidences has indicated that biodiversity is being depleted by fragmentation and loss of habitat, overexploitation of resources, expansion of human-sponsored ecosystems, pollution, invasive exotic species, diseases, shifting cultivation, and poaching of wildlife, besides climate change events. Estimates of species loss suggested that unless current trends are reversed, from one quarter to one half of the Earth's species will become extinct in the next 30 years. On the other hand, many of the tropical countries and islands offer a unique opportunity to conserve biodiversity that is increasingly under threat in the respective regions. At the same time, the pressures on natural ecosystems are likely to be exacerbated in the coming decade. Hence integrating the impacts of climate change on vulnerable species and ecosystems into immediate conservation planning measures will be a prudent approach in the long-term. In this context, a key challenge will be to address knowledge gaps both in terms of biodiversity status, strategy, and climate change impacts in comprehensive conservation planning in the tropical regions.

In the last few decades, the world has witnessed the active involvement of scientists, nature conservationist, and biodiversity supporters to drive towards some the international agreements such as the World Heritage Convention. This attempts to protect and support many hotspots of biodiversity in different parts of the world. Another treaty known as the Convention on International Trade in Endangered Species (CITES) intends to reduce the utilization of endangered plants and animals by controlling trade in their products. It is, therefore, important to develop more mechanisms and regimes of this kind, not only to prevent further degradation of the biodiversity resources but also to reverse the current trend of continuous loss of biological species and cultural assets.

References

Adams WM (2004) Against extinction: the story of conservation. Earth scan, London

Adams WM, Hutton J (2007) People, parks and poverty: political ecology and biodiversity conservation. Conserv Soc 5(2):147–183

Agardy MT (1994) Advances in marine conservation: the role of marine protected areas. Trends Ecol Evol 9:267–270

Allen GR (2007) Conservation hotspots of biodiversity and endemism for Indo-Pacific coral reef fishes. Aquat Conserv Mar: Freshwat Ecosyst. https://doi.org/10.1002/aqc.880

Anderson RC, Baldock M (2001) New records of birds from the Maldives, with notes on other species. Forktail 17:67–73

Anonymous (2011) Bangladesh red list reports. Bangladesh J Plant Taxon 18(1):81–91. www.iucn.org/news/bangladesh/201607/bangladesh-red-list-reports

Balick MJ, Mendelsoon R (1991) Assessing the economic value of traditional medicines from tropical rain forests. Conserv Biol 6(1):128–130

BDS/MoE and DNBG (2012) The National Red list of Sri Lanka. In: Weerakoon DK, Wijesundara S (eds) Conservation status of the fauna and flora. Ministry of Environment, GoS, Colombo, Viii+476

Biebighauser TR (2008) Build your own wetland. Ky Woodlands Mag 3(2):12–15

BMMRO (2009) Bahamas Marine Mammal Research Organisation, Annual report to The Bahamas Department of Marine Resources, Govt. of Bahamas

BSI (2011) Contribution towards preparation of status report on conservation of biodiversity in India II. Botanical Survey of India, Calcutta

Catarino L (2004) Fitogeografia da Guiné-Bissau. Dissertaçãodoutor. Universidade Técnica, Lisboa

Daltry JC (2007) An introduction to the herpetofauna of Antigua, Barbuda and Redonda, with some conservation recommendations. Appl Herpetol 4:97–130

FAO (2005) Global forest resources assessment 2005. Food and Agriculture Organization, Rome. http://www.fao.org/forestry/country/20807/en/guy
GBRMPA, WB, WCU (1995) A global representative system of marine protected areas
GoG (2014) Fifth national report to the convention on biodiversity. Government of Grenada, Grenada, p 75
GOJ (2010) NEPA. Fourth National Report for Jamaica to the Convention on Biological Diversity. www.earthendangered.com
Government of Seychelles (2014) Fifth National Report to the United Nations Convention on Biological Diversity. Ministry of Environment and Energy, P.O. Box 445, Botanical Gardens, Mont Fleuri, Victoria, Republic of Seychelles
Guzman HM (1991) Restoration of coral reefs in Pacific Costa Rica. Conserv Biol 5:189–195
Hodgson GA (1999) Global assessment of human effects on coral reefs. Mar Pollut Bull 38:345–355
ICBP (1992) Putting biodiversity on the map priority areas for global conservations. International Council for Bird Preservation, Cambridge, p 90
IPCC (2007) In: Gitay H, Suárez A, Watson RT, Dokken DJ (eds) Climate change and Biodiversity, Intergovernmental Panel on Climate Change Technical Paper V. IPCC, Geneva, p 85
Kaly UL, Jones GP (1998) Mangrove restoration: a potential tool for coastal management in tropical developing countries. Ambio 27:656–661
Keppel G, Morrison C, Meyer JY, Boehmer HJ (2014) Isolated and vulnerable: the history and future of Pacific Island terrestrial biodiversity. Pac Conserv Biol 20:136–145
Kinch J, Anderson P, Richards E, Talouli A, Vieux C, Peteru C, Suaesi T (2010) Outlook report on the state of marine biodiversity in the Pacific Island Region. Secretariat of the Pacific Regional Environment Programme, Apia
Kingsford RT, Watson JEM, Lundquist CJ, Venter O, Hughes L, Johnston EL, Atherton J, Gawel M, Keith DA, Mackey BG, Morley C, Possingham HP, Raynor B, Recher HF, Wilson KA (2009) Major conservation policy issues for biodiversity in Oceania. Conserv Biol 23:834–840
Kitalong (2010) The Republic of Palau statewide assessment of forest resources and resource strategy. Bureau of Agriculture, Ngerulmud
Mcneely Jeffrey A, Mittermeier RA, Brooks TM, Boltz F, Neville ASH (2009) The wealth of nature. CEMEX, Arlington
McNeill JR (1994) Of rats and men: a synoptic environmental history of the island Pacific. J World Hist 5:299–349
MEA (2005) Ecosystems and human well-being: biodiversity synthesis, Millennium Ecosystem Assessment. World Resources Institute, Washington, DC
Meerman JC (2005) Compilation of information on biodiversity in Belize. Forest Department, Belize, p 56
Myers N (1990) The biological challenge: 'hot-spots' in tropical forests. Environmentalist 10:243–256
NBSAP II (2015) National Biodiversity Strategy and Action Plan 2015–2020, National Commission for Monitoring and Evaluation, Ministry of Infrastructure, Natural Resources and Environment, São Tomé and Príncipe, 122
NRCBD (2015) Fifth National Report on the Convention on Biological Diversity. Ministry of Agro-Industry and Food Security, Port Louis, p 87
Posner S, Michel GA, Toussaint JR (2010) Haiti 118/119 biodiversity and tropical forest assessment, USAID, USDA Forest Service, pp. 88
Primack RB (2006) Essentials of conservation biology, 4th edn. Sinauer Associates, Sunderland
SCBC (2010) Secretariat of the convention on biological diversity, Global Biodiversity Outlook 3. Montréal
SER (2004) Society for Ecological Restoration International Science & Policy Working Group, The SER International Primer on Ecological Restoration (available from http//www.ser.org) accessed in November, 2017. Society for Ecological Restoration International, Tucson, Arizona
Sivaperuman C, Raghunathan C (2012) Fauna of the protected areas of Andaman and Nicobar Islands, Zoological Survey of India, Kolkata, p.26

Spalding MD, Ravilious C, Green EP (2001) World Atlas of coral reefs, Prepared at the UNEP World Conservation Monitoring Centre. University of California Press, Berkeley, p 432

SPREP (2012) Pacific environment and climate change outlook. Secretariat of the Pacific Regional Environment Programme, Apia

Teo DHL, Tan HTW, Corlett RT, Wong CM, Lum SKY (2003) Continental rain forest fragments in Singapore resist invasion by exotic plants. J Biogeogr 30:305–310

Thaman RR (2014) Agrodeforestation and the loss of agrobiodiversity in the Pacific islands: a call for conservation. Pac Conserv Biol 20:180–192

UNDP (2002) Human development report 2002. United Nations Development Programme. http://hdr.undp.org/en/reports/global/hdr2002/

Veron JEN, DeVantier LM, Turak E, Green AL, Kininmonth S, Stafford-Smith M, Peterson N (2009) Delineating the coral triangle. Galaxea 11:91–100

Vitousek PM, Mooney HA, Lubchenco J, Melillo JM (1997) Human domination of earth's ecosystems. Science 277:494–499

Whittingham E, Campbell J, Townsley P (2003) Poverty and reefs. DFID–IMM–IOC/UNESCO, p 260

Wilkinson C (2004) Status of Coral Reefs of the World, 2004, vol 1. Australian Institute of Marine Science, Townsville, p xiv + 301

Wilson EO (1988) The current status of tropical biodiversity. In: Wilson EO (ed) Biodiversity. National Academy Press, Washington, DC, pp 1–18

ZSI (2015) Animal discoveries new species and new records. In: Chandra K, Sheela S, Das D (eds) Zoological survey of India report. Government of India, Ministry of Environment, Forest and Climate Change, New Delhi, p 152

Alternative Farming Systems for Diversification and Conservation of Agro-biodiversity

15

Abstract

Food insecurity and loss of biodiversity are the two major juxtaposed challenges facing the humanity. The situation is further aggravated by resource degradation mostly as a result of intensive agriculture and climate change impact particularly in the coastal areas of tropical countries. This calls for alternative approaches to conventional intensive farming that should enhance agricultural production and ecosystem services besides achieving agricultural diversification to instil stability and sustainability to the production system. In this chapter the status of coastal agriculture is reviewed vis-à-vis challenges and threat to agricultural production with the focus on the impact of modern agriculture on biodiversity, ecosystem, and the society. The importance of diversification to ensure regional food security and conservation of agro-biodiversity is discussed in details. This chapter also attempts to justify the need for alternative farming systems, its definition, major types, and the major alternative farming systems suitable for the tropical coastal region. At the end the impacts of these alternative farming systems were compared in terms of its suitability, promotion of diversity, input use, ecosystem services, resilience capacity, labour engagement, and use of traditional knowledge biodiversity conservation.

Keywords

Agro-biodiversity · Farming Systems · Types · Tropical Region · Challenges

15.1 Introduction

With rapid increase in human population and decreasing natural resource base, it will be a great challenge to meet the growing demand for food and other agricultural products of 9–10 billion people by 2050 (Godfray et al. 2010) with greatest increases in tropical developing nations. Off this about 1.3 billon people lives in tropical

V. Ayyam et al., *Coastal Ecosystems of the Tropics - Adaptive Management*,
https://doi.org/10.1007/978-981-13-8926-9_15

coasts within 100 km of the coastline, primarily in developing nations (Sale et al. 2014). This rapid increase in population of coastal areas will lead to enormous increase in the requirement for food and other agricultural products posing a greater challenge for achieving efficient and productive agricultural land use while conserving biodiversity and sustaining the food production.

The traditional extensive production systems could not satisfy the food requirements, while modern-intensive, monoculture systems have increased the food production from 1.84 BT in 1961 to 4.38 BT in 2007 (Royal Society 2009) lifting millions out of poverty and provided a platform for rapid economic growth in many parts of the world. It led to worldwide losses of biodiversity (Tilman et al. 2001) caused by land use changes including the conversion of complex natural ecosystems to simplified managed ecosystems, intensification of resource use, including use of more agrochemicals, etc. It also resulted in degradation of land and water and inflicted damage to the functioning of several natural ecosystems, and the benefit is unevenly distributed.

The projections by FAO (2002) and OECD/FAO (2012) suggest that 80% global food demand will be met by crop intensification in the existing croplands, while 20% of production increase will come from new croplands by 2030 which require conversion of an area equivalent to South Africa, most of which occur in sub-Saharan Africa and South America. Though the rate of crop expansion is reducing, it occurred at the expense of natural habitats mostly in tropical forests (Koh 2007). At the same time, consistent side effect of agricultural expansion and intensification has been the fragmentation and isolation of natural habitats within an increasingly unsuitable matrix for many species (Donald et al. 2001). With increasing awareness and recognition on the environmental impacts of agricultural expansion globally, several initiatives were taken up to provide management options to farmers, to repair some of the environmental damages caused by agricultural intensification, to ensure a more sustainable exploitation of natural resources, and to improve the ecological status of agricultural landscape. As the comprehensive information on these aspects pertaining to tropical coastal areas is lacking and scattered in literature, this chapter focuses on alternative farming practices particularly suitable for coastal regions of the tropics to address the production issues and natural resource degradation challenges.

15.2 Tropical Coastal Areas

15.2.1 Regional Grouping

The coastal region plays a vital role in the global economy due to its rich natural resources, productive habitats, and biodiversity. Total coastline of the world is 356,000 km, and the coastal area covers more than 10% of the Earth surface. As we are focusing on tropical areas, it is important to understand what constitute a tropical country and tropical coastal area. The countries with at least one third of their land area between the Tropic of Cancer and Capricorn are defined as tropical countries. In this chapter, tropical countries with at least one third of land area under coastal region and all the island nations within the tropics are considered for this study. These tropical countries are grouped into different developing regions (FAO and World Bank 2001), viz.:

- South Asia
- East Asia and Pacific
- Sub-Saharan Africa
- Middle East
- Latin America and the Caribbean

The grouping was adopted to facilitate the analysis of agricultural status, types of existing farming systems, constraints and challenges for biodiversity conservation, diversification options for the nations, and islands located in the tropical region (Fig. 15.1). However, the land use statistics, farming systems, and agricultural production are not available separately for coastal areas rather the available information pertains to the administrative boundaries. Therefore, the data/available information need to be carefully integrated and generalized for the coastal areas.

15.2.2 Status of Agriculture

Agriculture in tropical regions is characterized by monsoonal weather systems having the majority of its rainfall during the warmer months of the year. Sometimes heavy downpour or cyclonic weather is experienced which creates potential for significant amounts of storm water runoff and flooding. Agriculture in the tropical world has been the backbone of the economy which is viewed as a holistic production system by the farmers rather than simple crop cultivation or rearing of animals. It shows variety of natural resources available to farm families and interlinks between them including the socio-economic aspects. These resources normally include different types of land, various water sources, and access to common

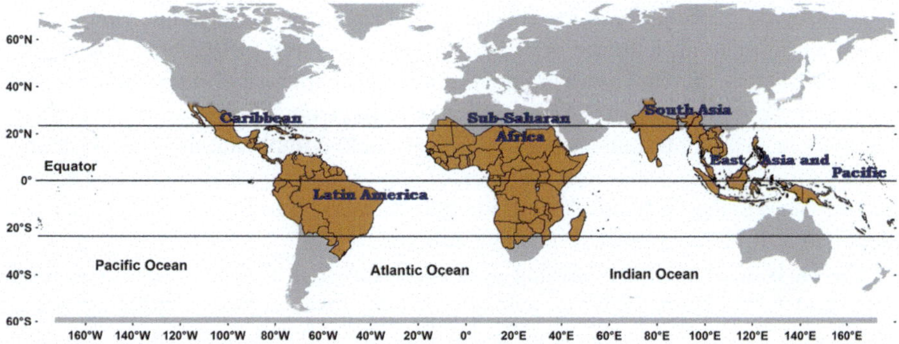

Fig. 15.1 Developing regions of the tropics

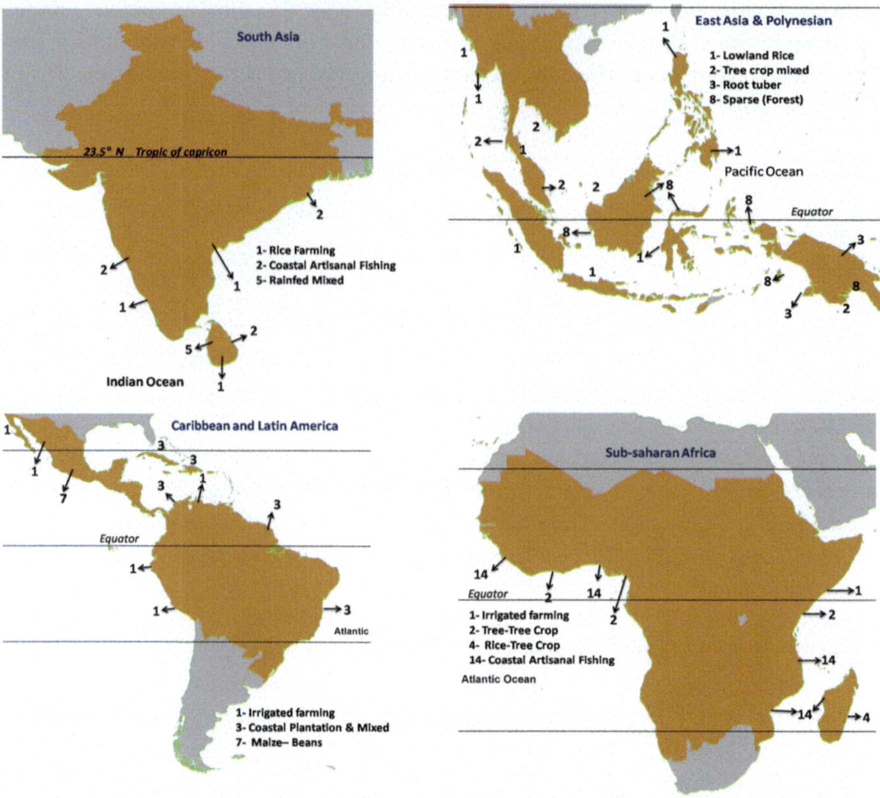

Fig. 15.2 Major farming systems practised in the tropical coastal areas. (Modified from FAO and World Bank 2001)

property resources—including ponds, grazing areas, and forest. To these basic natural resources may be added climate and biodiversity, as well as human, social, and financial capital which are essential for successful farming.

In this context, the delineation of the major farming systems practised in the tropical region particularly in the coastal areas provides a useful framework within which appropriate agricultural development strategies and technological interventions can be determined. However, identification of numerous, discrete, micro-level farming systems in each developing country in the tropical region could result in hundreds or even thousands of systems. That would complicate the understanding of the farming systems and interpretation of appropriate regional and global strategic responses to the problems. Therefore, only the major farming systems have been identified to find the alternative farming systems.

The classification of the farming systems in this region (Fig. 15.2) is based on (FAO and World Bank 2001) (i) available natural resource base, including water, land, farm, climate, grazing areas, and forest, and (ii) dominant pattern of farm activities and household livelihoods such as crops, livestock, trees, aquaculture, hunting and gathering, processing, and off-farm activities including intensity of production components. Based on these criteria, the following eight broad categories of farming system have been distinguished:

- Irrigated farming systems, embracing a broad range of food and cash crop production
- Wetland rice-based farming systems, dependent upon monsoon rains supplemented by irrigation
- Rainfed farming systems in humid areas of high resource potential, characterized by a crop activity (notably root crops, cereals, industrial tree crops—both small scale and plantation—and commercial horticulture) or mixed crop-livestock systems
- Rainfed farming systems in steep and highland areas, which are often mixed crop-livestock systems
- Rainfed farming systems in dry or cold low potential areas, with mixed crop-livestock and pastoral systems merging into sparse and often dispersed systems with very low current productivity or potential because of extreme aridity or cold
- Dualistic (mixed large commercial and small holder) farming systems, across a variety of ecologies and with diverse production patterns
- Coastal artisanal fishing, often mixed farming systems
- Urban-based farming systems, typically focused on horticultural and livestock production

It is reiterated here that the spatial mapping of farming systems represented in this chapter is only the major farming systems at the regional level. There may be other systems; subsystems within each system and sharp boundary rarely exist at field level.

15.3 Challenges of the Tropical Coastal Region

The tropical coastal regions are increasingly facing huge challenge emanating from population growth, developmental needs, and the burden of providing food to millions of people. Together with these anthropogenic factors, combinations of natural and hydrological factors are responsible for land degradation of different kinds. Therefore managing the increasing demand for coastal resources and the inevitable conflicts between competing users represents the major future challenge for sustainable management of tropical coastal region.

15.3.1 Population Growth

The coastal regions are affected by changing land use and land cover due to changes in the size, composition, and distribution of human population. The population density in coastal regions (80 persons per square kilometre) is twice that of world average. Migration and urbanization are the two other factors affecting coastal zones. The figures of migration are staggering in China and Southeast Asia where around 1000 people arrive to large coastal cities in China, Vietnam, and the Philippines. Besides, 40% of cities with one to ten million people are located near coastlines (Tibbetts 2002). The rapidly growing population in coastal areas by natural growth and migration is greater concern for many countries because population growth and the associated activities can degrade coastal and marine ecosystem. For instance, endemic coastal fish stocks reduced from 30% to 10% over a period of three decades by intensive fisheries. Similarly half of the world's wetlands as well as mangroves and nearly 60% of the coral reefs were seriously degraded by development and other human activities. The freshwater quality is affected by pollution from industry, agriculture, and urban areas (Creel 2003). These challenges are more acute for island countries, where coasts often comprise the entire territory. Overharvesting of sea-based biological resources for food can deplete many species and alter the biological structure of coastal ecosystems.

15.3.2 Climate Change

Climate change causes various impacts on coastal areas through extreme weather events, long-term changing averages in climatic variables, and increased weather variability. Because of their unique position, coastal areas are exposed to the influences of climate change either directly through sea level rise, storm surges, floods, and droughts or indirectly through modification of coastal ecosystems and habitats (Alcamo et al. 2007). The increase in mean annual atmosphere temperature may cause increased wave impact and increased storminess and storm impact causing flooding, erosion, and destruction of infrastructure.

Agriculture in coastal areas and deltas is susceptible to the impact of sea level rise (SLR) through inundation of land; erosion; salinization of rivers, bays, wells, and land; and modification or loss of ecosystems and habitats. The coastal aquifers will be affected by seawater intrusion rendering them unsuitable for drinking and agriculture and promote gradual salinization of coastal strips. This may cause habitat loss especially of wetlands and impact the abundance and distribution of sea grasses. SLR is of significant economic relevance as a number of productive areas are located near the coastline in the region. For example, about 25% of the coastal plain territory (142,500 ha) in Guyana would be affected by sea level rise, including the intensification of storm surges, of which 59% are dedicated to agriculture (Government of Guyana 2012). However, there is no comprehensive assessment of the systemic impact of sea level rise in agricultural areas in costal zones and deltas.

15.3.3 Impact of Modern Agriculture

With advent of 'Green Revolution' during the 1960s, an integrated system involving the use of synthetic compounds like pesticides, fertilizers, high-yielding varieties of crops, irrigation facilities, etc. has increased the food production tremendously in many parts of the world. But this was brought at the cost of more diversified farms and the following impact on the biodiversity and environment.

15.3.3.1 Loss of Biodiversity

Agriculture is one of the major anthropogenic activities altered the face of the Earth (Tilman et al. 2001). Habitat loss due to crop land expansion is one of the global threats to biodiversity (Green et al. 2005) which threatens to undermine the progress towards meeting the Aichi Biodiversity Targets (UNEP 2011). In recent years intensification of crop production has led to significant increase in food production than cropland expansion. The cropland expansion will account for only 20% of production increases in developing countries, while crop intensification accounts for the rest (FAO 2002). Much of this expansion is likely to be in sub-Saharan Africa, South America, and Tropical Asia. The greatest cropland expansion was reported in Nigeria, Sudan, Ethiopia, Brazil, and Indonesia over the period of 1999–2008. In several countries including India, Australia, Colombia, Mexico, and Thailand, the reported area of annual crops decreased which may be due to crop land degradation or conversion to other uses including perennial crops or increases in land use efficiency (Phalan et al. 2013).

The biodiversity of pristine ecosystems has been reduced not only by cropland expansion; intensively used agro-ecosystems have also witnessed drastic reduction in agro-biodiversity. Intensification is negatively correlated with the ecosystem services (Fig. 15.3). Further, intensification largely degrades the environmental quality and threatens the sustainability of food production by likely disruption of beneficial functions of biodiversity (Tscharntke et al. 2012). It is observed that only less than

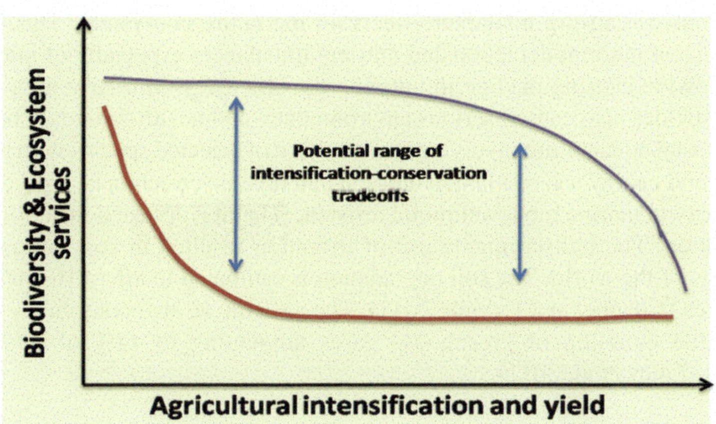

Fig. 15.3 The relationship between intensified agriculture and biodiversity

5% of the original plant species are retained in intensive non-shaded cacao production as that of pristine situation (Siebert 2002). Specialization or monoculture of crop production is an important aspect of conventional farming. Such type of specialization of agricultural production at landscape level also depresses ecosystem quality to the extent.

Besides cropland expansion and intensification, the spread of new modern high-yielding varieties was also responsible for loss of indigenous varieties or land races resulting in gene erosion. Although around 7000 species of plants are consumed, only three crops, viz. rice, wheat, and maize, account for about 60% of the calories and 56% of the protein people derive from plants. Mono-culturing of high-yielding varieties (HYV) of rice has reduced diversity including traditional varieties and fish species in Bangladesh, the Philippines, and India. In India almost half of the native varieties were replaced by HYV (Thrupp 2000). The loss of native species or varieties and mono-culturing of crops occurred not only in food grains but also in commercial crops and even in livestock. Many traditional strains of cattle have disappeared as the focus was on new breeds of cattle, pigs, sheep, and chickens all over the world. Many breeds were extinct, and exotic breeds were introduced in countries like India, Africa, etc. Though these changes have increased productivity, narrowing of genetic pool and homogenization of species increased the vulnerability to biotic and abiotic stresses. Further there has been a serious decline in soil organisms, beneficial insects, and other nontargeted organisms under intensive use of agrochemicals. Hence, there is an urgent need to conserve these traditional varieties or land races and native breeds of animals either on-farm or ex situ through alternative farming for productive genetic diversification of traditional agriculture and enhanced ability of the cropping system to evolve adaptations to environmental and economic changes (Wood and Lenné 1997).

15.3.3.2 Environmental Impact

The large-scale expansion of industrial agriculture involving monoculture not only in crops but also in cattle and fisheries resulted in contamination of land, soil, and water resources causing nontarget effects on the entire ecosystem. This has also resulted in soil nutrient depletion and nutrient imbalances especially of micronutrients worldwide causing decline in nutrient use efficiency. Similarly the excessive use of pesticides also caused serious environmental and health hazards. The expansion of coastal aquaculture involving monoculture of selected species with intensive chemical and energy use resulted in nutrient and organic enrichment, oxygen depletion, toxicity to marine biota, antibiotic resistance, habitat destruction, and sediment accumulation. The indiscriminate use of irrigation resulted in soil salinization in many parts of the world. The soil degradation is estimated to affect 16–40% of terrestrial area (Chappel and LaValle 2011). The agriculture also contributes substantially to the emission of greenhouse gases amounting to 14% of total global emissions (Foley et al. 2011).

15.3.3.3 Social Impact

The benefits of green revolution have not evenly spread across the world and even within a region. The continued expansion of industrial agriculture is witnessing rapid growth of land grabbing in recent years in response to food security concerns and food speculation especially in Africa, Asia, and Latin America. Since the year 2000, foreign investors have acquired 26.7 million hectares of land around the globe for agriculture; the largest acquisition is in Africa with 10 million hectares concentrated along important rivers such as the Niger and the Senegal and in East Africa (Land Matrix 2016). Land grabbing and its conversion to large-scale production for oil palm and sugarcane in Southeast Asian countries like Indonesia and Malaysia led to displacement of small holders, increasing monoculture, and abandonment of food cropping. This will not only deprive the land rights of small holder farmers but also threaten the food self-sufficiency and food sovereignty of these countries (Tscharntke et al.2012).

15.4 Diversification: The Concept

The demand for agricultural and food products caused by rising population and climate change situations will become most acute in the next few decades. Rising temperatures, altered rainfall patterns, and more frequent extreme events will increasingly affect crop production, often in those places that are already most vulnerable (Morton 2007). Notwithstanding the potential to adapt crops to changing environments, the need to mitigate climate change will increasingly challenge conventional, resource-intensive agricultural systems which depend on chemical inputs derived from fossil fuels and contribute significantly to greenhouse gas (GHG) emissions. The increase in food production is also limited by scarcity of natural resources. Consequently, attention is now being directed towards the development of crop production systems with improved resource use efficiencies and more benign effects on the environment without compromising food production (Foley et al. 2011).

In order to tackle the problems of crop expansion, intensification, and climatic vulnerabilities, there is an urgent need for identifying potential alternative farming strategies other than energy- and chemical-intensive one-crop farming. Alternative farming system, in this context, is defined as an activity on the farm with the potential for bringing in supplementary income which utilizes land, labour, capital, and entrepreneurial ability and which differs qualitatively from any other existing activity (Barlas et al. 2001). Alternatives include using animal and green manure rather than chemical fertilizers, integrated pest management instead of chemical pesticides, reduced tillage, crop rotation, alternative crops, diversification of the farm enterprise, etc.

In this context, promotion of agricultural diversity and ecological processes in agricultural production system is one of the four factors which will have positive impact on food and nutritional security (Caouette et al. 2002). Diversification can be facilitated by technological breakthrough, by changes in consumer demand or in government policy or in trade arrangements, and by development of irrigation, roads, and other infrastructures. On the contrary, it can be impeded by risks in markets and prices and in crop management practices, by degradation of natural resources, and by conflicting socio-economic requirements—perhaps for employment generation or for self-sufficiency or foreign exchange earning capacity in particular crops or livestock or fishery or forest products (Singh et al. 2002).

As discussed earlier the alternative farming should accommodate enough diversity to address these issues and provide stability to the production system. Indirectly it helps to conserve biodiversity and traditional wisdom/resources as it is a region specific. In general diversification is an adjustment to the farm enterprise pattern so as to increase farm income and stability or to reduce income variability. It exploits the potential of diversified products from diversified activities and the new market opportunities. Diversification may take the form of completely new enterprises or modification of part or may simply involve the expansion of existing, high value, enterprises. The addition or expansion of enterprises refers not only to crop and animal varieties/breeds and other inputs required for production but also to on-farm processing and other farm-based, income-generating activity.

The diversification can occur in many forms (genetic variety, species, structural) and over different scales (within crop, within field, landscape level), giving farmers a wide variety of options and combinations for the implementation of this strategy. The diversity can be created temporally as well as spatially through alternative farming systems, adding even greater functional diversity and resilience to systems with sensitivity to temporal fluctuations in climate. Diversification may be of different types and is aimed at increasing the farm income. Based on the level of diversification and resource endowments, the farming can be divided into four groups (Fig. 15.4), viz.:

- Low diversification and low per capita income
- High diversification but low per capita income
- Low diversification and high per capita income
- High diversification and high per capita income

The farming systems as practised in the tropical coastal areas may fall under any of the category or in different stages of transformation. Transformation into commercial fruit cultivation in some of the Polynesian islands and Caribbean countries is good example of market orientation. In South Asian countries, it is mostly aimed at meeting the food requirement of the burgeoning populations. In most of the locations, the farming falls under low per capita income with or without diversification. Agricultural diversity, which should be socially acceptable, enhances sustainability of farming systems in these areas and elsewhere. With the introduction of technology and capacity building in long run, farming in the coastal areas may transform itself and provide high per capita income and can reach level 4. It may also be possible to improve the existing farming systems which are climate resilient by technological interventions in certain areas.

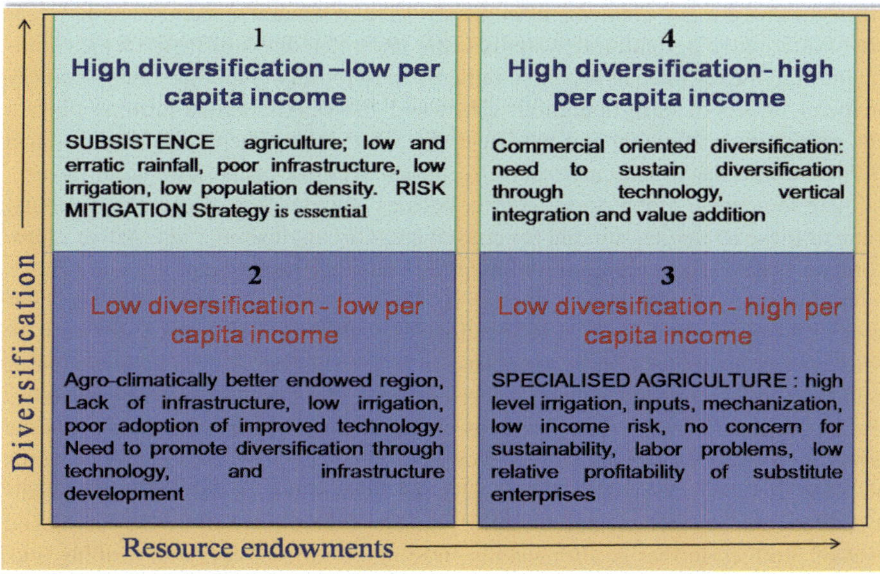

Fig. 15.4 Farming systems based on diversification and resource endowments

Thus these sustainable alternative farming practices will play a vital role in maintaining biodiversity through diversification, regulating climate change and energy use, addressing land degradation, conserving ecosystem services, optimizing productivity, and ensuring food security in the coastal areas of the tropics.

15.5 Alternative Farming Systems for the Coastal Region

The farming practices that avoid or at least minimize the use of nonrenewable production inputs such as synthetically compounded fertilizers, pesticides, and herbicides are broadly defined as alternative agriculture or farming (Anderson 1985). The alternative farming practices is viewed as a region-specific innovation. Any enterprise/activity which is not conventional to the region is considered to be an alternative enterprise as long as it either is entirely absent from that region or has been adopted by only a minority of farmers in that region (Ilbery et al. 1997).

The alternative farming practices encompass a variety of farming practices which might follow a common 'alternative belief system' characterized by ecological principals (Batie and Taylor 1991). They include sustainable agriculture, organic farming, natural farming, regenerative agriculture, low-input agriculture, agroecology, etc., and these practices are aimed at assuring the farm's viability and preventing the decline in farm income. The farming practices consistent with the philosophy of alternative agriculture vary in their details and utility to a specific region, but they have an important feature in common, i.e. they are designed to drastically reduce and preferably to eliminate the agrochemicals such as pesticides and inorganic fertilizers that are key elements of conventional agricultural systems. The key feature of alternative agricultural systems is the interrelatedness of their various components that makes them as a system rather than as loosely combined but essentially independent sets of farm operations (Harwood 1984). Alternative farming plays a vital role in diversification and biodiversity conservation. There are many on-farm alternative options which can ensure livelihood security of the individual farmers and help to achieve global food security besides conservation of agro-biodiversity. Some of these strategies suitable for coastal areas are inclusion of alternative crops, cropping systems, organic farming, integrated farming system, etc.

The alternative farming practices can be divided into three main categories: conversion, diversification, and extensification. The adoption of any crop or livestock which is new, unconventional, or unusual in a given region is called conversion. It includes significant changes in the nature of the crop or livestock like new varieties, new breed, etc. Diversification involves having different types of activities on the farm which includes activities other than agriculture as well and oriented towards the public, e.g. agro-tourism, recreation, crafts, farm shops, processing, value addition, direct marketing, etc. (Ilbery 1991). Finally, extensification describes a reduced level of production that is afforestation, free range farming, organic farming, etc. Some of the farming practices consistent with the philosophy of alternative agriculture suitable for the tropical coastal region/islands are given below.

- Organic farming
- Integrated farming systems
- Agroecological farming—diversified farming systems
- Natural farming
- Sustainable intensification
- Integrated aquaculture

15.5.1 Organic Farming

Organic farming systems seek to significantly reduce or to avoid entirely the use of synthetic fertilizers, pesticides, growth regulators, and other agricultural chemicals. The system relies instead on crop rotations, crop residues, animal manures, legumes, green manures, organic wastes, and non-chemical pest control to maintain the organic matter and tilth of soil, to supply nutrients, and to control insects, weeds, and diseases (Papendick 1984). It can be an effective alternative to conventional farming as it aims at ecosystem management and excludes agrochemicals like fertilizers and most pesticides (Stockdale et al. 2001). Organic farming can influence biodiversity either directly or indirectly (Rundlöf et al. 2016), the direct effects include the reduced exposure to agrochemicals, and indirect effects include the effects of changed farming practices which can result in increased local habitat diversity on organic farms. It helps to bring diversification by way of land improvement and providing opportunities for the cultivation of local/minor crops. Organic

Table 15.1 Region-wise organic farming potential and its effect on biodiversity

Sl. no	Region	Products exploited/produced by organic farming	Effects of organic farming on biodiversity
1.	South Asia	Rice including basmati rice, wheat, vegetables and fruits, cotton, soybean, tea, coffee, milk, sugar, medicinal plants, and aromatic herbs	Species abundance or richness, across wide range of taxa, tends to be higher on organic farms than conventional farms (Hole et al. 2005)
2.	Southeast Asia	Rice, vegetables and fruits, tea, spices, coconut	
3.	Pacific Islands	Beef, cocoa, coffee, fruit and fruit juices, herbs and spices, honey, Noni products, vanilla, vegetables, and virgin coconut oil (Vinning 2008)	Increased diversity or species richness of plant (Winqvist et al. 2011; Frieben and Kopke 1995), pollinating arthropods, viz. bees, hoverflies, butterflies (Schneider et al. 2014)
4.	Latin America and Caribbean	Banana, pineapple, pears, grapes, citrus fruits, vegetables, soybean, maize, wheat, coffee, cocoa, sugarcane, and meat (Garibay and Ugas 2010)	
5.	Sub-Saharan Africa	Banana, pineapple, spices, coffee, tea, cocoa, cotton, fresh vegetable, citrus, grapes, mango, avocado, tree nuts	Enhanced bacterial (Brossio et al. 1998) and fungal (Shannon et al. 2002) activity/ abundance
			Increased earthworm activity, species diversity (Brown 1999)
			Greater abundance and diversity of arthropods were also reported by Reddersen et al. (1997) compared to conventional system
			Brown 1999 observed greater activity levels of small mammals (wood mouse, bank vole, and common shrew) especially in field margins of organic fields
			Increased foraging activity of bats (Wickramasinghe et al. 2003), greater avian abundance, and species richness were also observed in organic farms (Hole et al. 2005)

farming is practised in all the regions with specialization in crops having high market potential, the details of which are given in Table 15.1.

Rice including basmati rice, wheat, cotton, soybean, tea, coffee, milk, sugar, medicinal plants and aromatic herbs, spices, and coconut are the major crops in South and Southeast Asian countries. The Pacific Islands are known for the organic production of cocoa, coffee, beef, fruits and vegetables, etc. The Latin America and Caribbean region fruits, vegetables, soybean, maize, wheat, coffee, and cocoa are produced organically and exported to the EU and the USA, while sub-Saharan

Africa is known for production of organic banana, pineapple, tea, cocoa, cotton, etc. Majority of the organic production from all the regions aimed at export market mainly for European Union and the USA.

15.5.1.1 The Effect on Biodiversity

The effect of organic farming on biodiversity in terms of species richness has been assessed in large number of studies. Bengtsson et al. (2005) observed that organic farming usually enhances species richness, most notably of plants, birds, and predatory insects. The effects of organic farming on species richness will be larger in intensively managed agricultural landscapes than in small-scale diverse landscapes with many non-crop biotopes. Similarly Tuck et al. (2014) observed 30% increases in local biodiversity by organic farming in their meta-analysis. However there is a large variation between studies and organisms which can be attributed to differential responses of organisms to changes in farming practice, soil conditions, differences in time, and the scale of conversion to organic farming, modifying effect of the heterogeneity of the surrounding landscape, and also the spatial distribution of organic farming (Fuller et al. 2005).

15.5.1.2 Organic Farming in Latin America and Caribbean: A Case Study

Latin American agriculture is in a state of flux, and organic production has gained momentum and increasingly opted by many farmers as they consider organic agriculture either as a way of life or as a good business opportunity to market fresh produce in local or foreign markets. Profitability alone is not the motive to produce organically, but it is considered as an alternative to maintain and protect their local resources, avoid environmental damage, and mitigate climate change effects.

The Latin American and Caribbean countries are known for production of banana, coffee, sugarcane, rice, ornamentals, and fruits on larger plantations as mono-crops under intense and predominant use of pesticides. Organochlorine compounds were heavily used during the 1970s and 1980s. But after their prohibition in 1988, several other agrochemicals were dominated the pesticide use which resulted in pesticide residues in soil, and many persistent compounds were also detected in tropical aquatic ecosystems.

Organic agriculture as a planned activity was started in many Latin American countries during mid-1980s with the production of vegetables for the local market; however, later it was supported by NGOs and multinational companies and boosted by national governments. In general, the organic movement in Latin America has grown on its own accord with some seed funding by international aid agencies especially from Germany, the Netherlands, Belgium, Switzerland, the USAID, etc. for extension and association building.

This type of agriculture has become more important because of a reduction in production costs and an increase in profitability per ha of cultivated land. The organic agricultural production has increased from 3.9 million ha in 2000 to 6.4 million ha representing nearly 1% of the total agricultural area. In addition, nearly

6000 ha of aquaculture in Ecuador and about 7.5 million hectares of wild collection mostly in Brazil and 0.5 million hectares certified bee keeping in Argentina are also reported (Garibay and Ugas 2010). Organic farming has been mainly used in cacao, banana, pineapple, coffee, fresh fruits and vegetables, sugar, meat, wines, and spirit stimulated by the foreign buying companies and fair trade agencies. It also encourages diversification and value addition. Costa Rica currently exports organic coffee, pineapple, banana puree and fresh banana, cocoa, blackberries, sugar, orange juice, mango, ginger, vanilla, and other aromatic spices (De La Cruz and Castillo 2010) mainly to Europe and the USA. Brazil sends apple and grapes; Chile exports kiwi, raspberries, and strawberries. Mexico, Columbia, Honduras, and the Dominican Republic export banana, pineapple, mango, and other tropical fruits. Mexico supplies apples, citrus fruits, and avocados in the world market, and Costa Rica and other Central American countries export fresh vegetables.

Organic banana is the major products in the international organic fruits market exported by Ecuador, Dominican Republic, Peru, Columbia, Brazil, Costa Rica, Honduras, Mexico, and Nicaragua. Mexico is the largest organic coffee producer and exporter followed by Guatemala and other Central American countries. It is mostly grown by small indigenous farmers and primarily defined by ecological forest management systems. So it is considered as an alternative to the deforestation taking place in this region. Most of these countries also produce and export organic cocoa which is an important source of income to small farmers throughout Central America and tropical areas of South America. Mexico, Nicaragua, Costa Rica, and Bolivia are producing organic chocolate to local markets as well.

15.5.1.3 Constraints

The major constraints for organic farming are:

- Competition from other certification standards like Rainforest Alliance, Fair Trade, etc. which permit use of chemical inputs. So the farmers need only fewer changes in the production rather than converting to organic farming.
- Increasing risk of production/crop loss due to climate change events. For instance, in Cuba, Haiti, and Jamaica, hurricanes have destroyed many crops incurring heavy losses to the farmers.
- Crop losses due to pest and diseases for many of which control measures were not developed, e.g. Monilia Pod Rot in cocoa (*Moniliphthora roreri*) which is a serious fungal disease causing severe damage in northwestern South America and southern Central America.
- The price of organic produce is sometimes doesn't cover the entire production cost discouraging the farmers from continuing with organic farming.
- Lack of processing and value addition in many of the Polynesian and Southeast Asian countries.

15.5.2 Integrated Farming System (IFS)

In recent times agricultural production has become unstable due to complex, dynamic, and interrelated factors such as climate, increasing food demand, scarcity of resources, volatile market forces, and national and international policies which are beyond the control of farming community. The pace, scale, and even the direction of such changes are hardly predictable (Thompson and Scoones 2009). Therefore new farming strategies should be developed by stakeholders incorporating innovations in objectives, organization, and practices adapted to changing production contexts and geographical location (Martin et al. 2013). As a consequence, farming system approach has gained momentum as an alternative to conventional monocropping to sustain the agricultural production especially of small and marginal farmers in developing countries mostly in tropical region.

15.5.2.1 IFS and Diversification

Farming system approach has been widely recognized and advocated as one of the tool for harmonious use of inputs and their compounded response to make the production system sustainable. A farming system is the result of complex interactions among a number of interdependent enterprises/components, where an individual farmer allocates certain quantities and qualities of factors of production, viz. land, labour, capital, and management to which he has access (Mahapatra 1992). Farming system is defined as a resource management strategy to achieve economic and sustained agricultural production to meet diverse requirements of farm livelihood while preserving resource base and maintaining a high level of environment quality (Lal and Miller 1990). The integration of different agricultural enterprises like cropping, animal husbandry, fishery, poultry, horticulture, forestry, etc. in the farming system mode has great potentialities to make farming more profitable and sustainable in the coastal region (Behera et al. 2004).

Historically, integrated farming systems were called mixed farming in which two or more commodities were produced separately in the same farm to use land efficiently and to increase the farmer's income (Djajdiredja 1980). An integrated farming system (IFS) consists of a range of resource-saving practices that aim to achieve acceptable profits and high and sustained levels of production. It promotes recycling of organic wastes within the farm which is very essential component of IFS. It helps to minimize the negative impacts of intensive farming and preserve the environment. It also promises to enhance food, nutritional, livelihood, and income of farmers apart from reducing risk of total failure of crops in monocropped areas due to weather vagaries and/or biotic factors. The promotion of agricultural diversity through integration of different enterprises will have positive impact on food and nutritional security at household level (Caouette et al. 2002). It provides such an opportunity for diversification of different agricultural activities and utilization of agro-biodiversity, and it can enhance the productivity and profitability of existing farming systems to ensure livelihood security (Ansari et al. 2014).

In IFS, output from one subsystem becomes an input to another subsystem resulting in a greater efficiency of output of desired products from the land/water area under the farmer's control (Edwards et al. 1988). This implies a reliance on on-farm resources for such recycling activities. El Titi (1992) defines integrated farming system as 'a holistic pattern of land use which integrates natural regulation processes with farming activities to achieve maximum replacement of off-farm inputs and to sustain farm income'. A broader definition of integrated farming also includes the use of off-farm resources, such as agro-industrial products and by-products (Little and Muir 1987). Taking a wider perspective, integration has been viewed as part of Integrated Resources Management with relevance at landscape and/or watershed level.

15.5.2.2 Components of Integrated Farming System

The major components of any integrated farming system include:

1. Crops, livestock, birds, and trees suitable for the region and agroclimatic conditions.
2. Crop may have subsystem like monocrop, mixed/intercrop, multitier crops of plantations, cereals, legumes (pulses), oilseeds, forage, etc.
3. Livestock components may be milch cow, goat, sheep, poultry, pig, and bees.
4. Tree components may include timber, fuel, fodder, and fruit trees which are mostly multifunctional.

15.5.2.3 Factors to Be Considered for Integration

While selecting the enterprises/components for integration, soil types, rainfall, and its distribution and length of growing season of the farm should be considered. The need, resource base of the farmers along with social considerations also plays a major role in selection of components for integration in any farm. Some of the examples are:

- Suitable grain crops: According to soil type and agroclimatic conditions, suitable crops should be selected, for example, in tropical and subtropical coastal region, rice is the major crop.
- Suitable forage crops: Based on soil type and water availability, forage crops should be selected. Sorghum and other fodders for rainfed areas.
- Suitable tree species: Based on the requirement and suitability, tree species need to be selected for ails, dykes, and road sides
- Suitable livestock and birds: Based on the need, goat, sheep, cattle, buffalo, poultry, and duck can be selected. In selecting the animals, social factors also play a major role, e.g. piggery is a major animal component in Nicobar Islands and Southeast Asian nations.

15.5.2.4 Advantages of Integrated Farming System

The benefits of IFS over conventional farming system are reported by many workers, and it is confirmed to reduce cost of production and thus increase productivity, income, nutrition, and overall welfare besides increasing personal savings and health of farmers (Yadav and Sharma 2013). Diversification of farming activities improved the utilization of labour and reduced unemployment in areas where there was a surplus of underutilized labour and provided a source of living for those households that operated their farm as a full-time occupation (Van et al. 2003). The advantages of integrated farming system can be seen from farm level as well as from regional perspectives. The major benefits include:

- Maximum farm yield and diversified products can be obtained from a minimum area.
- Efficient utilization of limited resources in terms of both space and time.
- Increased and beneficial spread of income throughout the year.
- Reduction of operating cost and market risks due to improved farming conditions and diversification of commercial activities.
- Slowing down migration processes and increasing social benefits through employment and income generation by diversification of farm activities and improving year-round labour demand.
- Food and nutritional security of farm family through on-farm production of diversified food products and increased farm income.
- Intensification of nutrient cycling and recovery of nutrients that have leached or drained to deeper soil layers, especially through tree and forage roots, and increase of soil organic matter through litter and decaying plant residues.
- Improved soil quality by increased soil cover from crops and pasture residues. This interaction prevents losses through erosion (soil, water sources, organic matter, and nutrients), stimulating the biota and its physical recovery.
- Improved microclimatic conditions due to the tree component reducing thermal amplitude, increasing air humidity, and lowering wind intensity results in increased animal wellbeing due to improved thermal comfort.
- Global warming mitigation through carbon sequestration especially by forest and forage components.
- Supporting biodiversity protection, especially due to the abundance of 'border effects' or interfaces, improving synergy through new niches and habitats for crop pollinators and natural enemies of pests and diseases

15.5.2.5 Main Challenges of Integrated Systems

- Farmers' traditionalism and resistance to adopt new technologies which require strong role for change agents.
- Higher financial investments on infrastructure on the integrated systems multiple components which is difficult for small holder farmers of the African and South Asian region.
- Returns in the medium to long term, especially with regard to forestry and or perennial tree components (coconut, spices, fruit trees, etc.).
- Long distances to final consumers or processing industries. In some regions, input purchase such as fertilizers, seeds, seedlings, agrochemicals, and animals is limited, and so it is for selling the outputs.
- Limited availability of skilled professionals necessary for farm management and components optimization especially with formal educational qualification.

15.5.2.6 Integrated Farming Systems for Tropical Coastal Zones

The integrated farming system has been in practice in different tropical countries in different ways depending on the agroclimatic condition, resource base, socio-economic milieu, constraints, and limitations of farming. The successful integrated farming systems developed and demonstrated in various tropical countries especially for coastal regions are summarized in Table 15.2. This provides existing farming systems, livelihood options, and its productive components and constraints in different regions.

Table 15.2 Existing farming systems and options for diversification

Major farming systems	Farming system character	Principal livelihoods	Constraints	Integrated farming systems (evaluated)
(A) South Asia				
Rice	Wetland rice—irrigated or rainfed	Wetland rice, vegetables, legumes, off-farm work	Water scarcity during dry months. Limited fodder production, declining soil fertility	Rice + fish
				Rice + *Azolla* + fish (Balusamy et al. 2003)
	Concentrated in Bangladesh, West Bengal, Tamil Nadu, and Kerala in India and Southern Sri Lanka			Rice-pulses/oilseeds/maize/vegetables + dairy (Swarnam et al., 2015)
				Rice-wheat-green gram + fish + duck + goat (Kumar et al. 2012)
	Cattle and small ruminants like goat are also found			Rice-cattle-small ruminants-poultry (Saadullah 1995)
Coastal artisanal fishing	Along the coast of Bangladesh, India, and around Maldives. Inshore fishing supplemented by rice, coconut, livestock, and vegetables	Fishing, coconuts, rice, legumes, livestock	Storms, cyclones, water logging, soil salinity, poor fertility, seawater intrusion and sea level rise	Rice-fish (Gupta et al. 1998; Purba 1998; Horstkotte-Wesseler 1999; Rothuis 1998)
				Fish-freshwater prawns in fruit/coconut orchards-rice (Hien et al. 1998)
				Composite carp culture (Nandeesha and Rao 1989)
				Poultry-fish (Edwards 1986)
				Duck-fish (Woynarovich 1980)
Urban based	Intensive cultivation of vegetables and dairy products in and around large cities	Horticulture, dairying, poultry, other activities	Price fluctuations and loss of produce due to lack of storage and value addition	

(continued)

Table 15.2 (continued)

Major farming systems	Farming system character	Principal livelihoods	Constraints	Integrated farming systems (evaluated)
(B) South East Asia and Pacific				
Lowland rice	Largely found in Thailand, Vietnam, Myanmar, the Philippines, Indonesia, and small areas are in Cambodia, Laos DPR, and Malaysia. Both livestock and off-farm income contributes to household livelihood	Rice, maize, pulses, sugarcane, oilseeds, vegetables, livestock, aquaculture, off-farm work	High mean temperature, irrigation water, shift in pest and diseases, declining soil fertility	'Forestry-fish' – cultivation of brackish water fish and shrimps in fenced-off mangrove forests (Johnston et al. 1999)
				Rice-fish (Sawah tambak), kang-kung-fish, genjer-fish, mendong-fish, duck-fish, fish-mangrove (Djajdiredja et al. 1980)
				Rice-duck-shrimp, rice-pig-duck
				Rice-melons-maize-fish (Othman 2004)
				Shrimp-mangrove (Binh et al. 2008)
				Poultry-fish
Tree crop mixed	Found in Malaysia, Indonesia, Thailand, Cambodia, the Philippines, Vietnam, and Papua New Guinea. Principally found on flat undulating terrain where paddy rice cannot be grown. The major trees are rubber, oil palm, coconut, coffee, tea, and cocoa with other associated crops like pepper and spices. Livestock and off-farm activities also contributes to household livelihood security	Rubber, oil palm, coconuts, coffee, tea, cocoa, spices, rice, livestock, off-farm work	Low profitability, decline in processing, moisture stress	Rubber-crops-cow-goat-poultry (Paris 2002)
				Coconut+ fodder-cattle-goat-poultry (Devendra et al.1997)
				Coconut-sheep-goat (Faylon and Alo 1995)
				Rubber/oil palm-sheep (Pillai et al. 1985)
				Rubber + legumes-goat/sheep (Tan and Abraham 1980)

(continued)

Table 15.2 (continued)

Major farming systems	Farming system character	Principal livelihoods	Constraints	Integrated farming systems (evaluated)
Root-tubers	Found mainly in Papua New Guinea and the Pacific Islands. They are typically in found in areas with low population density in both plains and hill landscapes. Often merges into the coastal artisanal fishing towards the coastline. It is the dominant farming system in many of the Pacific Islands. It is based on the use of root crops (yams, taro, sweet potato), vegetables and fruits (banana), coconut, livestock supplemented by hunting and gathering in the forest	Root crops (yam, taro, sweet potato), vegetables, fruits, livestock (pigs and cattle), off-farm work	Soil erosion, declining fertility, moisture stress	Root/tubers-banana-coffee-casuarina (Bourke 1985)

(continued)

Table 15.2 (continued)

Major farming systems	Farming system character	Principal livelihoods	Constraints	Integrated farming systems (evaluated)
Sparse (forest) farming system	Found in the major islands of Indonesia excluding Java and Papua New Guinea. In these Islands, small, scattered settlements (mainly tribals) depend on upland rice, root crops, and small and large ruminants, supplemented by gathering wild plants and hunting animals. In India such system exists in Nicobar group of islands; coconut is the major crop instead of rice		Shifting cultivation, loss of soil fertility, soil erosion	Coconut-pig-fish-cum-poultry, Home garden-pig/goat-backyard poultry (Swarnam et al. 2014, 2016)
Irrigated sugarcane	Found in wet tropical coasts of Australia. Monoculture in sugarcane is common. Sometimes it is rotated with N-fixing legumes	Sugarcane, raw sugar and by-products molasses, bagasse, and fibre		

(continued)

Table 15.2 (continued)

Major farming systems	Farming system character	Principal livelihoods	Constraints	Integrated farming systems (evaluated)
Latin America and Caribbean				
Coastal plantation and mixed farming	Mostly found in Caribbean Islands and coastal areas of northern Latin America. This system occupies some of the richest agricultural land in the region but also include mangrove swamps and isolated areas of tropical forests. There are two subsystems (i) small-scale family farms with mixed agriculture, in-shore fishing and off-farm activities. (tourism), (ii) large-scale plantations	Export crops/ tree crops, fishing, tubers, tourism	Salinity, waterlogging, cyclones	Mangrove-shrimp (Altieri 1999)
				Pig-poultry-tilapia (Paulin 1982)
				Pig-poultry-fish polyculture/ tilapia-prawns (Paulin 1982)
Irrigated	Found in coastal areas of Mexico. Have high degree of intensification with commercial production of rice, cotton, fruit, horticulture and vines	Horticulture, fruit, cattle		Pig-poultry-tilapia (Paulin 1982)
				Pig-poultry-fish polyculture/ tilapia-prawns (Paulin 1982)

(continued)

Table 15.2 (continued)

Major farming systems	Farming system character	Principal livelihoods	Constraints	Integrated farming systems (evaluated)
Maize-beans (Meso-American)	Found in central American countries stretching from Central Mexico to the Panama Canal mainly based on the production of maize and beans for subsistence. Historical loss of the valley lands to non-indigenous settlers and commercial operations	Maize, beans, coffee, horticulture, off-farm work		
Urban based	Specific peri-urban, intra-urban agricultural systems have developed to serve the major conurbations and population centres. Include horticulture, dairy and poultry. Off-farm income is major source of livelihood with many members engaged in agriculture as parttime basis	Horticulture, dairy, poultry		

(continued)

Table 15.2 (continued)

Major farming systems	Farming system character	Principal livelihoods	Constraints	Integrated farming systems (evaluated)
Sub-Saharan Africa				
Tree crop	Found in the humid regions in Western Africa and mainly runs from Cote d'Ivoire to Ghana and from Nigeria and Cameroon to Gabon with smaller packets in Congo and Angola. Cocoa, coffee, oil palm, and rubber are the major crops. Food crops are interplanted between tree crops and few cattle are raised	Cocoa, coffee, oil palm, rubber, yams, maize, off-farm work	Low profitability, declining soil fertility	Pig-fish (AIFP, 2005) Pig-fish-duck (Lally, 2000) Poultry-fish (Gabriel et al. 2007)
Rice-tree crop	Is located in Madagascar mostly in the moist subhumid and humid agroecological zones. Banana and coffee cultivation is complemented by rice, maize, cassava, and legumes	Rice, banana, coffee, maize, cassava, legumes, livestock, off-farm work	Water scarcity, loss of soil fertility, limited fodder availability	Rice-fish (Bodiguel 2016)

(continued)

Table 15.2 (continued)

Major farming systems	Farming system character	Principal livelihoods	Constraints	Integrated farming systems (evaluated)
Coastal artisanal fishing	In East Africa it includes the coastal areas of Zanzibar, Comoros, and Madagascar. In West Africa it stretches southwards from the Gambia and the Casamance region of Senegal, along the coast of Guinea Bissau, Sierra Leone, Liberia, Cote d'Ivoire, and Ghana, to Nigeria, Cameroon, and Gabon. Based on artisanal fishing supplemented by crop production. Multi-storied tree crop gardens with root crops under coconuts, fruit trees, and cashews and livestock mainly goat and poultry are also found. Off-farm activities are concentrated in tourism mainly in tourists resorts along beaches and with large tree crop estates	Fish, coconut, root crops, cashew, goat, poultry	Poor soil fertility, destruction of reefs, decimation of fish stock	Sea weed farming—Zanzibar, Tanzania, Mozambique, and Kenya
				Edible oyster (*Saccostrea cucullata*), black-lipped pearl oysters,
				Integration of Seaweeds and Abalone – (Ronnback et al. 2003)
				Integration of milkfish, rabbit fish, and mullets together with mollusks and seaweeds (Mmochi et al. 2001)
				Poultry-fish, duck-fish (Kapetsky 1994)

(continued)

Table 15.2 (continued)

Major farming systems	Farming system character	Principal livelihoods	Constraints	Integrated farming systems (evaluated)
Urban based	In some cities it is estimated that 10% or more of the population are engaged in urban agriculture. This farming system is very heterogeneous; ranging from small-scale but capital-intensive market-oriented commercial vegetable growing, dairy farming and livestock fattening, and part-time farming by the urban poor to cover part of their subsistence requirements	Fruits, vegetables, dairy, cattle, goats, poultry, off-farm work		
Middle East				
Highland mixed	Mostly in Yemen where rainfed cereal and legume cropping is done with tree crops like qat and coffee on terraces	Cereals, legumes, qat, and coffee	Moisture stress, soil erosion	

15.5.3 Agroecological Farming

Agroecological farming is knowledge, management, and labour intensive rather than external input-intensive monoculture farming. It aims to regenerate long-term agroecosystem properties by incorporating functional biodiversity leading to sustainable, resilient systems. It is based on the understanding that more than 70% of world food production is by small farmers in Asia, Africa, and Latin American countries having an average holding of less than 2 ha land. These small family farms are much more productive than large farms if total output is considered rather than yield from a single crop. Moreover, these small-scale farmers using agroecological methods have the potential to feed 15–20 people annually from 1 ha of land. These small farmers have more coping mechanism to climate change and extreme events through increased use of local varieties, mixed cropping, soil and water conservation measures, water harvesting, and a series of traditional techniques (Altieri and Koohafkan 2008; Altieri et al. 2012) which means highly diversified farming. Diversified farming is one of the earlier systems developed based on the agroecological principle, and it is discussed in detail here.

15.5.3.1 The Concept

Kremen et al. (2012) defines diversified farming systems as ‘Farming practices and landscapes that intentionally include functional biodiversity at multiple spatial and/or temporal scales in order to maintain ecosystem services that provide critical inputs to agriculture, such as soil fertility, pest and disease control, water use efficiency, and pollination’. The farmers manage this functional biodiversity to generate critical ecosystem services to agriculture. Diversification is achieved at different scales like field/plot, farm, and landscape levels (Fig. 15.5). There is a constant interaction between the components of agro-biodiversity and also with the physical environment to provide ecosystem services to the farming process, such as nutrient cycling, nitrogen fixation, water infiltration, etc. This type of farming primarily depends on inputs generated within the system. Diversification is achieved both spatially and temporally to create needed interactions and synergies between the components.

Though the DFS is similar to multifunctional, organic, sustainable, and ecological agriculture, it differs from each of the above in a subtle but fundamental way, i.e. the basic concept of DFS is to have different farm practices that support functional diversity across spatial and temporal scales, supplying the necessary ecosystem properties providing critical inputs to agriculture.

15.5.3.2 Level of Diversification

In DFS, diversification is achieved at different scales from plot to landscape level. At field/plot level, it can be done by inclusion of number of varieties of a single crop, mixed crops by way of intercropping, different breeds/species of livestock,

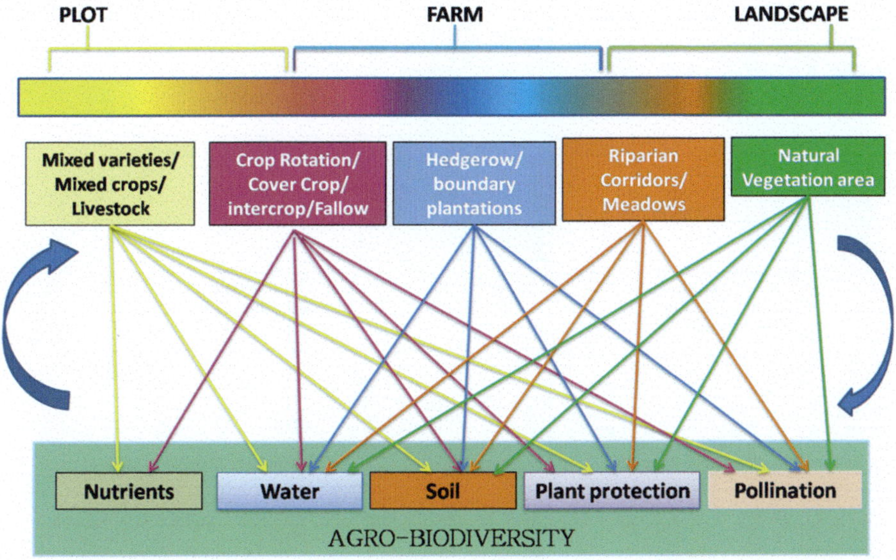

Fig. 15.5 Conceptual model of diversified farming system

etc. At farm level it is done by crop rotations, cover crops, strip cropping, hedge rows, boundary plantations, and integration of different farm enterprises like livestock, fisheries, etc. At landscape level inclusion of hedge rows, meadows, riparian strips, and other natural and seminatural habitats in the cropped area is done to have diversified farming system. These practices result in heterogenous farming systems at different scales which include intended biodiversity while supporting others unintentionally. Temporally, these farming practices along with the natural succession enhance agro-biodiversity and ecosystem services in a dynamic way (Kremen et al. 2012).

15.5.3.3 Diversification and Biodiversity Conservation: A Case Example of South Andaman, India

The Andaman Islands, India, is located in the Bay of Bengal between 11 and 14° N latitude and 92–94° E longitude. The islands witness tropical humid climate with an annual average rainfall of 3080 mm. Rice is the major annual crop grown in the coastal plains (lowlands) where water stagnation for more than 6 months prevents growing of arable crops during wet season. In contrast water scarcity during dry months forced the land to be left fallow. Hence, long-duration rice is grown under subsistence farming.

Crop diversification was initiated with the introduction of short-duration high-yielding rice varieties and growing of drought-tolerant pulse and oil seeds after rice harvest. With increasing population and advent of tourism, the demand for vegetables and other food products increased tremendously in the islands. At the same time, efforts were made to reclaim the tsunami-affected lands, and new techniques

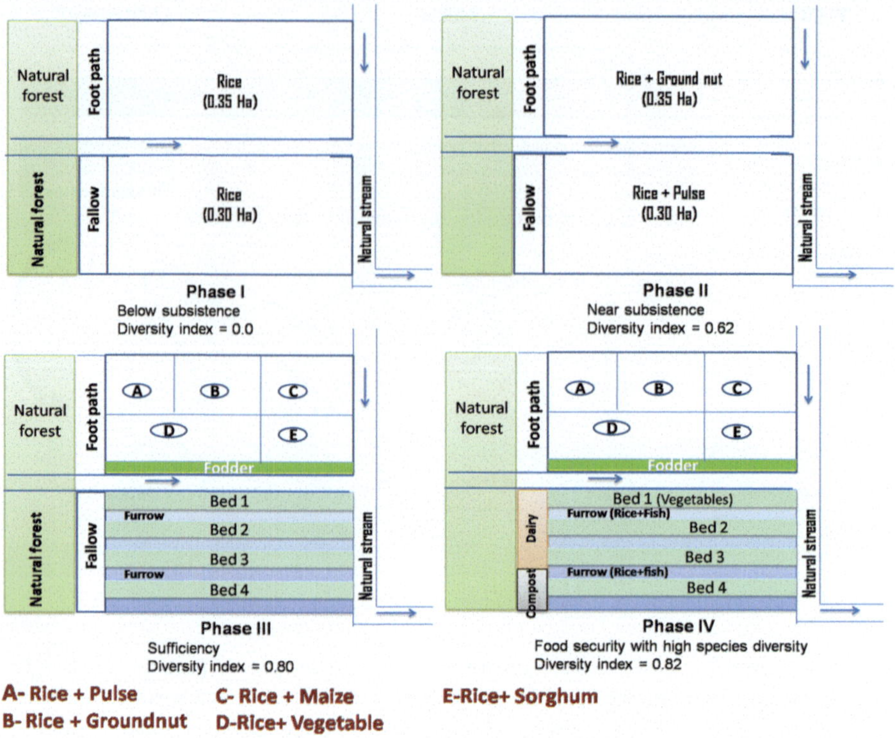

Fig. 15.6 Stages of diversification of a farm in tropical coastal lowland

such as broad bed and furrow systems were introduced. These techniques facilitated the diversification of crops in coastal lowlands by improving drainage and in situ water harvesting. Such efforts have helped to diversify the crop cultivation from rice monocropping to polyculture of several vegetables and flowers and also raising fish on the same piece of land.

In this context a model was evaluated for diversification in coastal lowlands in an area of 0.75 ha. Initially long-duration rice was the only crop grown during wet season, and the land was normally left fallow due to water scarcity during dry period (Fig. 15.6). At this stage due to monocropping, the diversification measured as diversification index (Simpson's diversity index- Kumar et al. 1994) is only 0.0 (no diversification). Later short-duration rice varieties were grown during wet season, while ground nut and green gram during dry months use residual soil moisture increasing the species diversity to 0.62. In the third phase with making of broad bed and furrow system in 0.30 ha area (land shaping), the crop diversity is further enhanced by growing seasonal vegetables in the beds and rice-fish in the furrows. In each bed crop rotation involving three vegetables was grown utilizing the harvested rainwater in the furrows. Thus a total of 12 crops were harvested in a year. In remaining area of 0.35 ha rice-based cropping systems, viz. rice-maize, rice-sorghum, rice-groundnut, and rice-pulse/green gram, were grown to meet the

Table 15.3 Measures of agro-biodiversity enhancement for food and nutritional security

Measures of diversification	Nature of diversification	Examples
Varietal diversification	Grow a mix of varieties or closely related species	Rice is the major cereal crop grown in the islands with dominance of long-duration varieties like C14-8. The other promising varieties like CSR 36, Jeya, CARI 5, CARI 7, Naveen, etc.
Diversification by crop rotation	Grow two or more crops within the field in different seasons	Pulses, oil seeds, and vegetables after the harvest of rice not only increase the yield but also biodiversity
		Crop rotation involving seasonal vegetables in beds and rice-fish in furrows of BBF
Farm-wide diversification	Agroforestry integrated with cropping, livestock, and fisheries	Diversification and integration of components result in residue recycling, increased farm production, and also agricultural diversity
Boundary plantations or field margin non-crop vegetations	Growing of green manure/fodder/fruit crops as biofence	Growing of tree crops (banana) or shrubs (*Gliricidia* and fodder trees) as biofence will improve the biodiversity

nutritional requirement of a farm family. As a result the diversity index has increased to 0.80, and this also witnessed increase in cropping intensity and farm production. At the final stage, farm and landscape level diversification was achieved by integrating dairy into the system along with residue recycling through composting. Besides, all along the field boundary fodder crops and fruit trees were grown to provide green fodder to the livestock unit. With farm diversification and integration of different components, the level of external input use was drastically reduced. Besides the direct effect on species diversity, the diversification indirectly increased the soil microflora and fauna. As a result the farm diversification has increased to 0.82. This system not only enhances farm diversity but also improves farm productivity and ensures food and nutritional security of small and marginal farmers of small islands with limited resource base. This diversified model suitable for coastal lowlands can be upscaled to similar agroclimatic and physiographic locations in other tropical countries.

It is also seen that the degree of complexity involved in increasing biodiversity ranges from diversifying varieties within a monoculture to landscape level diversification including non-crop and perennial vegetation. The details of such diversification involved in the diversification model are given in Table 15.3.

The DFS essentially enshrine the characteristics of multifunctional, organic, sustainable agriculture, but the reverse may not be true; nowadays organic agriculture is often practised on large-scale monocultures with external inputs and does little to foster biodiversity or sustain ecosystem services. Diversified farming systems are known to play a significant role in biodiversity conservation; control of arthropod pests, weeds, and diseases; pollination services; soil quality maintenance; energy-use efficiency; and reduction of global warming potential, resistance, and resilience of farming systems to extreme weather events and enhanced carbon sequestration and water holding capacity in surface soils. Long-term field trials on diversified farming practices demonstrated a strong positive impact on soil quality and increased nutrient uptake by crop biomass and/or soils, thus enhancing nutrient use efficiency which is particularly important in tropics where soils are inherently poor in available nutrients (Kremen and Miles 2012).

15.5.4 Natural Farming

15.5.4.1 The Concept

Natural farming (NF) is an alternative farming method which avoids manufactured inputs and equipment. It works on the philosophy of working with nature to produce healthy food and to keep the living beings and the land healthy (Devarinti 2016). It is an environmentally sustainable way of growing food based on the principle of building communion and kinship between farmer and nature. As a result, individual *methods* and *techniques* between natural farmers vary widely. It is also called as 'Do Nothing Farming', as the farmer is considered to be a facilitator, while the real work is done by nature herself. Unlike organic farming no manure or compost is used in natural farming. The basic principles of this farming method are:

1. No tillage: As soil is the home to rich biodiversity that contributes to soil health in some way, while tilling these organisms are affected and deplete the soil's fertility and create the conditions for addition of manufactured inputs.
2. No weeding or pest control: All living things must be respected. This does not mean 'never cut a weed' or 'never take actions to keep pests off plants', but it does mean that we should work to respect their lives and their roles as parts of this living nature. Promote/support nature to control itself through feedback mechanisms when a particular species/pest causing damage to the natural system proliferate.
3. No external input application: It is based on the conception that the land has the power to grow and regenerate itself and no external inputs such as fertilizers, pesticides, or literally anything else from outside the farm area are needed.
4. Allowing nature to decide what to grow: Instead of planning based on our own desire or forcing things to grow which are not suitable to a particular climate or area, let the nature show what should be grown and where.

15.5.4.2 Natural Farming in India

Natural farming is popular as Zero Budget Natural Farming (ZBNF) was evolved in Karnataka as an alternative to conventional farming mainly to end the reliance of farmers on external credit, thereby relieving them from the vicious cycle of debt by drastically cutting the production cost. The word 'budget' refers to credit and expenses; thus the phrase 'Zero Budget' means without using any credit and without spending any money on purchased inputs. 'Natural farming' means farming with nature and without chemicals. This method was evolved and propagated by Subhash Palekar. In ZBNF, soil is applied with microbial inoculums to accelerate the multiplication of soil flora, beneficial to soil enrichment. Indigenous pesticide decoctions from leaves with cow urine are used for pest control. The basic idea of natural farming is to nurture the growth of the beneficial microorganisms and maintain healthy soil environment without using external manure and chemical inputs. It involves four different kinds of preparations which act as the four pillars of farming (FAO 2007). The details of which are given below:

(i) **Jivamrita/jeevamrutha**: It is a fermented microbial culture. It provides nutrients as well as acts as catalysts to promote soil biological activity including microbial and earthworm activity. It also helps to prevent fungal and bacterial plant diseases, and it requires for the first 3 years of the transition, after which the system becomes self-sustaining.
(ii) **Bijamrita/beejamrutha**: It is used for treatment of seeds, seedlings, or any planting material against seed and soilborne pathogens. It is prepared from local cow dung, cow urine, lime, and soil.
(iii) **Acchadana (mulching)**: There are three types of mulching, viz. soil mulch, straw mulch, and live mulch.
    - Soil mulch protects topsoil during cultivation and does not destroy it by tilling. It promotes aeration and water retention in the soil.
    - Straw mulch uses dried wastes/residues of previous crop to cover the soil. Soil fertility can be improved by providing dry organic material which will decompose and form humus through the activity of soil biota activated by microbial cultures.
    - Live mulch (symbiotic intercrops and mixed crops) promotes multiple cropping patterns of cereals and other crops like pulses or oilseeds, etc. to supply all essential elements to the soil and crops. For example, legumes fix N, while other crops like rice and wheat supply other elements like potash, phosphate, and sulphur.
(iv) **Whapasa (moisture)**: The supporters of natural farming believe that plant roots need only water vapour, so encourage reducing irrigation, irrigating only at noon, in alternate furrows.

The other important principles in ZBNF are intercropping, contours and bunds, native earthworms, and cow dung. The major effect of this type of farming is not only agronomic but also social and economic. Majority of the farmers adopted this method of farming reported improvements in yield, soil conservation, seed diversity, quality of produce, household food autonomy, income, and health over the years. Most experienced reduced farm expenses and a reduced need for credit, one of the major problems plaguing Indian farmer.

15.5.5 Sustainable Intensification

To meet the increasing demand for food, several options have been proposed, viz. closing the yield gap, increasing the production potential of crops, reducing waste, and changing diets and expanded aquaculture, all of which should be coordinated in a multifaceted and linked global strategy to ensure sustainable and equitable food security (Godfray et al. 2010). The future increases in food supply should be met without increasing the agricultural area and without damaging the biodiversity, protecting our freshwater resources and reducing adverse impact on the whole range of ecosystem services (Firbank et al. 2011). While delivering the food security, there is a need to reduce the impact of climate change on food production, and its resilience to future environmental change should be improved. This can be achieved by 'sustainable intensification', which is regarded as an enhancement of current agricultural production but with increased efficiency. It is defined 'as the process of delivering safer, nutritious food per unit of input resource, whilst allowing the current generation to meet its needs without compromising the ability of future generations to meet their own needs' (Smith 2013).

The future increase in global food demand (100–110% increase from 2005 to 2050) can be met by moderate intensification of existing croplands of nations where yields are suboptimal through adaptation and transfer of high-yielding technologies with global technological improvement with minimal environmental effect (Tilman et al. 2011). The options for sustainable intensification range from adoption of new technology to improve the efficiency of current food production. They include genetic modification of living organisms, use of cloned livestock and nanotechnology, and closure of yield gaps by the rebalancing the distribution of inputs (fertilizers, water) from regions of overuse to under-used areas or where the input supply is limiting (Godfray et al. 2010; Mueller et al. 2012). Foley et al. (2011) showed 2.3 billion tonnes of additional production by redistribution of such imbalances and bringing yield of 16 important food and feed crops to 95% of their production potential.

15.5.6 Integrated Aquaculture Production

The coastal aquaculture plays an important role in livelihood security of coastal communities all over the world. It is considered as a sustainable alternative to fishing. There are many forms of traditional culture systems prevalent in countries like Java, Hawaii, West Africa, Mauritius, etc. where fish, shrimp, mullet, etc. were grown to supplement food production. Such systems are sustainable than modern-intensive, one species farming as they maintain the biodiversity, have less input use, and do not cause pollution by using little or no feed as they rely on natural production for feed. However, these extensive, traditional systems cannot satisfy the increasing food demand of the burgeoning coastal population (Brzeski and Newkirk 1997).

Though the modern-intensive, monoculture systems have increased the total production substantially, they require higher level of inputs such as water, feed, fertilizer, and chemicals besides large support area as in the case of shrimp farming. Cage farming of salmon is the most expensive production system requiring

Table 15.4 Types of integrated agriculture-aquaculture practices in tropical coastal areas

Sl. no	Type	Remarks
1.	Rice-fish	In flooded rice fields, living aquatic resources (LARs) such as fish, freshwater prawns and crabs, snails, mussels, and frogs occur naturally which are exploited. Sometime, rice fields are used as nursery for raising fingerlings before stocking in permanent ponds. The integrated fish-in-paddy field system functions through the feeding of fish on organisms (particularly insects and other possible rice pests) and weeds and the stirring of the sediment through their foraging action which leads to nutrient resuspension (Lightfoot et al. 1993; Cagauan 1995)
2.	Trenches in fruit orchards	Trenches were made within the fruit orchards by excavating the soil to raise the beds for fruit trees. These trenches were usually surrounded by lateral trench and connected to the adjacent rice field (Hien et al. 1998)
3.	Raised bed and sunken furrow system and paired bed system	They are land management techniques similar to the trenches made in fruit orchards. Here, vegetables are grown in the raised beds. The furrows are used for rice-fish or as fish nurseries to grow fingerlings (Velmurugan et al.)
4.	Forestry-fish	It is the practice of the cultivation of brackish water fish and shrimps in fenced-off mangrove forests in Malaysia, the Philippines, and Vietnam (Johnston et al., 1999)
5.	Pond-based IFS Poultry-cum-fish system	On the embankments of ponds, vegetables and fruits are grown using the mud or silt and water taken from the ponds. The wastes from processing of fruits and vegetables and droppings from the poultry shed are used for raising fish in the ponds (Zhong 1995)

25–50 kcal of energy per kcal of fish produced. Besides, the farm discharges pose severe environmental problems, viz. nutrient enrichment, oxygen depletion, loss of biodiversity and habitat destruction, threat to marine life, antibiotic resistance, human health hazards, etc. Hence, alternatives such as polyculture, integrated aquaculture practices were developed in many countries to augment the coastal food production without affecting the natural resource base (Folke and Kautsky 1992). A wide range of integrated agriculture-aquaculture (IAA) systems are in use and are mainly practised in Bangladesh, India, Indonesia, Malaysia, Thailand, Vietnam (Prein 2002), and sub-Saharan Africa. There are several integrated aquaculture systems practised in most of the tropical coasts, the details of which are given in Table 15.4.

The small-scale IAA systems play an important share of fish used for consumption and income generation by the rural poor in many of the tropical countries including Asia and Africa. They have greater potential for increasing the aquaculture production and ensuring food security among rural poor in these nations.

15.6 Effect of Alternative Farming Practices on Diversification and Agro-biodiversity

The effects of different alternative farming systems as discussed above are compared with each other and with conventional farming (Table 15.5). External inputs are very essential to meet the crop demand and for higher production as advocated in all the farming systems except natural farming. Similarly agrochemicals for various purposes and GMO are used in conventional farming, IFS, and sustainable intensification based on the need; in contrast in other systems, it is avoided. Plant and animal diversity is an indispensable aspect in all other farming systems aimed to harness the diversity to ensure stability and enhance the productivity, but it is not essential in conventional farming as it advocated monoculture to reap higher productivity.

Table 15.5 Comparison of effectiveness of alternative farming systems

	Conventional (intensive monoculture)	Organic farming	Diversified farming system	Integrated farming system	Natural farming	Sustainable intensification
Use of external input	Yes	Yes	Minimal	Yes	Nil	Yes
Use of agrochemicals (synthetic)	Yes	No	No	Yes	No	Yes
Use of HYV/GMO	Yes	No	Minimal	Yes	No	Yes
Plant and livestock species diversity	No	Minimal	Yes	Yes	Yes	Minimal
Promotion of non-farm species diversity	No	Sometimes	Yes	Yes	Rare	Rare
Integration of livestock	Rare	Yes	Yes	Yes	Yes	Rare
Spatial heterogeneity	No	Rare	Yes	Yes	Yes	Rare
Use of ecosystem services	Rare	Yes	Yes	Yes	Yes	Sometimes
Capability for resilience	Nil	Less	More	More	–	Less
Labour dependence	Low	High	High	High	Medium	Low
Use of traditional knowledge	Rare	Sometimes	Often	Sometime	Often	Rare

Agricultural production in the tropical coastal areas is subjected to large uncertainties imposed by climate and socio-economic challenges. Unlike the conventional farming, alternative farming systems are resilient and exploit the ecosystem services very well. In general, the traditional knowledge and labour are indispensable for alternate farming systems whereas it is rarely found place in the conventional farming. Notwithstanding anything in this discussion, it is widely reported that traditional farming in many of the coastal areas and islands in the tropical regions has evolved to adjust with the external changes. In continuous coping with extreme weather events and climatic variability, farmers of these regions have developed and inherited complex farming systems, the scientific potential of which can be evaluated and integrated into the alternative farming systems to address the location specific challenges. Unlike the conventional farming, these alternative farming systems have the potential to bring solutions to many uncertainties facing humanity in an era of climate change.

15.7 Conclusion

The current rate of degradation of natural resources in the coastal areas accelerated by anthropogenic activities is the serious challenge to achieve food security and conserve the agro-biodiversity. It has been increasingly realized that the commodity centric modern agricultural practices have inflicted heavy damage to the sustainability of agricultural production system. Therefore, alternative farming systems supporting diversification, biodiversity conservation, and sustainable food production are highly advocated for this region. With respect to the agro-biodiversity conservation, it is the biology of farm productivity with resilience to climate change. This provides scope for incorporation of traditional knowledge and region-specific biodiversity in the alternate farming systems. In this context, it is understood that without supporting human interest in the productive capacity of agro-ecosystems, on-farm survival of diversity is highly doubtful.

Among the different alternative farming systems identified for tropical coastal region, except natural farming, all other systems advocate use of external inputs. Use of agrochemicals and GMO/new varieties is continued to constitute the core of the conventional farming. However, such inputs are used based on need in IFS and sustainable intensification. With reference to the organic farming other than commercial farms of select commodities, many farmers in the tropical islands are farming organically simply by virtue of utilizing their traditional knowledge, available resources, and techniques because purchasing inputs is highly costly and lives in relative isolation. Several studies have pointed out that inclusion and proper utilization of plant and animal diversity are vital components in all other farming systems to ensure stability and enhance the productivity. On the other hand, conventional farming as practised in the coastal region largely depends on monoculture and intensive agriculture with agrochemicals to reap higher productivity.

The current trend and future climate change projections also point to the fact that we would be concerned more with food and natural resource conservation while meeting the demands of growing population particularly in the tropical coastal areas. At the same time, in continuous coping with extreme weather events and climatic variability, farmers living in these harsh environments like regions of Africa, Asia, and Latin America have developed and inherited complex farming systems that have the potential to bring solutions to many uncertainties facing humanity in an era of climate change. Several researchers concluded that alternative farming techniques can produce enough food to feed the world agricultural land base and aggravating the resource degradation (Koohafkan and Altieri 2010). In view of this, it is wise to evaluate and upscale such alternative farming systems and also to address the challenges posed by climate change and food demand. In summary, agricultural environmental imperatives must be simultaneously tackled and an ecosystem approach be adopted in agricultural production management as an alternate to the conventional farming systems.

References

AIFP (2005) Farming Nigeria waters. Newsletter of the Aquaculture and Inland Fisheries project of the National special Programme for food security in Nigeria. FAO Office Abuja, Nigeria 3(4):2–4

Alcamo J, Moreno JM, Novaky B, Bindi M, Corobov R, RJN D, Giannakopoulos C, Martin E, Olesen JE, Shvidenko A (2007) Europe. In: Parry ML, Canziani OF, Palutikof JP, van der Linden PJ, Hanson CE (eds) Climate change 2007: impacts, adaptation and vulnerability, Contribution of Working Group II to the Fourth Assessment Report of the Intergovernmental Panel on Climate Change. Cambridge University Press, Cambridge, pp 541–580

Altieri MA (1999) Applying agroecology to enhance productivity of peasant farming systems in Latin America. Environ Dev Sustain 1:197–217

Altieri MA, Koohafkan P (2008) Enduring farms: climate change, smallholders and traditional farming communities, Environment and Development Series 6. Third World Network, Malaysia

Altieri MA, Funes-Monzote FR, Petersen P (2012) Agroecologically efficient agricultural systems for smallholder farmers: contributions to food sovereignty. Agron Sustain Dev 32:1–13

Anderson J (1985) Farm programs and alternative agriculture. In: Proposed 1985 farm bill changes: taking the bias out of farm policy. Proceedings of the Institute for Alternative Agriculture Second Annual Scientific Symposium. Institute for Alternative Agriculture, Greenbelt

Ansari MA, Prakash N, Baishya LK, Punitha P, Sharma PK, Yadav JS, Kabuei GP, Levis Ch KL (2014) Integrated farming system: an ideal approach for developing more economically and environmentally sustainable farming systems for the Eastern Himalayan Region. Indian J Agric Sci 84(3):356–362

Balusamy M, Shanmugham PM, Baskaran R (2003) Mixed farming an ideal farming. Intensiv Agric 41(11–12):20–25

Barlas Y, Damianos D, Dimara E, Kasimis C, Skuras C (2001) Factors influencing the integration of alternative farm enterprises into the agro-food system. Rural Sociol 66:342–358

Batie S, Taylor D (1991) Assessing the character of agricultural production systems: issues and implications. Am J Altern Agric 6:184–187

Behera UK, Jha KP, Mahapatra IC (2004) Integrated management of available resources of the small and marginal farmers for generation of income and employment in eastern India. Crop Res 27(1):83–89

Bengtsson J, Ahnström J, Weibull A-C (2005) The effects of organic agriculture on biodiversity and abundance: a meta-analysis. J Appl Ecol 42:261–269

Binh CT, Phillips MJ, Demaine H (2008) Integrated shrimp- mangrove farming systems in the Mekong delta of Vietnam. Aquac Res 28(8):599–610

Bodiguel C, Wallemacq F, San M (2016) Rice-fish farming: a development lever for smallholder farming in Madagascar. IOC SmartFish, Mauritius, p 3

Bossio DA, Scow KM, Gunapala N, Graham KJ (1998) Determinants of soil microbial communities: effects of agricultural management, season, and soil type on phospholipid fatty acid profiles. Microb Ecol 36:1–12

Bourke MR (1985) Food, coffee and Casuarina: an agroforestry system from the Papua New Guinea highlands. Agrofor Syst 2:273–279

Brezeski V, Newkirk G (1997) Integrated coastal food production systems – a review of current literature. Ocaen Coast Manag 34(1):55–71

Brown RW (1999) Margin/field interfaces and small mammals. Asp Appl Biol 54:203–210

Cagauan AG (1995) Overview of the potential roles of pisciculture on pest and disease control and nutrient management in rice fields. In: Symoens J-J, Micha J-C (eds) The Management of Integrated Freshwater Agro-Pisciculture Ecosystems in tropical areas. Proceedings of an International Seminar, 16–19 May 1994, Brussels, Belgium. Technical Centre for Agricultural and Rural Cooperation (CTA), Wageningen, Holland and the Belgian Royal Academy of Overseas Sciences (ARSOM), Brussels, Belgium, pp 203–244

Caouette D, Taylor S, Trowell C, Buckles D (2002) Making the connections: small scale agriculture, food Security and nutrition. Research into the effectiveness of Sustainable Agriculture Projects. Discussion Paper by the NGO Food Security Program Effectiveness Group. http://www.interpares.ca/en/publications/pdf/making_the_connections

Chappell MJ, LaValle LA (2011) Food security and biodiversity: can we have both? An Agroecological analysis. Agric Hum Values 28:3–26. https://doi.org/10.1007/s10460-009-9251-4

Creel L (2003) Ripple effects: population and coastal regions. Making the Link. Population reference Bureau. https://www.prb.org/wp-ontent/uploads/2003/09/RippleEffects_Eng.pdf. Accessed on 16 July 2018

De La Cruz EM, Castillo LE (2010) The use of pesticides in Costa Rica and their impact on coastal ecosystems. In: Taylor MD, Klaine SJ, Carvalho FP, Barcelo D, Everaarts J (eds) Pesticide residues in coastal tropical ecosystems: distribution, fate and effects. CRC Press 2003

Devarinti SR (2016) Natural farming: eco-friendly and sustainable? Agrotechnology 5:147. https://doi.org/10.4172/2168-9881.1000147

Devendra C, Thomas D, Jabbar MA, Kudo H (1997) Improvement of livestock production in crop–animal systems in rainfed agro-ecological zones of South-East Asia. ILRI, Nairobi

Djajdiredja R, Jangkaru Z, Junus M (1980) Freshwater Aquaculture in Indonesia with special reference to small –scale agriculture – aquaculture integrated farming systems in West Java. In: Pullin RSV, Shehadeh ZH (eds) Integrated agriculture – aquaculture farming systems, Integrated agriculture-aquaculture farming systems. ICLARM Conference proceedings 4,258p. International Center for Living Aquatic Resources Management, Manila and the Southeast Asian center for Graduate Study and Research in Agriculture college, Los Banos

Donald PF, Green RE, Heath MF (2001) Agricultural intensification and the collapse of Europe's farmland bird populations. Proc R Soc Lond B Biol Sci 155(1462):39–43

Edwards P (1986) Duck/fish integrated farming systems. In: Farrell DJ, Stapleton P (eds) Duck production science and world practice. University of New England, Armidale, pp 267–291. 430 p

Edwards P, Pullin RSV, Gartner JA (1988) Research and education for the development of integrated crop–livestock–fish farming systems in the tropics, ICLARM Studies and Reviews 16. ICLARM, Manila

El Titi A (1992) Integrated farming: an ecological farming approach in European agriculture. Outlook Agric 21:33–39

FAO (2002) World agriculture: towards 2015/2030. 97 p

FAO and World Bank (2001) Farming systems and poverty- improving farmers' livelihoods in a changing world. FAO and World Bank, Rome

FAO (2007) Zero budget natural farming in India. http://www.fao.org/3/a-bl990e.pdf
Faylon PS, Alo AMP (1995) Research methodology for coconut-based small ruminant system. In: Devendra C, Sevilla C (eds) Crop–animal interaction, IRRI Discussion Paper Series No, vol 6. IRRI, Manila, pp 197–203
Firbank L, Bradbury R, Jenkins A, Ragab R, Goulding K, Harmer R, Hess T, Jenkins A, Pilgrim E, Potts S, Smith P, Storkey J, Williams P (2011) Enclosed farmland. In: UK National Ecosystem Assessment (ed) Understanding nature's value to society, Technical Report. UNEP-WCMC, Cambridge, pp 197–239
Foley JA, Ramankutty N, Brauman KA, Cassidy ES, Gerber JS, Johnston M, Mueller ND, O'Connell C, Ray DK, West PC, Balzer C, Benett EM, Carpenter SR, Hill J, Monfreda C, Polasky S, Rockstorm J, Sheehan J, Siebert S, Tilman D, Zaks DPM (2011) Solutions for a cultivated planet. Nature 478:337–342
Folke C, Kautsky N (1992) Aquaculture with its environment: prospects for sustainability. Ocaen Coast Manag 17:5–24
Frieben B, Kopke U (1995) Effects of farming systems on biodiversity. In: Isart J, Llerena JJ (eds) Proceedings of the First ENOF Workshop – biodiversity and land use: the role of organic farming. Multitext, Barcelona, pp 11–21
Fuller RJ, Norton LR, Feber RE et al (2005) Benefits of organic farming to biodiversity vary among taxa. Biol Lett 1:431–434
Gabriel U, Akinrotimi OA, Bekibele DO, Anyanwu PE, Onunkwo DN (2007) Economic benefit and ecological efficiency of integrated fish farming in Nigeria. Sci Res Essay 2(8):302–308
Garibay VS, UGas R (2010) Latin America overview. In: Willer H, Kilcher L (eds) The world of organic agriculture – statistics and emerging trends 2010. IFOAM, Bonn
Godfray HCJ, Beddington JR, Crute IR, Haddad L, Lawrence L, Muir JF, Pretty J, Robinson S, Thomas SM, Toulmin C (2010) Food security: the challenge of feeding 9 billion people. Science 327(5967):812–818. https://doi.org/10.1126/science.1185383
Government of Guyana (2012) Guyana Second National Communication to the United Nations Framework Convention on Climate Change
Green RE, Cornell SJ, Scharlemann JPW, Balmford A (2005) Farming and the fate of wild nature. Nature 307:550–555
Gupta MV, Sollows JD, Mazid MA, Rahman A, Hussain MG, Dey MM (1998) Integrating aquaculture with rice farming in Bangladesh: feasibility and economic viability, its adoption and impact, ICLARM Technical Report 55. ICLARM, Penang
Harwood R (1984) Organic farming research at the Rodale Research Center. Bezdicek, D., Power, J. F., Keeney, D. R., Wright, M. J.. Organic farming: current technology and its role in a sustainable agriculture. Agronomy Society of America, Crop Science Society of America, and Soil Science Society of America. Madison, 2–3
Hien TTT, Minh TH, Phuong NT, Wilder MN (1998) Current status of freshwater prawn culture in the Mekong River Delta of Vietnam. JIRCAS J 6:89–100
Hole DG, Perkins AJ, Wilson JD et al (2005) Does organic farming benefit biodiversity? Biol Conserv 122:113–130
Horstkotte-Wesseler G (1999) Socioeconomics of rice–aquaculture and IPM in the Philippines: synergies, potentials and problems, ICLARM Technical Report 57. ICLARM, Manila
Ilbery BW (1991) Farm diversification as an adjustment strategy on the urban fringe of the west midlands. J Rural Studies 7(3):207–218. https://doi.org/10.1016/0743-0167(91)90085-7
Ilbery B, Bowler I, Clark G, Crockett A, Shaw A (1997) Farm based tourism as an alternative farm enterprise. A case study from the Northern Pennines, England. Reg Stud 32:355–364
Johnston D, Clough XTT, Phillips M (1999) Mixed shrimp- mangrove farming systems in Ca Mau province, Vietnam. Aquac Asia 4(2):6–12
Kapetsky JM (1994) A strategic assessment of warm-water fish farming potential in Africa, CIFA Technical Paper, No. 27. FAO, Rome. 67 pp
Koh LP (2007) Potential habitat and biodiversity losses from intensified biodiesel feedstock production. Conserv Biol 21:1373–1375
Koohafkan P, Altieri MA (2010) Conserving our world's agricultural heritage. FAO, Rome. 2010

Kremen C, Miles A (2012) Ecosystem services in biologically diversified versus conventional farming systems: benefits, externalities, and trade-offs. Ecol Soc 17:40

Kremen C, Iles A, Bacon C (2012) Diversified farming systems: an agroecological, systems-based alternative to modern industrial agriculture. Ecol Soc 17(4):44. https://doi.org/10.5751/ES-05103-170444

Kumar MB, George SJ, Chinnamani S (1994) Diversity, structure and standing stock of wood in the Homegardens of Kerala in Peninsular India. Agrofor Syst 25(3):243–262

Kumar S, Subash N, Shivani S, Singh SS, Dey A (2012) Evaluation of different components under integrated farming system (IFS) for small and marginal farmers under semi-humid climatic environment. Exp Agric 48(3):1–15

Lal R, Miller FP (1990) Sustainable farming system for tropics. In: Singh RP (ed) Sustainable agriculture: issues and prospective, vol I. Indian Society of Agronomy, IARI, New Delhi, pp 68–89

Lally E (2000) Zambian farm finds a winning formula. Fish Farmer 14:–35

Land Matrix (2016) Agricultural drivers. November 2016. https://landmatrix.org/en/get-the-idea/agricultural-drivers. Accessed 26 July 2018

Lightfoot C, Roger PA, Cagauan AG, dela Cruz CR (1993) Preliminary steady-state models of a wetland ricefield ecosystem with and without fish. In: Christensen V, Pauly D (eds) Trophic models of aquatic ecosystems, ICLARM Conference Proceedings 26. ICLARM, Manila, pp 56–64

Little DC, Muir J (1987) A guide to integrated warm water aquaculture. Institute of Aquaculture, University of Stirling, Stirling

Mahapatra IC (1992) Farming systems research challenges and opportunities. East Indian Farming Syst Res Ext Newsl 6(4):3–10

Martin G, Martin-Clouaire G, Duru M (2013) Farming system design to feed the changing world. A review. Agron Sustain Dev. (2013 33:131–149. https://doi.org/10.1007/s13593-011-0075-4

Mmochi A, Mozes N, Kite-Powell HL, Dubi AM, Gordin H, Jiddawi N, Kissil G, Msuya F, Mwangamilo J (2001) Design and preliminary results of an integrated mariculture pond system at Makoba, Zanzibar, Tanzania. Paper presented at 20th Anniversary Conference of the Institute of Marine Sciences, University of Dar es Salaam. Zanzibar, 28th June–1st July, 1999, 19 pp.

Morton JF (2007) The impact of climate change on smallholder and subsistence agriculture. Proc Natl Acad Sci 104(50):19680–19685. https://doi.org/10.1073/pnas.0701855104

Mueller ND, Gerber JS, Johnston M (2012) Closing yield gaps through nutrient and water management. Nature 490:254–257

Nandeesha MC, Rao KG (1989) Recent developments in carp culture technology in India with special reference to the State of Andhra Pradesh. In: Huisman EA, Zonneveld N, Bouwmans AHM (eds) Aquaculture research in Asia: management techniques and nutrition. Pudoc, Wageningen, pp 203–210

OECD-FAO (2012) Agricultural outlook 2013–2022. OECD, Rome

Othman K (2004) Integrated farming system and multifunctionality of agriculture in Malaysia. Acta Horticulturae. https://doi.org/10.17660/ActaHorti.2004.655.36

Papendick R (1984) Alternative agriculture systems: benefits to soil and water resources. Proceedings of the Institute for Alternative Agriculture First Annual Scientific Symposium. Institute for Alternative Agriculture, Greenbelt, 3–4

Paris RT (2002) Crop–animal systems in Asia: socio-economic benefits and impacts on rural livelihoods. Agric Syst 71:147–168

Phalan B, Bertzky M, Butchart SHM, Donald PF, Scharlemann JPW et al (2013) Crop expansion and conservation priorities in tropical countries. PLoS One 8(1):e51759. https://doi.org/10.1371/journal.pone.0051759

Pillai KR, Thiagarajan S, Samuel C (1985) Weed control by sheep grazing under plantation tree crop. In: Proceeding of 9th annual conference of the Malaysian Society Animal Production, Serdang, Malaysia, p 43–52

Prein M (2002) Integration of aquaculture into crop–animal systems in Asia. Agric Syst 71(1):127–146

Pullin RSV (1982) Experimental integrated farming systems in Mexico. ICLARM, Mexico. https://www.worldfishcenter.org/.../experimental-integrated-farming-systems-mexico
Purba S (1998) The economics of rice–fish production systems in North Sumatra, Indonesia: an empirical and model analysis. Farming systems and resource economics in the tropics, vol 31. Wissenschaftsverlag Vauk, Kiel
Reddersen J (1997) The arthropod fauna of organic versus conventional cereal fields in Denmark. Biol Agric Hortic 15:61–71
Ronnback P, Bryceson I, Kautsky N (2003) Coastal aquaculture development in Eastern Africa and the Western Indian Ocean: prospects and problems for food security and local economies. AMBIO J Hum Environ 31(7–8):537–542
Rothuis A (1998) Rice–fish culture in the Mekong Delta, Vietnam: constraint analysis and adaptive research. PhD thesis, Katholieke Universiteit, Leuven, Belgium
Royal Society (2009) Reaping the benefits: science and the sustainable intensification of global agriculture. RS Policy document 11/09. London. P.72
Rundlöf M, Smith HG, Birkhofer K (2016) In: eLS (ed) Effects of organic farming on biodiversity. Wiley, Chichester. https://doi.org/10.1002/9780470015902.a0026342
Saadullah M (1995) Integrated farming systems for rural poor. Rural Poor Programme Task Force of BRDB, Dhaka. 92 pp
Sale PF, Agardy T, Ainsworth CH, Feist BE, Bell JD, PatrickChristie GOV, Mumby PJ, Feary DA, Saunders MA, Daw TM, Foale SJ, Levin PS, Lindeman KC, KaiLorenzen PRS, Allison EH, Bradbury RH, Sheppard RC (2014) Transforming management of tropical coastal seas to cope with challenges of the 21st century. Mar Pollut Bull 85(1):8–23
Schneider MK, Luscher G, Jeanneret P et al (2014) Gains to species diversity in organically farmed fields are not propagated at the farm level. Nat Commun 5:4151
Shannon D, Sen AM, Johnson DB (2002) A comparative study of the microbiology of soils managed under organic and conventional regimes. Soil Use Manag 18:274–283
Siebert FS (2002) From shade- to sun-grown perennial crops in Sulawesi, Indonesia: implications for biodiversity conservation and soil fertility. Biodivers Conserv 11(1889–1902):2002
Sing NC, Sparkes DR, Charleston C (2002) Sugar farming systems development and demonstration on the wet tropical coast. http://hdl.handle.net/11079/13241
Smith P (2013) Delivering food security without increasing pressure on land. Glob Food Sec 2:18–23
Stockdale EA, Lampkin NH, Hovi M (2001) Agronomic and environmental implications of organic farming systems. Adv Agron 70:261–327
Swarnam TP, Velmurugan A, George Z, Dam Roy S (2014) Integrated farming system for sustainable livelihood in tribal areas of Nicobar Island, India. J Andaman Scie Assoc 19(1):19–22
Swarnam TP, Velmurugan A, Subramani T, Swain S, Sankaran M, Shrawan Singh, Kundu MS, Kiruba Sankar R, Nanda BK and Jaisankar I (2015) All India coordinated research project on integrated farming systems. ICAR – CIARI Annual Report. 2014–15. p14
Swarnam TP, Velmurugan A, Subramani T, Kundu MS, Kirubasankar R, Awnindra K Singh, Jaisankar I and Dam Roy S (2016) Agrobiodiversity management in integrated farming system for food and nutritional security. Vegetos 29:9–13
Tan KM, Abraham PD (1980) Livestock production on rubber plantations. First Asian-Australasian of Animal Science Congress, Serdang, Malaysia
Thompson J, Scoones I (2009) Addressing the dynamics of agri-food systems: an emerging agenda for social science research. Environ Sci Pol 12:386–397
Thrupp LA (2000) Linking agricultural biodiversity and food security: the valuable role of sustainable agriculture. Int Affairs (R Inst Int Affairs)., Special Biodiversity Issue 76(2):265–281
Tibbetts J (2002) Coastal cities: living on the edge. Environ Health Perspect 110(11):674–681
Tilman D, Fargione J, Wolff B, D'Antonio C, Dobson A, Howarth R, Schindler D, Schlesinger WH, Simberloff D, Swackhamer D (2001) Forecasting agriculturally driven global environmental change. Science 292:281–284
Tilman D, Balzer C, Hill J, Befort BL (2011) Global food demand and the sustainable intensification of agriculture. Proc Natl Acad Sci U S A 108:20260–20264

Tscharntke T, Clough Y, Wanger TC, Jackson L, Motzke I, Perfecto I, Vandermeer J, Whitbread A (2012) Global food security, biodiversity conservation and the future of agricultural intensification. Biol Conserv 151:53–59

Tuck SL, Winqvist C, Mota F et al (2014) Land-use intensity and the effects of organic farming on biodiversity: a hierarchical meta-analysis. J Appl Ecol 51:746–755

UNEP (2011) Report of the tenth meeting of the conference of the parties to the convention on biological diversity. UN Environment Programme, Montreal. UNEP/CBD/COP/10/27. 353 p

Van Brakel ML, Morales EJ, Turingruang D, Little DC (2003) Livelihood improving functions of pond based integrated agriculture and aquaculture systems, MRC Fisheries Programme (FP). Institute of Aquaculture, University of Stirling, Scotland

Vinning G (2008) Marketing organics in the Pacific Islands. Suva, Fiji

Wickramasinghe LP, Harris S, Jones G, Vaughan N (2003) Bat activity and species richness on organic and conventional farms: impact of agricultural intensification. J Appl Ecol 40:984–993

Winqvist C, Bengtsson J, Aavik T et al (2011) Mixed effects of organic farming and landscape complexity on farmland biodiversity and biological control potential across Europe. J Appl Ecol 48:570–579

Wood D, Lenné JM (1997) The conservation of agrobiodiversity on-farm: questioning the emerging paradigm. Biodivers Conserv 6(1):109–129. ISSN 0960-3115

Woynarovich E (1980) Raising ducks on fish ponds. In: Pullin RSV, Shehadeh ZH (eds) Integrated agriculture aquaculture farming systems. 1980, ICLARM Conf. 4, p 129–134

Yadav CM, Sharma RK (2013) Crop-livestock integrated farming system for the marginal farmers in rainfed regions of Bhilwara District in Rajasthan. Indian J Dry Land Agric Res Dev 28(1):74–76

Zhong GF (1995) Integration of agriculture and fish farming (dyke ponds) in Canton Region (China). In: Symoens J-J, Micha J-C (Eds) The management of integrated freshwater agro-pisciculture ecosystems in tropical areas. Proceedings of the International Seminar, 16–19 May 1994, Brussels, Belgium. The Technical Centre for Agricultural and Rural Co-operation (CTA), Wageningen, Netherlands and the Belgian Royal Academy of Overseas Sciences (ARSOM), Brussels, pp 297–307

Agroforestry for Livelihood and Biodiversity Conservation

16

Abstract

Trees play a predominant role in protecting and sustaining the productivity of coastal and island eco-regions, yet they are facing greater challenge from harsh climate, extreme events, anthropogenic activities, and land degradations. Different tree species can be produced by the concurrent production of agricultural crops and tree species in different ways by the conscious and deliberate use of land known as agroforestry systems. Agroforestry has the potential to contribute to the improvement of rural livelihood, due to the capacity of its various forms to offer multiple alternatives and opportunities to small holders to enhance farm production and income while conserving the biodiversity and environment. It is observed that plantation crops integrated with livestock, poultry, and rice are main features of this region. The growing food insecurity and deteriorating livelihood situations call for concerted and consorted actions to take advantage of the high potential of agroforestry, among other systems, for promoting livelihood. The structure, function, and role of different agroforestry systems in providing livelihood are discussed in this chapter with case examples from different tropical countries. Further, there is an increasing recognition of the contribution of agroforestry across tropical countries to improve ecosystem services and livelihoods. Therefore agroforestry research should focus more on conservation and utilization of diverse priority indigenous useful trees along with crops and livestock in a region-specific agroforestry-based integrated model.

Keywords

Agroforestry types · Tropical coast · Plantations · Silvopasture · Conservation

V. Ayyam et al., *Coastal Ecosystems of the Tropics - Adaptive Management*,
https://doi.org/10.1007/978-981-13-8926-9_16

16.1 Introduction

Relationship of man with trees and the land use systems can be appreciated in the temporal and spatial dimensions across agroclimatic regions. This led to the evolution of tree-based systems or agroforestry which is an age-old practice to meet the human need and protect their surroundings. Tree has the dominant role to play in all the agroforestry systems particularly in the coastal areas where different land use systems are practiced due to its geomorphology and geographical locations. Agroforestry has the capability to increase the productivity and at the same time maintain the nutrient balance as well as protect the coastal environment (FAO 1976). Acceleration in human and livestock population growth particularly in the coastal areas necessitated acquisition of more and more land under cultivation. In the absence of reasonable scope for horizontal expansion in land use, in order to increase the food production and provide people livelihood opportunities, the multitier system came into existence. Further, variety of land use practices have evolved through settled agriculture in different agroclimatic conditions, improvement in animal husbandry and organised forestry aimed at providing livelihood, utilization of available land, and ecosystem services.

Consequently increased interest in tree planting activities in the last few decades beginning from the 1970s resulted in the emergence and popularization of several terms related to forestry such as social forestry, farm forestry, and community forestry. Tree planting and all these forestry practices are greatly influenced by the climate, edaphic, and social factors. In this context the major climatic parameters that determine the environment and flora of a location in the tropics are rainfall (quantity and distribution) and temperature regimes. Altitude is important on land relief characteristics which determine the species distribution and plant-soil interaction. One of the special features of the tropics that are not a consequence of its climate and ecology is its economic, social, and developmental status (Nair 1993). At the same time, coastal areas in many of the tropical countries are suffering from over-exploitation of natural forests, waste disposal, oil spills, erosion, seawater intrusion, and salinity (Dagar 2000). Further, recent advances in climate change studies and its consequences and adaptation options have stressed the need for paradigm shift in agricultural practices with the inclusion of tree component and integration of animals. This strongly supports the agroforestry solutions to the coastal production and conservation challenges. Therefore, documentation and upscaling of agroforestry measures have attracted the active involvement of several international organizations and national governments. In all these activities and planned developments, some forms of livelihood support and biodiversity conservation through people's participation are common. This chapter details the role of agroforestry practices in providing livelihood and conservation of biodiversity across the tropical world.

16.2 Agroforestry

Farmers and land owners in different parts of the tropical world integrate a variety of woody perennials in their crop and livestock production depending upon the agro-climates, land, and local needs. It is observed that most of these practices are location specific, and the utility in other locations is probably underexploited; hence, it needs to be documented and evaluated before its upscaling.

16.2.1 The Concept

Agroforestry is a collective name for land use system and technology whereby woody perennials are deliberately used on the same land management unit as agricultural crops and or animals in some form of spatial arrangement or temporal sequence (Lundgren and Raintree 1982). In practical sense, agroforestry is a very sustainable land use system that maintains or increases total yields by combining food crops (annuals) with tree crops (perennials) and/or livestock on the same unit of land either alternately or at the same time using management practices that suit the social and cultural characteristics of the local people and the economic and ecological conditions of the area (Fig. 16.1). In economic terms, agroforestry emphasizes the interactive association between woody perennials and agricultural corps and animals for multiple products and services. In an agroforestry system, there are both ecological and economical interactions between the various components. This means, agroforestry normally involves two or more species of plants in which at least one of which is a woody perennial. It is also a common observation that the cycle of agroforestry systems is more than 1 year (Nair 1993).

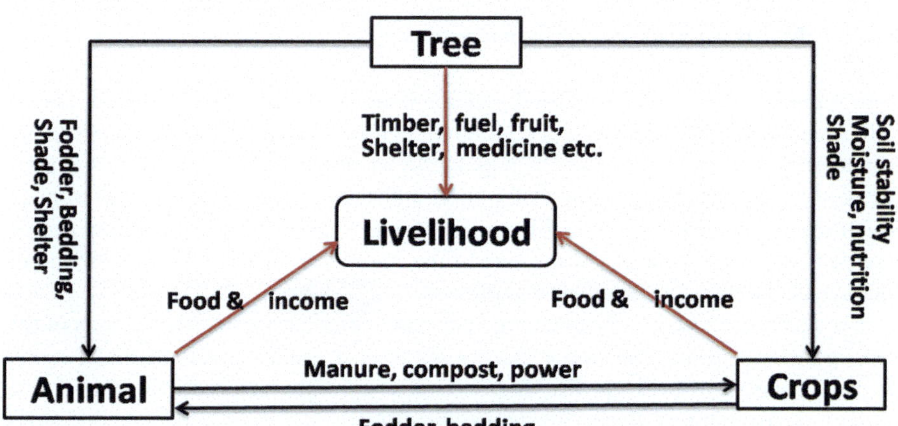

Fig. 16.1 Importance of agroforestry for providing livelihood

Therefore, invariably, the system is structurally, functionally, and socio-economically more complex than a monocropping system.

The National Forest Policy of different countries located in the tropical region envisages having 20–50% of the land area under the forest cover depending upon the land availability, socio-economic conditions, and geographical features for sound ecological balance. However, these policy documents are mostly silent on the spatial distribution pattern and to some extent time frame. On the other hand, positive aspects of many of these documents are recognition of the importance of improving the green cover by way of integrating tree components or increasing the tree outside the forest. This can be achieved to a large extent by different agroforestry systems in which coastal plantations are very important. Coastal plantations or agroforestry in the coastal areas is a vital component in the years to come to cope with the climate change. It is also important to note that many of the National Forest Policies state that as the natural forests serve as a gene pool resource and help in maintaining ecological balance, such forests will not be made available to industries. Forest-based industries should raise the raw material needed for meeting its own requirements, preferably by establishment of a direct relationship with a grower. In other words some of the forest-based industries should reorient or adjust by harnessing the potential of agroforestry. This means that agroforestry could provide livelihood as well as support the sustenance of the forest-based industries.

In summary, two fundamental attributes of all agroforestry systems are productivity and sustainability. The various productive functions of agroforestry are food, fodder, and fuelwood besides other products. The protective functions of agroforestry systems are windbreak, shelter belt, soil conservation, moisture conservation, providing shade, etc., for crops, man, and animals. The details are discussed in types of agroforestry systems in the following sections.

16.2.2 Need and Scope of Agroforestry

Agroforestry is not only the age-old practice in most of the tropical countries, but it is the need of the hour. It does not confine to regional, geographical, or agroclimatic boundaries. Owing to increase in population of human and cattle in the coastal areas, there is increasing demand for food and fodder particularly in the tropical countries. Each year farmers of the world have to feed 81 million more people irrespective of the weather. At the same time, there will be 15–19% decline in per capita crop land due to population explosion.

For decades, food insecurity and poverty have been hampering livelihoods of the rural poor and posing the major challenges to sustainable development of many tropical countries. Over the last four decades, agricultural production has come under increasing threat, due to the worsening climatic and environmental conditions and land and soil degradation, enhanced by a huge population pressure on limited resources. Because of that, agricultural production in these countries particularly in the coastal areas with increasing population has seldom matched the needs of the people. Further, it has been observed from the global climatic data records that coastal areas are most vulnerable to climate change and extreme events. Consequently the profound changes in farming systems, markets, and investment mechanisms are exposing small holders to increased vulnerability and often forcing them to change their traditional farming systems. This necessitates new management system that is capable of producing food from marginal agricultural land and is also capable of maintaining and improving quality of producing environment.

In many traditional farming systems, trees have always been a key element of sustainability, and farmers often conserve high percentage of trees on their lands to control soil erosion and usually adopt fallow systems to improve the soil fertility. Trees also provide farmers with supplementary benefits such as fruit for food, firewood, medicines, forages, etc. However, in the last two decades, the need to grow more food and to improve livelihoods has forced farmers to strive for maximum food production from fragile ecosystems using limited inputs, often resulting in the breakdown of the traditional farming system, and resource degradation. In this context, large-scale deforestation has taken place leading to decline in soil fertility and productivity, especially in the dry areas with low forest cover and in the humid tropics (Hosny El-Lakany 2004). In such situations agroforestry is more sustainable than monocropping. Well-planned and maintained agroforestry system can help to conserve biodiversity and promote agricultural diversifications in the coastal areas. This will ensure the livelihood of the coastal population who depends on primary activity in the tropical region.

16.3 Different Agroforestry Systems

The existing agroforestry systems across tropical coastal areas have got evolved or developed to suit to the different needs of the local populations which can sustain within the agroclimatic conditions. It means that it has specific structure, functions, and socio-economic role in different regions. Simultaneously agroforestry practices differ in their nature and arrangements of different components.

16.3.1 Arrangement of Components

The arrangement of components gives main priority to the plants which act as a base for other components to include and integrate. Even in agroforestry system involving animals, their management is according to a definite plan. For example, in the coastal grazing land, before selection of animals, it gives precedence to plants (fodder shrubs and grasses which can come up in the area) followed by animals. Such plant arrangements in multispecies combinations involve dimension of space and time. To some extent it also helps to conserve different species and maintain the diversity, but the economic interest is supporting livestock farming.

In spatial arrangement of plants in an agroforestry, mixture may result in dense stands (e.g. home gardens as practised in South and Southeast Asian countries) or in space mixed stands as in most systems of trees in pasture. The species or its mixture can be laid out in zone or strips of varying width based on the species characters and site characters. A common example of zonal pattern is hedgerow intercropping (alley cropping), and an extreme form is the boundary planting of trees on edges of plots and fields for a variety of purposes. The second method can lead to sole cropping system, but interactive association of different components can be used as criteria to decide limits between zonal agroforestry and sole crop plots.

Similarly the temporal arrangements of plants in agroforestry may also take various forms. A typical example is the conventional shifting cultivation cycle involving 2–4 years of cropping and more than 15 years of fallow cycle, when a selected woody species or a mixture of species may be planted. In the same context, some silvipastoral systems may involve grass alleys in rotation with some species of grasses remaining on the land for several years. These temporal arrangements of components in agroforestry are termed coincident, overlapping, separate, and interpolated.

16.3.2 Socio-economic Aspects

Based on the socio-economic criteria as scale of production and level of technology input and management, agroforestry systems are classified into three categories as commercial system, subsistence system, and intermediate system.

In commercial system the scale of production of the output usually a single commodity is the major aim of the system. The scale of operation is moderate to large, the ownership may be anyone, and the labour is paid or contracted. For example, the commercial production of plantation crops with pasture and animals or with food crops; commercial production of shade-tolerating crops such as coffee, tea, and cacao under shade trees; and commercial grazing and ranching under large-scale timber and pulp plantation.

Subsistence agroforestry systems are those wherein use of land is directed towards satisfying basic needs and is managed mostly by the owner or occupant and his family. Cash crops including sale of surplus production of commodities and all forms of traditional shifting cultivation are the best examples of this kind.

The third category is intermediate system which is between commercial and subsistence in terms of its scale of production and management. Production of perennial cash crops are undertaken in medium to small size farm wherein cash crops cater for cash needs; own land or with long-term tenancy rights to land, farmers reside and work themselves on the land, supplemented by paid temporary labour. Most of the agroforestry system in the tropical region is of this kind particularly those based on plantation crops providing livelihood and employment opportunities to the local population as large-scale labour and machine movement is not involved.

16.3.3 Major Agroforestry Practices

An agroforestry practice is a distinctive arrangement of the tree/shrub/crop components in space and time. In its evolutionary and economic perspectives, agroforestry shows the land use configuration with increasing intensity of management and inputs to increase output. While systems like home garden are climate based, all others also have their relation to rainfall, soil, and topography showing variations in farming systems.

Although several agroforestry systems have been recorded around the tropical world, considering various criteria having structural, functional, socio-economic, and ecological basis, these agroforestry practices are grouped into four major categories. They are agri-silvicultural, silvopastoral, agri-silvipastoral, and silviaqua systems. A generalized overview of the most common agroforestry practices of the tropics are given in Table 16.1. A closer examination of the distribution of these practices reveals that there is a clear relationship between climate, ecological, and socio-economic characteristics of a region on one side and the nature of the current agroforestry systems there.

16.3.3.1 Shifting Cultivation and Improved Fallows

The term *shifting cultivation* refers to farming or agricultural systems in which land under natural vegetation is cleared, cropped with agricultural crops for a few years, and then left untended while the natural vegetation regenerates. It is relatively primitive and subsistence in nature. It is found in most parts of the tropics, especially in the humid and subhumid tropics of Africa and Latin America and densely populated Southeast Asia. It is also known as Ladang (Indonesia, Malaysia), Jumar (Java), Ray (Vietnam), Tam-ray, rai (Thailand), Jhum (India), Chena (Sri Lanka), Tavy (Madagascar), and Milpa (Central America).

This can be described in two phases, viz. the cultivation and regeneration phase. The cultivation phase is usually short (2–3 years), but the regeneration phase, known as the fallow or bush fallow phase, is much longer (traditionally 10–20 years). The clearing is usually accomplished by the slash-and-burn method using simple hand tools (Nair 1993) either by individual or local society. In this method, the farmers usually take care that useful trees and shrubs are left standing and are sometimes lightly pruned. Other trees and shrubs are pruned down to stumps of varying height to facilitate fast regeneration and support for climbing species that require staking. After burning the ash or organic materials are mixed with the soil. The lengths of the cropping and fallow phases vary considerably in general; the fallow phase is several times longer than the cropping phase. The length of the fallow phase is considered critical to the success and sustainability of the practice because it helps to regain the fertility through the regenerative action of the natural vegetation.

In order to minimise the resource degradation in the traditional system of cultivation, Kang and Wilson (1987) developed a framework for a logical evolutionary pathway of traditional crop production systems in the humid tropics. This pathway highlights the major changes in cropping systems and indicates points at which intervention with planted fallows or other agroforestry methods could be introduced. It begins with a simple rotational sequence of temporal agroforestry. It is characterized by a very short cropping period followed by a very long fallow period which helps to restore soil productivity. In this method, at each of these successive stages, length of the cropping period extends progressively and that of the fallow diminishes correspondingly. This also advocates cover crops such as *Mucuna utilis*, *Pueraria phaseoloides*, *Centrosema pubescens*, and *Psophocarpus*

Table 16.1 Major agroforestry practices and their main characteristics

Sl. no.	Agroforestry practice	Arrangement of components	Major groups of components	Suitable areas
Agri-silvicultural systems (crops + trees of different nature)				
1.	Improved fallow	Woody species planted and left to grow during fallow phase	Fast-growing leguminous tree and common agricultural crops	Undulated terrain and shifting cultivation areas. Asia, East Africa and Central America
2.	Taungya	During the early stages of plantations, woody and agricultural crops are combined	Plantation forestry species and agricultural crops for food	Reforestation and afforestation areas mostly in Southeast and South Asia, Pacific islands
3.	Alley cropping	Agricultural crops are grown in between the hedges formed by the woody species	Fast-growing leguminous tree with good coppice with common agricultural crops	Subhumid to humid with population pressure and fragile soils. Southeast and South Asia
4.	Plantation crop combinations	Plantation crops in combinations Multistorey mixture of plantation crops Intercropping with agricultural crops	Plantation with shade trees or fuelwood or fruit trees and shade-tolerant species and agricultural crops	Humid to subhumid areas or highland and in small holder farm, South Asia and Tropical islands
5.	Home garden	Combination of different useful trees with intimate relation around homestead	Shade-tolerant agricultural crops along with fruit trees and creepers	In all the regions
6.	Trees for conservation	Trees with dense root on bund, terraces, etc. with grasses	Trees with good adaptability, multipurpose, and hardy crops	Sloppy land, sand dunes, and coastal degraded areas
7.	Multipurpose trees (MPT)	Trees scattered, in bunds, terraces, boundaries, etc.	MPT and crops	In all the regions and integrated with livestock
8.	Shelter belts, windbreak, and live hedges	Trees round the farmland	Combination of tall growing and dense trees	Coastal and wind prone areas, tropical islands
Silvipastoral systems (tree + pasture with or without animals)				
9.	Plantation with pasture and animals	Pasture under plantation with animals (free or captive)	Tree plantations, fast-growing grasses, or fodder trees with animals	Islands and South, S-E Asia
10.	Trees on rangeland	Trees scattered or in pattern in rangeland	MPT, high biomass fodder, and animals	Large grazing, sand dunes, and rehabilitated areas

(continued)

Table 16.1 (continued)

Sl. no.	Agroforestry practice	Arrangement of components	Major groups of components	Suitable areas
11.	Protein banks	Leguminous fodder trees in range/farm land for fodder	Leguminous trees with high biomass and herbaceous plants	In person: land ratio areas (land is limited in supply)
Agri-silvipastoral system (crops + tree + pasture/animals)				
12.	Home garden with animals	Various trees dominantly fruits and fodder with animals around homestead	Fruit trees, fast-growing fodder, and shade-tolerant herbs	In all ecological regions
13.	Apiculture with trees	Trees and beekeeping	Trees with high flowering nature/herbs with high nectar	Relatively less disturbed, rural areas
14.	Multipurpose woody hedgerows	Woody hedges with pasture cover	High coppicing woody trees like alley cropping with timber, fuelwood value	Sloppy land in humid, subhumid areas
Agri-/silviaqua system (tolerant crops/tree + aquaculture)				
15.	Aquaforestry	Trees around fish pond sometimes aquaculture in narrow furrows dug in between the trees	Trees and shrubs for shade and leaves used for feeding and manuring	Lowlands of humid and subhumid areas
16.	Home garden	Trees and crops are grown along with fodder and fish culture in the coastal areas	Multipurpose trees, fruits, fodder, and fish culture in the low-lying coastal areas	Lowlands of humid and subhumid areas

Source: Rocheleau et al. (1988), Nair (1989), and MacDicken and Vergara (1990)

palustris. Though the value and feasibility of planted fallows have been demonstrated experimentally (Webster and Wilson 1980), the practice has not become widespread due to various reasons besides gradual loss of biodiversity of woody perennials. Alternatively an improved tree fallow is a rotational system that uses preferred tree species as the fallow species as opposed to regeneration of natural vegetation, in rotation with cultivated crops as in traditional shifting cultivation. The reason for using such trees is production of an economic product, or improvement of the rate of soil amelioration, or both (Nair 1993).

16.3.3.2 Taungya

This is a modified form of shifting cultivation aimed at reforestation while it provides livelihood to the local population who depends on forest. The taungya system can be considered as a process of transformation from shifting cultivation to agroforestry. While shifting cultivation is a sequential system of growing woody species and agricultural crops, taungya consists of the simultaneous combination of the two components during the early stages of forest plantation establishment. The system was first practiced in Burma for afforestation of teak plantations and allows the local landless people to cultivate the forest land. In this method the labour is permitted to raise crops in an area but only side by side with the forest species, and they are responsible for the upkeep of the forest plantation. The practice consists of land preparation; tree planting; growing agricultural crops for 1–3 years, until shade becomes too dense; and then moving on to repeat the cycle in a different area. In some cases crops may be grown 1 year before the trees are planted. A large variety of crops and trees, depending on the soil and climatic conditions, are grown in India and South Asia (Nair 1993). The state owns the land and the trees. This system can be utilized to restore the forest land lying vacant and improve the productivity of land, while some livelihood is provided to the local people who render their labour. In Pacific Islands taro with *Anthocephalus* and *Cedrella* trees are grown in differnt ways. In East Asian countries, maize, rice, and pulses are grown in taungya system. However this system is not a welcome method in forest conservation and practiced in hilly and mountainous terrain which is normally located away from coastal areas.

16.3.3.3 Alley Cropping

In alley cropping with livestock, the hedgerows of trees are managed to provide high-quality fodder for sheep, goat, and/or cattle. There are two main types of systems that integrate alley farming principles with livestock production, viz. alley farming with crops and livestock and alley farming with fodder and livestock.

In the first system, the hedgerows are managed to provide fodder for the animals and mulch for the crops. The hedgerows can also be used as a stand for growing spices like black pepper. In the second system, both the trees and the intercropped fodder crops are fed to animals. Alley farming with food crops and livestock allows small holders to follow a low-cost pathway for improving both crop and livestock production, reducing reliance on external inputs, and increasing sustainable offtake from the land. Alley farming with grass and livestock is an innovative way of overcoming the nutritional constraints faced by livestock production in many areas of tropical Asia and Africa.

In the coastal areas and undulated island terrains, leguminous *Erythrina poeppigiana* and *Gliricidia sepium* are used for alley cropping (Fig. 16.2). They are propagated by large cuttings and well known for their ability to fix nitrogen and the relatively large amounts of crude protein (20–30%) in their leaves. The associated crops are maize, millets, cassava, and common beans. These trees used in hedgerows are periodically lopped, as supply of organic matter and nutrients to the associated crops, suppression of weeds, and action as mulch. The lopped branches and leaves could also be carried from the nurse trees to nearby crops. It is observed that *Erythrina* provides 8000-kg dry matter (with 3–4% nitrogen) per year and higher yield of maize than control.

In Andaman group of islands located in the Bay of Bengal, alley farming with food crops/grasses with livestock is the best option for hilly uplands. Food crops (maize) and vegetables (okra, brinjal, amaranth, etc.) are grown during wet season in the alleys of *Leucaena*, *Gliricidia*, and *Sesbania* formed across the slope or along the contours. The hedgerows are periodically lopped as green fodder for livestock

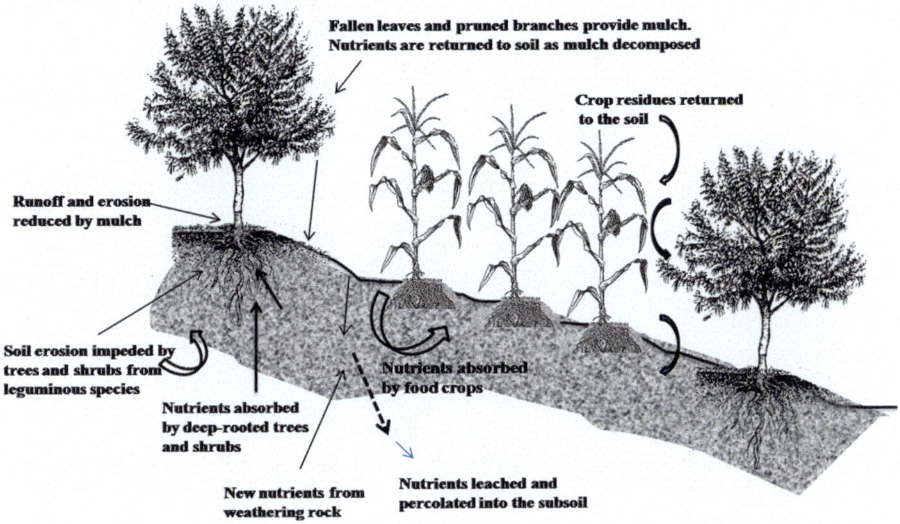

Fig. 16.2 Schematic representation of alley cropping and its ecological services

and green leaf manure or used as mulch for growing the black pepper which is trained on the hedgerows itself. As the hedges are formed across the slope, they prevent soil erosion and increase the water infiltration into the soil. The fodder crops grown in the alleys along with green fodder obtained from hedgerows can be used by the livestock component and thus can reduce the green fodder scarcity during dry months. The foliage from *Leucaena* and *Gliricidia* species contains over 20% crude protein (3.4% nitrogen) and hence provides a valuable high protein feed supplement for livestock. The leguminous trees should be used as supplements only to the extent of 20–25% of total feed intake. If it exceeds 30–40% of their total feed intake, *Leucaena* becomes toxic due to the high levels of mimosine which leads to hair loss and illness and can even be fatal.

Most of the tropical islands face shortage of green fodder in terms of its quality and quantity particularly during dry season resulting in low productivity of livestock. However, the multipurpose trees and shrubs used in alley farming have long tap roots and can access water at deeper soil layers than shallow-rooted crop plants. Thus, the MPTs can produce leaves and branches for forage throughout the year which can meet the green fodder requirement of livestock. The alley farming can be used both in hilly uplands and hill slopes where the problem of soil erosion is a serious concern and cultivation of annual crops is a serious problem due to erosion and water scarcity.

16.3.3.4 Homestead Garden

A homestead is an operational farm unit in which a number of crops including tree crops are grown with livestock, poultry, and/or fish production in the pond. Water harvesting resource recycling is very essential to minimize the production cost. It is an age-old, evolved practice in the coastal areas in which there are wide variations

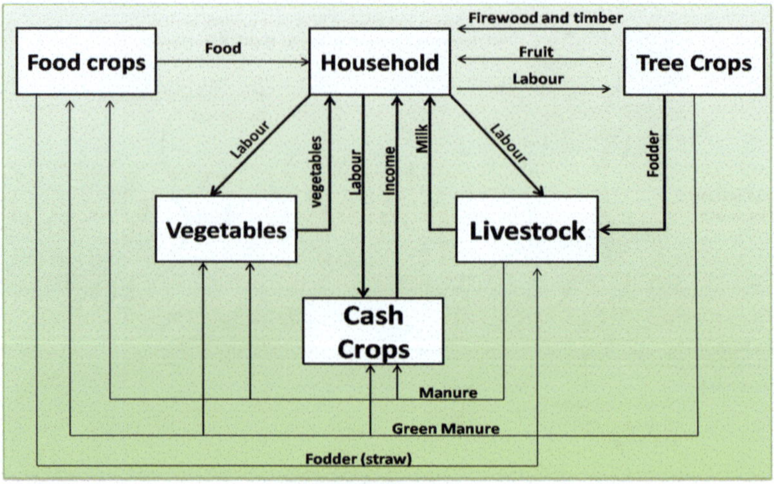

Fig. 16.3 Resource flow in homestead farming

in the intensity of trees, crops, and livestock based on the size of holding, needs of the people, and microclimate. Home gardens (also known as homestead and mixed gardens) are usually located, where they exist at all, close to the household as one of the more intensively cultivated parts of the overall farm. The present-day homestead garden is characterized by a mixture of several or many annual or perennial species grown in association, commonly exhibiting a layered vertical structure of trees, shrubs, and ground-cover plants, which recreates some of the properties of nutrient recycling, soil protection, and effective use of space above and below the soil surface to be found in forests (Fernandes and Nair 1986). Home gardens are widely used to supplement outputs from other parts of the farm through the cultivation of a variety of other subsistence and commercial crops.

In Java, home gardens have long existed as the principal farming system on dry lands, accounting for a substantial proportion of total land use, with irrigated rice cultivation forming the other main component of the farm system. The gardens are traditionally dominated by perennials rather than annuals and by woody rather than herbaceous growth. On farms with sufficient rice land to enable the household to meet its basic food needs, priority in labour and capital allocation is given importance in home gardens (Fig. 16.3). The garden areas in these larger farms are essentially forest gardens, with capital investment in trees of commercial value (Wiersum 1982).

In general, coconut, areca nut, guava, mango, citrus, jack fruit, papaya, banana, moringa, sesbania, custard apple, and many multipurpose trees are the major tree component found grown in home gardens. Cattle and poultry are the main components of homestead; therefore, forages like stylo (*Stylosanthes guianensis*, *S. hamata*), guinea grass, and Napier are also grown frequently. Irrespective of difference in components and structure, home garden across the tropical region is so intimately mixed in horizontal and vertical strata as well as in time that a complex

Gliricidia sp., *Sesbania grandiflora*

Fig. 16.4 Boundary planting as biofence, fodder, and food

interaction exists among soil, plants, other aboveground and below ground components, nutrition, and environmental factors. Home garden helps to conserve large variety of trees and shrubs when it is practiced by large number of people in a locality; it helps to advance genetic stock by selection. But identification of such advance plant is very important for its use. Mangroves form an essential part of homestead of backwater and bay. In island ecosystem coconut and pandanus (*P. tectorius*, *P. lirum*) can be commonly seen. Fish and shrimp culture in backwater channels and in association with mangroves is the main activity in homestead. In dry areas of South Asia, palmyrah palm is common. Thus home garden ensures diversity, sustainability, and livelihood to the farmers.

16.3.3.5 Boundary Planting

The MPTs can also be integrated into farming system by way of utilizing the border space as boundary planting and live fence. In valley areas, the border planting of *Leucaena*, *Sesbania*, *Gliricidia*, etc. will be highly suitable as they not only provide green fodder but also used as green leaf manure for growing rice (Fig. 16.4). Border planting is also highly suitable for homestead gardens in many of the tropical Islands, where the closer planting of above trees is used as green fodder for goat, as leaf manure, and as a starter in composting of crop residues. In most of the coastal areas, boundary plantations also help as shelter belts and windbreaks particularly in fruit orchards.

16.3.3.6 Windbreak

Windbreaks have been a traditional agroforestry practice which is characterized as linear plantings of trees and/or shrubs established primarily for environmental purposes such as reducing wind speed, snow accumulation, retaining soil moisture, and buffering extreme temperatures (Fig. 16.5). Integrated as part of crop or livestock production systems, the effects of windbreaks serve to enhance crop and animal production by protecting livestock and crops from harsh environmental conditions in addition to reducing soil erosion from agricultural fields. The height of a windbreak is directly related to the amount of leeward area protected by wind damage. These areas usually extend up to 10–15 times the tree height or up to

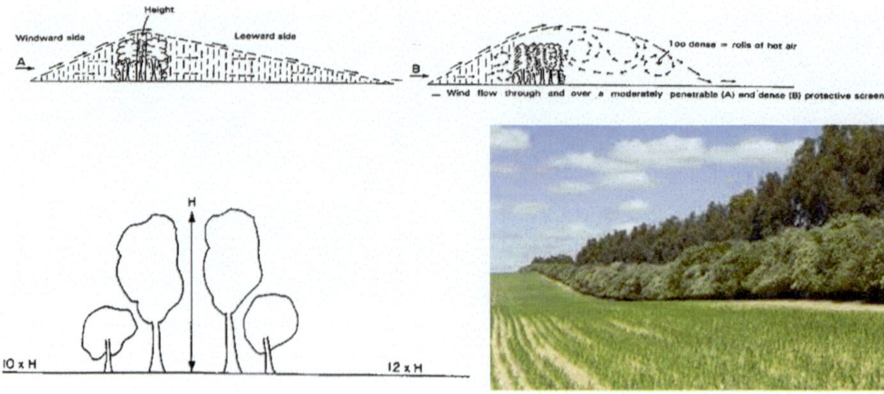

Fig. 16.5 Trees as windbreaks: A conceptual model and field photo

Fig. 16.6 Caribbean coffee plantation with shade trees

150-m with 10-m high trees. Some windbreaks system has shown to increase wheat yields by 15% and improve net economic benefits by 17%. Windbreaks may also serve additional benefits such as pollination sources for crops, wildlife habitat, and production of timber, posts, firewood, and other valuable resources. However, very little if any information is available on the economic impacts from these additional products from windbreaks. Some suitable trees and shrubs are *Acacia tortilis*, *Cassia auriculata*, *Pithecellobium dulce*, *Lawsonia inermis*, *Parkinsonia* aculeata, and *Prosopis chilensis.*

16.3.3.7 Shade Trees

Coffee and cacao are very valuable cash crops throughout Central America, Caribbean, Indian Ocean islands, and East Coast of India (Alvim and Nair 1986). The presence and utility of shade trees in these fields (Fig. 16.6) have therefore attracted considerable attention by agroforesters. It is now well recognized that the use of shade trees implies a multitude of biological, environmental, and socio-economic benefits. Although the amounts of nitrogen fixed are still a matter of discussion as large variations are observed, on an average of 30–40-kg N ha^{-1} $year^{-1}$ for *Inga* trees in coffee fields in Mexico was reported (Roskoski 1982). In the

Caribbean traditionally polyculture of shade trees with high density planting is maintained in the coffee garden, while in the commercial polyculture in recent times, shaded trees are judiciously planted to provide only required shade to the plantation.

16.3.3.8 Block Planting/Fodder Tree Banks

This system aims to integrate livestock through cultivation of fodder trees in the agricultural land. In this system, blocks of closely spaced trees or shrubs are planted in one corner of field. The trees are planted in hedgerows with 0.25-m spacing between trees and 1.0–1.5 m between rows. Such a layout maximizes protein yield on the available land area while allowing farmers to enter the blocks with ease. The blocks of closely spaced hedgerows are planted up to 10–30% of the total area based on the number of animals integrated into the system. They are managed as fodder tree banks or also called 'tree protein banks'. The blocks are fenced off for improved management, to allow only limited access by grazing animals (e.g. pregnant and lactating dams) and during the dry season.

16.3.3.9 Aquaforestry

Low-lying waterlogged marshy areas, flood plains, and ill-drained lands are common features in coastal areas. Mangroves and associated halophytes occupy swamps and river banks. Such conditions impose serious challenge to farming and livelihood. In those areas most of the farmers cultivate fish and prawn in saline water and growing coconut and other trees on bunds of ponds. These trees help in producing litter feed to fishery and generate extra income to farmers. Fish culture in mangroves is also advocated to provide livelihood to people which forms a rich source of nutrition to aquatic life and breeding ground for juvenile fish, prawn, and mussels. A well-balanced aquaforestry in the small ponds in home garden will increase moisture, energy, and nutrient use efficiency per unit area. The leaves of many leguminous trees, *G. sepium*, *L. leucocephala*, and *M. oleifera* have been found to serve as good fish feed when offered as pellets and improved its productivity. In South Asia, at certain places in the coastal areas, farmers are allowed to utilize the natural wetlands for aquaculture; in return they have to maintain or protect through local society. To certain extent this system helps to ease out pressure on natural wetlands or reduce it degradation.

16.3.4 Supporting Measures for Agroforestry Practices

Even forest grazing and natural silvipastures or natural savannas from where grazing and firewood collection benefits are obtained could be the part of agroforestry helping agricultural economy and livelihood need. Further, these systems are most suited for local conditions, with high adaptability and biodiversity. The stability and sustainability of these systems are high as compared to agriculture alone. These agroforestry practices fit into the socio-economic milieu of farmers in tropical region. While understanding the structure and deriving utility out of the agroforestry system, it is also vital to take suitable measures for successful development of agroforestry so as to have sustainable production and ensure long-term benefits.

- Soil conservation techniques such as contour farming, contour bunding, graded trenching, introduction of cover crops, shelter belt planting, runoff, recycling, etc. should be practised.
- Planting and growing of shade trees in plantation, e.g. *Erythrina*, *Grevillea robusta*, etc.
- Plant propagation techniques should be standardised. For example, plants may be propagated through seeds or by vegetative means, e.g. root suckers, coppice cuttings, etc.
- Total or partial weed control is required, and complete weeding, line weeding, interrow weeding, and spot-ringed weeding are required.
- Another important aspect of establishing successful agroforestry is cover crop which is a secondary crop which may be a small tree, grasses, herb, or shrub used to cover the soil below the main crop for minimizing soil erosion, etc., e.g. *Stylosanthes hamata*, *Centrosema pubescens*, cowpea, etc.
- To protect the area under plantation from grazing by cattle, live fencing through planting of *Leucaena*, *Leucocephala*, *Gliricidia sepium*, *Acacia villosa*, *Erythrina*, and *Murraya koenigii* can be practised which are highly promising. Growing of grass like vetiver grass and lemon grass reduces runoff velocity as well as soil erosion.
- Control of insect pests and other plant protection measures are unavoidable input measures.

16.4 Agroforestry in Livelihood and Biodiversity Conservation

16.4.1 Livelihood Through IFS and MPTs

A large number of tropical trees and shrubs are traditionally known and used for their fodder. In recent times, a lot of interest has been generated regarding the possibility of exploiting the fodder value of tropical shrubs for improved silvopastoral management, special attention being given to nitrogen-fixing species. In integrated farming system (IFS), various kinds of tree species are grown and integrated along with crops and animals along with composting and water harvesting, thereby providing huge scope for conservation of biodiversity. The major advantage of this system is production of food, fodder, and wood products with high sustainability for home consumption and sale for cash.

In the IFS system, trees are planted in various patterns such as alley cropping, homestead garden, multipurpose trees (MPT), multilayer garden, etc. The trees in agroforestry provide fruit, fuel, fodder, and wood and enhance soil fertility, conserve soil, and also act as live fence and windbreaks. A tree serving more than one purpose is a multipurpose tree, e.g. *Gliricidia sepium*. Growing of multipurpose trees can reduce the risk of total crop failure. The primary purpose is to maintain or increase crop yields by improvement of the soil and microclimate and weed control. Farmers may also obtain tree products from the hedgerows, including fuelwood, poles, food, medicine, and fodder, and on sloping land, the hedgerows and prunings may help to control erosion. The woody plants are cut regularly, and leaves and twigs are used as mulch on the cropped areas in order to reduce the evaporation losses of moisture from the soil, suppress weeds, and/or add nutrients and organic matter to the top soil. Where nitrogen is required for crop production, nitrogen-fixing plants are the main components of the hedgerows.

Dagar and Kumar (1992) stated that there is an urgent need to make a quantum improvement in the farming area for fodder production not only by seeding with legumes and other high yielding grasses and by controlled grazing but also by introduction of fodder tree species such as *Trema tomentosa*, *Erythrina variegata*, *Bauhinia*, *Sesbania*, *Leucaena*, *Pithecellobium*, etc. In general, about 10–15 fodder trees/ha with good coppicing ability may be optimum for coastal and island regions. Such a diverse system will not only provide stability but also sustain against climate change and its associated events than any other systems which are less diverse. Under IFS system, free grazing cannot be possible so that the stall feeding and cut-and-carry system is the best method to be followed. By cut and carry and pruning management, MPTs can provide high-quality fresh forage year round, even during the peak of the dry period in integrated farming system (Jaisankar et al. 2015). These trees can be propagated by seeds and stem cuttings. All other trees are planted with close spacing, while *Ficus*, *Bauhinia*, *Sapium*, and *Crypteronia* are planted in a wider spacing. The details are given in Table 16.2. These trees can produce 12–30 kg/tree of leaf biomass per year in 4–6 lopping.

In addition to biodiversity conservation, agroforestry provides food security and enhances farm income through enhancement in the yield of products and services from biodiversity rich agro-ecosystems. In Indonesia four million of agroforestry not only yield rubber of US $ 1.9 billion but also contain 250–300 spp. of plant (Leakey 1998; Mc Neely and Scherr 2001). Similarly majority of the farmers of Andaman and Nicobar Islands, India, practices coconut-based farming system which is the main source of economic return for them.

16.4.2 Plantation-Based Farming

Perennial plantation crops in the tropical region occupy about 8% of the total arable area in developing countries. With the realization of its huge potential and necessity for intensification of land use due to rapidly increasing populations, policymakers and farmers in tropical developing countries have turned their attention towards integration of plantation crops, annual crops, livestock production, and forestry to provide employment and livelihood. Some plantation crops (e.g. coconut) are more amenable to such integration than others (e.g. rubber) because of their growth habits as well as the methods of their cultivation. The West African coast is densely

Table 16.2 Growth and yield performance of important MPTs under island conditions

Sl. no	MPTs	Part (s) used as a fodder	Propagation	Method of planting	Suitable place for planting	No. of lopping per year	Number of trees/ha	Leaf biomass production (kg/tree/year)
1.	*Leucaena leucocephala*	Leaves, pod, flowers	Seed	High density planting	Boundary planting	5	200	20
2.	*Sesbania grandiflora* (L.) Poiret	Leaves, pods, flowers	Seed	High density planting	Alley planting/ pond bund	4	50	12
3.	*Mussaenda macrophylla* (Wall)	Leaves, flowers	Seed and stem cuttings	Close spacing	Border planting	4	40	25
4.	*Ficus hispida*	Leaves, fruits	Seed	Wider spacing	Terrace/alley planting	5	40	30
5.	*Trema tomentosa* Roxb.	Leaves, flowers, and fruits	Seed	Close spacing	Terrace planting/ pound bund	6	50	15
6.	*Bauhinia variegata* L.	Leaves, flowers, and pods	Seed	Wider spacing	Terrace planting	5	25	30
7.	*Sapium baccatum* Roxb.	Leaves and fruits	Seed and stem cuttings	Wider spacing	Intercropping	5	25	14
8.	*Gliricidia sepium* (Jacq.) Walp	Leaves, flowers, and pods	Seed and stem cuttings	High density planting	Boundary/alley planting	6	175	24
9.	*Crypteronia paniculata* Blume	Leaves and panicles	Seed	Wider spacing	Around poultry shed	6	15	24
10.	*Pithecellobium dulce* (Roxb.) Benth	Leaves, flowers, and fruits	Seed	Close spacing	Boundary/border	6	75	15

populated and the forests have been cleared to make space for palm, coffee, cacao, and rubber plantations, while coconut occupies predominant role during the colonisation history of Pacific Islands and fruits and spices have been introduced in recent decades.

In Trinidad, cacao is mainly a forest species, grown traditionally under shade trees, with no fertilizer or pesticide application. Many small holder rubber plantations in Southeast Asia are based on integrating rubber with a variety of crops, including soya bean, maize, banana, groundnut, fruit trees, black pepper, and coconuts. In Malaysia, poultry raising in rubber stands is also a common and remunerative practice (Ismail 1986). Notable examples of small holder systems in which coffee is integrated with other crops and/or livestock include banana particularly in Caribbean and West coast of India, Sri Lanka. Similarly coffee smallholdings of East Africa, the coffee and maize holdings at Jimma in the Ethiopian highlands, and the coffee and dairy milk production systems in Kenya exhibit same structure. Most of the coconut production in India, the Philippines, Sri Lanka, and the Pacific Islands comes from smallholdings in which the coconut palm is integrated with a large number of annual and perennial crops which provide additional income and stability to the farm income. In Sri Lanka, Lakshadweep, India, and the Pacific Islands, grazing under coconut is also common; thereby it helps to diversify the agricultural activities. Cashew grows in a wide range of ecological situations, including wastelands where few other species thrive. In India, Tanzania, Mozambique, and Senegal, small holders often grow cashew trees with other crops, planting the trees in a random way so that they appear scattered on the land. Grazing under cashew is also very common, particularly on smallholdings in East African coastal areas (Fernandes et al. 1984; Nair 1993).

Several studies indicated that sufficient amount of sunlight is transmitted to the lower profiles in palm stands but differ with age groups. Additionally, the rooting pattern of the palm in a managed plantation is such that most of the roots are found near the bole (Kushwah et al. 1973), and thus overlapping of the root systems of the palm and the intercrop species is minimal. Owing to the diverse conditions under which coconuts are grown, they can be interplanted with a large number of other economic species. At the same time, the species diversity is usually greater in less intensively managed holdings than at commercial palm cultivation. In well-maintained holdings, farmers exercise some care in the selection of the other species grown among coconuts, but, invariably, food crops that produce a reasonable yield under partial shade are a natural choice. For example, various tuber crops such as sweet potato, tapioca, and different species of yam, as well as several kinds of vegetables, are common choices. There are also other annuals such as ginger and turmeric and perennials such as banana, pineapple, cacao, clove, and cinnamon that grow well with coconuts. Where the population of palms per unit area is lower and other conditions are favourable, crops that require abundant sunlight, such as cereals and grain legumes, are also grown profitably. It can be seen that the intercrops range from staple food crops to cash and export crops. This system provides more stability and higher income besides employment opportunities and opens up possibilities for agro-tourism in tropical islands.

Table 16.3 Common pasture and forage species grown/occurring under coconuts plantation

Common name	Scientific name	Location and adaptability
Improved pasture grasses		
Carpet grass	*Axonopus compressus, A. affinus*	South East Asia
Palisade grass	*Brachiaria brizantha*	Wide adaptability
Para grass	*B. mutica*	
Signal grass	*B. decumbens*	Pacific islands
Koronivia grass	*B. humidicola*	East Africa
Congo grass	*B. ruziziensis*	
Alabhang	*Dicanthium aristatum*	
Rhodes grass	*Chloris gayana*	East Africa, India
Pangola grass	*Digitaria decumbens*	
Batiki blue grass	*Ischaemum aristatum*	
Molasses grass	*Melinis minutiflora*	Fiji
Guinea grass	*Panicum maximum*	Wide adaptability
Scrobic	*Paspalum commersonii*	
Napier grass	*Pennisetum purpureum*	Mostly as fodder
		Most of the coastal areas
Legumes		
Green leaf desmodium	*Desmodium intortum*	Wetter subtropics
Kaimi clover	*Desmodium canum*	Pacific Islands
Perennial soya bean	*Glycine wightii* fsyn. (*Wight & Arn.*)	East Indies
Leucaena	*Leucaena leucocephala*	Wide adaptability (except acid soils) for fodder
(Ipil-ipil)		
Siratro	*Macroptilium atropurpureum*	
Stylo (Brazilian lucerne)	*Stylosanthes guianensis*	Adaptable to infertile soils; also used as fodder

Modified from: Plucknett (1979)

16.4.3 Fodder Cattle Under Coconut

Grazing cattle on the pastures grown under coconuts (as opposed to the cut-and-carry system) is another major land use activity in coconut areas in many parts of the tropics. Cattle raising usually involves grazing on pastures composed of natural species, but, in some cases, special fodder plants are also cultivated which provides stability and additional income. In natural stands, the most important plants for grazing, as would be expected, are grasses and legumes, although many other types of plants that can be grazed are also found. A list of the common species occurring in the natural pastures in coconut areas is given in Table 16.3. Some of the species which are considered weeds in coconut gardens are also grazed. Moreover, cover crops such as kudzu (*Pueraria phaseoloides*), centro (*Centrosema pubescens*), and calopo (*Calopogonium mucunoides*) can also be found in natural pastures.

Table 16.4 Dry matter yields of pasture grasses grown under coconuts (Rajaguru 1991)

Species	D.M. kg ha^{-1} year^{-1}
B. miliiformis	15,426
B. mutica	10,666
Digitaria decumbens	15,824
Urochloa mosambicensis	11,598
Panicum coloratum	8919
Paspalum commersonii	10,812
P. notatum	13,193
P. plicatulum	12,824
Setaria anceps	14,268

Compilation of studies showed that inclusion of suitable grass species under coconut produced on an average 8–15 ton of dry matter per year in different tropical countries (Table 16.4). All these grasses could support maintenance of cattle for various purposes. In Fiji, *A. compressus* has been considered to be a useful feed under coconut plantations. In Thailand, Duryaprapan and Boonklinkajorn (1977) found that *Axonopus compressus* and *Paspalum conjugatum* produced green matter yields of 18 and 19 t ha^{-1}, respectively, compared with 24 t ha^{-1} for *Brachiaria brizantha*, under the shady conditions of a coconut plantation which supports livestock production along with coconut in the coastal areas. It was tolerant of grazing and shading, and nutritive value measurements indicate a high digestibility compared to *Stenotaphrum secundatum* (Samarakoon et al. 1990). The only similar or superior grass species under these conditions are buffalo couch (*S. secundatum*), Pemba grass (*S. dimidiatum*), and possibly Batiki Blue (*I. aristatum*). It has been shown that *A. compressus* was out yielded by *I. aristatum* at 40% and 31% and by *S. secundatum* at 40 and 20% light transmission (Smith et al. 1981). *A. compressus* may also be an indicator of low soil fertility or overgrazing (Steel et al. 1980) as demonstrated in Fiji where it invaded old *B. utica* pastures.

In Sri Lanka it was noted that *B. miliiformis* was more shade tolerant than *B. brizantha* and that *P. maximum* was a very competitive grass. However, it could successfully be used in the wet zone without any reduction of coconut yields provided both crops were adequately fertilized (Santhirasegaram et al. 1969). Apart from green fodder production, these grasses proved to be successful in introduction and sustenance of dairy units in many of the coconut-growing areas in the tropical countries.

16.5 Agroforestry for Biodiversity Conservation

Biodiversity is considered as cornerstone stability and basis of livelihood and sustainable development (MEA 2005). It provides farming systems and taxa the means to recycle nutrient; reduce insect, pest, and disease problems; control weed;

maintain good water and soil conditions; and handle climate stresses while producing commodities necessary for human survival. Farmers practice agroforestry for gaining livelihood, income generation, risk management, food security, and optimal use of available land, labour, and capital (Arnold and Dewees 1997). It is estimated that about 1.2 billion people (20%) of the world population depends directly on agroforestry products and services in developing countries which can provide goods and services that can offset 5–20% of deforestation (Leakey and Sanchez 1997; Dixon 1995).

Agroforestry systems are considered as diversity-enhancing land use system especially in the context of interspecies diversity as it brings together crops, shrubs, trees, and in some cases livestock on the same piece of land. The major physical parameters that determine the biodiversity and ecology of a location in the tropics are rainfall (quantity and distribution) and temperature regimes. Altitude also is important because of its influence not only on temperature but also on land relief characteristics. Many of the traditional agroforestry practices give focus on biodiversity aspects, while other modern systems give preference to commercial exploitation of combinations. Studies of the traditional agroforestry practices in the Pacific islands exhibited their diversity in both species and cultivars. In Papua New Guinea, there is wide diversity of cultivated plants in shifting cultivation system, on an average which ranges from six to ten cultivated species per 25 m^2. In some parts of Indonesia (Eastern islands), the agroforestry system contains 8 varieties of coconuts, 25 *Musa* spp., 13 *Colocasia* taro, 14 *Cyrtosperma* taro, and 25 breadfruit. In Hawaii nearly 200 varieties of sweet potato, 6 sugarcane, and 35 banana varieties were introduced and maintained in the garden (Manner 1981).

The biodiversity aspects of different agroforestry practices followed in tropical countries are given in Table 16.1. The selection of tree species in agroforestry should be such that cultivation of arable crops traditionally grown on that site can be continued under the new system. Guillerme et al. (2011) reported decline in diversity of indigenous multipurpose trees and shrubs and herbaceous components such as traditional vegetable crops and ornamental plants owing to the conversion of agroforestry systems (including home gardens or their parts) to monospecific production systems and introduction of exotic fast-growing multipurpose trees. It is estimated that 1400 m ha of croplands and agro-ecosystem may provide ecosystem services worth US$ 92 ha^{-1} per year in pollination, biological control, and food production amounting to total US $ 128 × 10^9 per year (Costanza 1997). In general, agroforestry plays five major roles in conserving biodiversity:

- Agroforestry conserves biodiversity within deforested, fragmented landscapes by providing habitats and resources for plant and animal species. Conservation of biodiversity helps in preserving germplasm of sensitive species which can be used in the future or used as a donor of important desirable traits.
- Indirectly biodiversity conservation through different agroforestry practices helps to reduce the rates of conversion of natural habitat by providing a more productive, sustainable alternative to traditional agricultural systems that may involve clearing natural habitats.

- It provides connectivity by creating corridors between habitat remnants and forest patches which may support the integrity of these fragmented forests and the conservation of area-sensitive floral and faunal species.
- In addition to plant diversity, agroforestry systems also play an important role in increasing microbial, avian, and faunal diversities. Kumar and Nair (2004) reported species richness of tropical home gardens varying from 27 (Sri Lanka) to 602 (West Java). The greater diversity of birds and insect in agroforestry systems provides the beneficial service of pest reduction to adjacent crops (Gillespie et al. 1995; Schultz et al. 2000).
- It also provides other ecosystem services such as erosion control and water recharge, thereby preventing the degradation and loss of surrounding habitat (Jose 2009).

In the coastal areas, different types of land forms are encountered mostly due to the interface of sea and land and its physiographic position. As a result different agroforestry models have evolved which act as a custodian of biodiversity particularly conservation of locally adapted species. A glimpse of different agroforestry models practised in the tropical region is given in Fig. 16.6. Though many scientific inputs are given to plantation crops and multiple cropping systems, quite large areas under coconut plantation are still neglected and remain open for grazing. This is particularly relevant in the Pacific and Indian Ocean tropical islands. These areas may be brought under multistoreyed cropping systems. Spices like clove and cinnamon may be planted as middle storey crops, and pineapple or forage grasses like hybrid Napier (*Pennisetum purpureum*), guinea (*Panicum maximum*), and legumes like *Stylosanthes guianensis* may be grown to support the livestock and diversification (Fig. 16.7).

16.6 Conclusion

Coastal areas in the tropics are characterized by low-lying waterlogged marshy areas, flood plains, elevated land, sand dunes, relict mountains surrounded by flood plan, or rock shores. In those areas inclusion of tree component in crop cultivation or adopting or improving the agroforestry practices is most suitable to utilize the available resources effectively, address the environmental degradation concern, and improve the productivity of the system. Thus agroforestry is a collective name for land use systems and technologies where woody perennials are deliberately used on the same land management units as agricultural crops and/or animals, in some form of spatial arrangement or temporal sequence. There are three main types of agroforestry systems practiced in the tropical region, viz. agri-silvicultural, silvipastoral and agri-silvipastoral systems. These practices judiciously combine tree species, crop, and animal components and help to effectively utilize the available resources though wide variation in tree species distribution and its performance are observed. Inclusion of fodder trees and silvipastoral systems enables maintenance of livestock in otherwise fodder scare coastal areas. At the same time, agroforestry

Fig. 16.7 Different agroforestry model providing livelihood and biodiversity conservation

Multilayer crops | Wind breaks

Aquaforestry in coastal areas | Silvipasture with livestock in coastal areas

Fig. 16.7 (continued)

is crucial to small holder farmers and other rural people of this region because it can enhance their food supply and income and helps in conservation of biodiversity. Further, it is evident that agroforestry systems offer more hope for conservation of plant and animal species than the monoculture crops. Traditional agroforestry system offers more potential towards the biodiversity conservation, while modern agroforestry technologies provide more income to the rural household. Thus integrated approach involving agroforestry systems is essential for the development and conservation of resources of coastal areas which can provide livelihood and can impart sustainability and save the fragile coastal ecosystem from degradation.

References

Alvim R, Nair PKR (1986) Combination of cacao with other plantation crops- an agroforestry system in Southeast Bahia, Brazil. Agrofor Syst 4:3–15

Arnold JEM, Dewees PA (1997) Farms, trees and farmers: responses to agricultural intensification. Earthscan, London, p 292

Costanza R (1997) The value of world's ecosystem services and natural capital. Nature 387:253–260

Dagar JC (2000) Agroforestry systems for coastal and island regions. Indian J Agrofor 2:59–74

Dagar JC, Kumar V (1992) Agro-forestry for Bay Islands. Indian Forester 18(6):411–415

Dixon RK (1995) Agroforestry systems: sources or sinks of greenhouse gases? Agrofor Syst 31(2):99–116

Duryiyaprapan S, Boonklinkajorn P (1977) Fertilizer experiment on guinea grass under coconuts. J Agric Sci 10:201–208

El-Lakany H (2004) Improvement of rural livelihoods: the role of agroforestry, first world agroforestry congress, 27 June-2 July 2004, Orlando
FAO (1976) A framework for land evaluation. FAO soils bulletin. FAO, Rome, p 32
Fernandes ECM, Nair PKR (1986) An evaluation of the structure and function of tropical homegardens. Agric Syst 21:279–310
Fernandes ECM, O'kting'ati A, Maghembe J (1984) The Chagga home gardens: a multistoried agroforestry cropping system on Mt. Kilimanjaro (N. Tanzania). Agrofor Syst 2:73–86
Gillespie AR, Miller BK, Johnson KD (1995) Effects of ground cover on tree survival and growth in filter strips of the corn belt region of the Midwestern U.S. Agric Ecosyst Environ 53:263–270
Guillerme S, Kumar BM, Menon A, Hinnewinkel C, Maire E, Santhoshkumar AV (2011) Impacts of public policies and farmer preferences on agroforestry practices in Kerala India. Environ Manag 48(2):351–364
Ismail T (1986) Integration of animals in rubber plantations. Agrofor Syst 4:55–66
Jaisankar I, Velmurugan A, Swarnam TP, Gangaiah B, Subramani T, Dam Roy S (2015) Multipurpose fodder trees in integrated farming system: scope for increasing livestock and agricultural production in Bay Islands, Technical bulletin, CARI/NRM TB-03, CIARI, Port Blair, India
Jose S (2009) Agroforestry for ecosystem services and environmental benefits: an overview. Agrofor Syst 76:1–10
Kang BT, Wilson GF (1987) The development of alley cropping as a promising agroforestry technology. In: Steppler HA, Nair PKR (eds) Agroforestry: a decade of development. ICRAF, Nairobi, pp 227–243
Kumar BM, Nair PKR (2004) Tropical home gardens: a time tested example of sustainable agroforestry. In: Advances in agroforestry, vol 3. Springer, Dordrecht, pp 355–370
Kushwah BL, Nelliat EV, Markose VT, Sunny AF (1973) Rooting pattern of coconut (*Cocos nucifera* L.). Indian J Agron 18:71–74
Leakey RRB, Sanchez PA (1997) How many people use agroforestry products? Agrofor Today 9(3):4–5
Leakey RRB (1998) Agroforestry for biodiversity in farming systems. In: Collins W, Qualset C (eds) The importance of biodiversity in Agroecosystems. Lewis Publishers, New York, pp 127–145
Lundgren BO, Raintree JB (1982) Sustained agroforestry. In: Nestel B (ed) Agricultural research for development: potentials and challenges in Asia. ISNAR, The Hague, pp 37–49
MacDicken KG, Vergara NT (1990) Agroforestry: classification and management. Wiley, New York
Manner HI (1981) Ecological succession in new and old swiddens of montane Papua New Guinea. Hum Ecol 9(3):359–377
Mc Neely J, Scherr SJ (2001) Common grounds common future: how ecoagriculture can help feed the world and save wild biodiversity. IUCN, Gand/Future Harvest, Washington, DC
MEA (2005) Millennium ecosystem assessment, ecosystems and human well-being: biodiversity synthesis. World Resources Institute, Washington, DC
Nair PKR (1989) Agroforestry systems in the tropics. Kluwer Academic Publishers, Dordrecht
Nair PKR (1993) An introduction to agroforestry. Kluwer Academic Publishers, Dordrecht, p 491
Plucknett DL (1979) Managing pastures and cattle under coconuts. Westview Press, Boulder
Rajaguru ASB (1991) Integration of livestock under tree crops in Sri Lanka. In: Proceedings of the FAO/MARDI international livestock-tree cropping workshop, 5–9 December 1988, Serdang, Malaysia, pp 101–111
Rocheleau D, Weber F, Field-Juma A (1988) Agroforestry in dryland Africa. ICRAF, Nairobi
Roskoski JP (1982) Nitrogen fixation in a Mexican coffee plantation. In: Robertson GP, Herrera R, Roswall T (eds) Nitrogen cycling in ecosystems of Latin America and the Caribbean. Martinus Nijhoff, The Hague, pp 283–292
Samarakoon SP, Wilson JR, Shelton HM (1990) Growth, morphology and nutritive quality of shaded *Stenotaphrum secundatum, Axonopus compressus* and *Pennisetum clandestinum*. J Agric Sci Camb 114:161–169

Santhirasegaram K, Ferdinandez DEF, Goonasekera GCM (1969) Fodder grass cultivation under coconut. Ceylon Cocon Planters Rev 5(4):160–166

Schultz RC, Colletti JP, Isenhart TM, Marquez CO, Simp-kins WW, Ball CJ (2000) Riparian forest buffer practices. In: Garrett HE, Rietveld WJ, Fisher RF, American Society of Agronomy. Madison, WI, USA (eds) North American agroforestry: an integrated science and practice. University Press, Cambridge, pp 189–281

Smith MA, MacFarlane DC, Whiteman PC, Shelton HM (1981) Technical report on regional trials, pasture species evaluation and soil fertility assessment. Aruliho land system – Guadalcanal; Ri Ri – Makira, Dala – Malaita, Alu – Shortland Island. Solomon Islands Pasture Research Project, University of Queensland, Mimeo report

Steel RJH, Smith MA, Whiteman PC (1980) A pasture handbook for the Solomon Islands. University of Queensland/ADAB, Ministry of Agriculture and Lands, Honiara, Solomon Islands

Webster CC, Wilson PN (1980) Agriculture in the tropics. Longman, London

Wiersum KF (1982) Tree gardening and taungya in Java: examples of agroforestry techniques in the humid tropics. Agrofor Syst 1:53–70

Land and Water Conservation: Dealing with Agriculture and Aquaculture Conflicts

17

Abstract

The tropical coastal environment has undergone rapid changes in recent times with the introduction of aquaculture to meet the rising demand for fish products and to supply nutritional requirement of growing population. The major concern in these changes is the way that natural habitats have been extensively cleared and converted to shrimp farming and agriculture and in recent times agriculture into aquaculture. This is driven by high demand for aquaculture products in global markets and, together with technological advances, has fuelled rapid expansion. These rapid and generally unplanned changes in the coastal areas have provoked conflicts between agriculture and shrimp farming and further resulted in degradation of land and water. This necessitates integrated management of coastal zone and proper regulations to deal with the conflicts and land use change. This chapter deals with the trend in land use change and its associated problems and approaches to manage agriculture-aquaculture conflicts aimed at combined production and resource conservation in the tropical coastal areas.

Keywords

Alternative aquaculture · Integrated fish farming · Coastal regulation · ICZM

17.1 Introduction

Coastal areas are home to more than one fourth of the world's population and support much of the world's food production besides industrial, transportation, and recreation needs. It encompasses very wide and important coastal ecosystems which deliver several vital ecosystem services. In spite of this, during the last century, the coastal environment has undergone rapid changes due to pressure from both anthropogenic activities and natural hazards. The scale of the stresses imposed on coastal environment poses a significant threat to the resilience of both natural and

V. Ayyam et al., *Coastal Ecosystems of the Tropics - Adaptive Management*,
https://doi.org/10.1007/978-981-13-8926-9_17

human systems. This buildup environmental pressure happens through land use change, intensification of resource exploitation, urbanization, industrial development, tourism, and recreational demand. Consequently the interlinked casual factor-induced changes affect the state of the coastal environment which include altered nutrient, sediment, and water fluxes; degradation of habitats and loss of biodiversity; and pollution of soils, groundwater, and surface water.

Among these changes the negative effect on land and water resources is the key issue as it affects the livelihood and future survival of coastal communities and ecosystem function and services which are centred on these resources (Velmurugan et al. 2015). In other words land and water are very vital resources required for agricultural production which in turn has to feed the increasing population. As the demand grows for agricultural land use, lands from other natural use are being diverted. At the same time, high-profit and protein-rich shrimp aquaculture has rapidly grown and occupied significant proportion of coastal land. Change in the state of the coastal environment may be multifaceted, but the major concern is the way natural habitats particularly mangrove forests and salt marshes have been extensively cleared and converted to agriculture, shrimp farming, and other uses.

To worsen the situation, market demand-driven expansion of shrimp farming onto agricultural lands has created imbalance in resource use. This is the second key issue for the sustainable coastal zone management and resource conservation. But it is also important to see both in the context of the range of conflicting demands of the different stakeholders who live within and depend upon the resource base in this environment (FAO 1997). Further, the competition for space has resulted in negative interactions with traditional and new coastal users (Dempster and Sanchez-Jerez 2008) and negative effects of other activities on aquaculture like discharge of agriculture and sewage (Díaz et al. 2012). With this background this chapter attempts to provide some insight into the functioning of land and water resources in the coastal region and land use change due to aquaculture and agricultural activities, conflicting interest, and its effect on land and water resources.

17.2 The State of Land and Water Resources

Land degradation is an important global issue because of its adverse impact on agricultural productivity, the environment, and food security. It has been estimated that 23% of all usable land (excluding mountains and deserts) has been affected by degradation to a degree sufficient to reduce its productivity (UNEP 1994). The main types of soil degradation are water erosion (56%), wind erosion (28%), chemical degradation (12%), and physical degradation (4%) (Oldeman et al. 1992). In the coastal region, salinization, accumulation of toxic substances, and erosion are the major process of land degradation (Fig. 17.1). Drainage congestion and seawater intrusion into the low-lying areas create waterlogging and salinization. Some places due to exploitation of groundwater resulted in the intrusion of sea water into the coastal aquifers.

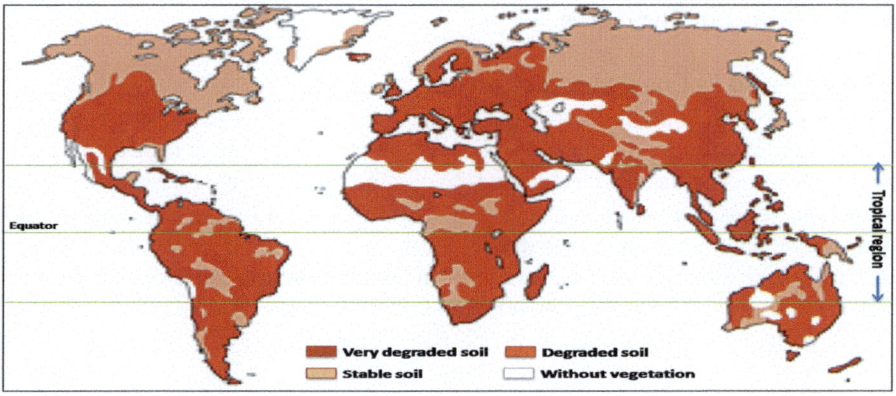

Fig. 17.1 Status of land degradation. (Modified from UNEP 1992)

Depending on their inherent characteristics and the climate, lands vary from highly resistant, or stable, to those that are vulnerable and extremely sensitive to degradation. However, human actions accelerate the natural process which also reduces resilience and increases the sensitivity. Fragility, extreme sensitivity to degradation processes, may refer to the whole land, a degradation process (e.g. erosion), or a property (e.g. soil structure). Stable or resistant lands do not necessarily resist change; rather they are in a stable steady-state condition with the environment. Under stress, fragile lands degrade to a new steady state, and the altered state is unfavourable to plant growth and less capable of performing environmental regulatory functions (Lal et al. 1997). In the tropical coastal areas, most of the soils are degraded due to various factors discussed above, while erosion is the major degradation process in the inlands.

As the agricultural production largely depends on the availability of land and water, degradation of these vital resources will adversely affect the food production system. In coastal regions due to excessive withdrawal of groundwater, high evapotranspiration, rise of saline groundwater, and seawater intrusion pose a major challenge not only to agriculture but other dependant sector as well. Further, degradation may lead to shift in native and exotic flora and fauna in the coastal region. Therefore, conservation of these precious resources is vital for the coastal region.

On the other hand, the complex interactions occurring among different users of coastal areas often leave little space for aquaculture. Additionally marine aquaculture requires coastal waters with specific environmental and water quality characteristics. Therefore, maintaining the water quality is essential to support the fisheries and other marine life in the nearshore region. The result is that suitable coastal areas for aquaculture are severely limited in many areas, and this has become a major barrier to expansion.

17.3 Trend in Land Use

17.3.1 Trend in Coastal Aquaculture

Aquaculture is a highly profitable venture for the coastal region all over the tropical world where good market demand and favourable climate are experienced. At the same time, it has increased the availability of nutritive food for the growing population. The advent of aquaculture is mainly attributed to depletion or standstill of capture fishery, availability of vast stretches of brackish water lands due to various reasons (Kagoo and Rajalakshmi 2002), and increasing market demand for selected fishery products. Owing to this push and pull factors, aquaculture has grown enormously prompting purchase of agricultural and fallow lands by entrepreneurs for setting up small- and large-scale aqua farms in many of the tropical coastal regions.

Though extensive shrimp farming has been practiced for a very long time in some of the tropical countries as part of the traditional livelihood system, recent strong demand in global markets together with technological advances has fuelled rapid expansion and intensification. It has seen annual growth rate of nearly 17% during 1970 and 2000 which was noticeably higher than that of other food production sectors (FAO 2003). Growth of aquaculture is seen as a success story by some researchers (Fegan 1999) having developed from a cottage industry based upon a backyard production system to a global industry in little more than 30 years. Conversely for some others (EJF 2003), this rapid growth is associated with serious negative environmental and social impacts.

In 2000, brackish water aquaculture particularly shrimp farming comprised 4.6% by weight of total aquaculture production but 15.7% by farm gate value. This clearly shows the economic value and market importance of shrimp farming. Moreover shrimp farming has emerged as a major source of foreign earnings and an important source of income and employment. Estimates for the main shrimp-producing countries put the total employment generated by shrimp farming at around two million people. The top shrimp-producing tropical countries and the production are given in Fig. 17.2. Of the major countries, South and Southeast Asia alone accounted for 90% of world production. Thailand and Indonesia alone produced nearly 450,000 mt annually accounting for nearly one third of total production. Proportionately these countries also witnessed significant expansion of aquaculture area in the coastal region. Based on the level of technology adopted, stocking density, and yield, shrimp farming methods are classified into extensive, semi-intensive, and intensive. The level of intensification and inputs largely decides the production environment and impact on the surroundings.

17.3.2 Rice Paddies

The coastal low-lying areas of the tropical regions suffer from the twin problem of waterlogging and moisture stress at different periods in the same year. High rainfall together with seawater intrusion is responsible for waterlogging. But the effect gets

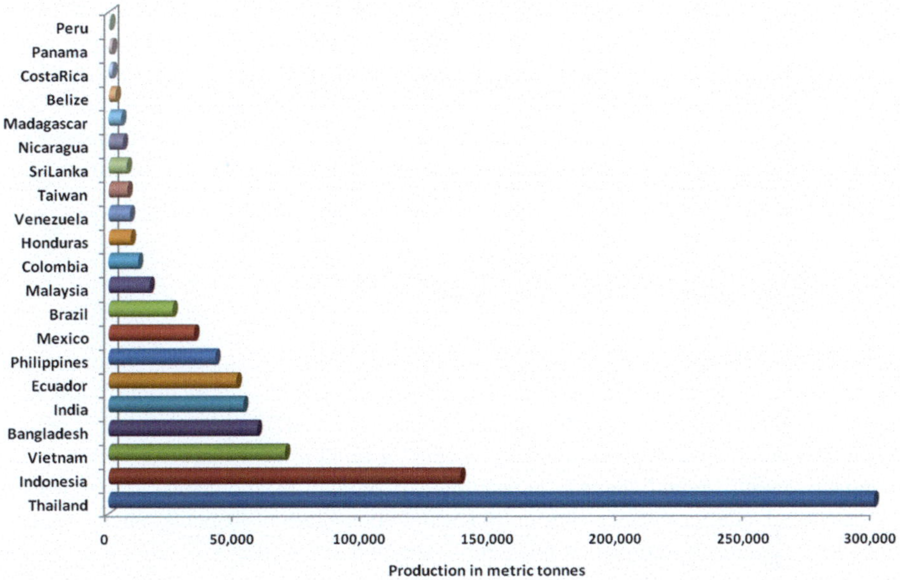

Fig. 17.2 Shrimp production in tropical countries. (Data source: FAO 2003)

diluted due to heavy rainfall during monsoon season (Velmurugan et al. 2015). However, increase in salinity during summer months restricts crop production, and crop production cannot be carried out without some irrigation. Similar situation is experienced in deltaic and estuarine environments because of seasonally varying freshwater inputs; however, the situation may get aggravated by upstream river basin management. Agricultural lands in the brackish water zone generally have lower productivity than those in the freshwater zone. As demand for shrimp increased, many farmers found that shrimp farming of such saline land has comparative advantage over crop production. For this reason many of the farmers from Southeast Asian nations have converted their rice fields into shrimp ponds. But it is difficult to estimate the extent to which brackish water shrimp farming has encroached onto agricultural land due to wide variations in methods and overlapping of enterprises adopted by them among other factors. Despite difficulties of having reliable estimate, the available data from a few countries illustrate the significance of the trend.

A large number of rice farmers in the Southeast Asian countries and coastal areas of India have converted paddy fields into shrimp ponds beginning from the 1990s (Szuster et al. 2003). In Thailand surveys suggested that shrimp farms operating within freshwater areas could have accounted for as much as 40% of Thailand's total farmed shrimp output (Limsuwan and Chanratchakool 1998). In Bangladesh, because of commercial interests of shrimp, many coastal polders constructed since the1960s to protect agricultural land from inundation of saltwater were turned into large shrimp culture *ghers* (ponds) during the 1990s. Shrimp area expanded from 51,812 ha in 1983 to 141,353 ha in 2002 (DoF 2003). In the coastal zone of the

Mekong River Delta of Vietnam, rice area decreased from 970,000 ha in 2000 to 800,000 ha in 2002, whereas corresponding increase in shrimp farming was recorded. In conclusion the available national level data clearly showed that the tropical coastal areas witnessed increase in paddy cultivation in the 1960s and 1970s by diversion of mangrove areas, while the same regions have witnessed change of paddy field into aquaculture pond due to loss of profitability of rice cultivation and to some extent soil salinization.

17.3.3 Diversion of Mangrove Area

In the intertidal areas along tropical and subtropical coasts, especially where large river systems deposit alluvial sediment and salinity is moderated by high freshwater discharge, luxuriant mangrove forests are found. Under such saline and waterlogged environment, plants require certain special adaptation to survive. Therefore, mangroves with special adaptation features surrounded by mangrove associates in the landward side represent a dominant natural ecosystem of tropical coastal zones. The global extent of mangroves has been estimated at 181,077 km^2 (Spalding et al. 1997), but this is under reported compared with their undisturbed extent. Historically, mangrove ecosystem has been exploited for forest products or converted to various alternative forms of land use such as rice field, salt pans, aquaculture ponds, recreation, and industrial development. In some of Southeast Asian countries, mangrove area has decreased by more than 50%.

Many researchers have attributed much of the global loss of mangroves to the expansion of shrimp farming, but the evidence for this assertion is not always strong (Janssen and Padilla 1999; Nickerson 1999; EJF 2003). For example, mangrove loss in Thailand over the last two decades has been exacerbated as expansion of shrimp farming. In Chakaria, Bangladesh, shrimp areas expanded from 10,000 to75,000 ha between 1967 and 1988, at the expense of decrease in mangrove area from 70,000 to 7000 ha. Southeast Asian countries show evidence that before the expansion of shrimp farming, large tracts of mangrove had already been converted to rice farming. In Africa, the extent of mangrove destruction is comparable with that in Asia even though coastal aquaculture in general, and shrimp farming in particular, is not widespread. Within Southern Asia, mangroves in India and Bangladesh have been heavily exploited for timber, fuelwood, and other forest products for centuries, and population pressure has led to serious forest degradation.

A comprehensive survey of 5000 shrimp farms in Asia (ADB/NACA 1997) showed that less than 20% of the area occupied by intensive and semi-intensive farms was former mangrove. Because of unavailability or unreliability of data on land use, comprehensive land conversion matrix could not be worked out which increased the uncertainty (Walters 2003). Nevertheless, conflicts do exist between shrimp farmers and mangrove resource users.

17.4 Conflict Assessment

In general tropical coastal zones support three distinct types of resource-dependent livelihood activity, viz. agriculture, aquaculture (shrimp farming), and fishing. Several studies on coastal land use change and resource assessment concluded with cautionary note about environmental and social problems arising from the impacts of land use change (Table 17.1). In the coastal areas, land use change has emerged as a key issue responsible for conflict of interest. Conversion of rice field and other freshwater bodies into aquaculture unit to culture shrimp results in salinization of the entire field, while it is not the case where shrimp farms are located in the brackish water zone, where salinity intrusion is a natural seasonal phenomenon and crop production is relatively nonremunerative as in the Mekong Delta of Vietnam and

Table 17.1 Impact of expanding coastal aquaculture and A-A conflicts

Sl. No.	Cause	Activity/problems	Impact on resources/ES function
1	Aquaculture	Destruction of mangrove, wetlands, and other sensitive aquatic habitats	Land degradation, loss of biodiversity, loss of coastal protection
2	Agriculture	Intensive irrigated agriculture, use pesticides and fertilizers in all season	Algal bloom in the natural wetland and marine ecosystems in shallow water are severely affected
3	A-A conflict	Release of aquaculture pond effluents, excessive use of bioactive chemicals, no proper planning/space for waste treatment or disposal	Water pollution, marine ecosystems in shallow water affected
4	Aquaculture	Drainage and seepage from intensive aquaculture ponds	Land and water degradation in the vicinity, groundwater pollution
5	Aquaculture	Spread of aquatic animal diseases	Native and wild population gets affected
6	A-A conflict of interest	Acidification arising from development of acid sulphate soils, marginal land, conversion of paddy field	Land and water degradation due to mixing of toxic compounds, natural wetland gets affected by acidity
7	Aquaculture	Negative effects on biodiversity caused by exploitation of wild shrimp larvae/brood-stock and destruction of habitat	Loss of natural habitat and biodiversity
8	A-A conflict of interest	Negative impact on vegetation cover and terrestrial livestock	Loss of biodiversity, sustainability, land degradation
9	Aquaculture and A-A conflict	Loss of subsistence fishery and indebtedness or coercion for aquaculture/agriculture	Social conflict and livelihood loss
10	A-A conflict	Restricted access to common property resources and increased inequity and social unrest	Environmental protection and public use are affected, public health problem, social harmony

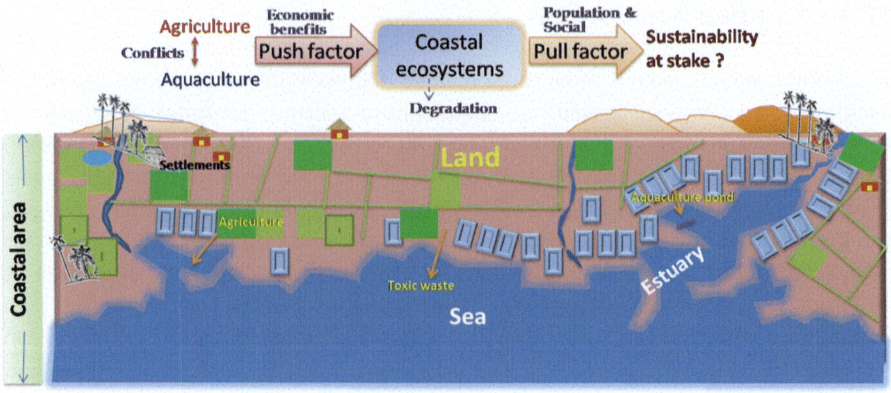

Fig. 17.3 Schematic representation of aquaculture-agriculture conflicts in the coastal areas

east coast of southern part of India. But the adoption of shrimp-rice production systems in the brackish water zone may also led to the encroachment of shrimp ponds onto homestead lands due to the economic benefit. A schematic representation of this conflicts and pollution in the coastal environment is given in Fig. 17.3. In Bangladesh before 1975, most of the shrimp farms (>80%) were located further than 500 m from homestead areas, but, in 1999, 46% of the shrimp ponds were found within 10 m of homestead land. Consequently multipurpose trees, traditional fruits, and many plant species have gradually decreased because of salinity and the shrinking of homestead areas. In addition this led to decline in grazing land and its vegetation cover.

Another major source of conflict is the discharge of high loads of nutrients and suspended solids which transfers the adverse effects onto the receiving waters, including the stimulation of algal blooms and the creation of anoxic conditions (Naylor et al. 1998). Burford et al. (2003) linked ecological processes in intensive shrimp ponds with impacts downstream. In contrast extensive and low-level semi-intensive shrimp farms do not cause appreciable chemical discharge pollution which provide scope for properly managed shrimp production with much less effect on the environment.

Incidentally at many places, aquaculture and agriculture developmental activities have taken up acid sulphate soils (ASS) which are associated with salt marshes, mangrove forests, and other estuarine wetlands. On exposure this pyrite (FeS_2)-rich soil gets oxidized into sulphate, which reacts with water forming sulphuric acid. As a result the soil pH becomes very low, and the concentrations of aluminium and iron increase. This condition is highly toxic to shrimps, agricultural crops, and the surrounding aquatic environment (Minh et al. 1998). Further these toxic substances move further into tidal creeks and estuarine through waters where mass mortalities of fish have been recorded with chronic long-term effects on coastal habitats (Sammut et al. 1996; Wilson et al. 1999).

At the same time, it is observed that the socio-economic impacts associated with change in resource-dependent livelihoods have often been underestimated or oversimplified. This can perhaps be explained in part by failures to recognize off-site effects, which occur when the boundaries of the system under consideration are not properly defined (Phillips et al. 2001). In many South Asian nations, salinization of land and shortages of fodder and fuelwood were the main identified problems for farming communities. Together with land use change, loss of employment and livelihood creates unrest and social conflicts (Patil and Krishnan 1997).

The foregoing analysis showed that aquaculture expansion and the A-A conflict resulted in the destruction of mangrove, wetlands, and other sensitive aquatic habitats, pollution due to release of wastewater from aquaculture pond, contamination of groundwater, spread of aquatic animal diseases, environmental degradation due to aggressive land conversion and development, social unrest, and loss of common property services as well. Though the focus here is on the negative impacts of land use change, it is wise to identify possibilities for co-existence of aquaculture and agriculture and creates win-win scenarios for future land and water resource use.

Bailey and Pomeroy (1996) argue that the complexity and high natural productivity of the environment provide many niches for accommodating these different activities. Sustainability to the aquaculture-agriculture production with conservation of resources can be brought in with eco-friendly technologies and co-existence rather than specialization. Lewis et al. (2003) similarly argue that the promotion of coastal aquaculture in an environmentally responsible manner requires 'adopting the principles of co-existence of mangroves and aquaculture'. They note that 'implying co-existence is possible and documenting its actual occurrence are two different things'.

17.5 Strategies for Conflict Resolution

17.5.1 Regulation

Aquacultural activities should be controlled or regulated to avoid environmental problems and conflicts with agriculture. The following measures are taken to regulate the activities.

- Specific effluent standards—In order to address the issues related to coastal aquaculture particularly shrimp farming, governments of tropical coastal nations have responded mainly with specific regulations such as effluent limits, design standards, best management practices, and codes of conduct. For example, Belize, Brazil, Ecuador, India, Mexico, Thailand, and Venezuela have made elaborate aquaculture effluent regulations, which are designed to prevent effluents from causing negative consequences on receiving waters. Though it is a welcome step, these farm-level measures have often been ineffective and depend on the local government's will to comply and enforce. Some non-governmental organizations have also proposed effluent standards for aquaculture.

Shrimp farmers should be motivated to follow the effluent standards so as to avoid direct conflicts with agriculture and degradation of water resources.

- Codes of conduct for aquaculture—Alternatively large number of governmental fishery agencies, international development organizations, producer associations, non-governmental organizations, and other stakeholders have formulated codes of conduct for aquaculture (Boyd et al. 2002). In its basic form, code of conduct is nothing but a set of guiding principles consisting of broad statements about how management and other operational activities should be conducted. Most aquaculture codes reference the Code of Conduct for Responsible Fisheries formulated by the Food and Agriculture Organization (FAO 1997) which has global applications. But most of the codes do not have any legal authority, and adoption is usually voluntary or depends on the individual countries to enact legislation.
- Best management practices—Most of the aquacultural ponds are operated by small farmhold or business houses which are scattered in the coastal areas; therefore, it is difficult to effectively regulate aquaculture effluents by applying traditional water quality standards. Collection and disposal are also difficult in several island countries. As an alternative to this problem, best management practices (BMPs) can be used as a tool. A BMP is the best available and practical means of preventing a particular environmental impact while still allowing production to be economically efficient. The best inducement is when adoption of BMPs clearly increases profit and also reduces pollution level. For example, if the BMPs recommend lower stocking rates and use better feed management, it will result in better water quality, less stress, faster growth, better feed conversion ratios, and less waste produced. This scenario will also increase efficiency and profits.
- Freshwater storage in the pond—Storage of rainfall in ponds will help to avoid overflow. Reduced or less overflow in turn reduces amount of water to be pumped into ponds to maintain water levels. A reduction in pump operation will also reduce costs and increase profits.
- Use of biofilter—The pond effluents can be discharge into a mangrove wetland which is used as a biofilter to remove suspended solids, lower BOD, and absorb nutrients in order to limit the risk of eutrophication of the adjacent waters (Robertson and Phillips 1995; Rivera-Monroy et al. 1999). This is another way of integrating shrimp ponds and mangrove areas. Corredor and Morell (1994) reported their effectiveness in removing nutrients from effluents. However, any general recommendation about an appropriate ratio of mangrove-shrimp pond area can be attempted only if the nutrient assimilation capacity of different kinds of sediments and plants is known.

17.5.2 Spatial Allocation

When there is conflict for space, and congestion of activities with contrasting interest, it is advisable to go for spatial allocation of different/contrasting activities providing operational space. In this context, 'zoning' implies bringing together the criteria for locating aquaculture, agriculture, and other activities in order to define broad zones suitable for different activities or mixes of activities. This largely excludes the conflicts and reduces the social disharmony in the coastal areas. Modern tools like GIS are well suited to facilitate this kind of task more accurately. Zoning may be used either as a source of information for potential developers to identify a particular location or as a planning and regulating tool, in which different zones are identified and characterized as meeting certain objectives.

Zoning of land (and water) for certain types of aquaculture development may help to control environmental deterioration at the farm level, reduce adverse social and environmental interactions, and serve as a focus for estimates of environmental capacity and a framework for providing or improving infrastructure to small-scale farmers. The strength of zoning of coastal areas for different activities lies in its simplicity, clarity, and potential in terms of streamlining procedures. What is allowed and what is not allowed are clear, and developers can go ahead with the plan. Monitoring of the whole coastal zone is desirable and possible rather than individual farms located in that zone.

17.5.3 Integrated Coastal Zone Management

In most of the tropical coastal countries, due to the complexity of ecosystems and interlinks of various problems, it is becoming more difficult to manage any one selected natural resource or promote one economic sector in isolation, in opposition to comprehensive, integrated framework for policy planning and coastal management. Therefore, a special and distinct management methodology is required, such as integrated coastal management programme. In common parlance integrated coastal area management refers to a special type of programme for coastal countries developed for the purpose of conserving coastal resources or environments (Sorensen and McCreary 1990) through proper management of developmental activities. The overall goal of ICAM programme is to ensure sustainable use of coastal natural resources and conservation of coastal ecosystems.

Based on a review of experience, GESAMP (2001) concluded that comprehensive ICZM may be effective as a starting point where coastal aquaculture is in the early stages of development, institutions for resource management are flexible or undeveloped, and appropriate legal and institutional frameworks are in place or can be developed rapidly. However it is often experienced that the available scientific and technical capacity in many developing countries is a constraint, and hence, there is a need to develop appropriate planning tools.

In spite of best efforts in planning, often it is undermined by the strength of economic and political interests. Evidence from Thailand (GESAMP 2001) suggests that more locally focused initiatives (e.g. relating to an estuary or lagoon system) may offer the most practical starting point, since they retain the benefits of integration but at a smaller scale. In the same context, the analysis of different situations and requirements of coastal countries necessitate that participatory coastal area development provides the best option for development. At the same time, involvement of local government and community accountability in coastal management will help to make these efforts more effective and sustainable.

17.5.4 Alternative Aquaculture

Aquaculture, including marine aquaculture (mariculture), plays a major role in meeting the rising demand for fish products and supplying healthy and nutritious protein to a growing global population(Larsen and Roney 2013). However, available space for new aquaculture development in coastal zones is now becoming increasingly limited, and it is facing conflict of interest with agriculture. Hence, it necessitates looking for new opportunities for aquaculture production within the limitation imposed by nature and available space. Some of the most important alternative aquaculture (other than coastal aquaculture pond) opportunities are discussed below:

17.5.4.1 Wetland Fisheries

Wetlands are dynamic ecosystems and are among the most precious natural resources on Earth. Broadly coastal wetlands include sand or pebble shores, estuarine lakes and lagoons, coastal floodplain, and lagoons. They are important nursery and feeding areas for fishes, crustaceans, and other aquatic life. Wetlands situated in the coastal areas (in the dry zone) towards landwards are normally marshy and excludes highly saline environment due to the inflow of water from the river and stream and base flow from inland. Low gradients, floods, and tectonic activities contribute to the genesis of this wetland.

The dynamic ecological character caused by the cyclic changes in the hydrology and morphometry, water chemistry, and sediment characteristics leads to growth of unique floral and faunal associations. These biological communities are adapted to aquatic and its surrounding environments. They play a vital role in the recruitment of fish population in the wetland ecosystems and provide breeding, nursery, and feeding grounds for several fish species. Due to high competition for land and geographical limitation in many tropical islands, these wetlands can be developed for sustainable fisheries production within the ecological boundaries. The species suitable for each region and wetland should be identified for culturing after due consideration for its ecosystem impact. Additionally rational management, conservation, and careful monitoring are needed to sustain this vital resource.

17.5.4.2 Estuarine Fisheries

Estuaries are the transitional zones between the rivers and the sea. It has specific ecological properties and biological composition. In the estuary the sea water is diluted with freshwater flow from the land drainage. The salinity of the estuarine waters varies between 0.5% and 35% and extremely variable, while marine and fresh water have distinctive stable salinities. Based on physicochemical characteristics, estuaries can be subdivided into three regions, tidal zone, mixing zone, and nearshore turbid zone. The estuarine environment is highly variable in terms of size, depth, and physical and chemical features. The tropical estuarine environment can be divided into four broad categories, viz. open estuary, estuarine coastal waters, blind estuary, and coastal lake.

These estuaries offer immense biological wealth characterized by the diversified rich flora and fauna including fisheries. This is because of its abundant availability of primary productivity, phytoplankton, phytobenthos, and green rotted plants. Thus there is huge scope of development of commercially important fisheries in this ecosystem. This is expected to provide livelihood to the coastal communities. However, this ecosystem faces severe problems due to reduced river water discharge, formation of sand bars, deposition of silt, barrier of fish migration, overfishing, and pollution. The estuarine environment should be judiciously exploited so that natural resources of fish and prawn and their recruitment levels are not damaged or destroyed.

17.5.4.3 Integrated Fish Farming

Integration of fish with agriculture in various combinations offers one the best solutions to avoid aquaculture-agriculture conflict and simultaneously contributing to the food basket and nutritional security. The integration of fish farming with agriculture and animal husbandry is considered as sustainable farming system, which offers greater efficiency in resource utilization, reduces risk by diversifying crop, and provides additional income and food for small-scale farming household. This system involves recycling of bio-resources or by-products and interconnected nutrient flows of one system as input for other, by maximizing the production from the same unit areas. The system is very much suitable for the tropical coastal countries experiencing population explosion and high demand for diversified food.

In traditional practices of integrated fish farming polyculture fingerlings, stocking of local and exotic carp has been practised in many tropical countries. By considering the compatibility for habitat preference and available natural fish food organisms, promotion of multi-species concept and diversified cultivable practices with medium and minor carps is identified as important regional species introduction to polyculture system. The carp species can also be cultured under mixed system of freshwater prawn (*Macrobrachium rosenbergii*, *M. malcolmsonii*) or catfish (*Clarias batrachus*, *Heteropneustes fossilis*) and murrels (*Channa marulius*, *C. striatus*) with proper composition and ratio for optimum utilization of the resources.

In general fishes can be integrated into different agricultural production components, viz. crops, livestock, or combination, aimed at increasing farm productivity and maximizing synergies between these components. Most profitable and sustainable combinations are paddy-fish system, horticulture-fish system, aquatic weed-fish, cattle-fish, pig-fish, goat-fish, poultry-fish, and duck-fish. In multi-utilization system in the pond area, more than two farming systems are planned to utilize the potential of pond water, available area, etc., resulting in extra income from the land area. Further, this combination of agri-aqua-animal husbandry at one place provides several opportunities for judicious recycling of nutrient linkages contributing to farming intensification, effective use of bio-resources, optimum production, agricultural diversification, and environmental sustainability.

All these outcomes significantly reduce the aquaculture-agriculture conflicts in the coastal region.

17.6 Conclusion

In the coastal areas the direct conflict between aquaculture and agriculture is intense. At the same time, the conflicting use along with other land uses gradually changes the overall coastal land scape and its use. Integrated management involving all stockholders is required to significantly reduce land use change from natural system to highly managed system which act as a source of environmental pollution. In several tropical coastal countries, restoration and conservation of land and water are necessitated by the negative impact of intensive shrimp farming and agriculture. Additionally, the coastal production system has to address the impact of climate change. The rational solution to these problems lies in avoiding conversion of natural habitat into aquaculture pond and agriculture to have consequential benefit from mangroves and other coastal wetlands. At the same time, conversion of agricultural land into semi-intensive aquaculture method and following best management practices in aquaculture production will considerably reduce the kind and amount of pollutants discharged into the environment. If fishermen and landless agricultural labourers were encouraged to practise aquaculture in alternative way, clashes and conflicts would be avoided.

Evidences suggested that sustainable food production through aquaculture can be achieved by carefully combining different types of aquatic and terrestrial cultivation systems in the coastal areas. An integrated fish farming with crop and livestock may help to curtail environmental hazards by using the wastes to increase production further and at the same time reduce the environmental impact. It is one of the best options to enhance production and productivity from existing agricultural land and water. Further, coastal water bodies other than sea should be targeted for intensive combined production. Strategy should be developed to trap the potential of large water bodies without affecting the ecosystem function and work out the harvestable potential and ways to improve its productivity.

References

ADB/NACA (1997) Aquaculture sustainability and the environment. Asian Development Bank and Network of Aquaculture Centres in Asia-Pacific, Bangkok

Bailey C, Pomeroy C (1996) Resource dependency and development options in coastal South-east Asia. Soc Nat Resour 9(1):191–199

Boyd CE, Hargreaves JA, Clay JW (2002) Codes of practice and conduct for marine shrimp aquaculture. Report prepared under the World Bank, NACA, WWF and FAO Consortium Program on Shrimp Farming and the Environment. p.31

Burford MA, Costanzo SD, Dennison WC, Jackson CJ, Jones AB, McKinnon AD, Preston NP, Trott LA (2003) A synthesis of dominant ecological processes in intensive shrimp ponds and adjacent coastal environments in NE Australia. Mar Pollut Bull 46:1456–1469

Corredor JE, Morell MJ (1994) Nitrate depuration of secondary sewage effluents in mangrove sediments. Estuaries 17:295–300

Dempster T, Sanchez-Jerez P (2008) Aquaculture and coastal space management in Europe: an ecological perspective. In: Holmert M, Black K, Duarte CM, Marbà N, Karakassis I (eds) Aquaculture in the ecosystem. Springer, Dordrecht, pp 87–116

Díaz R, Rabalais NN, Breitburg DL (2012) Agriculture's impact on aquaculture: hypoxia and eutrophication in marine waters. Organisation for Economic Co-operation and Development. www.oecd.org/tad/sustainableagriculture/49841630.pdf. Accessed 15 Feb 2018

DoF (2003) Fishery statistical yearbook of Bangladesh: 2002–2003. Department of Fisheries, Dhaka, 41pp

EJF (2003) Risky business: Vietnamese shrimp aquaculture – impacts and improvements. Environmental Justice Foundation, London

FAO (1997) Towards sustainable shrimp aquaculture development: implementing the FAO code of conduct for responsible fisheries. Food and Agriculture Organization of the United Nations, Rome

FAO (2003) Review of the state of world aquaculture. Food and Agriculture Organization, Rome, 95 pp

Fegan DF (1999) Research issues in sustainable coastal shrimp farming: a private sector view. In: Smith PT (ed) Towards sustainable shrimp culture in Thailand and the region, Proceedings of a workshop held at Hat Yai, Songkhla, Thailand. Australian Centre for International Agricultural Research, Canberra

GESAMP (2001) Planning and management for sustainable coastal aquaculture development. IMO/FAO/Unesco/WMO/WHO/IAEA/UN/UNEP Joint Group of Experts on the Scientific Aspects of Marine Pollution, Rome

Janssen R, Padilla JE (1999) Preservation or conversion? Valuation and evaluation of a mangrove forest in the Philippines. Environ Resour Econ 14(3):297–331

Kagoo EI, Rajalakshmi N (2002) Environmental and social conflicts of aquaculture. J Soc Econ Dev 4(1):13–26

Lal R, Blum WEH, Valentin C, Stewart BA (1997) Methods for assessment of land degradation. CRC, Boca Raton

Larsen J, Roney M (2013) Farmed fish production overtakes beef. Earth Policy Institute, Washington, DC. www.earthpolicy.org/plan_b_updates/2013/update114. Accessed on 15 Feb 2018

Lewis RR, Phillips MJ, Clough B, Macintosh DJ (2003) Thematic review on coastal wetland habitats and shrimp aquaculture. Report prepared under the World Bank, NACA, WWF and FAO Consortium Program on Shrimp Farming and the Environment. 81 pp.

Limsuwan CS, Chanratchakool P (1998) Closed recycle system for sustainable black tiger shrimp culture in freshwater areas. Proceedings of the Fifth Asian Fisheries Forum, 11–14 November 1998, Chiang Mai, Thailand

Minh LQ, Tuong TP, van Mensvoort ME, Bouma J (1998) Soil and water table management effects on aluminum dynamics in an acid sulphate soil in Vietnam. Agric Ecosyst Environ 68(3):255–262

Naylor RL, Goldburg RJ, Mooney H, Beveridge M, Clay J, Folke C, Kautsky N, Lubchenco J, Primavera J, Williams M (1998) Ecology: nature's subsidies to shrimp and salmon farming. Science 282(5390):883–884

Nickerson DJ (1999) Trade-offs of mangrove area development in the Philippines. Ecol Econ 28:279–298

Oldeman LR, Hakkeling RTA, Sombroek WG (1992) World map of the status of human-induced soil degradation, an explanatory note. ISRIC, Wageningen

Patil PG, Krishnan M (1997) The Kandaleru shrimp farming industry and its impact on the rural economy. Agric Econ Res Rev 10(2):293–308

Phillips MJ, Boyd C, Edwards P (2001) Systems approach to aquaculture management. Aquaculture in the third millennium. NACA/FAO, Bangkok

Rivera-Monroy VH, Torres L, Bahamon N, Newmark F, Twilley R (1999) The potential use of mangrove forests as nitrogen sinks of shrimp aquaculture pond effluents: the role of denitrification. J World Aquacult Soc 30(1):12–25

Robertson AI, Phillips MJ (1995) Mangroves as filters of shrimp pond effluents: predictions and biogeochemical research needs. Hydrobiologia 295:311–321

Sammut J, White I, Melville MD (1996) Acidification of an estuarine tributary in eastern Australia due to drainage of acid sulfate soils. Mar Freshwat Res 47(5):669–684

Sorensen JC, McCreary ST (1990) Institutional arrangements for management of coastal resources. Coastal Management Publication No. 1, (Rev. ed.) U.S.Nat. Pk. Serv./USAID Series. 194 p

Spalding M, Blasco F, Field C (1997) World mangrove atlas. The International Society for Mangrove Ecosystems, Okinawa

Szuster B, Molle F, Flaherty M, Srijantr T (2003) Socio-economic and environmental implications of inland shrimp farming in the Chao Phraya Delta. In: Molle F, Srijantr T (eds) Perspectives on social and agricultural change in the Chao Phraya Delta. White Lotus Press, Bangkok, pp 177–194

UNEP (1992) World atlas of desertification. United Nations Environment Programme, Edward Arnold, London

UNEP (1994) Land degradation in south Asia: its severity, causes and effects upon the People. INDP/UNEP/FAO. World Soil Resources Report 78, FAO, Rome, Italy

Velmurugan A, Swarnam TP, Lal R (2015) Effect of land shaping on soil properties and crop yield in tsunami inundated coastal soils of Southern Andaman Island. Agric Ecosyst Environ 206:1–9

Walters BB (2003) People and mangroves in the Philippines: fifty years of coastal environmental change. Environ Conserv 30(3):293–303

Wilson BP, White I, Melville MD (1999) Floodplain hydrology, acid discharge and change in water quality associated with a drained acid sulfate soil. Mar Freshwat Res 50(2):149–157

Conservation Agriculture for Rehabilitation of Agro-ecosystems

18

Abstract

Conservation agriculture (CA) is promoted as a solution for the problem of poor agro-ecosystem productivity, land degradation, and climate change impact in most sub-Saharan Africa, South Asia, Latin America, and many tropical island nations. It is characterized by continuous minimum mechanical soil disturbance, permanent organic soil cover, and diversified crop rotations in the case of annual crops or plant associations in the case of perennial crops. This chapter describes the benefits of CA, a suggested improvement over conventional tillage, where no-till, mulch, and rotations significantly improved soil properties, moisture retention, and other biotic factors. More significantly CA permits higher rates of carbon sequestration, organic matter accumulation, reduction in soil erosion, and its associated improvement in soil biological properties. Further reduction in greenhouse gas emission as a result of CA is also reviewed. Implementation of CA practices particularly in areas facing land degradation and depleting resources presents a win-win scenario due to improved crop yield and trends of increasing soil fertility and scope for climate change adaptation. This chapter presents evidences in favour of CA for many of the tropical countries to recuperate degraded soils and improve ecosystem services.

Keywords

Tropical coast · Degradation · Rehabilitation · No-till · Residue mulch · Soil cover

18.1 Introduction

The coastal regions are commonly attributed with lower geographic elevations and are often associated with higher population densities than that of the hinterland. The low-lying coastal land is subjected to inundation due to seawater intrusion and

V. Ayyam et al., *Coastal Ecosystems of the Tropics - Adaptive Management*,
https://doi.org/10.1007/978-981-13-8926-9_18

drainage congestion during heavy downpour. At the same time, the pressure on land for higher productivity is on the rise. In the humid and subhumid tropics particularly in the coastal region, the climate is relatively harsh in terms of humidity, rainfall, temperature, and violent storms. Due to high temperature and rainfall, the rate of weathering is fast, and the minerals released into the soil solution are quickly leached down by the runoff water. As a result the soils become deficient in major nutrients and get depleted over years leaving mostly iron and aluminium resulting in soil acidity in these areas.

The soils are also prone to erosion, while the rate of soil organic matter mineralization is usually high. Concerns about soil erosion affecting soil productivity have resulted in an emphasis on trying to stop negative effects on crop yields attributed to erosion and runoff. For example, in Brazil, the use of disc ploughs for growing maize, soybean, or sorghum has induced high levels of soil erosion in the first years of cultivation following forest clearing, contributing to the loss of up to 30–50% of the initial soil organic matter (SOM) (Resck et al. 2000). Under such practices, yield potentials declined sharply over time despite an increased use of chemical inputs and the introduction of crop rotations (Landers 2007).

In these conditions if the emphasis is shifted towards the soil as a habitat for roots and soil loss and runoff are recognized as consequences of prior damage to soil porosity, a different perception about the problem emerges which favours providing better conditions for the roots to grow and allow the infiltration of runoff. Key goals of improving and maintaining excellent soil conditions for and with roots as propounded by Food and Agriculture Organization (FAO 2012) include:

- Increasing the reliability of plant production in the face of unpredictable variations in the weather and other hazards of the environment
- Reducing production costs and raising net returns to producers
- Increasing the quality of the land and its resilience to extreme weather conditions

Conservation agriculture (CA) has the potential to bring about all these positive changes to soil in these regions. CA restores agro-ecosystem services by combating land degradation, increasing productivity and resilience, regulating potential greenhouse gas emissions, and reducing production costs. Originally, CA practices were promoted to combat soil degradation and erosion resulting from intensive agriculture based on tillage. In later years, CA helped to reduce production costs and raise productivity through increased yield and efficiency. In recent years CA has spread to South Asia and sub-Saharan Africa among small holder farmers, (Baudron et al. 2012) and almost 50% of the land area under CA are located in developing countries. The main focus of this chapter is to review the potentials and problems associated with the conservation agricultural practices in the tropical regions with the emphasis on rehabilitation of coastal agro-ecosystems.

18.2 The Concept

Though conservation agriculture was in practice in different names and methods, the term 'conservation agriculture' was adopted during the First World Congress on Conservation Agriculture, Madrid 2001, organized by FAO and the European Conservation Agriculture Federation (Saturnino and Landers 2002). Conservation agriculture (CA) is an approach developed to manage farmland for sustainable crop production while simultaneously conserving soil and water resources (Erenstein 2009). It aims to conserve, improve, and make more efficient use of natural resources through integrated management of available soil, water, and biological resources combined with external inputs. In the beginning, CA got evolved to reduce the effort and cost of farming in a way that protects and improves agricultural soils. It contributes to environmental conservation as well as enhanced and sustained agricultural production. It can also be referred to as resource-efficient or resource-effective agriculture.

CA is defined as a 'concept for resource-saving agricultural crop production that strives to achieve acceptable profits together with high and sustained production levels while concurrently conserving the environment' (FAO 2007). It is also viewed as 'the simultaneous application of minimal soil disturbance, permanent soil cover through a surface mulch of crop residues or living plants, and crop rotations' (FAO 2015). Conservation agriculture (CA) maintains a permanent or semi-permanent organic soil cover. This can be a growing crop or dead mulch. Its function is to protect the soil physically from sun, rain, and wind and to feed soil biota. The soil microorganisms and soil fauna take over the tillage function and soil nutrient balancing. Mechanical tillage disturbs this process. Therefore, zero or minimum tillage and direct seeding are important elements of CA. A varied crop rotation is also important to avoid disease and pest problems.

CA is based on three interlinked principles and other related practices (Fig. 18.1):

- Minimal soil disturbance or no-till, aimed to maintain the quality and productivity of the soil, which is the base on all CA-based farming practices and systems
- Continuous soil cover—with crops, cover crops, or a mulch of crop residues to reduce runoff, soil, and water erosion, improve infiltration, protect the soil from harsh climate extremes, and serve as a substrate for soil microorganisms and fauna
- Diversified crop rotation over time and space to strengthen the system resilience against biotic and abiotic stresses (FAO 2015)

There is a confusion regarding conservation tillage (CT) and conservation agriculture (CA) among the stakeholders. The conservation tillage is a set of practices that leave crop residues on the surface which increases water infiltration and reduces erosion. It is a practice used in conventional agriculture to reduce the effects of tillage on soil erosion. However, it still depends on tillage as the structure-forming element in the soil. Nevertheless, conservation tillage practices such as zero tillage or minimum tillage practices can be transition steps towards conservation

Fig. 18.1 Principles and agricultural practices in conservation agriculture

agriculture (Hobbs et al. 2008). In other words, CT uses some of the principles of CA but has more soil disturbance, and CA is a broader concept that requires compliance with the simultaneous application of all the three criteria, viz. diversified crop rotations and associations, permanent soil cover, and minimum mechanical soil disturbance continuously maintained over time. For better understanding, Hobbs et al. (2008) explain the differences between traditional tillage, conservation tillage, and conservation agriculture in Table 18.1.

18.3 Scope of CA

Persistent use of conventional farming practices based on extensive tillage, especially when combined with removal or in situ burning of crop residue, has magnified soil erosion losses, and the soil resource base has been steadily degraded. It has been estimated that human activity is responsible for the loss of topsoil which is 2.6 times the natural rate of soil degradation. Erosion has been estimated to cause USD $44 billion a year in damage to farmland, waterways, infrastructure, and health (Pimental et al. 1995). Similarly soil fertility degradation and emergence of soil compaction, soil salinity, low infiltration rate, buildup of pest and diseases, etc. have also reported from the intensively cultivated agricultural land throughout the tropical region. Another direct consequence of farmers' persistent use of conventional production practices is rapidly increasing production costs; the costs of inputs such as improved varieties and fertilizers continue to increase, and farmers make inefficient use of them. All these problems are being aggravated by the observed changes in climate, and the projection is expected to further worsen the declining farm productivity and instability in agro-ecosystem performance and services in the tropical coastal regions.

Table 18.1 A comparison of traditional tillage (TT), conservation tillage (CT), and conservation agriculture (CA)

Traditional tillage (TT)	Conservation tillage (CT)	Conservation agriculture (CA)
It disturbs the soil and leaves a bare soil surface	It reduces the soil disturbance and keeps the soil covered	CA envisages minimal soil disturbance and soil surface permanently covered
The wind and soil erosion is maximum	The wind and soil erosion is reduced significantly	The wind and soil erosion is the least of the three in CA
Soil health is poor due to compaction and frequent soil disturbance	Soil health is significantly improved under reduced tillage due to reduced compaction and better biological activity	The compaction can be a problem, but use of mulch and promotion of biological tillage help to reduce compaction. The soil biodiversity and biological properties are improved
The water infiltration is lowest after soil pores are clogged. There is a loss of soil organic matter due to its oxidization	The water infiltration is good and promotes soil organic matter buildup in the surface layers	The water infiltration is best, and soil organic buildup in the surface layers is better than CT
The weed control is complete and also causes more weed seeds to germinate	The reduced tillage also controls weeds and exposes other weed seeds for germination	The weeds are a problem especially in the early stages of adoption, but problems are reduced with time, and residues can help suppress weed growth
The energy in the form of diesel use is higher and also the cost of production	The diesel use is intermediate so as the cost of production	The diesel use is much reduced, and cost of production is lowest of the three

Source: FAO (2007) and Hobbs et al. (2008)

Conservation agriculture addresses these issues and brings other benefits to the soil and production system as a whole. These CA principles are applicable to a wide range of crop production systems from low-yielding, dry, rainfed conditions to high-yielding, irrigated conditions (FAO 2012). However, application of the principles of CA will vary from one situation to another. CA also contributes positively to environmental conservation and helps in climate change mitigation. The most prominent benefit of CA is checking land degradation and improving soil fertility status including physical condition, carbon sequestration, and biodiversity conservation. Figure 18.2 provides contrasting situations as to how land is being degraded under conventional agriculture and alternatively how CA benefits the land. In many environments, conservation agriculture can be considered the ultimate soil and crop management system. Conservation agriculture has been successfully implemented in both small-scale and large-scale (FAO 2000) farming, where it has given economic benefits as well as improved soil and water resources. It holds considerable promise for small holder farmers particularly in sub-Saharan Africa and South Asia to control soil erosion, reverse land degradation, give more stable yields, and reduce labour and fuel needs (Baudron et al. 2012).

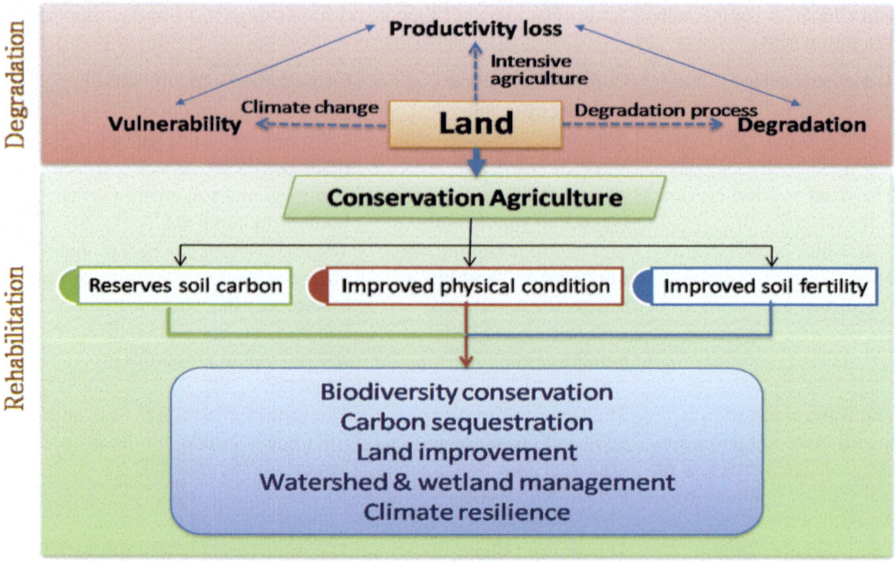

Fig. 18.2 Scope and ecosystem services of conservation agriculture

18.4 CA Methods and Practices

The traditional methods of intensive digging with hand hoes or ploughing have damaged the soil structure, reduced its ability to retain moisture, and caused wind and water erosions resulting in loss of nutrients and topsoil. In contrast, conservation agriculture (CA) is the integrated management of the available natural resources such as soil, water, flora, and fauna with partial outside inputs which increases the efficiency of natural resource use. It provides sustainability in farming production through maintaining the quality of natural resources by stable or semi-stable organic cover to soil. Zero or minimum tillage or no-till (NT) and minimum disturbance of soil along with varying rotations of crops are components of CA and must for future prospects. However there are wide variations in methods and tools required to accomplish the principles of CA.

18.5 CA in Different Types of Farms

Unlike conventional farming methods, CA practices disturb the soil as little as possible to plant their seed directly into the soil and for other intercultural operations as and when required. The task is accomplished by using several small tools and equipments which widely vary in different parts of the tropics which mostly depends on the socio-economic profile of the farmers and edaphic and climatic conditions of any region. A summary of tools and methods for different farming conditions is given in Table 18.2. It is inferred from the table that CA principles can be applied in

Table 18.2 Different tools and sources of power used in conservation agriculture

Options	Land preparation	Planting of main crop	Planting of cover crop	Weeding	Remarks
Simple hand tools (small holder, tribes, and vulnerable households)	Hand hoe, clod breaker	Stick, jab planter	Hand hoe, stick, or jab planter	Manual, hand hoe for shallow weeding	Farmers use different hoes and sticks. Slashing and spreading of crop and cover crop residues, followed by planting directly through residues. Rouging and shallow weeding done
Simple hand tools with herbicides (small holders with local market linkage)	Hand hoe to make planting pits	Hand hoe, seed drill, or stick	Hand hoe, stick, or jab planter	Weed wiper/brush cutter/ herbicides	Needs training to use weed wiper. Planting of cover crops is possible after applying herbicides
Animal-driven reduced tillage (small- to medium-scale farmers)	Ripper or subsoiler and herbicide application	Hand hoe, jab planter, seeder	Hand hoe, stick, or jab planter	Shallow weeding, herbicides	Cover crops might replace other weed control after few years. Subsoiling may be done once every 5 years. Herbicide application depends on weeds
Animal-driven no-tillage (small- to medium-scale farmers)	Herbicide application, knife roller/ pegs	No-till planter	Hand hoe, stick, or jab planter	Hand pulling (roguing), knife roller, herbicides	Knife roller is needed only for well-established soil cover. This is used from second year and beyond to crush residues before planting
Tractor-operated implements (large-scale farmers)	Subsoiler herbicide before planting	No-till tractor-mounted direct seeder	Tractor-mounted direct seeder	Hoe for shallow weeding, knife roller, herbicides	Subsoiling done once in every 4 5 years, small scale farmers can hire implements for farm operations

a wide range of conditions, and it can be practised in different types of farms, with different combinations of crops, tools, and sources of power. Some of the major farming practices are hoe farming and small holder farming with draught animals and mechanized farms.

Many African and tropical island farmers practice hoe farming by cultivating land using only hand hoes. Small planting holes are dug in lines leaving the rest of the soil unturned. If hoeing in previous years has produced a hardpan, the holes must be deep enough to break through this hard layer. The farmers put compost or manure in the holes to raise the soil fertility and the water holding capacity. Maize/millets or beans are sown in the holes and cover crops between the planting holes to protect the soil from erosion and to suppress weeds.

Farmers who own draught animals use them for pulling implements necessary for practising conservation agriculture. They can use a subsoiler to break up the hardpan (if there is one). If there is no hardpan, the farmers can use an animal-drawn ripper to open up a narrow furrow for planting seed. It is possible to use rippers and subsoilers that sow seeds and apply fertilizer at the same time, saving time and work. Farmers can use a hoe or knife roller to kill weeds or apply herbicide using a sprayer or wiper. They sow a cover crop with maize crop to smother weeds and to reduce evaporation from the soil surface. When harvesting the main crop, the farmers leave the residues and cover crop on the field.

Farmers with tractors can also easily practise conservation agriculture by replacing their mould board ploughs, discs, and harrows with rippers, subsoilers, and direct-drill planters to practise zero tillage. At the beginning of the season, farmers use a knife roller, sometimes in combination with herbicides, to kill the previous season's cover crop and weeds. In the first season of conservation agriculture, they may need to use a subsoiler to break up the plough pan. They then use a direct-drill planter to sow seeds. A post-emergence herbicide kills any weeds that come up after planting and before the crop canopy can cover the surface. After harvest, the crop residues are left in the field. The farmer then sows a cover crop to protect the soil until the next planting season. In practice, many farmers are able to use a combination of implements and different types of power. For example, a hoe farmer may be able to hire a tractor from neighbour to subsoil the field. Later various other options for intercultural operations are available so that full benefits of conservation agriculture can be exploited.

18.6 CA Methods in Different Regions of Tropics

Though the principles of CA remain same, different methods are used in different regions to achieve them. In subhumid and humid tropical areas, weeds and erosion are likely to be more of a problem. So, crops are planted at closer spacing, and cover crops help suppress weeds and protect the soil. In large-scale farms, sometimes herbicides are used to eliminate the dangerous weeds. On slopes different tools and methods are required, viz. terracing, contour grass strips, and other erosion control methods. CA can be directly practised if the terraces already exist. Coconut is one of the major crops in the coastal areas of humid tropics, and at several countries, it is practised as a multistoried cropping system. This can be brought under CA by employing cover crop, intercrop, and mulching. Some of the CA methods followed in the tropical coastal regions are given in Fig. 18.3. These include land preparation, sowing, mulching, crop rotation, cover crop, water harvesting, etc.

Peanut sowing with pole and covering with leg

Land preparation with hand hoe for inercropping

Animal traction no-till planter used for planting beans inbetween sorghum

Mechanical ripping

Direct sown rice (furrow) + vegetable sown with hand hoe and covered with stubbles (bed)

Water harvesting (furrow) and vegetables crop roration with ground cover (bed)

Coconut intercropped with ginger (cover)

Short duration pulse as cover crop

Fig. 18.3 Different conservation agricultural practices followed in the tropical coastal regions

Crop residue cover inbetween the crop row

Dibling of pulse in rice residues

Coconut + Black pepper with mulch

Soil cover using crop straw

Fig. 18.3 (continued)

A summary of various conservation practices followed in different tropical region is summarized in Table 18.3. It is inferred that CA practices vary in different regions, and even within the region, the farmers' ability to accommodate costly tools and use inputs decides the methods, crop rotation, and soil cover followed in different farm types. In general, tillage is performed either by hand tools or tractors, while in sloppy areas efficient hand tools should be used. Crop rotation is very important, but it should take care of local climatic and edaphic conditions. In most of the CA regions, cereals followed by pulses are the predominant system. Straws from maize, rice, millet, and sorghum are used to cover the soil which also adds organic matter to the soil. Coconut-based system is practised in the coastal areas of South Asia, Eastern Africa, and tropical islands. In the Latin American regions, large-scale CA farms are mostly mechanized and are known to use herbicides and high inputs. In spite of all these differences in CA practices, one common problem faced in all these regions is soil erosion due to high rainfall and frequent occurrence of tropical storms.

Table 18.3 Various conservation practices followed in different tropical regions

CA practice/type	Region	Tillage	Mulch	Rotation/soil cover	Farm type	Input	Main issues	References
Direct planting with hand tools	Subhumid, sub-Saharan Africa,	None. Use of pointed stick to plant (dibbling)	Little to no mulch of crop residues	Maize predominant—some legumes which covers soil	Small holder mixed crop-livestock farms (<3 ha)	Low level of fertilization no/limited use of herbicides	Water conservation and erosion control is limited, biomass goes as feed, rotation is limited, yield increase is condition for fertilizer application	Thierfelder et al. (2014)
	Indian Ocean tropical islands	Very less. Use of small hoe or stick to plant	Crop residues left on the surface (coconut, areca nut)	No rotation but multistorey cropping is practised in coconut	Small holder, mixed farming	Low to moderate	Traditional practices, rotation is not properly followed, moisture stress in summer	Swarnam et al. (2015)
				Natural plants cover the ground				
Direct planting by dibbling	South and Southeast Asia	Zero tillage for pulse only land preparation for rice	Rice stubbles and straw	Rice-pulse. Pulse act as cover crop, green manure is grown pre-rice	Small holder mixed type	Low	Uncertainty in rain creates waterlogging and heat stress	

(continued)

Table 18.3 (continued)

CA practice/type	Region	Tillage	Mulch	Rotation/soil cover	Farm type	Input	Main issues	References
Planting basins (conservation farming)	Semi-arid Southern Africa	Localized hoeing to make planting	Soil mulch Crop residue	Maize/sorghum/pearl millet predominant—some legumes	Marginal and small farmers	Low	Soil erosion, moisture stress	Twomlow et al. (2008) and Mazvimavi and Twomlow (2009)
Animal-driven reduced tillage	Subhumid Southern Africa	Use of ripper or subsoiler to make planting furrows	Crop residue	Maize predominant—some legumes	Small holder mixed crop-livestock farms (2–5 ha)	Low	Uncertainty of rain, pest infestation	Thierfelder et al. (2014)
Animal-driven no-tillage	Subtropical Southern tropical Brazil	Use of direct seeder for planting directly through mulch into soil	Mulch of crop residue and cover crops	Maize, soybean, and beans followed by winter wheat, black, oats, rye, and leguminous cover crops	Medium-sized mixed crop-livestock farms (20–50 ha)	Medium level of fertilization and use of herbicides	Control of soil erosion, longer growing period not utilized, soil compaction and invading weed species	Bolliger et al. (2006)

Tractor-operated no residue tillage (small tillage)	NW Indo-Gangetic plains (South Asia)	Use of no-till tractor-mounted direct seeder (locally manufactured)	Partial mulch of crop residue	Wheat crop, in irrigated wheat intercropped with mustard/gram-based double crop system, (e.g. rice-wheat), dhaincha before rice	Mixed crop-livestock farms (<20 ha)	Irrigation, fertilization, and use of herbicides	Reduced crop, yield effect due to timelines, crop residue use and handling	Erenstein and Laxmi (20082008)
	West Asia-North Africa	Reduced tillage	Limited mulch of crop residues	Wheat. Barley, legumes	Mechanized crop-livestock (sheep) farm	Medium use of fertilizer and herbicides	Proper soil and water conservation, increase in yield only if early or timely sowing	Kassam et al. (2012)
Tractor-operated direct seedlings (large scale)	North America (Canada and the Midwest)	Tractor-mounted direct seeder	Soil mulch	Maize-soybean	Mechanized large-scale farms (<500 ha)	Medium to high	Emergence of herbicide resistance-limited crop diversity	Hansen et al. (2015)
	Cerrado region Brazil (tropical, subhumid climate)	Reduced tillage/no-till	Plastics, soil, crop residues	Maize/vegetable-based irrigated system	Mechanized large-scale farms enterprises (500–5000 ha)	More input	Erosion control, herbicide resistance, and soil compaction	Bolliger et al. (2006)

18.7 Restorative Effect of CA on Degraded Lands

The conservation agriculture plays an important role in improving soil health including soil's physical, chemical, and biological properties providing short-term and long-term benefits. When evaluating an agricultural production system (agro-ecosystem) for sustainability and productivity, the question is centred on the health of the system and how it is measured. When rehabilitation of the degraded agro-ecosystem, maintaining the quality of land is the main focus of conservation agriculture while harvesting enough for the people at farm level, and it should be assessed using suitable indicators. Further, the assessment of the indicators should also show the negative consequences if any including the production system which will not exhaust the natural resources to make a meaningful comparison. The effect of soil cover, mulching, minimum disturbance, and crop rotation is felt on the soil quality, soil aggregate stability, infiltration, porosity, soil organic carbon, erosion, fertility, soil biodiversity, etc. Besides this, there are other benefits such as climate resilience and ecosystem series to the society. A summary of impact of CA on different quality parameters and its implication for rehabilitation of degraded land is given in Table 18.4. These indicators show the condition of soil under CA and the ability of the production system to support crop production. These aspects are discussed in detail in the following subsections.

18.7.1 Enhancement of Soil Quality

Soil quality gives an idea about the degree of fitness of a soil for specific use. Soil quality is defined as 'the capacity of a specific soil type to function, within naturally managed ecosystem boundaries, to sustain plant and animal productivity, maintain or enhance water and air quality, and support human health and habitation', or in other words it is the ability of the soil to maintain high productivity, without significant soil or environmental degradation. This is overall effect of different parameters.

Evaluation of soil quality is based on the physical, chemical, and biological characteristics of the soil. Management factors such as tillage and residue management can modify the soil quality besides climatic factors. A comparative soil quality evaluation of any management system in relation to alternatives indicates the performance and impact of such systems which has been in place over a period of time. In the context of conservation agriculture, majority of the studies indicated the improvement of soil quality over the period of time from its original state across different regions. But the impact may considerably vary based on the management and climatic factors.

18.7.1.1 Soil Structure and Aggregation

The soil structure and aggregate stability are important physical attributes deciding the quality of soils and indicate any significant improvement in the previously degraded soils. Soil structure is a key physical factor in soil functioning and is an

Table 18.4 Effect of conservation agricultural practices on agro-ecosystem restoration

Sl. no.	Soil quality indicator	Impact as measured	Rehabilitation effect	Reference
1	Soil structure	Soil structure improved	The soil is disturbed very little, so its structure stays intact, and no hardpan forms. Deep-rooted legumes can break the hard pan formed as a consequence of intensive agriculture. Organic matter builds up and improves the soil structure	Li et al. (2007) and Verhulst et al. (2010)
		After years of CA		
2	Aggregate stability and bulk density	The increased aggregate size and its stability, reduced bulk density	Due to binding of soil particles by organic matter, increased microflora more water-stable aggregates are formed	Lichter et al. (2008)
		The aggregate stability index was significantly higher for all size classes >106 μm in diameter from surface samples of no-till than of conventional tillage		Beare et al. (1994)
3	Infiltration and soil moisture	CA increased soil water content (25–55%) and crop yields (12–25%) in Kenya, Zambia, and Zimbabwe	Water can soak into the soil easily through pores and cracks. There is no hardpan, so water can percolate deep into the ground. Mulch and cover crops shade the soil surface, so less water evaporates resulting in more water available for crops	Gicheru et al. (2006) and Thierfelder and Wall (2010)
		Increased infiltration, reduced evaporation from soil, more soil moisture storage		Verhulst et al. (2010)
4	Soil organic carbon	Increase in C stocks under CA is approximately 17 t ha^{-1} higher than under TA, and 46–68% C gains occurs at 30–85 cm depth	Soil organic carbon is very important to restore the biological activity and retain the nutrients in organic farm. In the degraded soils, SOM helps to build soil aggregates, reduce soil erosion, increase infiltration of water, and improve soil moisture storage. For these reasons SOM is single most important indicator of rehabilitation of degraded land	Sisti et al. (2004)
		SOC accumulation ranges from 0.04–0.88 t ha^{-1} $year^{-1}$ up to 30 cm		Boddey et al. (2009)

(continued)

Table 18.4 (continued)

Sl. no.	Soil quality indicator	Impact as measured	Rehabilitation effect	Reference
		It is estimated that on an average Brazil has sequestered 12 million tons of carbon on 23.6 Mha of no-tillage land at a rate of 0.51 t ha^{-1} $year^{-1}$		Derpsch (2005)
		C storage rates under CA varying from 0.4 to 1.9 Mg C year ha for the 0 to 40 cm layer		Bernoux et al. (2006)
		The average carbon sequestration rate of legume-based cropping systems (with N fertilizers) was 1.42 t ha^{-1} $year^{-1}$ when considering the soil profile down to 100 cm depth		Scopel et al. (2005)
5	Soil fertility	C/N ratios of mineral-associated organic matter were higher in no-till than conventional tillage	As crop residues and cover crops stay on the soil, it adds organic matter, favours in situ composting, and reduces nutrient loss through runoff. Legumes fix atmospheric nitrogen and add to the soil. Soil microbial population and earthworm activity increases. All these have positive effect on soil fertility	Beare et al. (1994)
		Increase of N, P, and K particularly in the surface layers		Verhulst et al. (2010)
		The high organic matter contents under CA increased the cation exchange capacity of the topsoil		
		Increase in total N		Govaerts et al. (2007)

(continued)

Table 18.4 (continued)

Sl. no.	Soil quality indicator	Impact as measured	Rehabilitation effect	Reference
6	Erosion control	Runoff under CA reduced to 7.3–50% of rainfall as against 11–83.5% under conventional systems	Cover crops and mulch protect the soil surface from heavy rain. Roots bind the soil together, so it is less easily eroded, more infiltration, so less runs off, consequently erosion is reduced. As a result the degraded soil slowly recovers in its productivity	Scopel et al. (2004, 2013)
		CA reduces soil loss through erosion, only 0.1 t ha^{-1} soil loss was estimated in CA in Latin America		Derpsch and Moriya (1999)
		CA systems based on no-tillage could reduce soil erosion loss by a factor of 2–20 depending on the slope and soil texture		Scopel et al. (2013)
		Mean runoff co efficient and soil loss with CA plots were 45% less and 54% less than conventional agriculture plots		Ghosh et al. (2015)
7	Weeds and pests	CA increased the natural enemy and maintains low equilibrium level of pest population than CT	The cover crop or mulch smothers weeds and prevents them from growing quickly. In large CA farm, herbicides are carefully used	Govaerts et al. (2006)
			Crop rotation breaks the life cycle of pests and disease organisms as well. A healthy soil with crop rotation helps control pests and diseases buildup	
		Majority of the farmers in Latin America (57%) experienced/believe CA can't get rid of perennial weeds		

(continued)

Table 18.4 (continued)

Sl. no.	Soil quality indicator	Impact as measured	Rehabilitation effect	Reference
8	Crop diversity and resilience	Reduced the risk associated with the monocropping	Crop rotations and crop mixes produce a range of crops. They reduce the risk of ruin if one crop fails and provide a more diverse diet. Hence improves the land productivity	Vilela et al. (2004)
		There are reports of yield increase in CA (Knowler and Bradshaw 2007)		Knowler and Bradshaw (2007)
9	Soil biodiversity and biological fertility enhancement	Soil biological activity and diversity are higher in systems with reduced or no-tillage and with a surface mulch or cover crops	Improves the soil biodiversity by providing good habitat and organic matter which in turn improves the degraded land	Rabary et al. (2008)
		Threefold increase in soil macrofauna density under non-tilled CA systems compared to conventional systems		Brown et al. (2001) and Blanchart et al. (2007)
10	Environment and ecosystem	Positive carbon sequestration rates and net carbon differences of CA was higher than conventional system	Conservation agriculture is environment friendly, enhances agro-biodiversity, helps to conserves natural ecosystems, and acts as a 'carbon sink'	Nyborg et al. (1995)
11	Climate change regulation—greenhouse gas (GHG) emission	Reduction in GHGs	CA improves soil organic carbon level, less fuel consumption and release of GHG, thereby helps in mitigation of global warming	Lal (2001) and Verhulst et al. (2010)
		3% reduction in GHG was observed for rice-wheat system		Grace et al. (2012)
		CA reduced ammonia emission in N fertilized soil		Bhagat and Verma (1991)

Soil aggregates with clear structure under CA

A dense growth of legume roots with N-fixing nodules and earthworm in CA

Fig. 18.4 Effect of CA on soil structure and aggregates

important factor in the evaluation of the sustainability of crop production systems. It has been defined as the size, shape, and arrangement of solids and voids, continuity of pores and voids, their capacity to retain and transmit fluids and organic and inorganic substances, and the ability to support vigorous root growth and development (Fig. 18.4). Soil structure is often expressed as the degree of stability of aggregates, viz. water-stable aggregates. In general it is the ability of aggregates to remain intact when exposed to different stresses (Verhulst et al. 2010). In the context of CA, formation of a stable structure is highly desirable, and it shows the improvement of degraded land.

The soils under zero tillage with residue retention and crop rotation particularly with legume become more stable and less susceptible to structural deterioration when compared to conventionally tilled soils due to residue retention on the surface which reduces the raindrop impact; decomposing organic matter increases the aggregation process; no or minimum soil disturbance increases the presence of root fragments and mycorrhizal hyphae which are major binding agents for macroaggregates (Li et al. 2007; Bronick and Lal 2005). Besides, the conservation agriculture enhances the macrofauna population (Blanchart et al. 2007) such as earthworms which have positive effect on soil aggregate formation (Six et al. 2004) and soil porosity enhancement. The crop residue retention with no-tillage along with crop rotations eliminates the formation of surface sealing and crusting even in soils with low organic matter and enhances infiltration and water retention capacity of the soil (Cassel et al. 1995). Some cover crop species such as *Brachiaria* sp., *Eleusine* sp., *Andropogon* sp., and *Cynodon* sp. for tropical conditions (Blanchart et al. 2004; Mannetje 2007) are efficient in restoring the soil structure because of their substantial rooting systems in the surface soil layer (0–40 cm).

18.7.1.2 Soil Porosity

The main characteristic feature of the soil porosity is the size, shape, and continuity of the pores which influence the water infiltration, storage and drainage of water, the movement and distribution of gases, and the ease of penetration of soil by growing roots. The pores are created by abiotic (e.g. tillage and traffic, freezing and thawing, drying and wetting) and biotic factors (e.g. root growth, burrowing fauna). The changes in pore characteristics primarily reflect changes in the form, magnitude and frequency of stresses imposed on the soil, the placement of crop residues, and the population of microorganisms and fauna in the soil (Verhulst et al. 2010).

Total porosity is the sum of spaces occupied by water and air in soil. It is normally calculated from measurements of bulk density so the terms bulk density and total porosity can be used interchangeably. CA decreases bulk density and increases total porosity, both of which attains dynamic equilibrium in a given climatic conditions (Horne et al. 1992). But the effect of tillage and residue management on soil density is mainly confined to the topsoil (plough layer), and in deeper soil layers, soil bulk density is generally similar in zero and conventional tillage.

Total porosity of soils is distributed among different pore size classes, and different size classes fulfil different roles in aeration, infiltration, drainage, and storage of water and offer different mechanical resistance to root growth. In this context, a study reported higher volume of mesopores (0.2–30 μm) in the 0–3 cm layer in zero tillage with residue retention than in zero tillage without residue retention (Blanco-Canqui and Lal 2007). Water infiltration, retention, and flow not only depend on the quantity and size of pores but also on the interconnectivity and shape of pores. CA is known to improve these aspects as well. Hydraulic conductivity would be expected to be higher in zero tillage with residue retention compared to conventional tillage due to the larger macropore conductivity, which is the result of an increased number of biopores.

Soil management practices that increase the organic matter content in the soil could have a positive impact on soil water holding capacity because organic matter can withhold soil water more than its weight. Water holding capacity has been observed to increase with increases in soil organic matter, meaning CA has the potential to increase water holding capacity (Verhulst et al. 2010).

18.7.1.3 Runoff and Soil Erosion Control

In tropical region soil erosion is the major problem affecting farm production. In agriculture apart from topography and slope, water runoff and soil erosion occur due to surface and subsurface compactness, crusting and sealing, hardpans, and decreased macropores affecting the water infiltration. The high-intensive rainfall received in humid tropical regions further accentuates the rate of soil loss. The major benefit of CA is the control of soil erosion due to maintenance of soil cover, greater infiltration, and reduced runoff (Erenstein 2002), and early studies showed that CA systems based on no-tillage could reduce soil erosion loss by a factor of 2–20 depending on the slope and soil texture (Scolpel et al. 2013). High residue addition and growing cover crops decrease the impact of raindrops on soil surface, slow down the velocity of runoff water across the soil surface, provide more opportunity and time for infiltration, and act as surface storage of rainwater (Indoria et al. 2017). However, when no-till is practised in the absence of effective soil mulch cover, the effects can be disastrous with rapid surface sealing leading to increased runoff and accelerated soil erosion (Guto et al. 2011). Tillage is also known to induce soil erosion resulting in severe soil carbon loss and soil translocation in convex upper slope positions of cultivated, upland landscapes (Lobb et al. 1995). In summary, CA reduces erosion rates much closer to soil formation rates than conventional tillage and therefore could provide a foundation for sustainable agriculture.

18.7.1.4 Soil Water Balance

Despite inconsistent results on the effect of tillage and residue management on soil hydraulic conductivity, infiltration is generally higher in zero tillage with residue retention compared to conventional tillage and zero tillage with residue removal. This is probably due to the direct and indirect effects of residue cover on water infiltration. McGarry et al. (2000) found that time-to-pond, final infiltration rate, and total infiltration were significantly larger with zero tillage with residue retention than with conventional tillage. This was ascribed to the abundance of apparently continuous soil pores from the soil surface to depth under zero tillage, as opposed to a high-density surface crust in conventional tillage.

Soil evaporation is determined by two factors: how wet the soil is and how much energy the soil surface receives to sustain the evaporation process. Residue and mulches reduce soil water evaporation by reducing soil temperature, impeding vapour diffusion, absorbing water vapour onto mulch tissue, and reducing the wind speed gradient at the soil-atmosphere interface. The rate of drying of the soil is determined by the thickness of the residue together with the atmospheric evaporative potential.

Conservation agriculture can increase infiltration and reduce runoff and evaporation compared to conventional tillage and zero tillage with residue removal. Consequently, soil moisture is conserved, and more water is available for crops. Mulching helps conserve soil water in a season with long periods without rain. Soil water content increases with increases in surface cover. More soil water enables crops to grow during short-term dry spells. Thus, tillage and residue management may significantly affect crop yields in areas or seasons with poor rainfall distribution.

18.7.1.5 Soil Organic Carbon and Soil Fertility

Soil organic matter plays a crucial role in maintaining soil health and its productivity potential. However, most of the world's agricultural soils have become depleted in organic matter due to tillage, which accelerates the decomposition of soil organic matter. Conservation agriculture, a system avoiding or minimizing soil disturbance, combined with soil cover and crop diversification, is considered to be a sustainable production system that can also sequester carbon unlike conventional agriculture based on tillage. The C storage potential of conservation agriculture depends on antecedent soil C concentration, cropping system, management duration, soil texture, slope, and climate (Govaerts et al. 2009).

There are inconsistent reports on soil carbon storage and C sequestration by conservation agriculture. However, surface soil C may accumulate under CA due to lack of soil mixing (Powlson et al. 2014), significant contribution from plant roots (Rasse et al. 2005), and increased biomass production and crop residue retention by growing more than a crop. Residue retention leads to increase in soil organic matter which can induce changes in nutrient transformations in the soil as well as improvement in soil physical properties. Besides, the reduced soil erosion is partly responsible for increases in soil C (Scopel et al. 2005). Nyborg et al. (1995) showed positive carbon sequestration rates and net carbon differences of CA relative to TA.

At the same time, it has been reported that lack of proper balances of the nutrients, viz. nitrogen (N), phosphorus (P), and sulphur (S), often limits soil organic matter buildup (Richardson et al. 2014). The carbon accumulates in the soil only when the nitrogen balance of the system is positive, i.e. when the N input either through N fixation or external addition is higher than the nitrogen exported with harvested produce plus the amount lost by leaching or in gaseous forms (Bayer et al. 2000). In tropical region, the accumulation of organic C in the soil under CA systems observed to occur only when nitrogen is not limiting (Scopel et al. 2013) and rates of organic returned to the soil are more than 12 Mg dry matter ha^{-1} (Corbeels et al. 2006). The selection of crops in crop rotation also plays an important role in SOC accumulation in soil under CA. The type of crops, intensity of cropping, and duration of the cropping systems determine the amount of inputs and the ability of CA to store more C. Some studies showed negative SOC accumulation rates under CA mainly associated with specific rotations, i.e. with fallow-, barley-, soybean-, and maize-based rotations (Angers et al. 1997; Yang and Kay 2001). Intensification of cropping systems with high above- and belowground biomass may enhance CA systems for storing C relative to CT (Luo et al. 2010). Moreover, inclusion of legumes and green manure crops with high annual above-ground biomass production is essential to keep the nitrogen balance of the crop rotation positive and to achieve significantly greater carbon stocks.

Maintenance of soil organic matter is a key to sustainability of agricultural soils which requires N, P, and S not C inputs alone. This necessitates the use of fertilizers and N-fixing legumes and pastures besides crop residues as mulch to maintain soil organic matter and crop productivity (Da Silva et al. 2014). However, the long-term accumulation of soil C and increases in soil bulk density due to the lack of soil tillage can lead to excess water retention (Rusinamhodzi et al. 2011) and yield declines.

Increased soil C accumulation in the soil surface layers along with mulch plays an important role in enhancing the soil water supply for crop growth. The enhanced soil structural stability and the protection of the soil surface from direct raindrop impact lead to greater infiltration and capture of rainfall for crop growth (Giller et al. 2015) and increased water use efficiency (Unger 1990). The presence of mulch protects against volatilization of ammonia from fertilizer N (Bhagat and Verma 1991) and slower decomposition of surface placed residues prevent loss of nutrients through leaching (Kimmell et al. 2001) and increased nutrient use efficiency particularly N (Jat et al.

2012). No-till and residue retention under CA result in concentration of available potassium (K) and phosphorus (P) in the surface soils due to higher concentration of K in cereal residues and reduced P sorption resulting from higher SOM. However, concentration of nutrients in the surface soil layers with CA practices may lead to problems in years of prolonged drought as the crop roots could be distributed more superficially where nutrients are concentrated (Paul et al. 2003), but this may be offset by higher soil moisture in CA. They also reduce soil evaporation which results in overall yield benefits in some dry climates (Pittelkow et al. 2015; Wilhelm and Wortmann 2004). Eliminating tillage and retaining residue increased soil water content and crop yields in African countries like Kenya, Zambia, and Zimbabwe (Naab et al. 2017).

Studies on CA have found that the pH of the topsoil was lower under zero tillage than conventional tillage. This is because of greater soil organic matter accumulation in the topsoil and release of organic acids during decomposition. Further there is possibility for acidification effect from the superficially applied nitrogen and phosphorus fertilizers under zero tillage than under conventional tillage. Similarly many studies on conservation agriculture showed decrease in soil salinity particularly reduction in salt accumulation in the root zone due to decrease in salt movement in vapour form and dilution effect of soil moisture on the surface and (Choudhury et al. 2014) reported amelioration of sodic soils.

18.7.2 CA and Climate Regulation

It is suggested that by adopting improved management practices on agricultural land (use of NT and crop residues), food security would not only be enhanced but also offset fossil fuel emissions at the rate of 0.5 Pg C year (Lal 2005). With changes in the intensity of mechanical tillage, reduced irrigation, less N fertilizer application, and the associated reduced use of fossil fuels CA have considerable impacts on greenhouse gas emissions compared to conventional agriculture (Pathak 2009). For instance, in the IGP, resource-conserving technologies like reduced tillage and residue burning along with increased N use efficiency in the rice-wheat cropping systems could save 50–60 l of diesel ha^{-2} which will significantly reduce release of CO_2 to the environment. In tropical regions due to high temperature and moisture and intensive runoff after rains, CO_2 emission related to heterotrophic microbial activity is greater in CA systems than in conventional systems (Metay et al. 2003). However, C losses through decomposition are largely compensated by higher biomass returns produced in a two-crop CA system than in conventional systems with one crop per year (Bustamante et al. 2006). A conceptual model of mitigation effect of conservation agriculture is presented in Fig. 18.5.

Nitrous oxide is a long-lived GHG which is produced in soils through microbial mineralization processes. The effect of CA on N_2O emissions will depend on the crop rotation, type, and amount of crop residues returned under the system compared to conventional agriculture. The retention of crop residues and higher soil C in surface soils under CA play major roles in these processes. When the soil is saturated, higher concentration of soluble carbon or readily decomposable organic

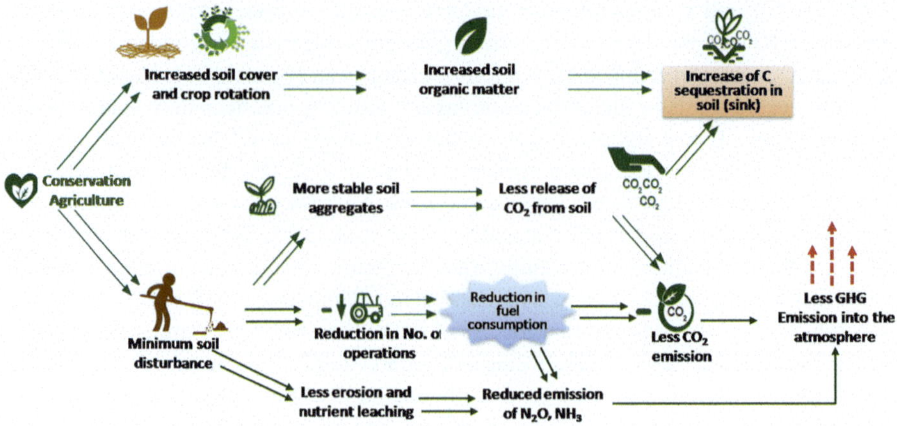

Fig. 18.5 Mitigation effect of conservation agriculture

matter favours denitrification under anaerobic conditions releasing N_2O (Palm et al. 2014). CA practices themselves can sometimes increase the nutrient use efficiency of N fertilizer (Jat et al. 2012), thus providing a synergy between CA and N fertilizer management which could contribute to decreased GHG emissions. Another important GHG is the methane, which has a global warming potential 25 times that of CO_2 over a period of 100-year horizon. Flooded rice is the largest contributor of CH_4 emission from agriculture. The overall impact of reduced tillage under this system appears to be relatively minor where only 3% reduction was observed with widespread implementation of CA in rice-wheat systems of Indo-Gangetic Plains (Grace et al. 2012).

18.7.3 CA and Soil Biodiversity

Changes in tillage, residue management, and crop rotation practices induce major shifts in soil fauna and flora, including both pests and beneficial organisms. CA alters below and above-ground species differentially than conventional agricultural practices (McLaughlin and Mineau 1995). This alteration drives wide range of responses among different groups of organisms enhancing greater abundance and diversity (Palm et al. 2014). Reduced tillage and residue retention provides a stable environment with less physical disruption and higher soil moisture content favouring earthworms, which are found to have higher abundance and diversity with CA compared to conventional systems. Enhanced activity of earthworms can alleviate excessive buildup of soil organic matter in the surface horizons (Singh et al. 2015) and also play a significant role in water infiltration and water regulation (Verhulst et al. 2010).

The residue retention combined with reduced tillage may provide suitable habitat for soilborne pathogens including a diversity of necrotrophic fungal pathogens. For example, surface crop residues infected with grey leaf spot (*Cercospora zeae-maydis*) provide an early-stage inoculum for the next maize crop resulting in acute infection (Thierfelder et al. 2014). The detritivore earthworm *Lumbricus terrestris* was shown to reduce the biomass of *Fusarium culmorum* in wheat straw, thus compensating for the negative effects of no-till due to crop residue accumulation (Wolfarth et al. 2011). Adopting suitable crop rotations may avoid the buildup of undesirable organisms including weeds, insects, and diseases. In CA systems intensive crop rotations are essential component to enhance microbial diversity and their functional roles. The potential for biological control of pathogens, insect pests, and weeds can be enhanced when crop rotations in CA involve number of diversified crops which can break the buildup of populations (Vilela et al. 2004). Govaerts et al. (2006) observed creation of conditions favourable for the development of antagonists and predators and fosters new ecological stability by zero tillage with crop residue retention on a long run. Thus, the potential exists for generally a higher suppression of pathogens in direct-seeded soils with crop residue retention.

18.8 Constraints and Challenges in CA

CA is a long-term strategy for sustainable intensification of arable lands that is essential to advance food and nutritional security, alleviate poverty, restore degraded soils, reduce GHG emissions from agriculture, minimize nonpoint source pollution and contamination of surface water bodies, conserve biological diversity, preserve ecosystem services, and make the farming systems more sustainable and resilient to climate change events (Lal 2015). However, adoption of CA systems is often limited due to the fact that they are quite different from traditional tillage agriculture. There are several challenges faced by the small holder farmers in tropical countries even though they are willing to shift to conservation agriculture.

- CA is knowledge intensive, sensitive to equipment and expertise, and requires external inputs including herbicides, seeds of improved varieties, and fertilizers which are limited by financial capacity of small holder farmers particularly in Africa and South Asia.
- There are numerous reports indicating significant reduction in agronomic yields with no-tillage. This decline sets-in-motion a vicious cycle of decreasing biomass causing low SOC pools, lower yields, lower biomass input, lower soil quality, and, finally, even lower yields (Lal 2015).
- The more visible effects of CA including soil loss and increased soil organic matter content and yield benefits accrue on a long-term basis which means mostly after 5 years which is a major hurdle for adoption (Giller et al. 2009).
- There is a strong competing uses for crop residues as fodder limits adoption by small holder farmers in Africa and South Asia where mixed crop-livestock farming is predominant.
- Weeds have severe consequences on crop productivity under tropical conditions which limits the profitability and increases the use of herbicides. Increase in

labour, when no-till is practised in manual low-input systems as in Africa, is also a major factor limiting its adoption.

18.9 Conclusion

It is widely agreed that over the years, most of the world's agricultural soils have been depleted in organic matter and soil health under plough-based modern agriculture compared with their original state under natural vegetation. Accelerated erosion and unjust agricultural practices have further damaged the soils resulting in soil degradation of varying types and intensities. With the increase in population, the existing arable lands with alarming rate of degradation could not cater production demand. Research on conservation agriculture, organic farming, natural farming, etc. as an alternative to the existing practice proved that this degradation process is reversible to a large extent, and the major ways to improve soil health seems to be (i) keeping the agricultural soil disturbed at absolute minimum, (ii) leaving crop residues as carbon source on the soil surface and cover the soil with suitable crop, and (iii) using effective crop rotations and associations suitable for the location. This is the three basic principles in conservation agriculture as concluded by several researchers across the globe. Improving soil fertility through integrated nutrient management (INM) for healthy crop growth and biochemical transformation of biomass C into SOM or humus was also included as a component of conservation agriculture (Lal 2014). The implementation of these principles through suitable practices/methods, as evidenced from several studies pertaining to the tropical regions, can help to rehabilitate the degraded agro-ecosystem to a productive and sustainable state.

There is significant diversity in local weather, edaphic condition, socio-economic status, land resources, and its degradation status in different areas in the tropical belt. Therefore, the conservation agriculture should represent a diversified set of options offering small holder farmers the possibility to achieve sustainable and intensive crop production under a wide range of agro-ecosystems. Evidences from CA experiments indicated that CA improves the soil physical, chemical, and biological properties considerably when compared to the conventional system which has been aimed largely at productivity maximization. Further, CA also increases the carbon sequestration in soil, reduces GHG emission into the atmosphere, and restores several agro-ecosystem services. In the coastal areas, CA helps to make suitable adjustments to cope with the emerging trend in climate thereby provide for climate change adaptation. Similarly long-term adoption of CA in a sloppy area can help to control soil erosion while allowing farmers to produce their staple crops.

Even though some authors reported significant increase in soil fertility and microbial activity soon after transition to CA, fuller advantages of CA in terms of soil health and its productive capacity can usually be observed only in the medium to longer term, when CA practices and soil biological processes become well established within the farming system. There are counterarguments against the positive impact of CA particularly its ability to produce more than conventional systems and sufficient quantity to feed people. The metadata analysis by several researchers positively indicated that CA can offer an opportunity for arresting and reversing the downward spiral of resource degradation, decreasing cultivation costs, and making agriculture more resource-use efficient, competitive, and sustainable. However, the success of the CA rests in identifying suitable site-specific components based on economic, ecologic, and sociocultural acceptability, integrating all the four basic components into a system and finally developing, and implementing suitable package of farm operations and cultural practices specifically designed for a CA system in a given agro-ecosystem. Though the crop productivity may not be comparable with the level obtained in intensive irrigated agricultural system, certainly CA helps to rehabilitate the degraded agro-ecosystem in the tropical region particularly in Latin America, sub-Saharan Africa, South Asia, and several tropical island nations and provide farmers the best alternative for production system management which is more sustainable and climate resilient.

References

Angers DA, Bolinder MA, Carter MR, Gregorich EG, Drury CF, Liang BC, Voroney RP, Simard RR, Donald RG, Beyaert RP, Martel J (1997) Impact of tillage practices on organic carbon and nitrogen storage in cool, humid soils of eastern Canada. Soil Tillage Res 41:191–201

Baudron E, Tittonell P, Corbeels M, Letourmy P, Giller KE (2012) Comparative performance of conservation agriculture and current smallholder farming practices in semi-arid Zimbabwe. Field Crops Res 132:117–128. https://doi.org/10.1016/j.fcr.2011.09.008

Bayer C, Mielniczuck J, Amado TJC, Martin-Neto L, Fernandes SBV (2000) Organic matter storage in a sandy clay loam Acrisol affected by tillage and cropping systems in southern Brazil. Soil Tillage Res 54:101–109

Beare MH, Hendrix PF, Coleman DC (1994) Water-stable aggregates and organic matter fractions in conventional and no-tillage soils. Soil Sci Soc Am J 58:777–786

Bernoux M, Cerri CC, Cerri CEP, Siqueira Neto M, Metay A, Perrin AS, Scopel E, Blavet D, Piccolo MC, Pavei M, Milne E (2006) Cropping systems, carbon sequestration and erosion in Brazil, a review. Agron Sustain Dev 26:1–8. https://doi.org/10.1051/agro:2005055

Bhagat RM, Verma TS (1991) Impact of rice straw management on soil physical properties and wheat yield. Soil Sci 152:108–115

Blanchart E, Albrecht A, Chevallier T, Hartmann C (2004) The respective roles of biota (roots and earthworms) in the restoration of physical properties in vertisol under a *Digitaria decumbens* pasture (Martinique). Agric Ecosyst Environ 103:343–355

Blanchart E, Bernoux M, Sarda X, Siqueira Neto M, Cerri CC, Piccolo M, Douzet JM, Scopel E, Feller C (2007) Effect of direct seeding mulch-based systems on soil carbon storage and macrofauna in Central Brazil. Agric Conspec Sci 72:81

Blanco-Canqui H, Lal R (2007) Impacts of long-term wheat straw management on soil hydraulic properties under no-tillage. Soil Sci Soc Am J 71:1166–1173

Boddey RM, Jantalia CP, Conceicao PC, Zanatta JA, Bayer C, Mielniczuk J, Dieckow J, Dos Santos HP, Denardin JE, Aita C, Giacomini SJ, Alves BJR, Urquiaga S (2009) A carbon accumulation at depth in Ferralsols under zero-till subtropical agriculture. Glob Chang Biol 16(2):784–795

Bolliger A, Magid J, Amado JCT, Skora Neto F, Ribeiro MFDS, Calegari A, Ralisch R, De Neergaard A (2006) Taking stock of the Brazilian "Zero-Till Revolution": a review of landmark research and farmers' practice. Adv Agron 91:47–110

Bronick CJ, Lal R (2005) Soil structure and management: a review. Geoderma 124:3–22

Brown GG, Pasini A, Benito NP, De Aquino AM, Correia MEF (2001) Diversity and functional role of soil macrofauna communities in Brazilian no-tillage agroecosystems: a preliminary analysis. In: International symposium on managing biodiversity in agricultural ecosystems, Montreal, Canada, 8–10 November 2001

Bustamante MMC, Corbeels M, Scopel E, Roscoe R (2006) Soil carbon storage and sequestration potential in the Cerrado region of Brazil. In: Lal R, Cerri CC, Bernoux M, Etchevers J (eds) Soil carbon sequestration and global climate change: mitigation potential of soils of Latin America. Harworth, Binghamton, pp 285–304

Cassel DK, Raczkowski CW, Denton HP (1995) Tillage effects on corn production and soil physical conditions. Soil Sci Soc Am J 59:1436–1443

Choudhury SG, Srivastava S, Singh R, Choudhuri SK, Sharma DK, Singh SK, Sarkar D (2014) Tillage and residue management effects on soil aggregation, organic carbon dynamics and yield attribute in rice-wheat cropping system under reclaimed sodic soils. Soil Tillage Res 136:76–83

Corbeels M, Scopel E, Cardoso A, Bernoux M, Douzet JM, Siqueira Neto M (2006) Soil carbon storage potential of direct seeding mulch-based cropping systems in the Cerrados of Brazil. Glob Chang Biol 12:1–15. https://doi.org/10.1111/j.1365-2486.2006.01233.x

Da Silva FD, Amado TJC, Ferreira AO, Assmann JM, Anghinoni I, Carvalho PCDF (2014) Soil carbon indices as affected by 10 years of integrated crop-livestock production with different pasture grazing intensities in Southern Brazil. Agric Ecosyst Environ 190:60–69. https://doi.org/10.1016/j.agee.2013.12.005

Derpsch R (2005) The extent of conservation agriculture adoption worldwide: implications and impact. In: Proceedings of the third world congress on conservation agriculture: linking production, livelihoods and conservation, Nairobi, Kenya, 3–7 October 2005

Derpsch R, Moriya K (1999) Implications of soil preparation as compared to no-tillage on the sustainability of crop production: experiences from South America. In: Reddy MV (ed) Management of tropical agro-ecosystems and the beneficial soil biota. Science Publishers, Enfield, pp 49–65

Erenstein O (2002) Crop residue mulching in tropical and semi-tropical countries: an evaluation of residue availability and other technological implications. Soil Till Res 67(2):115–133

Erenstein O (2009) Adoption and impact of conservation agriculture-based resource conservation technologies. In: 4th world congress on conservation agriculture held on 4–7 February (2009) in New Delhi, organised by the National Academy of Agricultural Sciences, vol Lead Papers. p 439–444

Erenstein O, Laxmi V (2008) Zero tillage impacts in India's rice-wheat systems: a review. Soil Till Res 100:1–14. https://doi.org/10.1016/j.still.2008.05.001

FAO (2000) Conservation agriculture. Food and Agriculture Organization of the United Nations, Rome. http://www.fao.org/ag/ca/ (accessed on 18 June 2017

FAO (2001) Conservation agriculture – case studies in Latin America and Africa. FAO Soil Bulletin 78, Rome, Italy

FAO (2007) Tropical crop–livestock systems in conservation agriculture. The Brazilian experience, Integrated Crop Management, vol 5. Food and Agriculture Organization of the United Nations, Rome, p 106

FAO (2012) What is CA"? "Principles of CA", "Benefits of CA. Available at: http://www.fao.org/ag/ca/index.html

FAO (2015) Conservation agriculture. http://www.fao.org/ag/ca/index.html

Ghosh BN, Dogra P, Sharma NK, Bhattacharyya R, Mishra PK (2015) Conservation agriculture impact for soil conservation in maize–wheat cropping system in the Indian Sub-Himalayas. Int Soil Water Conserv Res. https://doi.org/10.1016/j.iswcr.2015.05.001

Gicheru PT, Gachene CKK, Mbuvi JP (2006) Effects of soil management practices and tillage systems on soil moisture conservation and maize yield on a sandy loam in semi-arid Kenya. J Sustain Agric 27:77–92. https://doi.org/10.1300/J064v27n03_06

Giller KE, Witter E, Corbeels M, T ittonell P (2009) Conservation agriculture and smallholder farming in Africa: the heretics' view. Field Crop Res 114:23–34. https://doi.org/10.1016/j.fcr.2009.06.017

Giller KE, Andersson JA, Corbeels M, Kirkegaard J, Mortensen D, Erenstein O, Vanlauwe B (2015) Beyond conservation agriculture. Front Plant Sci 6:870. https://doi.org/10.3389/fpls.2015.00870

Govaerts B, Mezzalama M, Sayre KD, Crossa J, Nicol JM, Deckers J (2006) Long-term consequences of tillage, residue management, and crop rotation on maize/wheat root rot and nematode populations in subtropical highlands. Appl Soil Ecol 32:305–315

Govaerts B, Fuentes M, Mezzalama M, Nicol JM, Deckers J, Etchevers JD, Figueroa Sandoval B, Sayre KD (2007) Infiltration, soil moisture, root rot and nematode populations after 12 years of different tillage, residue and crop rotation managements. Soil Till Res 94:209–219

Govaerts B, Verhulst N, Castellanos-Navarrete A, Sayre K, Dixon J, Dendooven L (2009) Conservation agriculture and soil carbon sequestration: between myth and farmer reality. Crit Rev Plant Sci 28:97–122

Grace PR, Antle J, Ogle S, Paustian K, Basso B (2012) Soil carbon sequestration rates and associated economic costs for farming systems of the Indo-Gangetic Plain. Agric Ecosyst Environ 146:137–146

Guto S, Pypers P, Vanlauwe B, de Ridder N (2011) Tillage and vegetative barrier effects on soil conservation and short-term economic benefits in the Central Kenya highlands. Field Crop Res 122:85–94

Hansen NC, Tubbs S, Fernandez F, Green S, Hansen NE, Stevens WB (2015) Conservation agriculture in North America. In: Farooq M, Siddique KHM (eds) Conservation agriculture. Springer International Publishing, Dordrecht, pp 417–441

Hobbs P, Sayre K, Gupta R (2008) The role of conservation agriculture in sustainable agriculture. Phil Trans R Soc B 363:543–555

Horne DJ, Ross CW, Hughes KA (1992) 10 years of maize oats rotation under 3 tillage systems on a silt loam in New-Zealand. 1. A comparison of some soil properties. Soil Till Res 22:131–143. https://doi.org/10.1007/s13593-012-0106-9

Indoria AK, Srinivasa Rao C, Sharma KL, Sammi Reddy K (2017) Conservation agriculture – a panacea to improve soil physical health. Curr Sci 112(1):52–61

Jat RA, Wani SP, Sahrawat KL (2012) Conservation agriculture in the semi-arid tropics: prospects and problems. Adv Agron 117:191–273

Kassam A, Friedrich T, Derpsch R, Lahmar R, Mrabet R, Serraj R, Basch G, Gonzalez-Sanchez EJ (2012) Conservation agriculture in the dry Mediterranean climate. Field Crops Res 132:7–17

Kimmell RJ, Pierzynski GM, Janssen KA, Barnes PL (2001) Effects of tillage and phosphorus placement on phosphorus losses in a grain sorghum-soybean rotation. J Environ Qual 30:1324–1330. https://doi.org/10.2134/jeq2001.3041324x

Knowler D, Bradshaw B (2007) Farmers' adoption of conservation agriculture: a review and synthesis of recent research. Food Policy 32(1):25–48

Lal R (2001) Managing world soils for food security and environmental quality. Adv Agron 74:155–192

Lal R (2005) Enhancing crop yields in the developing countries through restoration of the soil organic carbon pool in agricultural lands. Land Degrad Dev 17:197–209. https://doi.org/10.1002/ldr.696

Lal R (2014) Societal value of soil carbon. J Soil Water Conserv 69(6):168A–192A

Lal R (2015) Sequestering carbon and increasing productivity by conservation agriculture. J Soil Water Conserv 70(3):55A–62A

Landers JN (2007) Tropical crop–livestock systems in conservation agriculture. The Brazilian experience, Integrated crop management, vol. 5. Food and Agriculture Organization of the United Nations, Rome. pp 106

Li HW, Gao HW, Wu HD, Li WY, Wang XY, He J (2007) Effects of 15 years of conservation tillage on soil structure and productivity of wheat cultivation in northern China. Aust J Soil Res 45:344–350

Lichter K, Govaerts B, Six J, Sayre KD, Deckers J, Dendooven L (2008) Aggregation and C and N contents of soil organic matter fractions in a permanent raised-bed planting system in the highlands of Central Mexico. Plant Soil 305:237–252

Lobb D, Kachanoski RG, Miller MH (1995) Tillage translocation and tillage erosion on shoulder slope landscape positions measured using 137Cs as a tracer. Can J Soil Sci 75:211–218

Luo Z, Wang E, Sun OJ (2010) Can no-tillage stimulate carbon sequestration in agricultural soils? A meta-analysis of paired experiments. Agric Ecosyst Environ 139:224–231

Mannetje L (2007) Climate change and grasslands through the ages: an overview. Grass Forage Sci 62:113–117. https://doi.org/10.1111/j.1365-2494.2007.00574.x

Mazvimavi K, Twomlow S (2009) Socioeconomic and institutional factors influencing adoption of conservation farming by vulnerable households in Zimbabwe. Agric Syst 101:20–29. https://doi.org/10.1016/j.agsy.2009.02.002

McGarry D, Bridge BJ, Radford BJ (2000) Contrasting soil physical properties after zero and traditional tillage of an alluvial soil in the semi-arid subtropics. Soil Till Res 53:105–115

McLaughlin A, Mineau P (1995) The impact of agricultural practices on biodiversity. Agric Ecosyst Environ 55:201–212

Metay A, Oliver R, Douzet JM, Bernoux M, Feller C, Feigl B, Rodrigues A, Moreira JA, Scopel E (2003) Short-term evolution of C-CO emissions for a Brazilian Oxisol: effect of tillage and rainfall. In: Producing in harmony with nature. II World congress on Sustainable Agriculture proceedings, Iguaçu, Brazil, 10–15 of August

Naab JB, Mahma YG, Yahaya I, Prasad PVV (2017) Conservation agriculture improves soil quality, crop yield, and incomes of smallholder farmers in North Western Ghana. Front Plant Sci 8:996

Nyborg M, Solberg ED, Malhi SS, Izaurralde RC (1995) Fertilizer N, crop residue, and tillage alter soil C and N content in a decade. In: Lal R, Kimble J, Levine E, Stewart BA (eds) Soil management and greenhouse effect. Lewis Publishers/CRC Press, Boca Raton, pp 93–99

Palm C, Blanco-Canqui H, DeClerck F, Gatere L, Grace P (2014) Conservation agriculture and ecosystem services: an overview. Agric Ecosyst Environ 187:87–105. https://doi.org/10.1016/j.agee.2013.10.010

Pathak H (2009) Greenhouse gas mitigation in rice-wheat system with resource conserving technologies. In: Fourth world congress on conservation agriculture, February 4–7, New Delhi, India, pp 373–377

Paul KI, Black AS, Conyers MK (2003) Development of acidic subsurface layers of soil under various management systems. Adv Agron 78:187–214

Pimental D, Harvey C, Resosudarmo P, Sinclair K, Kurz D, McNair M, Crist S, Shpretz L, Fitton L, Saffouri R, Blair R (1995) Environmental and economic costs of soil erosion and conservation benefits. Science 267:1117–1123

Pittelkow CM, Liang X, Linquist BA, van Groenigen KJ, Lee J, Lundy ME et al (2015) Productivity limits and potentials of the principles of conservation agriculture. Nature 517:365–368

Powlson DS, Stirling CM, Jat ML, Gerard BG, Palm CA, Sanchez PA et al (2014) Limited potential of no-till agriculture for climate change mitigation. Nat Clim Chang 4:678–683

Rabary B, Sall S, Letourny P, Husson O, Ralambofetra E, Moussa N, Chotte JL (2008) Effects of living mulches or residue amendments on soil microbial properties in direct seeding cropping systems of Madagascar. Appl Soil Ecol 39:236–243

Rasse DP, Rumpel C, Dignac MF (2005) Is soil carbon mostly root carbon? Mechanisms for a specific stabilisation. Plant Soil 269:341–356

Resck DVS, Vasconcellos CA, Vilela L, Macedo MCM (2000) Impact of conversion of Brazilian Cerrados to cropland and pastureland on soil carbon pool and dynamics. In: Lal R, Kimble JM, Stewart BA (eds) Global climate change and tropical ecosystems. Advances in soil science. CRC, Boca Raton, pp 169–196

Richardson AE, Kirkby CA, Banerjee S, Kirkegaard JA (2014) The inorganic nutrient cost of building soil carbon. Carbon Manage 5:265–268. https://doi.org/10.1080/17583004.2014.923226

Rusinamhodzi L, Corbeels M, van Wijk M, Rufino MC, Nyamangara J, Giller KE (2011) A meta-analysis of long-term effects of conservation agriculture on maize grain yield under rain-fed conditions. Agron Sustain Dev 31:657–763

Saturnino HM, Landers JN (eds) (2002) The environment and zero tillage. Associação de Plantio Direto no Cerrado, Brasília, p 144

Scopel E, Triomphe B, Ribeiro MF dos S, Séguy L, Denardin JE, Kochhann RA (2004) Direct seeding mulch-based cropping systems (DMC) in Latin America. In: New directions for a diverse planet. Proceedings of the 4th international crop science congress, 26 September–1 October, Brisbane, Australia

Scopel E, Triomphe B, Affholder F et al (2013a) Conservation agriculture cropping systems in temperate and tropical conditions, performances and impacts. A review. Agron Sustain Dev 33:113. https://doi.org/10.1007/s13593-012-0106-9

Scopel E, Triomphe B, Affholder F et al (2013b) Agron Sustain Dev 33:113

Singh P, Heikkinen J, Ketoja E, Nuutinen V, Palojärv IA, Sheehy J et al (2015) Tillage and crop residue management methods had minor effects on the stock and stabilization of topsoil carbon in a 30-year field experiment. Sci Total Environ 51:8–51., 9: 337–344. https://doi.org/10.1016/j.scitotenv.2015.03.027

Sisti CPJ, dos Santos HP, Kohhann R, Alves BJR, Urquiaga S, Boddey RM (2004) Change in carbon and nitrogen stocks in soil under 13 years of conventional or zero tillage in southern Brazil. Soil Tillage Res 76:39–58

Six J, Bossuyt H, Degryze S, Denef K (2004) A history of research on the link between (micro) aggregates, soil biota and soil organic matter dynamics. Soil Till Res 79:7–31

Scopel E, Findeling A, Guerra EC, Corbeels M (2005) Impact of direct sowing mulch-based cropping systems on soil carbon, soil erosion and maize yield. Agron Sustain Dev 25:425–432

Thierfelder C, Wall PC (2010) Investigating conservation agriculture (CA) systems in Zambia and Zimbabwe to mitigate future effects of climate change. J Crop Improv 24:113–121

Thierfelder C, Rusinamhodzi R, Ngwira AR, Mupangwa W, Nyagumbo I, Kassie GT, Cairns JE (2014) Conservation agriculture in Southern Africa: advances in knowledge. Renew Agric Food Syst 30:328

Twomlow SJ, Urolov JC, Jenrich M, Oldrieve B (2008) Lessons from the field—Zimbabwes conservation agriculture task force. J SAT Agric Res 6(2008):1–11

Unger PW (1990) Conservation tillage systems. Adv Soil Sci 13:27–67. https://doi.org/10.1007/978-1-4613-8982-8_3

Verhulst N, Govaerts B, Verachtert E, Castellanos-Navarrete A, Mezzalama M, Wall P, Deckers J, Sayre KD (2010) Conservation agriculture, improving soil quality for sustainable production systems? In: Lal R, Stewart BA (eds) Advances in soil science: food security and soil quality. CRC Press, Boca Raton, pp 137–208

Vilela L, Martha JGB, Barcellos A, De O Barioni LG (2004) Integração lavoura/pecuària:a sustenibilidade do Cerrado. XXV Congresso de Milho e Sorgo, 29th August–9th September, Cuiabà-MT Brazil

Wilhelm WW, Wortmann CS (2004) Tillage and rotation interactions for corn and soybean grain yield as affected by precipitation and air temperature. Agron J 96:425–432. https://doi.org/10.2134/agronj2004.0425

Wolfarth F, Schrader S, Oldenburg E, Weinert J, Brunotte J (2011) Earthworms promote the reduction of Fusarium biomass and deoxynivalenol content in wheat straw under field conditions. Soil Biol Biochem 43:1858–1865. https://doi.org/10.1016/j.soilbio.2011.05.002

Yang XM, Kay BD (2001) Impacts of tillage practices on total, loose- and occluded particulate, and humified organic carbon fractions in soils within a field in southern Ontario. Can J Soil Sci 81:149–156

Part V

Enhancing the Productivity of Coastal Region

Water Management for More Crops per Drop in the Coastal Areas

19

Abstract

The availability and distribution of fresh water has steadily emerged as an important global issue in agriculture as it is directly linked with the food production system. The recent changes in rainfall pattern, projected decrease in per capita availability of water, water pollution due to increased human activities, and the ever-increasing demand for multiple use of water exert pressure on the already constrained water resources. Consequently rainwater harvesting and its judicious use should form the centre of any developmental strategy to increase agricultural production. Different methods are employed for rainwater harvesting and its efficient use to increase which are based on the local conditions. Protected cultivation, micro irrigation, land leveling, and use of water-efficient crops improve water productivity. The details are discussed in this chapter with some case examples and illustrations.

Keywords

Water crisis · Water productivity · Rainwater harvesting · Land leveling

19.1 Introduction

Water is the most important resource on Earth and has been a central issue on the international agenda for several decades. As the demand for water grows while the water resources are being polluted, many areas of the world are affected by water scarcity (Alcamo et al. 1997). The projected increase of the world population growth rate (UNDESA 2013) suggests that higher food demand is expected in the future, with a direct effect on agricultural water usage. In addition, as a result of the increased water scarcity and due to climate change (Jimenez Cisneros et al. 2014), extensive water withdrawal from the available freshwater resources for irrigation and other purposes is expected to occur in the context of increasing competition

V. Ayyam et al., *Coastal Ecosystems of the Tropics - Adaptive Management*,
https://doi.org/10.1007/978-981-13-8926-9_19

between agriculture and other sectors of the economy. Irrigation water is also required to combat climate change-related parameters, apply nutrients that are dissolved in the water, improve the physical properties of land, remove excess salinity from the soil, and modify the microclimate to promote high-value crops.

Agriculture in the tropical coastal region whether irrigated or rainfed directly or indirectly depends on amount and pattern of rainfall. Another important hydrological character of this region is that it experiences excess water during the rainy season, whereas water deficit/scarcity along with salinity is observed in the post monsoon period. Technological gap in improving water use efficiency is also a major concern for enhancing cropping intensity and extending irrigation facilities. This poses severe threat to agricultural diversification, enhancing agricultural production as well as productivity.

Since the balance between water demand and water availability has reached critical levels in many regions of the world particularly in the coastal areas, sustainable approach to water resource management in agriculture is essential to produce more crop per drop of water used. The sustainable water management concept refers to all practices that improve crop yield and minimize non-beneficial water losses. In addition, biofuel production is increasingly competing with food production, and this trend could reduce water resources and food availability (Nhantumbo and Salomao 2010). In order to cope with future projected water shortages, some measures aimed at streamlining and optimizing the efficiency of water consumption in the agricultural sector are critical. In this context, the aim of this chapter is to provide a broad review of several water-related issues and to highlight the potential solutions to cope with water scarcity for food production.

19.2 Production System Constraints Linked with Water Resources

The success of agriculture in the coastal and island ecosystem largely depends on performance of monsoon. It is characterized by excess water during the monsoon period followed by water scarcity in the post monsoon period. Apart from erratic rainfall pattern, the poor and inefficient water management practices result in about 70% loss of the rainwater, mostly as surface runoff to the seas, thereby reducing the per capita availability of fresh water to the coastal people. West coast of India presents a typical example where despite more than 3000 mm annual rainfall, the per capita effective rainwater availability is less than even some dry parts of India (Than desert, Rajasthan) mainly due to heavy surface runoff loss because of steeply sloping topography, excessive deforestation, and road construction. Moreover, the growing competitive demand for good-quality water from industry, power, and household sectors is further compounding the water crisis. The excess runoff results in eroding the top fertile soils of the region and also leads to poor groundwater recharge due to lesser ponding time. The indiscriminate and excessive withdrawal of groundwater in this region causes lowering of water table in several place and also ingress of sea water, which, in turn, leads to salinization of land and water. The poor drainage is another key factor limiting the agricultural production and low agricultural diversification in this region. The excess rainwater available during the monsoon period needs to be conserved through reduced runoff and storage in ponds so as to reuse it as life-saving irrigations to the crops during the lean periods. However, it is important to optimize the land and water allocation for maximizing the net profits. The problems related to productivity in coastal areas may be summarized as:

- Inundation of agricultural lands by tidal water and seawater ingress in aquifers.
- Impeded surface and subsurface drainage conditions.
- High soil salinity due to shallow saline groundwater in winter/summer months.
- High and erratic rainfall causes low availability of water for irrigation in dry periods and for second crop apart from poor irrigation infrastructure.
- Unscientific farming system reduces the crop and water productivity.
- Conflict between farmers practicing agriculture and brackish water aquaculture.
- Sea level rise, extreme weather events, and cyclones/storms in climate change scenario increase the vulnerability of coastal and island agriculture.

19.3 Rainwater Management Strategy

Integrated natural resource management is good for planning strategy, watershed management is good for programme implementation, and farming system approach is more suiting to beneficiaries. These terms are often used synonymously though their dimension of area varies from basin to farm scale (Fig. 19.1). The integrated approach revolves around rainwater management dealing with storage of excess rainwater during monsoon season and recycling it for irrigation of crops during deficit periods in dry season with the objective to introduce multicropping and multiple uses of water in the otherwise predominantly monocropped rice areas. However, looking to the small land holdings in coastal and island ecosystem, farming system approach seems to be more appropriate. As region faces the severe threat in the form of sea level rise, extreme weather events, storm/cyclonic disasters, and coastal erosion in climate change scenario, it calls for development of integrated strategies based on natural resource management for enhancing agricultural productivity to ensure livelihood suiting to different coastal subsystems.

The other aspects of the strategy should be increasing the water productivity of available water in various agro-ecosystems. High water productivity may not be a suitable target, where the concurrent opportunity costs or forgone values, like the alternative potentials of the saved or lost water, are not taken into account (Zoebl 2006). Therefore, understanding the concept of agricultural water productivity and exploring opportunities for its enhancement in irrigated, rainfed, and waterlogged areas through integrated and multiple uses assume greater importance. Agriculture with improved water management system and/or integrated farming and multiple water use systems, i.e. using the available water source(s) for more than one production system (crops, horticulture, livestock, fisheries, etc.) is the viable strategy to produce ‘more food with less water’.

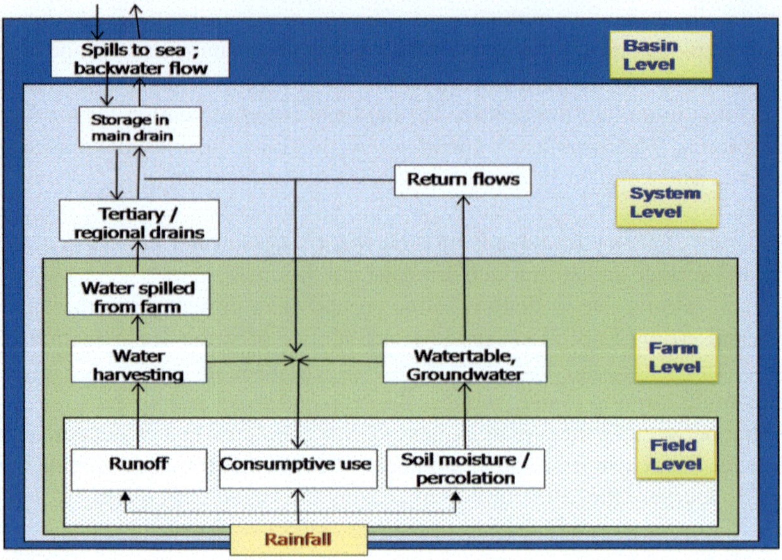

Fig. 19.1 A comprehensive framework of hydrological processes at different scales

19.4 Water Harvesting Methods

19.4.1 Land Shaping for Rainwater Harvesting

In humid coastal and island regions, surface waterlogging during *kharif*, non-availability of water for irrigation, and severe soil moisture stress during *rabi* are the major problems. In these areas, the agricultural development is centred on rainwater harvesting. Land configuration/shaping method can be a viable option for rainwater harvesting besides addressing the concern of waterlogging. Land shaping is defined as the articulation of land arrangement so as to overcome hydrologic problems in agricultural area for potential crop cultivation. Some of the water harvesting techniques suitable for coastal and island regions are farm pond, broad bed and furrow, paddy-fish, and three-tier systems (Fig. 19.2). A comparison of total amount of rainwater harvested per unit area, additional irrigation benefit, and impact on the cropping intensity is given in Table 19.1.

19.4.2 Subsurface Water Harvesting

In coastal areas due to scarcity of water during post monsoon season, the crop productivity is low; however, the rainfall is not enough to support the second crop, or surface moisture is lost at faster rate. At the same time below the surface, lighter fresh water floats above the heavier saline water in these areas. This could be tapped through subsurface water harvesting structures to meet the irrigation demand of *rabi*

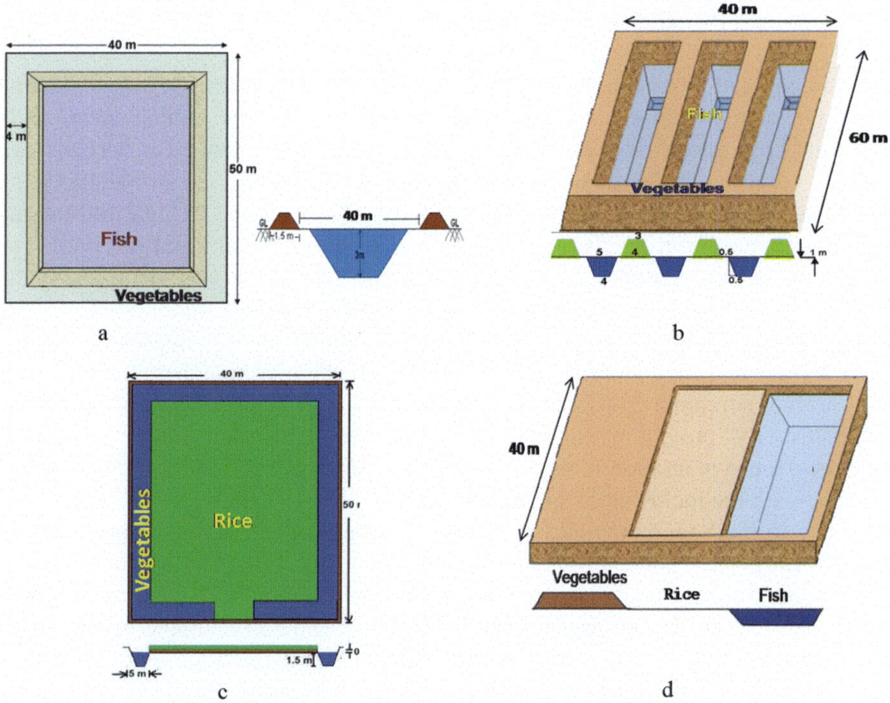

Fig. 19.2 Land shaping methods suitable for coastal and island regions. (**a**) Farm pond and its side view. (**b**) BBF system. (**c**) Paddy-fish system. (**d**) Three-tier farming

Table 19.1 Rainwater harvesting and its impact in different land shaping methods

Sl. No.	Land shaping method	Rainwater harvested (m^3 ha^{-1} $year^{-1}$)	Cropping intensity (%) Before intervention	Cropping intensity (%) After intervention	Additional area irrigated (ha)
1	Farm pond	8000	100	200	2.5
2	BBF	3812	100	240	1.0
3	Paddy cum fish	3564	100	200	1.0
4	Three-tier system	4509	100	220	1.5
5	Paired bed	3750	100	240	1.0

crop. Subsurface water harvesting structure varies from 200 to 1500 m^3 depending on the size of the holding, and the depth can vary from 2 to 4 m. The depth should be restricted to sandy zone, and only 1–2 hp pumps are recommended to lift water so as to avoid intrusion of saline water. The system also helps to mitigate early drought in *kharif* crops. The average water productivity of this system is Rs.36 per m^3 of water (Chowdhury et al. 2016).

19.4.3 Lined Ponds for Hilly Areas

In the sloppy and mountainous terrain, if a part of runoff, which goes waste, could be stored, supplementary irrigation can be provided to the annual crops and plantation crop during dry season. This will provide ample scope for introducing multiple cropping in the monocropped areas and enhancing crop productivity in such areas where irrigation facilities are not available. But due to high infiltration rate and seepage from ponds, the storage level in unlined tanks and reservoirs in these hilly terrains reduces at a faster rate. The seepage losses are quite high particularly in the island region as the lower strata comprises fractured stones or gravels. Further, high evaporation rate and gravitational flow increases the water loss. For these reasons annual crops could not be cultivated in the slopes alternatively; in these areas perennial crops such as coconut and areca nut are grown under rainfed conditions. By providing suitable rainwater harvesting and micro irrigation structures, these plantation monocropped areas can be converted into multistoried cropping with the inclusion of spices and fruits.

Owing to the huge potential for enhancing crop production and diversification through rainwater harvesting, several sealants and brick lining have been recommended for minimizing seepage losses from the ponds. However, it did not find favour with the farmers due to its low efficiency and cost among other reasons. Pooranchandan et al. (1978) have recommended a 3-mm-thick bitumen spray, but this also proved to be ineffective due to seasonal influences. Similarly it is quite difficult for brick lining as the availability of bricks are limited, and tanks are constructed by mechanical excavator due to shortage of manpower. The side slopes of these tanks are made in a ratio of 0.5:1 due to the cutting by mechanical excavator. This makes difficult for lining by bricks and cement concrete. Alternatively lining of tanks with LDPE film, covered with soil with a side slope of 3.5:1, was also suggested. Rao et al. (1980) have suggested a 2:1 side slope with steps having film on bottom and sides covered with soil for effective storage of rainwater in the hill slopes.

A study conducted at ICAR-CIARI (2007) revealed that the daily water losses from different lined ponds constructed by using different lining and cover materials were observed in the range of 3–6 mm/day, whereas daily evaporation was observed in the range of 2.5–5.5 mm/day. Silpaulin was found to be a cost-effective and best lining material suitable for lining (Fig. 19.3). Water can also be collected from roofs of polyhouses or other structures. A tank of suitable dimension (15 × 7 × 2.5 m for a 1 ha land) size constructed at hill top or slopes can be lined with silpaulin for effective storage of water and provide irrigation to crops. The bottom bed and side slope area of the pond have to be well dressed and compacted to remove any angular projections. The film is spread and anchored on sides, and the film at the bottom is buried under 15 cm layer of soil. Tiles and cement blocks can also be used above the silpaulin lining in order to prevent the film getting exposed to the UV light from the sun. Since most of the tanks are on the top of the hillock or on the slope, the outlet should be designed properly so that a gravity flow is possible without any damage to the lining work.

Fig. 19.3 Lining of tank—dug out tank and reinforced plastering

The construction cost varies from Rs. 50,000 to 70,000 depending on the materials used and site characters. It was found that by supplying supplementary irrigation to areca nut plantations during the dry period, yield increased by 1.5 times than the yield from a rainfed areca nut plant.

19.4.4 Micro Tube Wells in Areas Having Saline Groundwater

In the coastal and island conditions where saline groundwater exist below 5–10 m, use of tube wells is not advisable due to the risk of saline water intrusion. In such condition a micro tube well can be used to extract shallow fresh water and avoid saline water intrusion. The depth of the tube well should not exceed 10 m or just above the saline water. The diameter of the well should be only 5 cm, and the capacity of the pump should be within 2 hp capacity. It can support 1–1.5 ha areas for *rabi* season crops. When combined with micro irrigation, it can greatly enhance crop productivity and diversification in otherwise rainfed areas (Chowdhury et al. 2016).

19.4.5 Traditional Method of Small Lined Tank (*Jal Kund*)

This is a kind of rainwater harvesting structure practised in rainfed areas with uncertain rainfall. It is known in different local names depending on where it is used. The most popular Konkan Jal Kund is an underground tank that stores rainwater with the help of polyethylene sheet. It is also known as polyethylene-lined tank. A tank of size 4 × 1 × 1 m or 2 × 2 × 1 m is excavated depending on the land size and quantity of water to be stored. After lining the pit with rice husk, a polyethylene sheet is laid on it to develop a proper-shaped water storage tank. The evaporation losses can be minimized by covering the tank with panels made from locally available bamboo and rice husk or dry grass. This meets the crop water requirement during dry season and long break in monsoon particularly in a small holders farm. This decentralized approach will be highly useful and practicable in such areas where developments of large- to medium-scale irrigation structures are not feasible due to topographical limitations.

19.5 Efficient Water Use

19.5.1 Land Leveling and Drainage

In the undulated and gently sloping areas, land leveling will improve the movement and drainage of water at faster rate. Further, land leveling takes care of several on site/farm terrain deformation thus facilitate efficient use of irrigation water in crop production. Besides, land leveling minimizes the changes of waterlogging in the field. On the other hand, most of the coastal low-lying areas are frequently inundated due to ingress of sea water through tidal estuaries. These estuaries are slowly silting up due to high silt load carried along with the river flows causing congested drainage system and thus flooding of agricultural fields. It has been suggested that earthen embankments, preferable brick pitched, having side slope of 3:1 on the river side and 2:1 on the country side, with 1 m free board above the high tidal level, are appropriate for protection against flooding (Rao and Dhruvanarayana 1979). In order to facilitate quick removal of excess rainwater and to stop the ingress of sea water in low-lying areas, construction of peripheral embankments with provision of one-way sluice gates and construction of drainage network will be necessary.

19.5.2 Precision Farming and Improved Water Use

Precision farming or site-specific farming is an emerging technology that allows farmers to have most efficient use of inputs and agronomic practices. It has the potential not only to reduce cost of cultivation through more efficient and effective application of crop inputs but also to protect the fragile environment. Precision farming is very essential because inputs in agriculture such as fertilizers and pesticides and based on nonrenewable source of energy. This calls for efficient use of inputs when the distribution of rainfall is not uniform throughout the year. A careful water balance analysis of annual precipitation and potential evapotranspiration is essential to determine the water surplus/deficit periods beforehand, so that all the agricultural operations can be adjusted accordingly. This will help to save the crops from water stress (excess/deficit) during its critical growth stages and also help the farmers in the decision-making regarding crop choice as well as its acreage depending upon the water availability. The concept of harvesting rainwater in on-farm reservoir (OFR) may be effectively utilized by applying exact quantity for growing pulses, oilseeds, tuber crops, vegetables, spices, etc. after *kharif* rice.

Precision farming enables judicious use of water for irrigation or fertigation; for this purpose the rainwater harvesting can meet large portion of crop water requirements. Limited information is available on water harvesting and irrigation methods for the coastal ecosystems, where, in general, the soils are to a large extent saline and medium to heavy texture in nature. Under such conditions, micro irrigation, especially in the form of drip, should be a potential irrigation method for coastal environment (Fig. 19.4). Drip irrigation system can prove beneficial particularly to irrigate the high-value crops such as plantation crops of coconut,

Fig. 19.4 Drip irrigation system

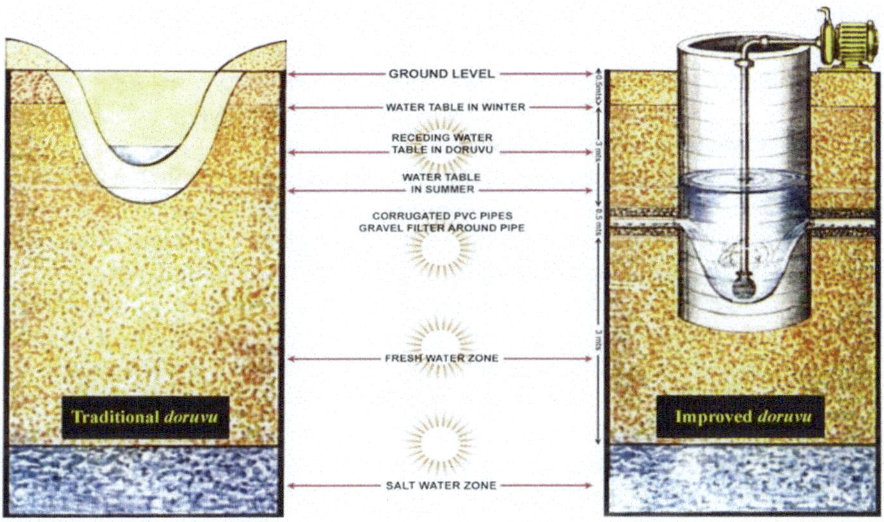

Fig. 19.5 Improved *Doruvu* for freshwater skimming

areca nut, oil palm, cocoa, and cashew and spices such as black peeper, cardamom, ginger, turmeric, coriander, and cumin. The controlled use of water and fertilizer directly in the root zone through drip irrigation not only increases the input use efficiencies but also reduces possible negative interactions in the soil profile. Besides, this provides scope of using poor quality water in conjunction with fresh water for irrigation while it increases the crop productivity.

Another improved technology to utilize the surface water is life-saving irrigation by digging *doruvus* (Fig. 19.5) which is a conical pit dug to collect seeped in water. This method is suitable particularly for sandy loam soil where fresh water floats over saline groundwater. This can be skimmed horizontally through tile drains at the rate of 18 lps, which would be sufficient for operating six sprinklers continuously. The technology has been successfully used in Andhra coast and can be upscaled for similar situations.

19.5.3 Proper Choice of Crops, Cropping and Farming System

The yield of several crops and cropping systems can be substantially improved if proper choice is made with regard to the season and location. Crop varieties tolerant to salinity and waterlogged conditions can yield better than susceptible crops, thereby increases the water productivity than the normal varieties.

19.5.3.1 Crops for Island Conditions

In general, in the low-lying waterlogged and salt-affected soils, farmers grow tall, extremely long-duration traditional paddy cultivar. It covers about 60–70% of the rice-cultivated area in Andaman Islands. In its place, farmers can use long-duration high-yielding paddy varieties, viz. Ranjeet and Varshadhan, in waterlogged areas which are suitable for such conditions. Some of the salt-tolerant rice varieties and suitable vegetable/fruit crops (Table 19.2) cultivated at proper season in this region resulted in higher yield and enhanced the water productivity.

19.5.3.2 Coastal and Saline Areas

The coastal saline and waterlogged soils are monocropped with low-yielding, traditional rice varieties during the monsoon season from June to December. Cropping intensity in these areas can be increased by proper choice of crops, and their varieties fit into an efficient cropping system. It is often observed that most of the saline areas in the coastal region remain fallow in the dry season because of high soil salinity and the lack of good-quality irrigation water. A combination of on-farm storage of surface water, to prolong freshwater availability beyond the end of the rainy season, together with the proper selection of rice varieties, can increase cropping intensity and productivity of these areas. In the wet season, the traditional rice varieties should be replaced by short-duration, high-yielding varieties (HYV), which can be harvested earlier, about 30–45 days before traditional varieties. This opens up opportunities for early establishment (November) of short-duration HYV of rice during the dry season. The stored water can be used to irrigate rice from February to the end of March as supplemental irrigation.

Table 19.2 Vegetables and fruit crops with their salinity tolerance for coastal areas

Vegetable crops	Varieties	Salinity levels (dS/m)
Brinjal	Pusa hybrid 6 and Pusa Uphar	8.0
Tomato	Pusa Ruby, Pusa 120, and Pusa Rohini	8.0
Chilli	CA 960, Suryamukhi and Pusa Jwala	7.0
Watermelon	Sugar sweet and sugar baby	6.0
Sapota	Badami, cricket ball, and Kalapatti	10.0
Guava	KG, Kashi, and L-49	6.0

Table 19.3 Biomass yield (t/ha) of promising forage grasses under soil salinity

	Biomass yield (t ha^{-1})					
Soil salinity range (dSm^{-1})	Hybrid Napier (DHN-6)	Hybrid Napier (DHN-9)	Guinea grass	Grazing guinea	Para grass	Rhodes grass
<4	3.89	3.30	5.48	6.20	7.58	18.45
4–6	3.56	3.21	4.89	5.57	6.89	16.78
6–8	2.85	2.71	3.53	5.10	5.74	14.63
8–10	2.89	2.62	3.83	4.00	4.68	12.74
10–12	2.04	1.94	2.08	3.68	3.56	11.46
12–14	1.17	1.20	1.34	2.58	2.56	10.85
14–16	1.10	0.92	1.16	2.12	2.45	9.60
>16	0.85	0.80	0.95	1.34	1.50	8.34

19.5.3.3 Use of Overaged Rice Seedlings in Waterlogged Areas

In low-lying areas, crop yield is affected due to unpredictability of extent of waterlogging. In such situations use of overaged rice seedlings is promising in avoiding seedling mortality and low productivity. Taller, overaged (40–50 days old) seedlings at 20 × 20 cm spacing should be practised. Phosphorus and potassium at 40:40 kg/ha should be applied during sowing time, and nitrogen is applied through mudball in split doses.

19.5.3.4 Integrated Farming System and Inclusion of Grasses

The land use statistics and remote sensing image of the coastal and island regions indicated lot of culturable wasteland, whereas the pressure on land for higher productivity has been increasing. At the same time, the non-availability of grasses limits the expansion of livestock sector and production. Both these constraints can be converted into opportunity by growing salt-tolerant grasses in the wasteland. In these areas land allocation for forage crops and integration of animal component with crop production will ensure not only increased income but also stability to the production system. There are several grass species found suitable for cultivation in the saline areas which perform better than several food crops. Among several grasses, Rhodes grass perform better even in high salinity of more than 10 dSm^{-1} followed by Para grass (Table 19.3). The income and productivity from integrated farming system model are two to four times higher than single crop cultivation in the coastal and island regions particularly in saline and acid-saline conditions.

19.6 Enhancing Water Productivity

Given the growing water scarcity that is resulting from increased water demand, the challenge is to increase water productivity for food and livelihoods, thereby usher in an era of the 'evergreen revolution'. It can help alleviate poverty through improved access to water for the poor and also have multiplier effects on food security, employment, and livelihood. Besides the physical availability of water, yet another cause for concern in obtaining the maximum agricultural output, or value for every drop of water used in agriculture, is the economical accessibility of water. This necessitates the efficient and low-cost technologies to enhance the water productivity which are acceptable to farmers.

19.6.1 Understanding Water Productivity

Water productivity is used exclusively to denote the amount or value of product over volume or value of water used or depleted or diverted (Kijne et al. 2003). Water productivity will depend on many factors other than the quantity of water applied or depleted. Though water is only one of the factors of agricultural production and cannot be meaningfully separated from the others, an estimate of its productivity and knowledge about the factors which influence the productivity will help understand the pathway to improving water productivity.

High water productivity may not be a suitable target, where the concurrent opportunity costs or forgone values, like the alternative potentials of the saved or lost water, are not taken into account (Zoebl 2006). Understanding the concept of agricultural water productivity and exploring opportunities for its enhancement in irrigated, rainfed, and waterlogged areas through integrated and multiple uses assume greater importance due to the perceived climate change. It is very pertinent to the coastal and island regions as these areas suffer from both excess and deficit rainfall and moisture availability, but in different seasons in a year.

19.6.2 Assessment of Water Productivity

Agricultural water productivity takes into account multiple water users, including conventional crops, horticulture, forestry, livestock, fisheries, environment, etc. It means that if all water users are taken into account and a concept of recycling and reuse of water is considered in an agricultural production system, then agricultural output per unit of total water input is referred to as agricultural water productivity. Since agricultural water productivity assessment considers the multiple uses of water, its value represents a composite or an integrated picture that is higher than the crop water productivity (Sikka et al. 2008).

Agricultural water productivity may be computed for the crop period or for the whole year considering the production value—first from crops only (crop water productivity) and then considering other water users including trees, livestock, and fish in case of multiple uses per unit of water inflow (including rainfall, groundwater, and canal water) as well as water used (total inflow excluding runoff) in the field, farm, irrigation system, basin, and landscape. The values of denominator, i.e. water input, may be different for different situations. In the estimation of water productivity, we are interested in water inflows (rain plus irrigation or just rainwater in rainfed agriculture) and water depletion (evaporation and transpiration). The amount of water depleted through evapotranspiration by crops will not be considered again while accounting for depletion in case of crop residues/straw for livestock, to avoid double accounting. Similarly, in case of fisheries, water depletion through storage losses is to be considered as input only when water storage in the pond/ tank is exclusively maintained for breeding fish.

Agriculture with improved water management system and/or integrated farming and multiple water use systems, i.e. using the available water source(s) for more than one production system (crops, horticulture, livestock, fisheries, etc.), is one of the responses to produce 'more food with less water'. In a case study of agricultural water productivity assessment, it was computed considering the output of various water uses like cereal crops, horticulture, trees, fisheries, and livestock. Water productivity was computed during kharif (monsoon) and rabi (winter) seasons considering the production value—first from crops alone and then considering other water uses including trees, livestock, and fish (Sikka et al. 2008). Crop water productivity (Rs/m^3) in relation to applied water ranged from 4.79 to 8.39, and in tube well commands, the applied water productivity ranged from 14.03 to 29.61. Similarly, agricultural water productivity (Rs/m^3) of applied water varied from 5.28 to 10.66 (Table 19.4).

The analysis indicates that both in the canal commands and tube well commands, agricultural water productivity of applied as well as total water inflow (including rainfall) taking into account other water users like trees, fodder, livestock, fish, etc. provides a better picture of the actual productivity of water. Multiple water use-based farming system with on-dyke horticulture and fish-prawn-poultry system in farmers' field in Orissa provided an excellent opportunity to productively use waterlogged area. The farmers converted 2.47 ha of waterlogged area into 1.64 ha of pond and 0.83 ha of raised embankment. While the pond area was utilized for fish and prawn

Table 19.4 Crop and agricultural water productivity

Water productivity (Rs/m^3)	Outlet 4 head reach	Outlet 17 middle reach	Outlet 27 tail reach	Tube well 2 land consolidation	Tube well 11-fragmented landholding
Area (ha)	30.61	43.68	4.65	18.74	13.21
Crop WP per unit of irrigation water applied	4.79	4.95	8.39	29.61	14.03
Crop WP per unit of water inflow including rainfall	2.42	2.73	3.11	2.81	2.39
Agricultural WP per unit of irrigation water applied	5.28	5.90	10.66	38.73	18.09
Agricultural WP per unit of water inflow including rainfall	2.67	3.25	3.96	3.68	3.09

Source: Sikka et al. (2008)

culture, 21-m-wide embankment was used for planting mango, teak, areca nut, coconut, banana, papaya, pineapple, mushroom, etc. Net water productivity of multiple use system was estimated to be Rs.7.5/m^3 against Rs.0.95/m^3 for lowland rainfed paddy alone and Rs.6.0/m^3 with vegetable production (Samra et al. 2003).

19.6.3 Paradigm Shift in Approach to Improve Water Productivity

A comprehensive assessment of water management in agriculture taken up by the International Water Management Institute (2001) calls for abandoning the 'obsolete divide between irrigated and rainfed agriculture' and to acknowledge rainfall as the key freshwater resource (Fig. 19.6).

The present paradigm operates on either of the extremes of rainfed or irrigated. Groundwater recharge and erosion control are the dominant elements of public investments in watershed development in rainfed areas, while a 'full-irrigation' centricity dominates the other end. Several studies suggest that better management of rainwater, soil moisture, and supplemental irrigation holds the key to high water productivity. It is estimated that a 35% increase in water productivity could reduce additional crop water consumption from 80% to 20% (Molden 2007). A study of over 500 districts in India concludes that provision of critical irrigation during the mid- and terminal drought periods has the potential to improve the yields by 29–114% for different crops (Sharma et al. 2010). The high levels of marginal productivity of water (through supplemental irrigation) in rainfed agriculture offer immense potential for national food security.

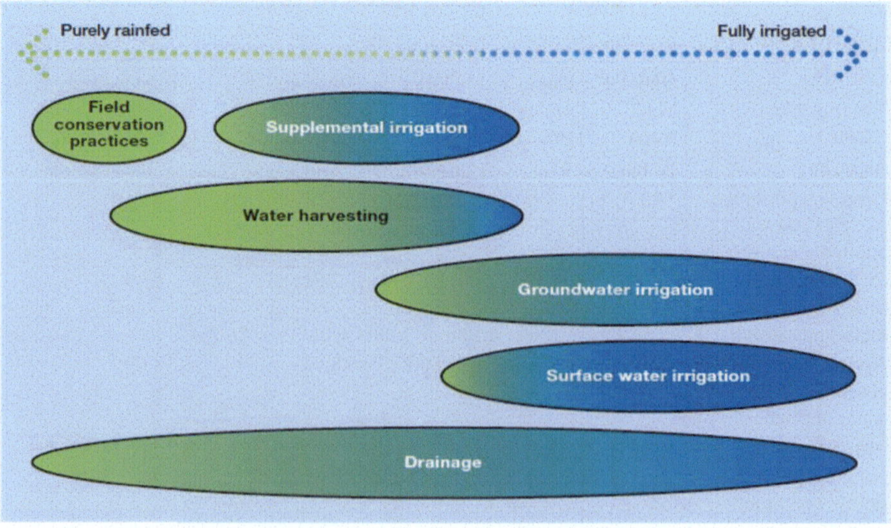

Fig. 19.6 Diverse options for agricultural water management along the spectrum

19.7 Water Management in the Tropical Region: A Case Study of South Andaman, India

Andaman Islands receive an average annual rainfall of about 3080 mm. Of that, nearly 95% rainfall is received during May to December, and the remaining 4 months during January to April is almost dry period. During dry period, agriculture suffers badly due to moisture stress. As rainwater is the only source of the freshwater availability, its harvesting and management forms the most important strategy of natural resource management in these islands (Ambast et al. 2008). Construction of drop spillway, an energy dissipation structure, in a series on a seasonal stream (nallah) at appropriate locations not only helps in preventing soil erosion but also acts as recharge structure when integrated with small reservoirs. These structures when integrated with small storage reservoir in the upstream area can also serve the purpose of rainwater harvesting during the period of excess water so as to utilize the stored water during the dry season.

19.7.1 Design of Drop Spillway

Drop spillways are abrupt decreases in channel-bed elevation with a downstream stilling basin used to dissipate hydraulic energy in open channels. Drop spillways often combine both hydraulic jump and impact features, although not all design situations are associated with hydraulic jump (Fig. 19.7). The energy dissipation occurs via floor blocks, end sill at downstream of basin, and turbulence in the tail water. The design of drop spillway thus involves design of stilling basin, floor blocks, and side and wing walls using empirically determined relationships (Donnelly and Blaisdell 1965).

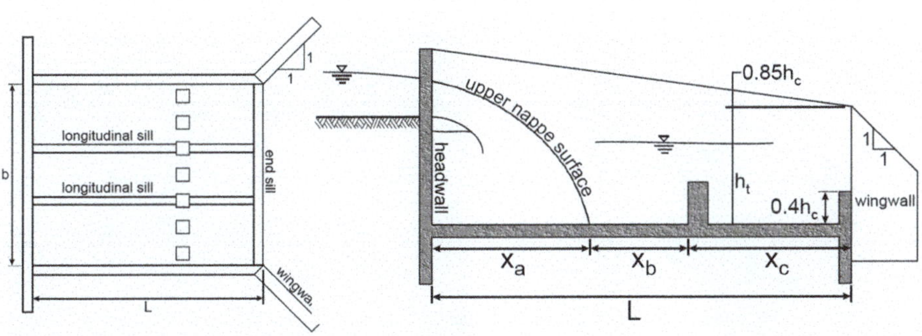

Fig. 19.7 Plan and side view of a drop spillway structure

19.7.2 Rainwater Management in Watershed

A water resource development model was developed for a small watershed. Spatial information techniques have been used to delineate watershed boundary using digital elevation model. The subsurface strata in the watershed is hard rock with localized pockets of moderate aquifer of fine sand at depths between 4 and 8 m which can be used for drawing the groundwater from shallow aquifers. In order to harvest the rainwater, a series of recharge structure-cum-well have been made in a model watershed (kajunallah). In order to study the impact of recharge structure on well yield and on water resource availability in the watershed, monthly recuperation test for seven wells were conducted during dry season (Fig. 19.8). The wells 1, 2, and 3 are located close to the bed of nallah where longitudinal slope of the nallah is more compared to the wells 4, 5, 6, and 7, which are located in the valley area. It has been observed that the well yield reduced significantly over the dry season particularly in the wells located at higher longitudinal slopes in comparison to wells located in the valley areas where longitudinal slope was less. It has also been observed that wells 1 and 5 which were supported by check dam/dugout storage had higher well yield (Fig. 19.9). However, the significant reduction in yield of well 1 was observed which was located in higher longitudinal slope. Thus it is suggested that recharge structure-cum-well system should be created at relatively less longitudinal slope along with recharge structure or dugout storage to get significantly enhanced and consistent well yield during the dry season (Ambast et al. 2010).

Further estimates have been made on surface and groundwater development with reference to water resource potential, realizable water potential, water requirement of crops, and livestock and human consumption to assess the water balance in the watershed. It indicated that development of about 10% of the realizable potential will meet the water requirement in the watershed, but at present, there is a shortfall in water resource development by about 21% of the water requirement.

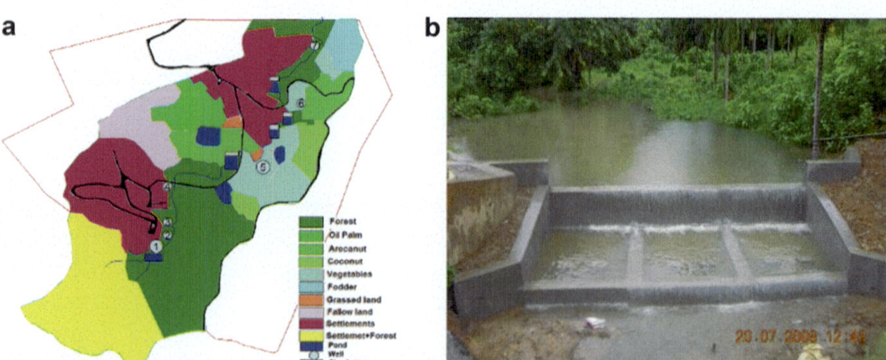

Fig. 19.8 (**a**) A model watershed showing the land use map and (**b**) view of a check dam located in the watershed

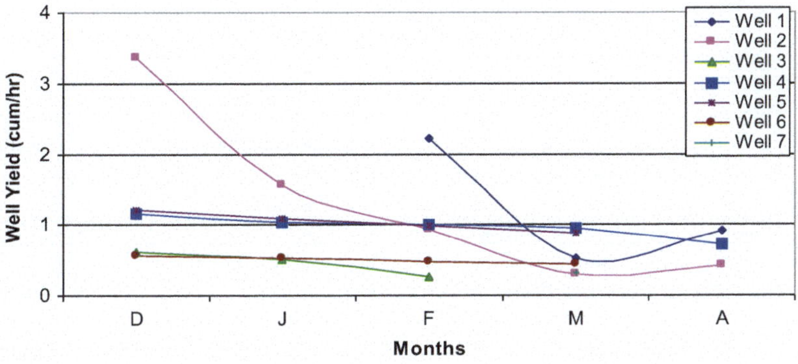

Fig. 19.9 Well yield in wells located downstream of check dam during dry period

19.8 Climate Change and Future Thrust Areas for Water Management

The major economic activities in coastal areas are agriculture, animal husbandry, fishery, and tourism. These areas entirely depend on rainfall for fresh water, which is severely scarce during dry period and not available for irrigation purposes. Any change in the pattern and quantum of rainfall, rise in temperature, and sea level as a consequence of climatic change would lead to scarcity/excess of water in the form of frequent droughts/floods resulting damages to land, crops, and animals affecting livelihood options in the coastal regions and tropical islands.

19.8.1 Management of Location-Specific Problems

In this context, emphasis should be given to location-specific problem identification with respect to water harvesting and its subsequent efficient use. A summary of different methods are given in Table 19.5 but is not exhaustive. Therefore, new methods and studies should focus on developing technologies for improving the water productivity in these regions.

19.8.2 Good Policy Measures to Improve the Water Management

Improvement in water management also depends on the policy support and its effective implementation at ground level. Some of the required policy and related supportive activities are listed below (IWMI 2007):

- Approach water management not just as a technical issue but also as a social and political issue. This would require meeting the multiple water needs of poor women and men—for growing food, for drinking, for enabling hygiene and sanitation, and for generating income through a range of activities.

Table 19.5 Summary of different methods of water harvesting and its efficient use

S.No	Location/management system	Details of method/techniques
	Water harvesting	
1	Hilly and top of hills	Lined tanks designed based on rainfall and substrata, terracing, and basin around trees
2	Island ecosystem/undulating terrain	Recharge structure-cum-well system
		Check dams
3	Island and coastal lowlands with waterlogging condition	Land shaping methods: BBF, RF, three-tier farming, farm pond
4	Coastal areas (East and West)	Properly designed farm pond and recharge of aquifers, subsurface water harvesting
5	Coastal Andhra Pradesh	Improved *Doruvu* for freshwater skimming and recharge of aquifers
6	Saline, aqua pond, and other degraded areas	Land leveling, open or subsurface drainage, storing rainwater in farm pond and use of one-way sluice gate
7	Coastal areas of Orissa, Andhra Pradesh, and Tamil Nadu	Micro tube wells with small pump sets
8	Coastal and island with gentle sloppy/ midland	Shallow pond is made in SRI field to collect excess water during rainy season and use during dry season
	Efficient use to produce more crops	
9	Plantation areas in coastal and island regions, rice monocropped areas in the coastal plains	IFS and multistorey cropping (for higher water productivity)
10	Coastal and degraded land	Protected cultivation
11	Rainfed and dry season in all areas	Use of drip/sprinkler irrigation
12	Using harvested water in saline areas and dry season	Supplemental irrigation and scheduling of irrigation
13	In the lowlands	Crops and cropping intensity
	Minimize losses/conservation to produce more	
14	Mulching—crop residue and plastic	To reduce moisture loss from soil and enhance the chances of dry season crop
15	Paired row planting	It minimizes water loss
16	Zero tillage for *rabi* crop	In waterlogged areas, second crop is grown after rice under zero tillage condition

- Institutionalize the development, coordination, and negotiation mechanisms among the state, civil society, and private organizations in water development and management and in related sectors for effective delivery and continuity beyond the political will.
- Agricultural water management is not an isolated issue of providing water alone, rather it requires integrated approach. Examples include managing water to enhance ecosystem services in addition to crop production, incorporating livestock and fisheries into water management, improving rainwater management and encouraging investments to upgrade rainfed production, and supporting systems and services that encompass multiple water uses, safe reuse of wastewater, and joint use of surface water and groundwater.

- Revamp the effectiveness of the state, particularly in its regulatory role, and find the right balance between action by the state and by other institutional actors.
- Create incentives and encourage water users, government agency, and NGOs to improve the equity, efficiency, and sustainability of water use.
- Build coalitions among government, civil society, and private and community users—and harness market forces for successful reform in water harvesting and management. At the same time, empower women, poor, and other stakeholders in the rural areas by making water their basic rights and involve them in water management to progress towards the development goals.

19.8.3 Future Research Need

Coastal areas are highly fertile and rich in natural resources but are fragile due to natural and anthropogenic reasons. The uncertainty in weather and likely extreme events are expected to further affect the unstable agricultural productivity in coastal region. The following water-related issues require more focus in future research to enhance the water productivity.

- Database on crop/farm/system productivity for coastal districts are important to set the priority areas of concern in coastal ecosystem.
- Application of modern techniques such as remote sensing and geographical information system may be used in future coastal agricultural research. However, application of microwave remote sensing may be explored in combination of optical remote sensing due to continuous cloud cover in the coastal regions.
- Monitoring of planned surface drainage improvement schemes would help in identifying the reasons of their poor performance and suggestive measures for their improvement.
- Different farming system options, available in each coastal subregion, may also be evaluated for water productivity apart from land productivity and their economics.
- Land shaping interventions should be seen as mitigation measures in the wake of sea level rise due to climate change and need to be evaluated at different subregions for suitable modification and adoption.

19.9 Conclusions

Enhancing water productivity in agriculture is an appropriate response to meet the food requirement and counter the challenge imposed by climate change. There are several ways by which rainwater can be harvested in the coastal and island regions. Protected cultivation, micro irrigation, land leveling, and IFS-based production system significantly enhance the agricultural production. Land shaping methods are best suited for utilizing low-lying areas as well for rainwater harvesting. The estimation of agricultural water productivity takes account of different water uses and production systems; it presents a better and composite picture of water productivity obtained in farming systems. The two case studies presented propose the way to assess the suitable option for rainwater harvesting in coastal/island areas. In Sundarbans delta, it is suggested to convert 20% of the farm area into on-farm reservoir to harvest excess rainwater during *kharif* season in Sundarbans delta. Storage of excess rainwater in on-farm reservoir may lead to reduction in surface waterlogging, and optimal land allocation indicated a cost-benefit ratio around 2, which justifies the investment. In island situations, drop spillway, an energy dissipation structure in open channel, is proposed for rainwater harvesting and in combination with well in the downstream can provide supplemental irrigation to crops during dry season.

References

Alcamo J, Doll P, Kaspar F, Siebert S (1997) Global change and global scenarios of water use and availability: an application of WaterGAP 1.0. University of Kassel, Kassel

Ambast SK, Srivastava RC, Raja R, Meena BL, Chand S (2008) Planning, augmentation and efficient utilization of water resources in Kajunallah watershed, Annual Report, 2007–08. Central Agricultural Research Institute, Port Blair

Ambast SK, Srivastava RC, Subramani T, Subhash C, Ravisankar N (2010) Design of drop spillway in undulated terrains for rainwater management in Kajunala watershed of A&N Islands. J Indian Soc Coast Agric Res 28(2):51–55

Chowdhury R, Rautary S, Panda SK, Das RK, Srivastava RC, Ambast SK (2016) Technological options for agricultural water management in Eastern regions of India. ICAR-IIWM, Bhubaneswar, p 39

Donnelly CA, Blaisdell FW (1965) Straight drop spillway stilling basin. ASCE J Hydraul Div 3:101–131

ICAR-CIARI (2007) Studies on effective storage of water in ponds, Annual Report 2007–08. Central Island Agricultural Research Institute, Port Blair, p 43

IWMI (2001) IWMI climate and water atlas. International Water Management Institute, Colombo

IWMI (2007) Water for food, water for life: a comprehensive assessment of water management in agriculture. Earthscan, and Colombo: International Water Management Institute, London. www.earthscan.co.uk

Jiménez Cisneros BE, Oki T, Arnell NW, Benito G, Cogley JG, Doll P, Jiang T, Mwakalila SS (2014) Freshwater resources. In: Field CB, Barros VR, Dokken DJ, Mach KJ, Mastrandrea MD et al (eds) Climate change: impacts, adaptation, and vulnerability. Part A: global and sectoral aspects. Contribution of working group II to the fifth assessment report of the intergovernmental panel on climate change. Cambridge University Press, Cambridge, pp 229–269

Kijne J, Barker R, Molden D (2003) Improving water productivity in agriculture: editors' overview. In: Kijne J, Barker R, Molden D (eds) Water productivity in agriculture: limits and opportunities for improvement, comprehensive assessment of water management in agriculture. CABI Publishing in Association with International Water Management Institute, Wallingford

Molden D (2007) Water for food water for life: comprehensive assessment of water in agriculture. Easthscan, London

Nhantumbo I, Salomao A (2010) Biofuels, land access and rural livelihoods in Mozambique. Institute for Environment and Development, London

Pooranchandan G, Chinnameni S, Jaykumar M (1978) Preliminary studies on the effect of different sealant materials for the storage tanks to the seepage losses in the Nilgiris. Agric Eng 20:7–9

Rao KVGK, Dhruvanarayana VV (1979) Hydrology and drainage condition of coastal saline soils. In: Proceedings first conference on problems and management of coastal saline soils, CSSRI, RRS, Canning, pp 29–52

Rao DH, Verma B, Narain P (1980) A case study on farm pond and its lining in semi- arid tract of south eastern Rajasthan. Paper presented in national symposium on soil conservation and water management in 1980's. March 12, 13 1980 at Dehradun

Samra JS, Sahoo N, Chowdhury S, Mohanty RK, Jena SK, Verma HN (2003) Sustainable integrated farming system for waterlogged areas of eastern India, Research Bulletin No.14. Water Technology Centre for Eastern Region (Indian Council of Agricultural Research), Bhubaneswar

Sharma BR, Rao KV, Vittal KPR, Ramakrishna YS, Amarasinghe U (2010) Estimating the potential of rainfed agriculture in India: prospects for water productivity improvements. Agric Water Manag 97:23–30

Sikka AK, Bhatnagar PR, Haris AA, Kaushal DK, Prasad LK, Dey A, Redy AR, Kumar U (2008) Productive utilization of waterlogged areas through multiple water use systems in Eastern India. Bulletin No. R-24/PAT-14

United Nations Department of Economic and Social Affairs (UNDESA) (2013) Population Division. World population prospects: the 2012 revision, volume II, demographic profiles, ST/ESA/SER.A/345. United Nations, New York

Zoebl D (2006) Is water productivity a useful concept in agricultural water management? Agric Water Manag 84:265–273

Approaches in Land Degradation Management for Productivity Enhancement

20

Abstract

Tropical coastal lands are vulnerable to different processes of land degradation such as salinization, waterlogging, drainage congestion, erosion, and acid sulphate formation. This is a serious impediment for agricultural growth and sustainable utilization of natural resources in the coastal region. Rehabilitation of degraded land is vital for improving the coastal ecosystems and provisioning of ecosystem services to support livelihood of coastal communities. Implementing innovative land management measures such as permanent raised beds and furrow; rice-cum-fish; three-tier land management; ponds with broader dykes, deep furrow, and high ridge; ridges and furrow systems; and brackish water aquaculture in combination with productive utilization of socio-economic opportunities of the coastal areas will help to address the land degradation concerns. Besides, these methods also enable to harvest rainwater and diversification of agriculture by favouring integration of fish and livestock components with crop cultivation providing greater stability to farm income. These land shaping methods also help to cope with the weather variations and projected impact of climate change.

Keywords

Coastal lowland · Degradation process · Salinization · Land shaping · Raised beds

20.1 Introduction

Land degradation will remain an important global issue for the twenty-first century because of its adverse impact on agronomic productivity, the environment, and its effect on food security and the quality of life. Land degradation is often related to decline in land quality, caused through its misuse by humans which results in decline in productivity (Eswaran et al. 2001). It can be defined as the rate of adverse changes

V. Ayyam et al., *Coastal Ecosystems of the Tropics - Adaptive Management*,
https://doi.org/10.1007/978-981-13-8926-9_20

in soil quality resulting in decline in productive capacity of land due to processes induced mainly by human intervention (UNEP 1992). In other words land degradation results from a mismatch between land quality and land use (Beinroth et al. 1994). Soil degradation is only a part of land degradation which refers to a decline in the soil productivity through adverse changes in nutrient status and soil organic matter, structural attributes, and concentration of electrolytes and toxic chemicals (Lal and Stewart 1994).

On a global scale, the productivity of prime land has declined by 50% due to soil erosion, desertification, and other causes of land degradation. Yield reduction in Africa due to soil erosion ranged from 2% to 40%, with a mean loss of 8.2% for the continent. In South Asia, annual loss in productivity is estimated at 36 million tons of cereal equivalent valued at US$5400 million by water erosion and US$1800 million due to wind erosion (Eswaran et al. 2001). However, when subject to the degradation process, all the land doesn't get degrade with the same intensity and time sequence. Depending on their inherent characteristics and the climate, lands vary from highly resistant, or stable, to those that are vulnerable and extremely sensitive to degradation. In this context fragility and resistant are used to describe the land degradation potential. Fragility is extreme sensitivity to degradation processes which may refer to the whole land and a degradation process (e.g. salinization) or soil property (e.g. soil structure). Stable or resistant lands do not necessarily resist change. They are in a stable steady-state condition with the new environment. Under stress, fragile lands degrade to a new steady state, and the altered state is unfavourable to plant growth and less capable of performing environmental regulatory functions. The former is mostly subjected to anthropogenic dominated changes, while the latter is more vulnerable to climate change events. In both cases coastal land is susceptible and results in degradation.

Out of several causes of land degradation in the coastal areas, salinization, waterlogging, and drainage congestion besides the climate-imposed constraints are identified as the major processes of land degradation. The condition is more aggravated by various human activities influenced by population growth and urge for development. Therefore, understanding the process and types of land degradation occurring in a particular site, land form, and agroclimatic conditions besides human needs could assist in developing management strategies aimed at productivity enhancement and its effective implementation. This is more pertinent to the coastal and island regions particularly in the tropical areas which are at various stages of land degradation highly vulnerable to climate change.

20.2 Causes of Land Degradation

Land degradation is a process driven by the factors/causes of degradation. These factors and its intensity are causing land degradation to exhibit spatio-temporal variations. In general the factors causing land degradation in the coastal areas are grouped into biophysical (land use including deforestation and land management like tillage methods), socio-economic (e.g. land tenure, marketing, institutional support), and political (policy incentives) (Eswaran and Dumanski 1994). In terms of the nature of these drivers, land degradation can be divided into natural, direct, and underlying causes which results from human-nature interactions and its effect while deriving some utility from the land. Otherwise the natural landscape is not described as degraded because natural system has its own structure and function.

- Natural causes of land degradation are otherwise called as natural hazards. This is defined as the conditions of the physical environment which lead to the existence of a high degradation hazard, for example, steep slopes as a hazard for water erosion, sea wave action on the coastal areas, seawater intrusion, and sea surges resulting in salinization.
- Direct causes of land degradation relate to unsuitable land use and inappropriate land management practices, for example, the cultivation of steep slopes without measures for soil conservation. This is an anthropogenic cause as a result of human activities.
- Underlying causes are the reasons why these inappropriate types of land use and management are practised. For example, the slopes may be cultivated because the landless poor need food while conservation measures could not be adopted because of lack security of tenure.

There is a distinction, although with overlap, between unsuitable land use and inappropriate land management practices. Understanding these terms assumes significance in the context of land degradation management. Unsuitable land use is the use of land for purposes for which it is environmentally unsuited for sustainable use. An example is forest clearance and arable use of steeply sloping areas which would have more value to the community as water sources, managed under a protective forest cover. Inappropriate land management practices refer to the use of land in ways which could be sustainable if properly managed but where the necessary practices are not adopted. An example is the failure to adopt soil conservation measures where these are needed. For example, some coconut gardens in the coastal regions are maintained without soil conservation measures. Another simple example is selecting inappropriate time for planting or harvest operation for tuber crops in the slopes. In some of the tropical islands, harvesting of tubers coinciding with the rainy season causes soil loss from the slopes which could have been prevented by proper crop planning. It can also refer to land use which is ecologically sustainable under low intensity of use but in which the management becomes inappropriate at higher intensities. Examples are shifting cultivation and the grazing of semi-arid rangelands.

20.3 Land Degradation Processes

Basically the processes leading to land degradation are triggered by excessive pressure on land to meet the competing demands of the rapidly growing population for food, fodder, and fibre. This leads to over-exploitation of natural resources with little consideration for maintaining the eco-balance sustainability. Mechanisms that

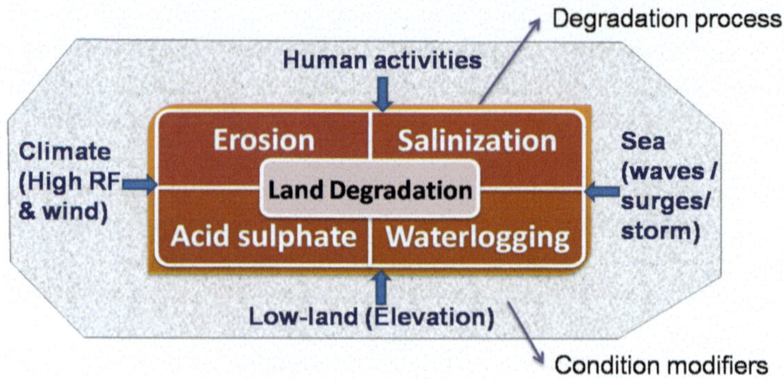

Fig. 20.1 Schematic representation of land degradation process and condition modifiers

initiate land degradation are grouped into physical, chemical, and biological processes (Lal 1994). These processes are triggered by human activities; sometimes activities accelerate the rate of degradation.

Different anthropogenic activities, such as introduction of large-scale irrigation projects, canals, over-exploitation of groundwater, and deforestation, led to the accelerated soil degradation through the process of salinization, flooding, drought, erosion, and waterlogging. It is very difficult to conclude the process of soil degradation in a particular area because its impact is a complex consequence of physical, chemical, social, political, and economic circumstances. The processes of land degradation in the coastal and island regions can be grouped into five major categories. They are salinization, erosion, soil fertility decline, waterlogging, and acid sulphate soils. These processes are modified (intensity and progress) by the modifying forces, viz. climate, physiography, human activities, and sea wave action (Fig. 20.1). As we intend to focus more on the land degradation management options, land degradation process are briefly described below:

20.3.1 Erosion

Erosion is a process of removal, transport, and depositing of soil particle from its original place to another place by the agents of erosion. In the tropical coastal region, only water and air are relevant. Water erosion is one of the most important geologic phenomena which accounts in large part for the levelling of mountains and the development of plains, plateaus, valleys, river flats, and deltas. Water erosion covers all forms of soil erosion by water, including sheet and rill erosion and gullying. Water erosion is mostly manifested in the form of ravines, human-induced intensification of land sliding caused by vegetation clearance, and washing out of large quantity of soil due to road construction in the coastal region. Sea wave action on the unprotected coastal land is the most important causes of soil erosion in coastal and island regions (Fig. 20.2). The major determinants are presence of elevated land and energy of sea wave action. In the future the projected sea level rise will have negative impact on land degradation by erosion and inundation.

Clearing of forest for cultivation in the slopes   Coastal erosion by sea wave action

Fig. 20.2 Causes of erosion

Wind erosion refers to loss of soil by wind, occurring primarily in dry/arid regions. Wind is not generally an important cause of erosion in humid areas except on unprotected sandy soils and tracts of drained and cultivated organic soils. At the same time, it is to be noted that high wind and cyclonic circulations will be highly destructive to the coastal vegetation besides inundation of large areas in the coastal region. Consequently it may trigger land degradation by accelerating water erosion. In regions of low rainfall, wind erosion can be widespread, especially during periods of drought. The wind erosion hazard increases with the removal or decrease in vegetation cover and most importantly the protective function of this vegetation as shelter belts/bio-shield will be reduced. At many places in the coastal region, the natural cover has been replaced by other land uses which are highly influenced by human activities. This is more prominently visible in the Gujarat plains, India, and coastal areas of eastern topical Africa and Caribbean coast due to accelerated land degradation and anthropogenic modification of the natural cover.

20.3.2 Soil Fertility Decline

This is used to refer to what is more precisely described as deterioration in soil physical, chemical, and biological properties which supports plant growth. While decline in fertility is indeed a major effect of erosion, the term is used here to cover effects of processes other than erosion. The main processes involved are:

- Lowering of soil organic matter due to lack of organic matter application and erosion from the slopes, with associated decline in soil biological activity. Heavy runoff from the sloppy and hilly terrain can lead to loss of soil organic matter though the terrain is covered with dense tropical forest.
- Consequent to the loss of organic matter from soil, degradation of soil physical properties (structure, aeration, water holding capacity) can happen.

- Adverse changes in soil nutrient resources including reduction in availability of the major nutrients (nitrogen, phosphorus, and potassium), onset of micronutrient deficiencies, and development of nutrient imbalances results in soil degradation.
- Buildup of toxicities, primarily acidification through incorrect fertilizer use, presence of large amounts of iron and aluminium as in the case of coastal acid soils of Andaman Islands and certain packets of Kerala coast, India.
- Loss of soil biodiversity especially beneficial organism due to pollution, excessive pesticide use, deforestation, and inappropriate soil management practices also results in fertility degradation.

20.3.3 Waterlogging

In the coastal lowlands, waterlogging is linked to high rainfall, drainage congestion, rising of groundwater, and low elevation. In the coastal region, many times waterlogged condition gets aggravated when high tide coincides with heavy rainfall. During rainy season in the humid areas, water table rises up to the surface resulting in ponding of water in the depression and surface. Under waterlogged condition biological oxygen demand rapidly increases due to depletion of oxygen. It also results in accumulation of several toxic compounds, soluble form of aluminium, iron, and other organic compounds. At the same time, it depletes or makes the plant nutrients unavailable under such conditions. Together these conditions adversely affect the crop growth and reduce the land productivity. The normal pattern of waterlogging in coastal lowlands of humid tropics during normal rainfall year is given in Fig. 20.3 (Velmurugan et al. 2015a). It is observed that maximum

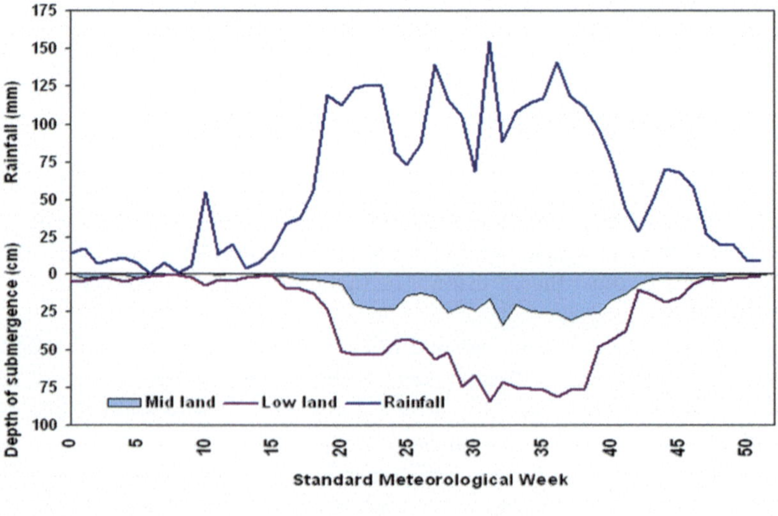

Fig. 20.3 Rainfall and waterlogging relationship in a coastal area

waterlogging occurs primarily during 20–40th standard meteorological weeks, and the depth of waterlogging reaches a maximum of 80 cm in the cultivated land. During dry season waterlogging is also associated with salinization due to seawater intrusion into the coastal lowlands.

20.3.4 Salinization

In general salinization refers to all types of soil degradation brought about by the accumulation of soluble salts in the soil. This occurs by seawater intrusion, raise of saline groundwater, and capillary rise of salts when evaporation exceeds precipitation during the dry season. The increase in salt accumulation is reflected as crop failure and gradual appearance of wild salt-tolerant plans, and the soil salinity is measured by ECe of soil. Salts including sodium accumulate and increase at the soil exchange sites by replacing other ions. As human-induced processes, this may occur mainly through incorrect planning and management of irrigation schemes. This covers saline intrusion, the incursion of sea water into coastal soils, and excess application of irrigation water resulting in rise of groundwater which bring soluble salts to the surface and over-exploitation of groundwater in the coastal areas (Fig. 20.4). Sometimes acid-saline soils occur due to the presence of sulphate and/or release of iron and aluminium from weathering. As the demand for fresh water increases, use of groundwater gets accelerated. This results in seawater intrusion into the freshwater aquifers and rise of salts through capillary pores to the soil surface which also gets accumulated in the root zone. This causes salinization, degradation of entire island resource, and further ecological consequences. It is brought about through tube well pumping of groundwater for irrigation and domestic use exceeding the natural recharge capacity. This occurs in areas of non-saline (sweet) groundwater located close to the sea and in small islands. Pumping for urban and industrial use is a further cause. In most of the places in the coastal areas, salinization occurs along with waterlogging causing severe degradation and lowering the productivity of land.

20.3.5 Acid Sulphate Formation

Acid sulphate soils are formed and deposited in areas that are, or once were, coastal environments. Coastal acid sulphate soils are commonly found in mangrove forests, salt marsh, floodplains, and salt water and freshwater wetlands. It is a serious but localized form of degradation, which may occur on drainage of coastal swamps. Acid sulphate soil is referred to the soils that contain metal sulphides and consequent low pH (<4.5). In an undisturbed and waterlogged state, these soils may pose no or low risk. However, when disturbed or exposed to oxygen, acid sulphate soils undergo a chemical reaction known as oxidation. Oxidation produces sulphuric acid, and as a result the soil pH is lowered. Sometimes these soils may contain soluble salts with high EC, when these soils contain both acidity and salinity called acid-saline soils (pH <5.5, EC >4.0). While inland acid sulphate soils form naturally

Fig. 20.4 Coastal degraded lands. (**a**) Supratidal iron staining, with salt deposit on the surface. (**b**) Seawater intrusion and waterlogging resulted in destruction of coastal plantations

in aquatic ecosystems, human activities may also increase their formation and accumulation. Water bodies become brown to dark reddish colour due the presence of iron (Figs. 20.4 and 20.5). Sulphate is one of the 'raw ingredients' required for the formation of acid sulphate soil, and salts already present in the landscape are a good source of sulphate. The salts are transported by surface and groundwater flows, and this movement has been greatly altered by human-induced changes to land use and hydrology.

Fig. 20.5 Plume of dissolved iron in a coastal pond located in Mithakari, S. Andaman, India

There are other minor/associated land degradation processes in the coastal and island regions which include forest degradation, rangeland degradation, soil pollution, soil destruction, and effect of global climate change.

20.4 Extent of Land Degradation

20.4.1 Global Scenario

Globally, land degradation affects about one sixth of the world's population, 70% of all dry lands, amounting to 3.6 billion ha and one quarter of the total land area of the world. An estimated 950 million ha of salt-affected lands occur in arid and semi-arid regions which is nearly 33% of the potentially arable land area of the world. Productivity of irrigated lands is severely threatened by buildup of salt in the root zone. On the contrary, soil acidity and the resultant toxicity of high concentrations of Al and Mn in the root zone are serious problem in sub-humid and humid regions (Eswaran et al. 1997). The current rate of soil degradation is estimated at 5–7 million hectares per year, and the annual rate may climb to 10 million hectares within a decade (FAO/UNEP 2012). The regional assessment of land degradation types and its integration at national and global level is still lacking. There are differences in assessment method, scale, and purposes. This is more pertinent to the coastal region where systematic assessment with uniform scale and method has not been carried out.

Global Assessment of Soil Degradation (GLASOD) project (Oldeman 1994) distinguishes four types of soil degradation, viz. water erosion, wind erosion, chemical degradation, and physical degradation which are presented in Table 20.1. It identifies four major anthropogenic activities as a factor for the different types of degradation. Globally intensification agriculture resulted in waterlogging, lowering of water table, and soil fertility degradation.

Table 20.1 Causes of degradation in the world as given in the GLASOD assessment

	Percentage area of degradation type			
Type of degradation	Deforestation	Overgrazing	Agricultural activities	Overcutting of vegetation
Water erosion	61	67	2	44
Wind erosion	21	46	1	98
Soil fertility decline	25	0	75	0
Salinization	34	30	14	87
Waterlogging	0	0	85	33
Lowering of water table	12	22	65	34
All types	37	46	15	63

About 24% of land in the world has been affected by various forms of land degradation. This degraded area is equivalent to the annual loss of about 1% of global land area, which could produce 20 million Mg of grain each year, or 1% of global annual grain production. About 1.5 billion people and 42% of the very poor people live on the degraded land (Nkonya et al. 2011). However, more detailed estimates on a regional scale or at climatic zone level is lacking. Alternatively access and compilation of data to project a global scenario are inadequate except for a few global level estimates.

20.4.2 Salt-Affected Soils of India: A Case Study

The coastal areas of India mostly suffer from salinization, erosion, and waterlogging the three major process of land degradation. Degraded lands particularly salt-affected soils are found in different physiographic units including the Great Plains, peninsular plain, and coastal plains of India. Each location has different nature of salt-affected soils, its severity, and causes. The data presented in Table 20.2 indicated that nearly 9.7% of the total salt-affected areas of India lie in the coastal and island regions. Out of 0.65 million ha of salt-affected soils found in the coastal region, it is distributed in the coastal plain of Gujarat (70.6%), Orissa (12.1%), Andhra Pradesh (8.6%), Kerala (3.1%), Tamil Nadu (2.6%), Andaman and Nicobar Island (2.6%), and Maharashtra (1%).

According to the naval hydrographic charts, the Indian mainland consists of nearly 43% sandy beaches, 11% rocky coast with cliffs, and 46% mudflats and marshy coast. Oscillation of the shoreline along the Indian coast is seasonal. Fifty percent of the beaches that do not regain their original shape over an annual cycle undergo net erosion. At present, about 23% of shoreline along the Indian main land is affected by erosion. In general, western coast of India is experiencing more of soil erosion due to steep slope, whereas eastern coast is gaining due to accretion and also facing erosion (Fig. 20.6) (Mandal et al. 2010). In the island region, Middle and North Andaman and Car Nicobar islands are mostly experiencing coastal erosion due to sea wave action.

Table 20.2 Distribution of salt-affected soils in the coastal and Island eco-region

S.No.	Name of the state and UT	Area (ha)
1	Gujarat	4,62,315
2	Maharashtra	6713
3	Karnataka	586
4	Kerala	20,000
5	Tamil Nadu	16,805
6	Andhra Pradesh	56,423
7	Orissa	79,543
8	Andaman and Nicobar islands	12,366
	Total	6,54,751

Source: Mandal et al. (2010)

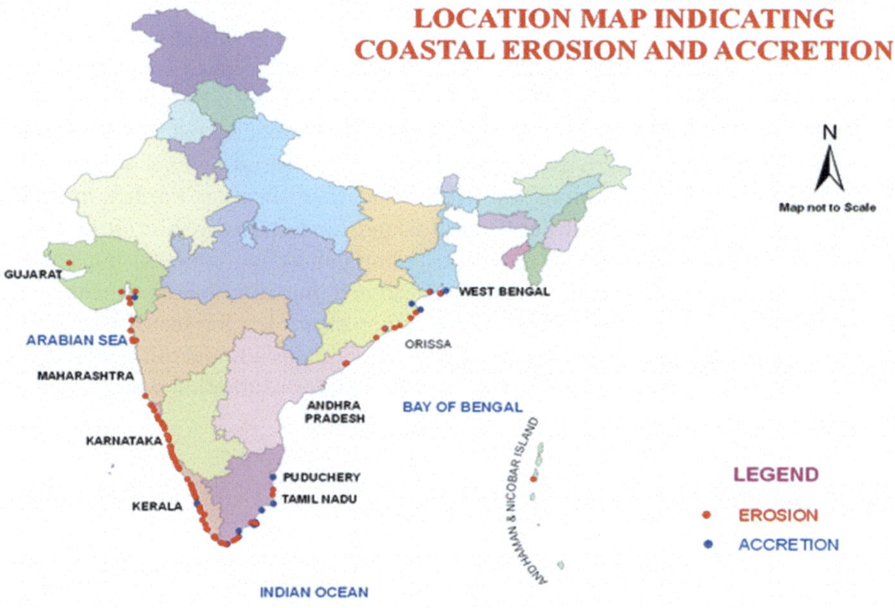

Fig. 20.6 Distribution of coastal erosion and accretion

As per a survey conducted by the Ocean Engineering Division of National Institute of Oceanography, Goa, about 23% of India's mainland coastline is affected by erosion. As much as 1380 km of the shoreline is getting eroded, all along the coast with 480 km of the 569 km shoreline of Kerala is the largest area affected by the phenomenon. In Karnataka, the erosion was marked over 250 km out of the state's total coastline of 280 km with the problem relatively more severe in south Kannada and Udupi coasts where about 28% of the total stretch is critical. Again, the coastal erosion is marked over 263 km of the 653 km shoreline of Maharashtra, while 108 km out of 476 km of Orissa's coastline has already been affected.

Shoreline erosion in the northern regions of Chennai, Ennore, Visakhapatnam, and Paradip ports has resulted due to construction of breakwaters of the respective port.

20.5 Rationale for Land Degradation Management

The coastal region plays a vital role in the global economy due to its rich natural resources, productive habitats, and biodiversity. Total coastline of the world is 356,000 km, and the coastal area covers more than 10% of the earth surface. The degradation of which will not only affect the land productivity but also the livelihood of the people who depend on it. Global estimates of the extent and severity of soil degradation and vulnerability to degradation processes are alarming (Oldeman 1994; Reich and Eswaran 2004).

The economic impact of land degradation is extremely severe in densely populated South Asia and sub-Saharan Africa particularly in the coastal areas and small islands. There are also serious (20%) productivity losses caused by erosion in Asia, especially in India, China, Iran, Israel, Jordan, Lebanon, Nepal, and Pakistan (Dregne 1992). The productivity of some lands in the coastal regions of India has declined by 40–60% as a result of soil erosion and desertification. Further, the land is limited in supply in most of the tropical islands and coastal region. It is in the context of these global to regional level economic and environmental impacts of land degradation, and numerous functions of value to humans, that management of land degradation is most relevant. It is also important to develop technologies and its larger implementation for reversing land degradation trends and mitigating its ill effects on the environment through land and ecosystem restoration which is more pertinent to coastal and island ecosystem of the tropics.

20.6 Management of Degraded Lands

Land resource in the coastal region is vulnerable to degradation due to combinations of natural, hydrological, and anthropogenic factors. Land degradation has both on-site and off-site effects. On-site effects are the lowering of the productive capacity of the land, causing either reduced outputs (crop yields, livestock yields) or the need for increased inputs. Off-site effects occur through changes in the water and wind regime, human activities in the catchment areas, and sedimentation of river beds and reservoirs. These effects are spatially variable. Therefore, proper understanding of the dominant factors and process is necessary for proper land management.

Although there are several constraints and climate-related challenges, there are tremendous opportunities to improve the production potential of the degraded land and water in the coastal region. Degraded land can be improved and put into productive use by two broad approaches (Fig. 20.7). The first method focuses on removal or modification of the conditions that favour the casual factors of land degradation or further aggravation of the existing situation by suitable methods.

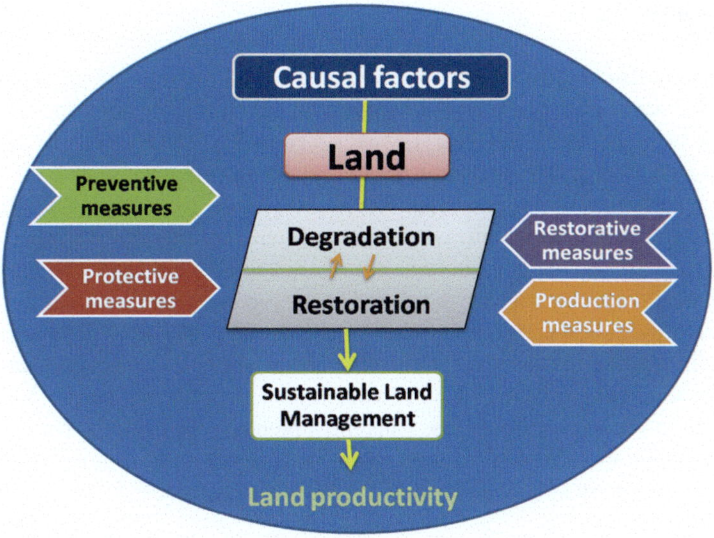

Fig. 20.7 Schematic representation of land degradation management in coastal areas

This could be supported by certain preventive and protective measures of land degradation which could be significant in certain cases. For example, avoid excessive use of irrigation, land levelling, and improving drainage system to avoid water stagnation, unjust agricultural practices, etc. Protective measures which are not directly related to agricultural production such as bio-shield/shelter belts (reduce wind effect), sluice gate (prevent seawater intrusion), bunding of coastal land (reduce wave action), etc. can help to reduce the land degradation. Secondly, productive and restorative measures could not only help to improve the quality of degraded land by removing/reducing the toxic or degradation process but restore and sustain the land productivity in the long run. While the production measure employs technology/methods to cope with the existing situation without further degradation of the land such as terracing, cover crops, improved agronomic practices, organic manure application, etc., restorative measures are special practices employed to restore/adjust with the current land condition and help to improve land quality in the long run in a given agroclimatic conditions. In most of the situations, both the approaches are employed to improve or restore the productivity of land. The main purpose is to employ suitable land management/restorative practices which address key challenges like land and water degradation (salinity), drainage congestion, and scarcity of fresh water for irrigation, all could enhance agriculture production and livelihood security of people in coastal region. The following subsections deal with innovative land management practices, termed as land shaping techniques for overall improvement of the situation besides brief mention of other conventional methods to address land degradation.

20.6.1 Saline Soils

Salt-affected soils contain considerable amounts of soluble salts accumulated where evapotranspiration greatly exceeds precipitation or weathering of parent materials containing high amounts of salt. However, secondary salinization is caused by human interventions such as inappropriate irrigation practices, e.g. with salt-rich irrigation water or insufficient drainage, rising of saline groundwater due to excess irrigation. In some areas of the coastal and island regions, due to inundation or intrusion of saline seawater, soluble salts mostly sodium, calcium, and magnesium salts of chloride and sulphate are deposited in the top soil. In these areas the amount of crop yield reduction depends on such factors as crop growth, the salt content of the soil, climatic conditions, etc. In extreme cases where the concentration of salts in the root zone is very high, crop growth may be entirely prevented. To improve crop growth in such soils, the excess salts must be removed from the root zone. The term reclamation of saline soils refers to the methods used to remove soluble salts from the root zone by any means. They are discussed below:

20.6.1.1 Removal of Salts

Methods commonly adopted to remove the salts include scraping, flushing, and leaching.

- Scraping is the removal of salts that have accumulated on the soil surface by mechanical means. However, it has only a limited success even though many land managers have resorted to this procedure. This method is simple and might temporarily improve crop growth, but the ultimate disposal of salts still poses a major problem.
- Flushing means washing away the surface accumulated salts. Flushing water over the surface is sometimes used to desalinize soils having surface salt crusts, because the amount of salts that can be flushed from a soil is rather small. The effectiveness can be improved by bunding around the field to facilitate collection and storage of rainwater for overnight. The salts get dissolved in the water and flushed out. This method was effectively used in the coastal areas to remove salts after December 2004 Indian Ocean tsunami.
- Leaching of soluble salts atleast up to the root zone is simple and effective method. Saline soils cannot be reclaimed by chemical methods, and it can only be reclaimed by removing soluble salts from the plant root zone. In some cases, selecting salt-tolerant crops may be needed in addition to managing soils. There are three ways to manage saline soils. First, salts can be moved below the root zone by applying more water than the plant needs. This method is called the leaching requirement method. The second method, where soil moisture conditions dictate, combines the leaching requirement method with artificial drainage. Third, salts can be moved away from the root zone to locations in the soil, other than below the root zone, where they are not harmful. This third method is called managed accumulation. However, leaching is possible only in elevated land or land with some gradient. Leaching is very difficult in coastal lowlands. But, the

high rainfall of these areas can be used to dilute the salts present in the soil which normally happens during rainy season. Leaching of salts can also be facilitated by raising of beds on which crops are grown. The details are discussed separately in land shaping methods.

20.6.1.2 Residue Management

Crop residue at the soil surface reduces evaporative water losses, thereby limiting the upward movement of salt (from shallow, saline groundwater) into the root zone. Evaporation and thus salt accumulation tend to be greater in bare soils. Fields need to have 30–50% residue cover to significantly reduce evaporation. Under crop residue, soils remain wetter, allowing fall or winter precipitation to be more effective in leaching salts, particularly from the surface soil layers where damage to crop seedlings is most likely to occur. In recent time plastic mulches are widely used with drip irrigation to effectively reduce the salt concentration by reducing evaporation.

20.6.1.3 Pre-plant Irrigation

Most crop plants are sensitive to salt injury during germination and at seedling stage. An early-season application of good quality water, designed to fill the root zone and leach salts from the upper 6–12 in. of soil, may provide good enough conditions for the crop to grow through its most injury-prone stages. The seedling of vegetable crops can be raised in portrays and transplanted in the main field after pre-plant irrigation. This is very useful for vegetable cultivation in the coastal and island regions particularly during the dry season (January–April) after the harvest of first-season rice crop (June–November).

20.6.1.4 Irrigation Frequency Management

Salts are most efficiently leached from the soil profile by irrigation given more frequently (short irrigation intervals) wherever irrigation facilities are available. Keeping soil moisture levels higher between the irrigation events effectively dilutes salt concentrations in the root zone, thereby reducing the salinity hazard. Most surface irrigation systems (flood or furrow systems) may not be suitable, but drip irrigation can be effectively used to improve the irrigation frequency with limited quantity of good quality water. This is mostly used for high-value and vegetable crops during dry season. Thus the water productivity can also be improved.

20.6.2 Alkaline Soils

20.6.2.1 The Nature

Alkali soils are formed due to concentration of exchangeable sodium resulting in high pH. Because of high alkalinity resulting from sodium carbonate and dissolution of certain portions of organic matter, the surface soil is discoloured to black; hence the term black alkali is also used. The main reasons are:

- The excessive irrigation of uplands containing sodium salts results in the accumulation of salts in the valleys.
- In arid and semi-arid areas, salts formed during weathering are not fully leached
- In coastal areas if the soil contains carbonates, the ingression of sea water leads to the formation of alkali soils due to formation of sodium carbonates.
- Irrigated soils with poor drainage.

Dispersion of soil particles due to high exchangeable 'Na^{+}' leads to poor physical condition of soil and low permeability to water and air, tends to be sticky when wet, and becomes hard on drying. It shows the presence of dispersed organic matter on the surface. Columnar structures present in the subsoil, and the presence of high exchangeable sodium decreases the availability of calcium and magnesium to plants.

20.6.2.2 Amelioration

It consists of two steps, viz. conversion of exchangeable sodium into water soluble form and leaching out the soluble sodium from the field. Amendments used for reclamation of alkali soils are:

- Gypsum ($CaSO_4$): It is slightly soluble in water; therefore it should be applied well in advance. For every 1 m.e of exchangeable Na per 100 gm of soil, 1.7 tons of gypsum/acre is to be added. It is advisable to work out the gypsum requirements as per the soil conditions.
- Pyrites ($Fe\ S_2$): Sulphur present in pyrites causes decrease in pH of soil due to formation of sulphuric acid (H_2SO_4). The other chemical method is application of sulphur or molasses to the field and provision of drainage channels around the field. However, this cannot be applied in acidic or acid sulphate soils.
- Growing and incorporation of green manure crops into the field.
- Growing of suitable crops such as sorghum, pearl millet, sugarbeet, rice, maize, red gram, green gram, cotton, sugarcane, mustard, sunflower, linseed, sesame, tomato, cabbage, cauliflower, cucumber, pumpkin, bitterguord, beetroot, guava, asparagus, banana, spinach, coconut, grape, and pomegranate. However, the selection of crops depends on climate and other soil conditions.

20.6.3 Waterlogged Soils

Wet/waterlogged areas often shows symptoms of other problems like soil compaction, impervious layer, etc. However, there are some soil types which, because of their position in the landscape or their texture and subsoil, are naturally poorly drained. Moreover, in the coastal lowlands, as discussed earlier, waterlogging occurs as a result of seawater intrusion during high tide and geographical limitation such as low elevation and drainage congestions. The problem is more pronounced

Poor establishment of crop field Stressed and dead plants

Fig. 20.8 Symptoms of waterlogging

during rainy season in the tropics along with salinity (Velmurugan et al. 2015a). Without proper attention and management, these poorly drained soils can develop other problems such as soil structural damage from tillage, planting, and harvest operations. Further wet soils are colder, and the slow warming of these soils can result in reduced yields.

20.6.3.1 Field Symptoms of Waterlogging

In many areas drainage problems also arise because of the accumulation and stagnation of rainfall or excess irrigation water on the soil surface. Surface drainage problems usually arise due to slopes that are too flat or slow water penetration because of structural instability of the soils or to uneven land. Temporary water stagnation in standing crops results in problems of aeration, disease, weed control, and nutrient supply (Fig. 20.8).

- Water is lying on the fields which is very prominent in the costal low-lying areas.
- Side hill seepage is evident.
- Crops are yellow or dead in areas of a field.
- Plants are stressed and more insect and disease damage are evident.
- Root growth is concentrated at a shallow depth.

20.6.3.2 Management

Proper land shaping and provision of surface drains are needed to solve the problems of surface water stagnation. Land shaping is discussed separately as it is most important to manage soil salinity and acid-saline soil as well:

- Install drainage and/or surface drains where ever gradient is available.
- Use crop rotations and include deep-rooted crops such as alfalfa, clover. etc.
- Grow crops suited to wetter soil conditions or crops that are planted later in the growing season, i.e. rice, soybeans, winter wheat, pasture or agroforestry.

Eucalyptus tereticornis

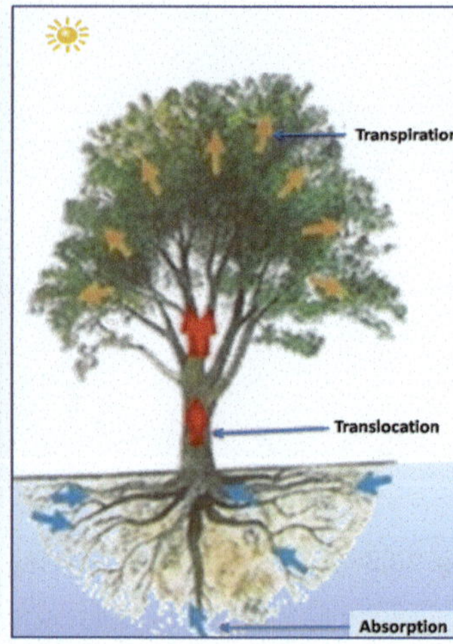

Fig. 20.9 Bio-drainage

- Use disease-resistant/disease-tolerant crop varieties.
- Use tillage carefully to expose soil to the air for evaporation and soil warming.
- Encourage earthworm populations for macropore development, by leaving residue on the soil surface wherever possible.
- Use timely tillage and field operations and minimize tillage passes to reduce compaction.
- In the coastal lowlands, land shaping is very effective to manage waterlogging and salinity which is discussed separately in the subsequent section.

20.6.3.3 Biological Drainage

This is nothing but removal of soil excess water by standing trees through fast transpiration for which the trees have certain special characters. Pumping of excess soil water/free water happens by deep-rooted plants using their bioenergy. With intensity of solar radiation, the transpiration from leaf surface also increases. *Casuarina* and *Eucalyptus* are best known examples for bio-drainage which are used in several agroforestry models (Fig. 20.9). In the east coastal areas of India, particularly in the deltaic region of Orissa and Andhra Pradesh, due to flat topography and high rainfall during monsoon, drainage is very difficult. In these areas the use of bio-drainage vegetations is promising. In the coastal Orissa, 30-cm-tall *Casuarina* saplings were tried in a raised mound with 4 × 4 m spacing. Rice was grown in *Kharif* and watermelon, pulses, and groundnut during *rabi* season. *Casuarina* withstands salinity and waterlogging. It reduces the water table and subsurface moisture by fast evapotranspiration. Rice grown in between the tree rows gave 3.5 t/ha, and other crops performed better than non-biodrained areas. Pruning of trees for fuelwood purpose is also worth Rs. 5500/ha (Roy Chowdhury et al. 2016).

20.6.4 Soil Conservation

In general soil and water conservation are discussed together because these are just two sides of the same problem. The water conservation measures directly conserve the soil from erosion. Soil conservation aims at sustained level of production by maintaining soil loss below a threshold level. In the coastal and island regions, soil loss by water erosion is considerable in the slopy areas particularly during the monsoon and heavy summer showers coinciding with land preparation/harvesting of tubers. In order to manage erosion, there are three basic principles which must be addressed. First, we must encourage as much of the precipitation as possible to run into the soil rather than of it. Second, the sediment burden of the runoff water should be kept low. Lastly, it is essential to control the accumulation of the runoff water into channels, thereby minimizing its cutting action and the sediment burden it can carry. Some of the structures used in water conservation are diversions, terraces, and waterways which are aimed at reducing slope length, to slow runoff velocity, to divert excess water safely, and to avoid runoff over fields. Soil conservation measures are broadly divided into agronomic/biological methods and mechanical methods. We have not provided more details on these aspects as we intend to focus on land shaping methods.

20.6.5 Land Shaping Interventions for Saline and Waterlogged Soils

In humid coastal and island regions, surface waterlogging during *kharif* and nonavailability of water during post monsoon season (December–April) are major problems for crop production. In addition, surface waterlogging coupled with varying degree of salinity is a common feature due to inundation/seepage of sea water. In these areas land shaping is a viable option to address the issue of waterlogging and salinity which helps to reclaim and use the land for production purpose. Land shaping is nothing but the articulation of land arrangement so as to overcome hydrologic problems in agricultural area for potential crop cultivation (Ambast et al. 2011). Some of the land shaping methods suitable for coastal plains are raised beds and furrow system, rice-cum-fish, pond with broader dykes, three-tier farming, and ridge and furrow systems. The same design could also be used for rainwater harvesting and fish culture.

20.6.5.1 Raised Beds and Furrow System (RBF)

It involves making of broad raised beds and furrows alternatively in rice fields. The ideal location is the flat low-lying area, where floodwater stagnates and only rice crop can be grown. Broad beds are made in the shape of inverted trapezium by digging soil from either side of the bed and raising the bed area using the soil by cut and fill method (Ravisankar et al. 2008). Rainwater is collected in the excavated portion called furrow which is used for rice or fish cultivation and the raised broad bed area which is above the water level of the paddy fields are used for cultivating

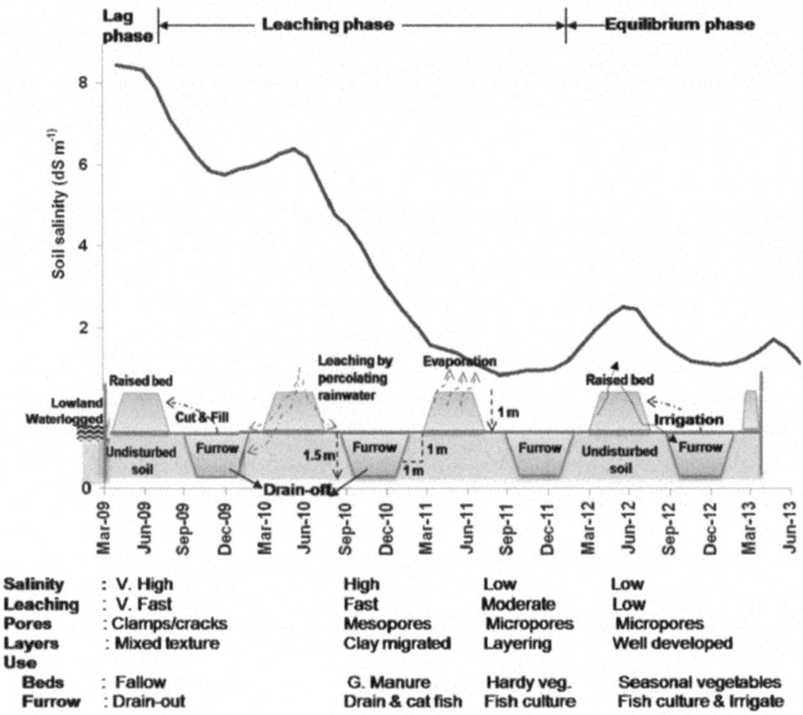

Fig. 20.10 Temporal changes in the leaching of soluble salts from the raised beds. (Source: Velmurugan et al. 2015b)

seasonal vegetable. The beds of 4–5 m wide and furrows of 5–6 m wide were found suitable for this system of cultivation (Fig. 20.10). The length can be of any suitable size according to the land availability and site conditions. The important aspect of making raised beds and furrow system is leaching of soluble salts and other toxic compounds from the raised beds using the rainwater at the same time it provides ideal crop growing conditions without waterlogging.

The observation of temporal changes in surface soil salinity in RBF system showed significant reduction in salinity from its initial high value to relatively low value after 2 years (Velmurugan et al. 2015b). In general, three fairly distinct phases are observed when leaching proceeds from high salinity to relatively lower value based on the rate of reduction in soil salinity, viz. lag phase, leaching phase, and equilibrium phase (Fig. 20.9). After the formation of raised beds by cut and fill method, the salt concentration on the beds used to be high. The beds are characterized by clamps of large size and lots of cleave between them. With no appreciable amount of rainfall for the initial 2 months, during the dry season there is little chance for downward movement of salts and clay migration which is recognized as 'lag phase'.

During the dry season, after the harvest of rice (June–Nov), soil salinity level increases. Mondal et al. (2001) also reported increasing trend in salinity during dry period from January to April in the coastal rice soils. However with the onset of monsoon in June, leaching proceeds at a faster rate in the raised beds for the next 2 years except for slight rise in salinity during the intervening summer. On an average, ECe decreases from the initial high value of 8.45–1.34 dS m^{-1} after 2 years recording 86% reduction from the initial level. This phase is termed as 'leaching phase' as it records maximum decrease in soil salinity. The duration of leaching phase may vary in other locations depending on the amount of rainfall, initial salt concentration, and area of water conducting pores. The rapid decrease in soil salinity from the raised beds is due to removal of soluble salts by the percolating rainwater as evidenced by the significant decrease in ionic concentration and improved drainage conditions created by RBF system (Velmurugan et al. 2015b). The improvement in drainage is attributed to higher percentage of macropores created by piling of soil lumps in the 1st year thereafter, decomposition of root system of the original vegetation resulting in higher percentage of mesopores (Minh et al. 1997).

Subsequently soil salinity attains dynamic equilibrium with the rainfall pattern as the possibility for significant salt addition from outside the RBF system is minimized. This process of detoxification also significantly increases the soil pH (6.36) of the beds than that of the surrounding soils (pH 5.80) treated as control. This is 'equilibrium phase' during which the salinity only fluctuates between wet and dry season around a mean value. Meanwhile, the leachate collected in the furrows along with the rainwater should be drained off by providing drainage pipes at the lower end of the slope. This helps to ease out the salt concentration in the furrow water and in turn from the RBF system.

Planting two rows of hybrid napier on both sides of the bed stabilizes the broad beds. At the end of the furrow on descending side, a fish shelter or pond is to be excavated. In this pond, the fish farming can be along with rice. This is added to the profit as well as livelihood of the farmer. From the bund of the furrow, two PVC pipes of 2 m length may be provided for draining out excess water. The system increases the cropping intensity from 100% in the rice monocropped area (surrounding) to 300 in the beds and 200 in the furrows of the broad bed-furrow system. The initial cost incurred on land manipulation is easily returned back from vegetables produced in one season. The technology helps in crop diversification by replacing 40% of the rice area with vegetables and fodder in the coastal degraded land. Different crops grown in the BBF system is shown in Fig. 20.11 This technology can also be extended to similar coastal plain areas in the tropics including the tsunami-affected lowlands of South Asian countries.

20.6.5.2 Paddy Cum Fish System (P-F System)

The P-F system standardised for degraded coastal lowlands involves excavation of trenches of about 5 m width and 1.5 m depth around a rice field and making 4-m-wide bund around the field using the excavated soil. The central land is used for paddy cultivation, while the trench made around the central rice field is used for harvesting rainwater and fish culture. Vegetables/fruits can be grown throughout the year on the raised bunds as in the case of RBF system. The leaching process in the bunds of paddy cum fish system is also similar to the one explained in Fig. 20.8.

Cultivation of multiple vegetables in BBF
Cultivation of Banana+rice+fish in BBF
Cultivation of vegetable+fish in BBF
Marigold cultivation in the beds of BBF

Fig. 20.11 Design and use of BBF under island condition

This system also provides the scope for in situ rainwater harvesting and thus enables growing of second crop during dry seasons (Fig. 20.12).

The plots utilised for paddy cum fish system in small holder farms of the tropical coastal areas are mainly based on organic fertilization with varieties of animals excreta such as poultry dropping, pig excreta, cow dung, and wastes and composts of plants such as rice husks, weeds, stalks, and ashes from household burnt and remains of burnt straws after the harvest is over. The lowland rice field of the coastal areas can be utilized for P-F system in the following two ways. Fishes can be reared from the month of May to September when the paddy crops grow in the field. The fish culture can also be taken up from the month of November to February after harvesting of paddy crops is completed and transplantation for the next season begins. The culture of fishes in paddy fields, which remain flooded even after the paddy is harvested, may also serve as an occupation for the unemployed youths. Paddy field is suitable for fish culture because of having strong bund in order to prevent leakage of water, to retain up to desired depth, and also to prevent the escape of cultivated fishes during floods. This is very much suitable for waterlogged and coastal saline soils which enable to retain the paddy as major food crop at the same time promote crop diversification by leaching out salts in otherwise monocropped areas.

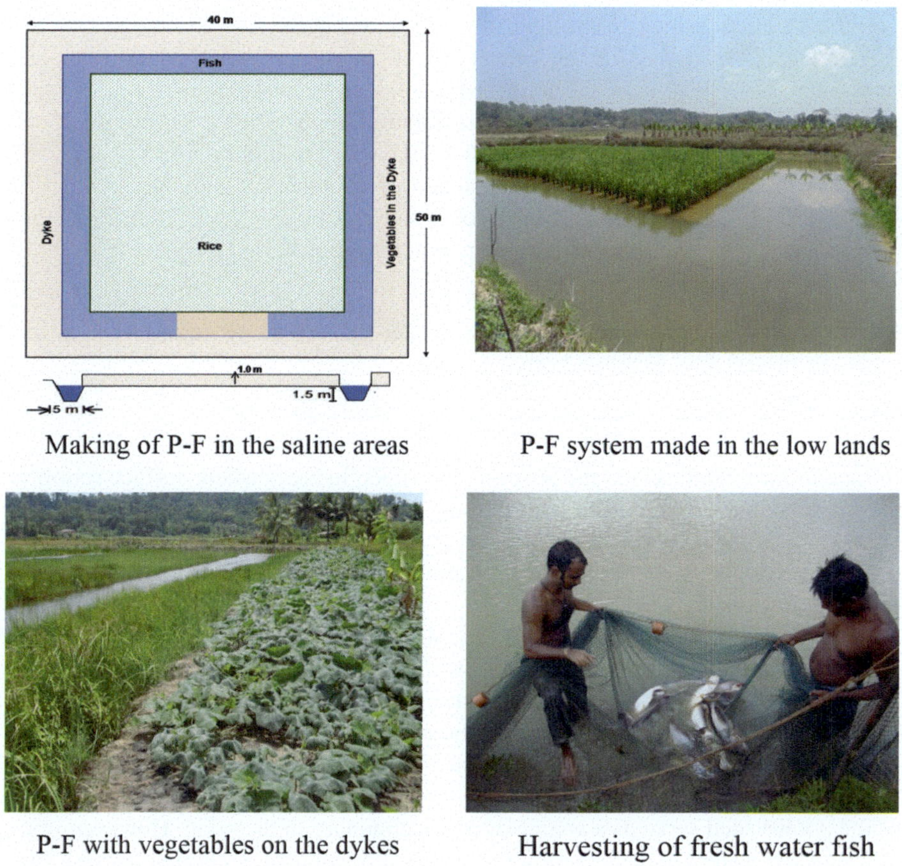

Fig. 20.12 A layout and a typical view of paddy-fish system

20.6.5.3 Pond with Broader Dykes

Farm ponds are common in several coastal region, but the bunds are small and not enough for vegetable cultivation. In the modified farm pond, the dykes are broader; pond is 2.5–3.0 m deep with 1:1 slope which is one of the suitable land shaping options which form the centre of integrated farming system. The steps involved in constructing a suitable farm pond includes estimation of rainwater availability, crop water requirements, design dimension of farm pond, location of the farm pond, and lining requirement for seepage control. It is made by cutting the soil up to 2.5–3.0 m deep and raising the bund 1.5 m away from the pond. The bund should have at least 3–4 m width so as to enable crop cultivation. The salts and other toxic substances present in the raised bunds get leached away in the subsequent rainy season. Besides this, the raised bund facilitates proper drainage which results in ideal condition for crop growth (Fig. 20.13). The system stores in situ rainfall or harvest surface runoff from surrounding areas depending upon the available rainfall in a region. In high rainfall areas where average annual rainfall is about 3100 mm, even in situ rainwater

Farm pond made in an acid soils | Vegetable cultivation in the dykes

Fig. 20.13 Schematic design and use of farm pond in different soil types

storage in farm pond serves the purpose. However, in areas where surface runoff is the main source of water, the contributing drainage area or watershed should be large enough to maintain desired water level in the farm pond. The raised and broader dykes also prevent the entry of saline sea water into the pond from the surrounding lowlands.

Once the rainwater is harvested in a suitable farm pond, it can be used during dry season for crop cultivation in the dykes while fish along with duck/poultry can be integrated with the farm pond. Studies have shown that the size of the pond should be 15% of the land holding to irrigate the remaining 85% area for summer crops (Ambast et al. 1998). Integration of poultry farming enhances plankton growth in ponds which ultimately leads to enhanced fish growth as the droppings and wastes of the poultryshed serve as a source of organic manure to the pond. However, optimal land and water allocation among several components of integrated farming system is the most important for maximizing the farm profit.

20.6.5.4 Ridges and Furrow System (R&F system)

In low-lying areas, soil remains under saturated conditions for most part of the year due to high rainfall, drainage congestion, and seawater intrusion during high tides. But the situation is not so severe to recommend raised beds and furrow system as the land is having gradient. Further, the pH is not less than 6.0 and not much degraded. In those areas the land can be converted into series of ridges and furrows during the dry season to facilitate drainage and support crop diversification (Fig. 20.14). The paddy land is converted into ridges by scooping the soil and making the ridge. The width of the ridge and furrows is 1.0–1.5 m and height of up to 0.5 m to facilitate drainage and provides the root some aeration during heavy rainfall. The earth cutting and filling operation could be accomplished manually in summer. At the time of filling, the soil should be compacted to prevent the soil erosion.

Salts from the ridges (mainly from rooting zone) are leached out, and the free water (excess of field capacity) drains into the furrow; thereby it facilitates better growing conditions in the root zone. Consequently the ridges are used for plating of

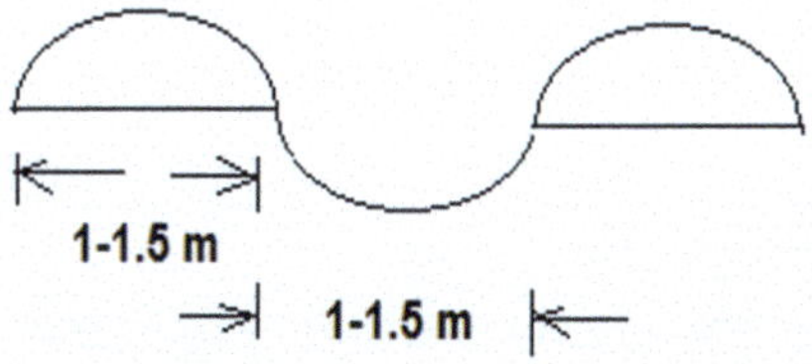

A layout of R & F system

Cucurbits in R & F made in the low land

Fig. 20.14 Layout and use of ridge and furrow system in the low-lying areas

vines, areca nut, vegetables, and banana. In coastal areas rice can be grown during rainy season in the furrows or used to drain out excess water. This system is semi-permanent type, which could be lasted for 3–5 years. Regular maintenance is required for proper shaping of ridges.

20.6.5.5 Three-Tier Farming System

In this method the narrow and long farm land is divided lengthwise into three equal parts. The basic idea in this system is to retain one third of the central portion of the land in its original state for rice cultivation and rising the one third of the land at one end using the soil excavated from the remaining one third portion located in the lower side of the original land. The one third part of the land in the lower side of the landscape is excavated to a depth of 2.5 m or more as a fish cum water harvesting pond depending up on the site-specific conditions. The dugout soil is taken to the other extreme of the field so as to raise this part by at least 1 m, while some soil can be used for making bunds around the pond (Fig. 20.15). The middle part of the field remains at its original ground level. Salts and other toxic materials present in the raised portion of the one third land (bed) is leached out in the subsequent rainy season as in the case of RBF explained earlier. The system helps to cultivate vegetables throughout the year in the raised part, whereas in the middle part, rice is grown. Fish can be integrated in the pond, and the rainwater harvested is used to grow vegetables in the remaining two third parts during the dry season.

20.6.5.6 Deep Furrow and High Ridge

In the deltaic areas of Sundarban (India and Bangladesh) and coastal areas of Southeast Asia, waterlogging during monsoon season is a major problem for agricultural diversification which gets aggravated by seawater intrusion. Rainwater harvesting and freshwater fish culture holds greater promise by providing stability to the farm income. In this method about 50% of the farm land is shaped into alternate ridges (1.5 m top width and1.0 m height) and furrows (3 m top width and 1.0 m depth). Dug out soil from furrows is used for making ridges. About 1875 m^3 ha^{-1} of

Fig. 20.15 Vegetable cultivation in the upper portion and fish culture in the lower portion

rainwater is harvested in the deep furrows and is used for fish cultivation and irrigation during post monsoon season and dry season (December to April). The ridges are used for cultivation of vegetables and other horticultural crops/multipurpose tree species (MPTs) round the year. Remaining portion of the farmland including the furrows is used for growing more profitable paddy + fish cultivation during monsoon season (June to October). During *rabi*/summer/dry season, farm land (non-furrow and non-ridge area) is used for low water requiring crops (Burman et al. 2015). Thus the system effectively brings together food crops, vegetables, fodder, fuelwoods, and fish cultivation centred on rainwater harvesting in the coastal degraded areas.

20.6.5.7 Impact of Land Shaping Methods

The experiences of all these land shaping interventions carried out in the tsunami-affected coastal lowlands in Andaman, India, revealed that these land shaping methods provide large scope for land degradation management and crop diversification integrated with pisciculture (NAIP 2014). These methods significantly improved the soil quality and helped in checking land degradation. In the tsunami-affected areas of Indonesia, bunding of coastal areas improved the soil conditions and helped to leach out the surface salts by using rainwater (Rachman et al. 2005). Soil MBC significantly increased both in bunded areas (193.6 mg kg^{-1}) and BBF systems (210.3 mg kg^{-1}) made in the tsunami-affected areas when compared to those immediately after tsunami (124.7 mg kg^{-1}). Increase in MBC even in tsunami-affected areas (140.5 mg kg^{-1}) may be attributed to reduction in salinity and improvement in

Table 20.3 Effect of different land shaping interventions on soil properties

Parameters	Tsunami affected (February 2005)	Interventions (March 2013) Bunding	BBF	No intervention
pHs	7.24 (0.16)[a]	5.98 (0.11)[b,c]	6.36 (0.06)[b]	5.88 (0.10)[c]
ECe (dSm^{-1})	22.14 (0.94)[a]	1.72 (0.10)[b]	1.18 (0.06) [b]	1.90 (0.08)[b]
BD (g cm^{-3})	1.47 (0.04)[a]	1.38 (0.03)[a]	1.35 (0.02) [b]	1.42 (0.03)[a]
O.C (g kg^{-1})	7.12 (0.20)[a]	11.46 (0.13)[b]	11.73 (0.11) [b]	7.59 (0.09)[a]
Na^+ (me l^{-1})	173.20 (2.02)[a]	38.80 (0.96)[c]	33.20 (0.91)[c]	55.40 (1.93)[b]
Ca^{2+}+Mg^{2+} (me l^{-1})	26.06 (0.71)[a]	15.80 (0.75)[c]	16.20 (0.74)[c]	20.45 (0.62)[b]
SAR	48.11 (0.54)[a]	13.91 (0.50)[c]	11.75 (0.43)[c]	17.43 (0.74)[b]
HCO_3^- (me l^{-1})	2.38 (0.11)[a]	0.30 (0.02)[b]	0.36 (0.04)[b]	0.50 (0.05)[b]
Cl^- (me l^{-1})	142.80 (2.00)[a]	25.64 (0.66)[c]	30.80 (0.88)[c]	47.00 (0.92)[b]
SO_4^{2-} (me l^{-1})	62.40 (2.89)[a]	14.40 (0.62)[d]	17.46 (0.85)[c]	33.60 (0.73)[b]
MB (mg kg^{-1} soil)	124.7 (2.64) [a]	193.6 (2.1)[c]	210.3 (1.8)[c]	140.5 (1.95)[b]

Mean values, $n = 10$; SE in parentheses; different letters within one row indicate a significant difference at $p < 0.05$

Source: Velmurugan et al. (2015a)

drainage over time. Significant reduction in MBC by increase in soil salinity was also reported by Ouni et al. (2013). Some of the effects and comparison of BBF and bunds made in the tsunami-affected coastal areas are given in Table 20.3.

Studies on effect of land shaping methods revealed that significantly higher O.C and lower B.D, Na^+, Ca^{2+}, and SAR values were observed in the beds of BBF compared with both the initial value and the surrounding areas under treated control. Soil organic carbon and its influenced soil properties are enhanced by improving the land environment and soil quality, which help rapidly restore soil functions, enabling different ecosystem processes interlinked with soil, water, and plants to be sustained (Shrestha and Lal 2006). Crop cultivation can be taken up on the raised beds of land shaping methods. Besides vegetable cultivation on the bed, fruits and fodder crops can also be grown on the slopes of raised bed in addition to fish in the furrows (Velmurugan et al. 2015a). Thus, it is evident that these methods not only improve the land conditions but also ensure enhanced farm income and improvement of livelihood security of small holder farmers of the tropical coastal regions.

20.7 Conclusion

Agriculture and land productivity of tropical coastal region are severely affected by land degradation primarily by waterlogging coupled with acid/saline conditions. Seawater intrusion into the coastal agricultural land and groundwater reinforced by the projected climate change is the major challenge of this region. At the same time, soil conservation measures are imperative in the hilly and undulated terrain. Reclamation and conservation of soil are very important as the availability of land is limited and the productivity of land is declining. Saline and alkaline conditions can be reclaimed by traditional chemical and cultural methods, but the same methods cannot be directly used for the coastal lowlands. In those lowlands land shaping is an innovative method for addressing the key challenges like land degradation (salinity), drainage congestion, and scarcity of fresh water for irrigation. This in turn has the potential for enhancing production, productivity, income, and employment. This is not only the most important land degradation management strategies but also climate change resilient technology as well. In addition, these techniques are financially viable and attractive proposition for the tropical coastal region. However, major constraints for adoption of land shaping techniques are marginal land holdings, high initial investment, presence of acid sulphate soils, and natural hazards. Hence proper policy support, development of viable farmer-inclusive institutions, and orchestrating national and international collaborations for sustainable coastal land management are some of the very important strategies required to support land degradation management aimed at enhancing agricultural productivity.

References

Ambast SK, Sen HS, Tyagi NK (1998) Rain water management for multiple cropping in Sundarbans delta (W.B.), Bulletin No 2/98. Regional Research Station, Central Soil Salinity Research Institute, Canning Town, p 69

Ambast SK, Ravisankar N, Velmurugan A (2011) Land shaping for crop diversification and enhancing agricultural productivity in degraded lands of Andaman and Nicobar Islands. J Soil Salinity Water Qual 3(2):83–87

Beinroth FH, Eswaran H, Reich PF, Van den Berg E (1994) Land related stresses in agroecosystems. In: Virmani SM, Katyal JC, Eswaran H, Abrol IP (eds) Stressed ecosystems and sustainable agriculture. Oxford and IBH, New Delhi

Burman D, Mandal S, Bandopadhyay BK, Maji B, Sharma DK, Mahanta KK, Sarangi SK, Mandal UK, Patra S, De S, Patra S, Mandal B, Maitra NJ, Ghoshal TK, Velmurugan A (2015) Unlocking production potential of degraded coastal land through innovative land management practices: a synthesis. J Soil Salinity Water Qual 7(1):12–18

Dregne HE (1992) Degradation and restoration of arid lands. Texas Technical University, Lubbock

Eswaran H, Dumanski J (1994) Land degradation and sustainable agriculture: a global perspective. In: Bhushan LS, Abrol IP, Rama Mohan Rao MS (eds) Proceedings of the 8th International Soil Conservation Organization (ISCO) Conference, vol 1. Indian Association of Soil and Water Conservationists, Dehradun, pp 208–226

Eswaran H, Reich PF, Beinroth FH (1997) Global distribution of soils with acidity. In: Moniz AZ, Furlani AMC, Schaffert RE, Fageria NK, Rosolem CA, Cantarella H (eds) Plant-soil interactions at low pH. Sustainable Agriculture and Forestry Production. Proceedings of the 4th International Symposium on Plant-Soil Interactions at Low pH, Belo Horizonte, pp 159–164

Eswaran HR, Lal R, Reich PF (2001) Land degradation: an overview. In: Bridges EM, Hannam ID, Oldeman LR, Pening de Vries FWT, Scherr SJ, Sompatpanit S (eds) Responses to land degradation, Proceeding of the 2nd international conference on Land Degradation and Desertification, Khon Kaen, Thailand. Oxford Press, New Delhi

FAO/UNEP (2012) Guidelines for the control of soil desertification. FAO, Rome. http://www.fao.org/ag/AGL/agll/prosoil

Lal R (1994) Tillage effects on soil degradation, soil resilience, soil quality and sustainability. Soil Tillage Res 27:1–8

Lal R, Stewart BA (1994) Land degradation. Advances in soil science, vol 11. Springer, New York

Mandal AK, Sharma RC, Singh G, Dagar JC (2010) Computerized database on salt affected soils in India, Technical Bulletin No.2. CSSRI, Karnal, p 28

Minh LQ, Tuong TP, Van Mensvoort MEF, Bouma J (1997) Contamination of surface waters as affected by land use in acid sulfate soils in the Mekong River Delta, Vietnam. Agric Ecosyst Environ 61:19–27

Mondal MK, Bhuiyan SI, Franco DT (2001) Soil salinity reduction and prediction of salt dynamics in the coastal rice lands of Bangladesh. Agric Water Manag 47:9–23

NAIP (2014) Strategies for sustainable management of degraded coastal land and water for enhancing livelihood security of farming communities, NAIP final report, Sub-project code: (2104), Component-3. Central Soil Salinity Research Institute Regional Research Station, Canning Town, p 106

Nkonya E, Koo J, Marenya P, Licker R (2011) Land degradation: land under pressure. In: Global food policy report. International Food Policy Research Institute, Washington DC, pp 63–67

Oldeman LR (1994) The global extent of land degradation. In: Greenland DJ, Szabolcs I (eds) Land resilience and sustainable land use. CABI, Wallingford, pp 99–118

Ouni Y, Lakhdar A, Scelza R, Scotti R, Abdelly C, Barhoumi Z, Rao MA (2013) Effects of two composts and two grasses on microbial biomass and biological activity in a salt-affected soil. Ecol Eng 60:363–369

Rachman A, Wahyunto H, Agus F (2005) Integrated management for sustainable use of tsunami affected land in Indonesia. Workshop on Sustainable Use of Problem Soils in Rainfed Agriculture, 14–18 April (2005) Khon Khaen

Ravisankar N, Ambast SK, Srivastava RC (2008) Crop diversification through broad bed and furrow system in coastal regions. Central Agricultural Research Institute, Port Blair, p 180

Reich P, Eswaran H (2004) Soil and trouble. Science 304:1614–1161

Roy Chowdhury S, Rautary SK, Panda RK, Das M, Srivastava RC, Ambast SK (2016) Technological options for agricultural water management in Eastern regions of India. ICAR-IIWM, Bhubaneswar, p 39

Shrestha RK, Lal R (2006) Ecosystem carbon budgeting and soil carbon sequestration in reclaimed mine soil. Environ Int 32:781–796

UNEP (1992) World Atlas of desertification. Edward Arnold, London

Velmurugan A, Swarnam TP, Lal R (2015a) Effect of land shaping on soil properties and crop yield in tsunami inundated coastal soils of Southern Andaman Island. Agric Ecosyst Environ 206:1–9

Velmurugan A, Swarnam TP, Ambast SK, Meena RL, Subramani T (2015b) Soil salinity dynamics in Raised Bed and Furrow (RBF) system and its effect on alleviating waterlogging in the coastal lowlands. J Soil Salinity Water Qual 7(2):90–97

Biosaline Agriculture

21

Abstract

In view of rapidly increasing human population and demand for diverse food items crop production must increase substantially if food security is to be ensured. At the same time, arable land and good-quality irrigation water resources are being depleted at faster rate particularly in the arid, semi-arid, and tropical regions. Over the years the salinization of soil and water has steadily increased due to various causes, and the increase in food production essentially depends on this degrading resources. In this context plant-based approaches of using salt-affected land and water are based on the salinity ranges of soil, water and other associated factors termed as biosaline agriculture. This chapter gives an overview of biosaline agricultural strategies as a solution to use salt-affected soils and water which primarily involves selection of suitable halophytes for direct use and production of new genotypic material capable of growing in the saline environment. The biosaline agriculture not only helps in halting further deterioration of marginal lands but also has direct commercial uses such as food, forage/fodder, livestock, medicinal uses, wood, biofuel and bioenergy. Further research is essential to explore the usefulness of several halophytes and its possible introduction into coastal saline areas, improvement of cultivated species, monitoring of salinity level in land and water, and its effective use through innovations.

Keywords

Halophytes · Tropical coast · Saline soils · Salinity tolerance · Food security

21.1 Introduction

The emerging situations arising out of population explosion in many of the developing countries in the tropics are posing major challenge in terms of demand for food and other products. At the same time improvement and adjustment in agriculture to

V. Ayyam et al., *Coastal Ecosystems of the Tropics - Adaptive Management*,
https://doi.org/10.1007/978-981-13-8926-9_21

face these challenges need to take into consideration the type of land and water available in different region (FAO 1992; Khan and Duke 2001). Global level assessment showed that millions of hectares of agricultural land are lost every year mostly during the last 50 years of agricultural development due to unsustainable intensive agricultural practices besides diversion of agricultural land for non-agricultural purposes. This has created enormous demographic and economic pressure which is expected to rise further due to changing needs of people for diversified products and globalization. Therefore, agriculture strategies for feeding the population from the available resources represent one of the most important challenges in the twenty-first century.

In addition, there are other factors which further aggravates this demand, viz. (a) the increase of land occupation for non-agricultural purpose and biofuel supply that diverts arable soils from food crops (Prins et al. 2011), (b) the challenges posed by increasing incidences of extreme environmental conditions and natural disasters, and (c) the raise of soil degradation and/or the increase of saline soils in the post green revolution era which is growing dramatically, reducing the area of arable land. Presently on an average the food grain production increases by 3–6% in different region as the soils get exhausted and inherent weaknesses have increased. Consequently the factor productivity is declining and global as well as region food production has fast approaching the plateau.

The projection given by FAO indicated increase of food requirements by 20% in developed countries and 60% in developing countries. It is therefore recognised that food requirements are increasing quicker than crop production, and researchers and policymakers increasingly highlight the urgent need to improve alternative agricultural strategies (FAO 1992; Yamaguchi and Blumwald 2005). In other words it has become necessary to invest in better technologies within the scopes of the Millennium Development Goals strengthening practices leading to sustainable agriculture without destroying lands and natural resources (Galvani 2007).

Among the many reasons affecting agricultural productivity in the tropical region, salinity and associated factors, like waterlogging and/or drought exaggerated by climate change, have contributed significantly. The increase in saline areas has been directly attributed to both water and soil salinity problems. In the coastal areas inundation of low-lying areas by sea water and seawater intrusion into the freshwater aquifers contribute to the coastal salinization. A number of approaches have been developed to combat the salinity problems and increase the food grain production, based on specific types of site and regional and global problems. These include soil reclamation, water management, and plant-based approaches. The first approach of soil reclamation is dealt in separate chapter, but it requires high financial investment if carried out in large areas. The other two approaches are management strategies using appropriate irrigation and drainage systems and selection of appropriate agriculture production systems, suitable for specific edaphic and climatic conditions. These approaches, when applied in an integrated form, constitute the backbone of biosaline agriculture, which are the main focus of this chapter.

21.2 Status of Saline Land and Water

Salinity is one of the most widespread soil degradation processes affecting the arable land on the Earth. Though the process of occurrence of salt-affected soils is different, they are found practically in all climatic belts from the humid tropics to the polar circle. They can be found in different altitudes, from territories below sea level, e.g., the district of the Dead Sea, to mountains rising over 5000 m as the Tibetan Plateau of the Rocky Mountains (Szabols 1995). Older estimates (Szabolcs 1989) suggest that 10% of the total arable land is affected by salinity and sodicity, extending over more than 100 countries occupying different proportions of their territory. The description of the types of salt-affected soils and causes of formation and hypothetical salinization cycle has been reported by many researchers (Shahid and Rahman 2011).

Currently the exact extent of salt-affected soils is unknown due to unavailability of updated information or lack of compilation of regional level assessments. Based on the FAO/UNESCO Soil Map of the World, Massoud (1977) made an estimate of 880 million ha (M ha) of salt-affected soils, of which 316 M ha are in developing countries. These are the potential areas where land can be leased for food security or alternate energy sources. Balba (1980) gave a global estimate of only 600 M ha as salt-affected soils (Africa 30 M ha, Asia 340 M ha, Australia 140 M ha, Europe 1 M ha, North America 26 M ha, and South America 60 M ha). Pessarakli and Szabolcs (2011) have quoted 954.8 M ha of salt-affected soils available worldwide. These estimates may include soils which are not currently in use and have the potential for practising biosaline agriculture to meet the food demands.

Similar to that of soils, the availability of fresh water is a major limiting factor in sustainable agriculture. In certain regions of the world, the water crisis is so severe than the availability of land. Global assessment of water availability and the future projections by several authors found decreasing water availability in these developing regions of burgeoning population pressure (Alcamo et al. 2000; IWMI 2007). The water scarcity is grouped into physical water scarcity, approaching physical water scarcity, and economic water scarcity to understand the water shortage which includes all purpose of water (Fig. 21.1). Physical water scarcity means water resource development is approaching or has exceeded sustainable limits. In this more than 75% of river, flows are withdrawn for agriculture, industry, and domestic purposes. This definition, relating water availability to water demand, implies that dry areas are not necessarily water scarce. In approaching physical water scarcity, more than 60% of river flows are withdrawn, and these basins will experience physical water scarcity in the near future. The situation may get worsen with more withdrawal to produce more food. Whereas in economic water scarcity even though water in nature is available locally to meet human demands, factors such as human, institutional, and financial capital limits access to water. The tropical developing region mostly faces the challenges of water scarcity. Economic water scarcity is the major limits for production in sub-Saharan Africa, while physical water scarcity limits the production in South Asia.

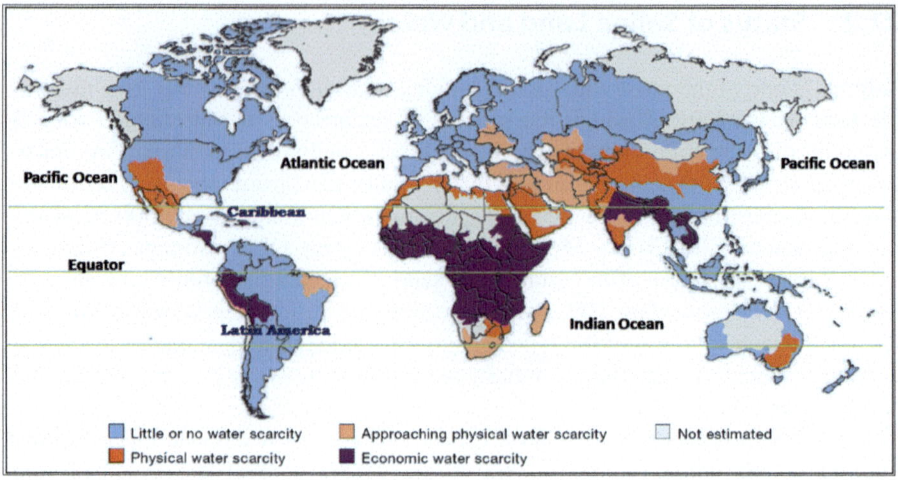

Fig. 21.1 Areas of physical and economical water scarcity at the basin level in 2007. (Modified from IWMI 2007)

In contrast to the crisis, some countries have developed technologies to utilize the saline water. For example, in Israel farmers carry out crop production with unconventional water resources irrigation, and the use of brackish water deserves particular attention. Also, in Tunisia strong investigations were addressed to this issue, and one can find agriculture practices based in alternative plant species, most of them are halophytes, which are able to tolerate high temperatures and/or low water availability (Lakhdar et al. 2009). Similar attempts are being made in some of the South Asian countries to meet the challenges, but in many cases these are at experimental stage.

The availability of fresh water for farming is an essential condition for achieving satisfactory and profitable yields, both in terms of unit yields and quality. In coastal regions due to excessive withdrawal of groundwater, high evapotranspiration, rise of saline groundwater, and seawater intrusion pose major challenge. Increase in groundwater pumping results in the intrusion of sea water into the freshwater aquifers. Furthermore, freshwater aquifers have been exhausted, and hence overlying saline water layers mix with fresh water, resulting in the increase of salinity in the groundwater. The most common reasons for the increase in salt-affected lands are the mismanagement of irrigated areas. Inadequate or absence of drainage systems and high rates of irrigation water application have resulted in the movement of salts in the soil profile, especially in the dry areas, where these salts are brought to the surface as a result of high evapotranspiration rates. As a result nearly 5–10% of the existing freshwater resources are getting salinized. The critical values of renewable freshwater resources and economic water scarcity and salinization indicate the necessity for regional water use policy coordination, appropriate water management strategies at the national and site levels, and international collaboration for protecting water resources of international significance.

21.3 Concept and Scope of Biosaline Agriculture

The emerging situations in soil and water availability and salinization particularly in the tropical coastal areas are imposing serious challenge on agricultural production systems. This means that agriculture will face the challenge of having to do with limited water at times with poorer quality and have to use saline or acid-saline soils. There is no choice left to manoeuvre but to use the land according to its production potential with required level of inputs and enhance the productivity by technological innovations. An innovative strategy for enhancing land and water availability for crop production is the use of saline soils and water, in a strategy designated as saline agriculture. This strategy is not new as the use of sea/saline water for crop production in coastal, arid, and deserts has already been suggested in the last several decades but practised to a limited extent (Epstein et al. 1980; Glenn et al. 1999). The saline water such as sea water, brackish water, drainage water from other plantation irrigations, drainage water from humanized areas, or even water derived from aquaculture waste can also be used to irrigate halophyte crops.

The concept and applicability of biosaline agriculture have changed due to new innovation in salinity management. In this context biosaline agriculture is relatively a new way of dealing with salinity in agriculture which got evolved in response to the shrinking land and freshwater resources. It develops specific cropping systems for saline environments, using the capacity of certain plants to grow under saline conditions in combination with the use of saline soil and water resources with improved soil and water management methods. As salinity is influencing our environments more and more, there is urgency in developing cropping systems that can produce in saline conditions. From the foregoing discussion, biosaline agriculture is defined as improved agricultural practices using saline land and irrigation water with the purpose to achieve better production through a sustainable and integrated use of genetic resources (plants, animals, fish, insects, and microorganisms) avoiding expensive soil recovery measures.

This is necessitated by the emerging situation in salinization of soil and freshwater resources. Soil salinization has been worldwide recognized as being among the most important problems for crop production in arid, semi-arid, and rainfed regions and now spreading to the coastal areas of the tropics as well (FAO 2008). As reported above, some of the emerging regions in risk of increasing levels of salinization of their soils are, for example, the Mediterranean Basin (Nedjimi et al. 2006), Central Asia, the Middle East, Northern and Eastern Africa, and South Asia (Yensen 2006; FAO 2008).

Experiences of salinity-tolerant trees grown in the saline land have demonstrated its usefulness to recover salt-affected land. Vegetation over saline soils tends to decrease salt concentration in the topsoil due to increased infiltration of rainwater and reduced capillary rise of water. This approach can allow farmers to get instant economic returns by growing field crops and also immediate economic benefits from saline wasteland with the help of salinity-tolerant trees. The imperative impact of trees on the microclimate, soil erosion, and floods is well known and are highly useful in the coastal areas. In addition, saline agriculture can also be a potential strategy for reducing CO_2 in the atmosphere from degraded salt-affected areas. Other studies refer the use of halophyte crops to reclaim saline soils since these plants can reduce the salt content of soil over time (Keiffer and Ungar 1997: Glenn et al. 1999). Such planting is preferred to various expensive engineering methods since it is cheaper and lasting.

21.4 Understanding Biosaline Technology

21.4.1 Components of Saline Agriculture

It is estimated that the agricultural production increase will need around 202 million ha of arable soil in developing countries, but approximately 93 million ha seems available for crop production (Khan and Duke 2001). The situation gets worse as millions of ha of agricultural lands are lost every year, mostly during the last 50 years of agricultural development due to unsustainable irrigation practices, excessive use of fertilizers, soil contamination, urban pressure, climate changes, etc. For these reasons it is vital to develop technologies to use saline soils and water and explore the potential of new plants which can tolerate salinity. Strategies to develop, utilize, and promote such technologies should be given high priority to sustain the production system and meet the food demand. Biosaline technology is based on three major components, namely (Fig. 21.2):

- The knowledge of salt-tolerant plants and halophytes in various saline environments and how to improve their productivity by plant breeding and agronomics
- Use of saline soils by innovative management practices for crop production
- The knowledge of salinity management in various saline environments including infrastructure for irrigation and drainage adapted to the needs of specific salt tolerant and halophytic crops

Fig. 21.2 Basic components of biosaline agriculture

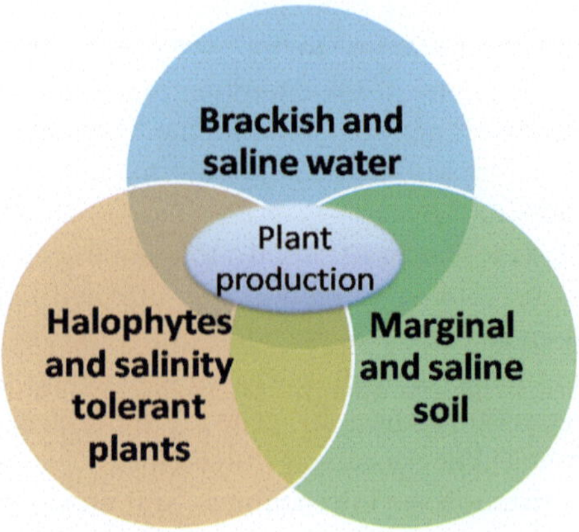

21.4.2 Desirability of Halophytes and Salt-Tolerant Plants

The potential applications of halophytes in phytoremediation, desalination, secondary metabolite production, medicine, food, and saline agriculture have been documented by several researchers (Glenn et al. 1999; Khan and Duke 2001; Yensen 2006; Ladeiro 2012). In the context of biosaline agriculture, a potential halophyte is cultivated in salt-contaminated soil using the available water; as a consequence slowly the severity of soil salinity also gets reduced. The harvested halophyte can have food, industrial, and other value; and later on, rehabilitated soil can be utilized for agriculture purpose.

Plant species used in biosaline agriculture are salt-tolerant plants which can grow even in higher salinity levels of soil and water wherein normal agricultural crops fail to grow and halophytes perform better than any other crops under saline environment. For halophytes succeed as irrigated crops, four basic conditions should be satisfied (Glenn et al. 1999). They are:

- High yield potential to justify its cultivation.
- The irrigation needs must not exceed the conventional crops; able to use saline water and be harmless to the soil.
- The products from halophyte crops must be able to replace the conventional crop products and meet human needs.
- High-salinity agriculture must be applicable to the existing agricultural infrastructure without much burden of capial investments.

The sustainable use of halophytes has multiple purposes from source of food to ecosystem protection. Though 2600 halophytic species are known, only few are extensively studied for their potential in agriculture and as biological resources with economical potential as sources of oils, flavours, gums, resins, oils, pharmaceuticals, and fibres (Galvani 2007). In addition, they have environmental potential for protection and conservation of ecosystems particularly in saline environment. For example, improvement of soil structure and fertility, habitat for wildlife, and source of biomass for the production of biodiesel will help in conservation and protection (Zhang et al. 2007). In this chapter we discuss some of the potential halophytic species to use as emergent or already used crops in arid, semi-arid, and tropical coastal regions, in a perspective of sustainable development.

21.4.3 Mechanism of Plant Adaptation to Salinity Stress

Molecular biological studies have shown that salt tolerance in halophytes depends on a range of adaptations. The multifactorial response includes many aspects such as gas exchange, water relations (osmotic adaptation), selective transport and uptake of ions, ion compartmentalization (homeostasis), osmolyte production, enzyme activities, ion excretion, and genetic control. The ability of plant cells to maintain low cytosolic sodium concentrations is an essential process for the ability of plants to grow in salty habitats that depend on several plant-specific interactions. Unfortunately, there are few investigations that combine studies of growth with the individual partially intracellular plant characteristics. Such joint investigations are the basis for the discovery of traits that present the ability to produce cash crops in saline environments (Koyro et al. 2008). Moreover the capacity to tolerate salinity is widely variable at intra- and interspecific level. It is known that variations of photosynthetic performances and related parameters (e.g. leaf area) are mainly responsible of growth reduction under salt stress. Generally studies have shown that salinity limits CO_2 availability at the carboxylation sites. A reduced assimilation rate is often accompanied by a more than proportional decrease in transpiration rate (Monteverdi et al. 2008). Other than genetic control, the adverse effects of salt stress on cereals such as maize and wheat can be alleviated by the exogenous application of chemicals such as glycinebetaine (GB) which in turn enhances photosynthetic capacity and modulates activities of antioxidant enzymes. However, effectiveness of such chemicals on regulation of photosynthetic and antioxidant capacities is species specific (Ashraf et al. 2008).

21.5 Biosaline Agricultural Technologies

21.5.1 Direct Use of Halophytes and Salinity-Tolerant Crop Plants

The use of salinized soils in alternative agriculture because of increase in salinized soils and reduction of arable land usable by conventional agriculture is an important strategy to cope with food demand (Shekhawat et al. 2006). Experiments involving potential halophytes have shown that profitable agriculture is possible using saline land and saline irrigation water avoiding expensive soil recovery measures. But for halophytes to succeed as irrigated crops, it should satisfy certain basic conditions as discussed in the previous section. This section highlights the use of such plants in combination with other technologies.

Encouraging results have been obtained for improving halophytes into new, salt-resistant crops and sometimes used directly after evaluation for desirable characters. On the other hand, when the economical production of conventional crop decreases due to salinity, halophytes could be used as a source of genes for conventional crop improvement. A summary of halophytes or salinity-tolerant plant species and their economic significance is given in Table 21.1. It can be used as food crops, oils seeds, genetic material in breeding, fuelwood, timber, source of chemicals, environmental protection, etc. Seeds of various halophytes, such as *Suaeda fruticosa*, *Arthrocnemum macrostachyum*, *Salicornia bigelovii*, *S. brachiata*, *Halogeton glomeratus*, *Kochia scoparia*, and *Haloxylon stocksii*, possess sufficient quantity of high-quality edible oil with unsaturation ranging from 70% to 80%. Seeds of *Salvadora oleoides* and *S. persica* contain 40–50% fat and are a good source of lauric acid—a potential substitute for coconut oil. The succulent *Salicornia bigelovii* (Chenopodiaceae) is a typical halophyte which is cultivated for its oil (both for human and animal use) and straw. The residual seed meal is very rich in protein

Table 21.1 Salinity-tolerant plant species/halophytes for coastal and saline inland

Sl. No	Economic benefit	Species	Purpose	References
1	Food crops	*Oryza indandamanica* Black Burma	Cereal crop	Velmurugan et al. (2016)
		Morinda citrifolia	Fruits are used	Velmurugan et al. (2016)
		Ardisia solanacea Roxb. and *A. andamanica* Kurz.	Minor fruits	Velmurugan et al. (2016)
		Phoenix paludosa	Fruits and salad	Velmurugan et al. (2016)
		Aster tripolium	Used as delicacy	Brock et al. (2007)
		Salicornia bigelovii	Protein, oil, and forage	Glenn et al. (1992)
		Diplotaxis tenuifolia	Salad and forage	Guerra (2008)
2	Genetic material	*Knema andamanica*	Used as root stock, for improving	Velmurugan et al. (2016)
		Annona glabra (Pond apple)		
3	Oils seeds	*Salvadora oleoides* and *S. persica*	40–50% fat and good source of lauric acid	Ladeiro (2012)
		Suaeda fruticosa, *Arthrocnemum macrostachyum*, *Halogeton glomeratus*, *Kochia scoparia*, and *Haloxylon stocksii*	Possess high-quality edible oil with unsaturation ranging from 70% to 80%	Ladeiro (2012)
		Salicornia bigelovii, *S. brachiata*,	Edible oil	Jaisankar et al. (2015)
		J. curcas (37%), *J. gossypiifolia* (40%), and *J. podagrica* (35%).	It contains 37–40% oil which has high biomedical use	Jaisankar et al. (2015)
		Aphanomixis polystachya (38%), *Calophyllum inophyllum* (51%), *Pongamia pinnata* (36%), *Sapium baccatum* (49%), and *Simarouba glauca* (53%)	Oil is used for various purposes including medicinal value. These trees grow in littoral region and some of them provide timber as well	Jaisankar et al. (2015)
4	Fuelwood, timber and multipurpose	*Prosopis*, *Tamarix*, *Salsola*, *Acacia*, *Suaeda*, *Kochia*, *Capparis*,*Casuarina*, *Pithecellobium*, *Parkinsonia*, and *Salvadora*	Fuelwood, construction material	Ladeiro (2012)
		Dalbergia sissoo, *Pongamia pinnata*, *Populus euphratica*, and *Tamarix spp.*	Good quality wood	Ladeiro (2012)
		Rhizophora, *Ceriops*, *Avicennia*, and *Aegiceras*	Good fuelwoods and contribute to charcoal production	

(continued)

Table 21.1 (continued)

Sl. No	Economic benefit	Species	Purpose	References
5	Environmental protection	*Spartina alterniflora*, *Spartina maritime*, and *Avicennia marina*	Coastline protection and restoration of coastal ecosystem	Callaway and Zedler (2004)
6	Source of chemicals	*Suaeda*, *Salicornia*, *Salsola*, and *Haloxylon*	Used in soap making	Ladeiro (2012)
		Annona glabra	Seeds used us insecticide	Khan et al. (2006)

(approximately 33–34% crude protein) which is fed to the cattle. In addition, *Diplotaxis tenuifolia* is a promising species for saline agriculture, as it lives naturally in saline/dry ecosystems, or with strong influence of sea. This plant has a potential to be used as salad (food) and forage for animals (Ladeiro 2012).

High oil prices in recent times are creating new markets for agricultural commodities that can be used as feedstock for the production of biofuels. The advantage of developing biofuel from halophytes as opposed to other types of biomass is that saltwater plants are not dependent on fresh water, which is in increasingly short supply, and can instead be irrigated using plentiful seawater supplies. Alternatively the results open the possibility of using saline land and water for carbon sequestration; biofuel production thereby helps in climate change mitigation. Further a kind of soda is obtained in large quantities from *Salicornia*, *Salsola*, and *Haloxylon* species, used in soap making and in glass industry. It is also interesting to note that seeds of *Annona glabra* is a source of botanicals (biological compounds used as insecticide) like *Azadirachta indica* which is known for its insecticidal properties (Khan et al. 2006). Halophytes and salinity-tolerant plants can also be sued as source of timber, fuelwood, coastal protection, and improvement of degraded soils and to provide some ecosystem services.

Some success with halophyte cultivation has been demonstrated in environmentally degraded soils, and it is imperative that large-scale adoption of halophytes, as potential candidates, can be accorded top priority for rehabilitating contaminated soils, which can pave way for sustainable agriculture (Ganesh et al. 2018). Yet more studies need to be conducted on the potential use of halophytes in saline agriculture and their use and incorporation in the consumers' diet. Also, the adequate conditions for increasing to industrial level production of halophyte species deserve more attention.

21.5.2 Biotechnological Approach

In recent years plant genetic engineering science has successfully risen to the challenge of producing plants tolerant to several abiotic stresses such as salt stress, moisture deficit, heat stress, low oxygen, etc. This approach works in harmony with the classical plant breeding approach. A comprehensive approach for utilization of

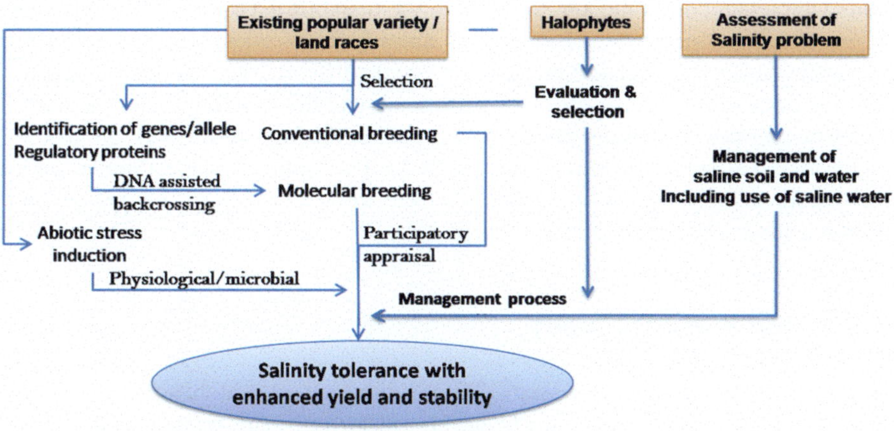

Fig. 21.3 Integrated approach for biosaline agriculture

halophytes and other salinity-tolerant wild relatives of cultivated plants using biotechnological tools is given in Fig. 21.3. Molecular control mechanisms for abiotic/salinity stress tolerance are based on the activation and regulation of specific stress-related genes. A host of genes encoding different structural and regulatory proteins has been employed for production of a range of abiotic stress-tolerant transgenic plants (Grover et al. 1998). The appreciation is growing for the use of regulatory genes which is more effective approach for producing stress-tolerant plants particularly salinity tolerant than other approaches. This is based on the observations that single regulatory gene leads to altered expression of a number of different downstream structural genes, thus leading to a wide-arrayed altered response (Pardo et al. 1998).

With the progress in plant genomics, more and more information is being gained about genes that respond to different stresses. Microarray analyses of transcriptome regulation under salt stress conditions have uncovered the complex gene networks involved in mechanisms of sensing, signaling, and short-term response. Most of this knowledge has been derived from shock-stress experiments conducted on one genotype under laboratory conditions. But crop response may vary under field codition which necesitates addressing the long-term acclimation to salt stress. The genes responsible for the variability of salt tolerance could be valuable resources in breeding programs, but they are difficult to identify in typical microarray experiments (Beritognolo et al. 2008). The genes revealed by transcriptome analyses of salt-stressed plants are often common to other stresses and other species but do not explain the heritable variation.

The work on identification, isolation, and cloning of abiotic stress-related regulatory genes is still in progress. However, a number of different stress-responsive genes have been identified which provide scope for development of transgenic plants with high level of adaptation to abiotic stress conditions (Dhaliwal et al. 1998). The advent of plant transformation may have placed within the grasp the possibility of engineering greater abiotic stress tolerance particularly salinity in plants which can contribute significantly to yield stability. However, the nature of the genetically complex mechanisms of abiotic stress tolerance, and potential detrimental side effects, makes this task extremely difficult The road to engineering such tolerance into sensitive species will be long, and the lack of public acceptance of genetic engineering means that other approaches raising tolerance to salt need to be considered at present (Munns et al. 2002). In this context it is most important to note that for salinity-tolerant transgenic plants to succeed in saline soils, proper land and water management is also very essential.

21.5.3 Tree-Borne Oil Seeds

Wide diversity of tree-borne oil seeds (TBOs) with high oil content and adaptability to marginal and coastal areas has been reported from many tropical islands and coastal regions (Ladeiro 2012; Jaisankar et al. 2015). A study conducted in Bay Islands, India, revealed that *Jatropha* is one of the most important species with high diversity which are widely distributed with varying amount of oil of economic importance, viz. *J. curcas* (37%), *J. gossypiifolia* (40%), and *J. podagrica* (35%). Apart from this, *Aphanomixis polystachya* (38%), *Calophyllum inophyllum* (51%), *Pongamia pinnata* (36%), *Sapium baccatum* (49%), and *Simarouba glauca* (53%) are other potential oil-yielding TBOs (Jaisankar et al. 2015). Similarly *Calophyllum soulattri (49%)* is identified as a potential TBO from Nicobar Islands, India, which are traditionally used by the tribals for its medicinal value. *Jatropha* has also been reported from Southeast Asian countries used for similar purpose. Other than these reported TBOs, there are wide diversity of TBOs known to the locals and tribals (traditional knowledge) which has the potential to be exploited commercially grown on saline and other degraded lands in a mixed stand or along with shelter belts in tropical coastal ecosystem.

21.5.4 Agri-Silvi-Pastoral System

An important agroforestry measure suitable for tropical coastal and island conditions is growing multipurpose tree species along with grasses and after reclamation with seasonal crops. This system is probably specific to the coastal ecosystem rather than inland saline areas. The climatic conditions and physiography of tropical coastal regions and islands offer ample scope for growing fodder trees, shrubs, and grasses together in silvipastoral system with appropriate silvicultural management to tolerate salinity. Silvopastoralism is one such agroforestry practice that intentionally integrates trees, forage crops, and livestock into a structural practice of planned interactions (Clason and Sharrow 2000). The primary role of this system is production of green fodder to support the livestock production on a sustainable basis without much constraint on soil resources and environmental degradation. Some of the suitable multipurpose trees employed in agroforestry for coastal saline conditions are (Table 21.2) *Tamarix*, *Prosopis*, *Salvadora*, *Acacia farnesiana*,

Table 21.2 Suitable trees and grasses for various problem sites

Strees condition	Suitable trees/shrubs or grasses
Very high salinity (Ece > 35 dS/m)	*Hibiscus* sp., *Pongamia pinnata*, *Desmodium umbellatum*, *Barringtonia asiatica*, *Manilkara littoralis*
High salt tolerant (Ece 25–35)	*Casuarina*, *Thespesia populnea*, *and Cocos nucifera (on specific sites)*
Tolerant (Ece 15–25)	*Casuarina sp.,, Pongamia pinnata*, *Eucalyptus camaldulensis*, *Albizia lebbeck*, *Ziziphus mauritiana*, *Parkinsonia aculeata*
Moderately tolerant (Ece 10–15)	*Casuarina*, *Azadirachta indica*, *Dendrocalamus strictus*, *Butea monosperma*, *Leucaena leucocephala*, *Tamarindus indica*, *Balanites roxburghii*
	Cenchrus ciliaris, *C. setigerus*

Casuarina (*glauca*, *obesa*, *equisetifolia*), *Acacia tortilis*, *A. nilotica*, *Pongamia pinnata*, *Albizia lebbeck*, *Ziziphus mauritiana*, and *Parkinsonia aculeata* (Dagar 1995). Some of these species are highly valued for its timber. It is highly valuable to note that species such as *Hibiscus* sp., *Pongamia pinnata*, *Desmodium umbellatum*, *Barringtonia asiatica*, and *Manilkara littoralis* can tolerate very high salinity and act as a shelter belts along with mangrove species.

21.5.5 Use of Salinity-Tolerant Microbes

One of the basic problems for saline agriculture is the poor germination and seedling establishment. Salts in higher concentration and some even at low levels are toxic to the crop plants, interfere with the water and nutrient uptake which adversely affects growth and development of crop plants, and are responsible for low agricultural production. The depressive effect of salinity on germination could be related to a decline in endogenous levels of hormones (Afzal et al. 2006). However, incorporation of certain salinity-tolerant microorganisms by seed biopriming in many cereal and vegetable crops has resulted in increased levels of plant growth hormones and improved seed performance (Howell 2003).

Several studies have shown that root colonization by *Trichoderma harzianum* increased the level of plant enzymes, including various peroxidases, chitinases, β-1,3-glucanases, lipoxygenase-pathway hydroperoxidelyase, and compounds like phytoalexins and phenols, to provide durable resistance against stress (Hoitink et al. 2006). Accumulation of some compatible solutes has also been observed under salt stress conditions and has been suggested as part of mechanism(s) that controls salt tolerance in plants. Proline is one of the best known solutes which accumulate due to *Trichoderma* inoculation under saline condition in brinjal, maize, and pulses. Another stress-related chemical malondialdehyde production increases as salinity stress increases in plants. This is an indicative of oxidative stress which serves as an index of lipid peroxidation. Peroxidation damage of the plasma membrane leads to leakage of contents, rapid desiccation, and cell death (Scandalios 1993).

Salinity-tolerant native strains of *Trichoderma* and bacterial antagonists were identified and used for the management of biotic stress in vegetable crops and inducing salinity resistance in brinjal (Velmurugan et al. 2015). Significant increase in seed germination and shoot and root growth in treated plants (Th-5) was noticed than untreated control under saline conditions. This was possibly due to the accumulation of proline and phenolic compounds in treated plants under salinity stress (6 dS m^{-1}) conditions.

21.5.6 Saline Water Resources and Use

Globally 98% of the water volume is sea water, 1% is fresh water, and 1% is brackish water. Agricultural use is currently responsible for 70% of all freshwater consumption but wide difference between various regions. Competition between the various users of fresh water is growing fast. The growing population of the world needs more food and agricultural crops for energy and industrial uses. This increasing demand for water resources in the world is uneven in its spatial distributions. Higher demand for water is experienced in the arid and semi-arid, tropical, and seasonal demand even in humid regions. It is highly unlikely however that there will be more water available for agriculture.

Under such situations, there are only two methods to solve the problem: (i) technology should be developed or evaluated to increase the efficiency of the current water use under saline conditions and (ii) start using other water resources, viz. brackish water and sea water. This has also forced farmers to use low-quality water for irrigation such as agricultural drainage water and marginal quality water, industrial wastewater. This necessitated the study of use of such water on cultivated crops. It is known that among the cultivated vegetables, tomato is moderately tolerant to salt. However, under stress conditions, antioxidative defence mechanisms in tomato are activated. The effects of diluted sea water on the antioxidant capacity, namely, ascorbate, tocopherols, and cellular redox status, have been evaluated in three tomato cultivars. The study showed that the three cultivars had different salt tolerance, and salad-type cultivar (Naomi) showed the best adaptive response due to its increased antioxidant pool after salinization (Incerti et al. 2008). Similarly the effects of irrigation with diluted sea water (20% and 30%) on main morphological characteristics of sunflower plants were studied. The results showed increase in Cl^- and Na^+ significantly but decreased K^+ and Ca^{2+}. However at 20% seawater dilution, the ecotypes produced achenes, probably due to micronutrients which increased the antioxidative defences (Izzo et al. 2008). These studies clearly established the fact there is wide variation in salinity tolerance among crop plants and even within cultivars. However, opportunity for identifying and utilizing such salinity-tolerant cultivated species with saline irrigation water are still wide open. Some of the important methods of using saline water in agriculture are:

- Irrigate the salinity-tolerant crop/halophytes with poor-quality or saline water with less amount, but frequency may be increased to meet the water demand. In some cases sea water can be used (salicornia, mangroves, seawater grasses); however, brackish water offers a higher potential in salt-tolerant species that can flourish on it.

- The irrigation with these waters does not always produce high yields. Mixing low-quality water with good-quality water to keep the salinity of the irrigation water below the threshold of the target crop is an acceptable practice which increases the crop yield than its direct application (Gawad et al. 2005).
- Another practical way of using the available water resources is the use of good-quality water during the sensitive stages of plant growth and the poor-quality water during the non-sensitive stages.
- Recently, seawater utilization has been explored to assess the possibility of obtaining a reasonable yield and quality of products from crops (Damico et al. 2004). But, seawater salinity delayed germination of tomato seed and reduced germination percentage especially with increasing salinity level (Hajer et al. 2006). Therefore salinity-tolerant varieties should be promoted in water-scarce areas.
- Selection of an appropriate irrigation method and the use of cultivars suitable for growth and providing economic yield even under saline conditions could improve water-use efficiency, thereby reducing the demand for fresh water and constituting a solution to minimize the effects of salinity.

21.6 Conclusion

Salinization of soil and water happens due to natural causes, pedological reasons, and anthropogenic activities. Sometimes in the coastal areas, the microscopic salt particles are carried by the monsoon wind from the oceans. Salts are mostly released by weathering of parent materials containing salts which get leached down by runoff water and get accumulated in the depression or low-lying areas. Sometimes the salts present in the soil deep below the surface are brought to the surface by water vapour during evaporation. But salinization is aggravated by excessive irrigation and withdrawal of groundwater termed as secondary salinization.

Several studies on resource degradation and agricultural production have concluded that the increase in demand for food by the burgeoning population constrained by the climate change events decreasing availability of arable land and fresh water led to the problem of degradation of agricultural production system and decline in food production. With these perspectives, saline agriculture is more pertinent and appropriate method for utilizing these degraded land and water to meet the food demand. From the summary of research findings on the use of salinized land through drainage/irrigation without using high-quality water it is infered that use saline water to harness the potential of suitable halophytes which are emerging as promising crops. Comparative genomics which is based on the comparison of genotypes differing in phenotypical behaviour is a promising approach to identify genes that control the heritable genetic variation of salt tolerance. Such valuable genes can be transferred to cultivated species having low salinity tolerance. There are several halophytes and salinity-tolerant plant species which can be used as a source of food, oil, timber, fuelwood, chemicals, and environmental protection. At the same time, it also requires changing the habits of mankind to incorporate halophytes and salinity-tolerant new crops in daily diet.

It is therefore clear that technological advances enable the use of saline soils and water to address the human population pressure for increased food grain production through biosaline agriculture. At the same time, without a comprehensive and long-term strategy adaptable to the prevailing climatic, edaphic, hydrogeologic, and socio-economic conditions of the coastal and other saline areas, it may not be possible to sustain the biosaline agriculture to meet the future challenges in food production. Therefore, research on potential use of halophytes in saline agriculture and suitable technology for halophyte species production at farm and industrial levels should continue.

References

Afzal I, Basra SMA, Farooq M, Nawaz A (2006) Alleviation of salinity stress in spring wheat by hormonal priming with ABA, salicylic acid and ascorbic acid. Int J Agric Biol 8:23–28

Alcamo J, Henrichs T, Thomas R (2000) World water in 2025, Global modeling and scenario analysis for the world commission on water for the 21st century. Centre for environmental systems research, University of Kassel, Germany

Ashraf M, Nawaz K, Akthar H, Raza SH (2008) Growth enhancement in two potential cereal crops, maize and wheat, by exogenous application of glycinebetaine. In: Abdelly C, Ozturk M, Ashraf M, Grignon C (eds) Biosaline agriculture and high salinity tolerance. Birkhäuser Verlag, Basel, pp 21–35

Balba AM (1980) Minimum management programme to combat world desertification. UNDP Consultancy Report on Advances in Soil Water Research, Alexandria

Beritognolo I, Sabatti M, Brosché M, Mugnozza GS (2008) Functional genomics to discover genes for salt tolerance in annual and perennial plants. In: Abdelly C, Ozturk M, Ashraf M, Grignon C (eds) Biosaline agriculture and high salinity tolerance. Birkhäuser Verlag, Basel, pp 273–286

Brock J, Aboling S, Stelzer R, Esch E, Papenbrock J (2007) Genetic variation among different populations of Aster tripolium grown on naturally and anthropogenic salt-contaminated habitats: implications for conservation strategies. J Plant Res 120(1):99–112

Callaway JC, Zedler JB (2004) Restoration of urban salt marshes: lessons from southern California. Urban Ecosyst 7:107–124

Clason TR, Sharrow SH (2000) Silvopastoral practices. In: Garrett HE, Rietveld WJ, Fisher RF (eds) North American agroforestry: an integrated science and practice. American Society of Agronomy, Madison, pp 119–147

Dagar JC (1995) Agroforestry systems for the Andaman and Nicobar Islands. Int Tree Crop J 8(2–3):107–128

Damico ML, Navari-Izzo F, Sgherri C, Izzo R (2004) The role of lipoic acid in the regulation of the redox status of wheat irrigated with 20% seawater. Plant Physiol Biochem 42:329–334

Dhaliwal HS, Kawai M, Uchimiya H (1998) Genetic engineering for abiotic stress tolerance in plants. Plant Biotechnol 15:1–10

Epstein E, Norlyn JD, Rush DW (1980) Saline culture of crops: a genetic approach. Science 210(4468):399–404

FAO (1992) The use of saline waters for crop production, FAO irrigation and drainage paper 48, Rep. Food and Agriculture Organization, Rome. www.fao.org/3/a-t0667e.pdf. Accessed 20 Sept 2018

FAO (2008) Land and plant nutrition management service. http://www.fao.org/ag/ AGL/public.stm

Galvani A (2007) The challenge of the food sufficiency through salt tolerant crops. Rev Environ Sci Biotechnol 6(1–3):3–16

Ganesh CN, Ashish KS, Sablok G, Girdhar KP, Tukaram DN, Suprasanna P (2018) Identification and validation of reference genes for quantitative real-time PCR under salt stress in a halophyte, *Sesuvium portulacastrum*. Plant Gene 13:1–18

Gawad GA, Arslan A, Gaihbe A, Kadouri F (2005) The effects of saline irrigation water management and salt tolerant tomato varieties on sustainable production of tomato in Syria (1999–2002). Agric Water Manag 78:39–53

Glenn EP, Hicks N, Riley J, Swingle S (1992) Seawater irrigation of halophytes for animal feed. In: Halophytes and biosaline agriculture. Marcel Dekker, New York

Glenn EP, Brown JJ, Blumwald E (1999) Salt tolerance and crop potential of halophytes. Crit Rev Plant Sci 18(2):227–255

Grover A, Pareek A, Singla SL, Minhas D, Katiyar S, Ghawana S, Dubey H, Agarwal M, Rao GU, Rathee J (1998) Engineering crops for tolerance against abiotic stresses through gene manipulation. Curr Sci 75:689–696

Guerra C (2008) Salt effects on growth, nutrient and secondary compound contents of diplotaxis tenuifolia [M.S. thesis], University of Aveiro, Aveiro

Hajer AS, Malibari AA, Al-Zahrani HS, Almaghrabi OA (2006) Responses of three tomato cultivars to seawater salinity. Effect of salinity on the seedling growth. Afr J Biotechnol 5:855–861

Hoitink HAJ, Madden LV, Dorrance AE (2006) Systemic resistance induced by Trichoderma spp.: interactions between the host, the pathogen, the biocontrol agent, and soil organic matter quality. Phytopathology 96:186–189

Howell CR (2003) Mechanisms employed by Trichoderma species in the biological control of plant diseases: the history and evolution of current concepts. Plant Dis 87:4–10

Incerti A, Izzo R, Belligno A, Navari-Izzo F (2008) Seawater effects on antioxidant production in berries of three cultivars of tomato (*Lycopersicon esculentum* Mill.) pp.43–51

IWMI (2007) Water for food, water for life: a comprehensive assessment of water management in agriculture. International Water Management Institute Earthscan, London

Izzo R, Incerti A, Bertolla C (2008) Seawater irrigation: effects on growth and nutrient uptake of sunflower plants. In: Abdelly C, Ozturk M, Ashraf M, Grignon C (eds) Biosaline agriculture and high salinity tolerance. Birkhäuser Verlag, Basel, pp 61–69

Jaisankar I, Velmurugan A, Swarnam TP, Dam Roy S (2015) Diversity of tree borne oil seeds in Andaman and Nicobar Islands. In: Roy D (ed) Extended summaries of national seminar on harmonizing biodiversity and climate change: challenges & opportunity, 17–19th April. CIARI, Port Blair, p 38

Keiffer CH, Ungar IA (1997) The effect of extended exposure to hyper saline conditions on the germination of five inland halophyte species. Am J Bot 84(1):104–111

Khan MA, Duke NC (2001) Halophytes—a resource for the future. Wetl Ecol Manag 9(6):455–456

Khan MA, Ansari R, Gul B, Qadir M (2006) Crop diversification through halophyte production on salt-prone land resources. CAB Rev 1:1–9

Koyro HW, Geißler N, Hussin S, Huchzermeyer B (2008) Survival at extreme locations: life strategies of halophytes – the long way from system ecology, whole plant physiology, cell biochemistry and molecular aspects back to sustainable utilization at field sites. In: Abdelly C, Ozturk M, Ashraf M, Grignon C (eds) Biosaline agriculture and high salinity tolerance. Birkhäuser Verlag, Basel, pp 1–20

Ladeiro B (2012) Saline agriculture in the 21st century: using salt contaminated resources to cope food requirements. J Bot (1–7). Article ID 310705, https://doi.org/10.1155/2012/310705

Lakhdar A, Rabhi M, Ghnaya T, Montemurro F, Jedidi N, Abdelly C (2009) Effectiveness of compost use in salt-affected soil. J Hazard Mater 171(1–3):29–37

Massoud FI (1977) Basic principles for prognosis and monitoring of salinity and sodicity. In: Proceedings of the international conference on managing saline water for irrigation. Texas Tech University, Lubbock, pp 432–45416–20 Aug (1976)

Monteverdi CM, Lauteri M, Valentini R (2008) Biodiversity of plant species and adaptation to drought and salt conditions. Selection of species for sustainable reforestation activity to combat desertification. In: Abdelly C, Ozturk M, Ashraf M, Grignon C (eds) Biosaline agriculture and high salinity tolerance. Birkhäuser Verlag, Basel, pp 197–206

Munns R, Husain S, Rivelli AR, James RA, Condon AG, Lindsay MP, Lagudah ES, Schachtman DP, Hare RA (2002) Avenues for increasing salt tolerance of crops, and the role of physiologically based selection traits. Plant Soil 247:93–105

Nedjimi BY, Daoud, Touati D (2006) Growth, water relations, proline and ion content of in vitro cultured Atriplex halimus subsp. schweinfurthii as affected by $CaCl_2$. Commun Biometry Crop Sci 1(2):79–89

Pardo JM, Reddy MP, Yang S, Maggio A, Huh GH, Matsumoto T, Coca MA, Paino-D'Urzo M, Koiwa H, Yun DJ, Watad AA, Bressan RA, Hasegawa PM (1998) Stress signaling through Ca^{2+}/calmodulin-dependent protein phosphatase calcineurin mediates salt adaptation in plants. Proc Natl Acad Sci U S A 95:9681–9686

Pessarakli M, Szabolcs I (2011) Soil salinity and sodicity as particular plant/crop stress factors. In: Pessarakli (ed) Handbook of plant and crop stress. CRC Press/Taylor and Francis Group, Boca Raton, pp 3–21., 496

Prins AG, Eickhout B, Banse M, van Meijl H, Rienks W, Woltjer G (2011) Global impacts of European agricultural and biofuel policies. Ecol Soc 16(1):49–55

Scandalios JG (1993) Oxygen stress and superoxide dismutase. Plant Physiol 101:7–12

Shahid SA, Rahman K (2011) Soil salinity development, classification, assessment and management in irrigated agriculture. In: Pessarakli M (ed) Handbook of plant and crop stress. CRC Press/Taylor and Francis Group, Boca Raton, pp 23–39

Shekhawat VPS, Kumar A, Neumann KH (2006) Bio-reclamation of secondary salinized soils using halophytes. In: Biosaline agriculture and salinity tolerance in plants, pp 147–154

Szabolcs I (1989) Salt-affected soils. CRC Press, Boca Raton, p 274

Szabolcs I (1995) Global overview of sustainable management of salt-affected soils. In: Proceedings of the international workshop on integrated soil management for sustainable use of salt-affected soils, Bureau of soils and water management, Diliman, Quezon City pp 19–38

Velmurugan A, Sakthivel K, Swarnam TP, Rachael SR, Dam Roy S (2015) Assessment of the plant growth promotion and phosphorus Solubilization by rhizosphere Bacteria isolated from *Troporthents* soils of Bay Island. Trends Biosci 8(11):2888–2892

Velmurugan A, Dam Roy S, Dagar JC, Swarnam TP, Jaisankar I (2016) Innovative technologies to sustain Saline Island agriculture in the scenario of climate change: a case study from Andaman Islands, India. In: Dagar J, Sharma P, Sharma D, Singh A (eds) Innovative Saline agriculture. Springer, New Delhi, pp 387–417

Yamaguchi T, Blumwald E (2005) Developing salt-tolerant crop plants: challenges and opportunities. Trends Plant Sci 10(12):615–620

Yensen NP (2006) Halophyte uses for the twenty-first century. In: Khan M, Ajmal W, Darrell J (eds) Ecophysiology of high salinity tolerant plants. Springer, Berlin, pp 367–396

Zhang JY, Wang ZY, Jenks MA, Hasegawa PM, Jain SM (2007) Recent advances in molecular breeding of forage crops for improved drought and salt stress tolerance. In: Advances in molecular breeding towards salinity and drought tolerance. Springer, Dordrecht, pp 797–817

Aquaculture-Based Systems for Harmonious Development of Coastal Region

22

Abstract

Aquaculture is an increasingly important food-producing sector and provides livelihood to millions of coastal people in the tropical coastal region. However unplanned development and unsustainable intensive practices have created competition for space and environmental concerns. In the process of aquacultural development, traditional polyculture which was once considered an environmentally sound practice based on optimum utilization of farm resources, including farm wastes, is now counted among potential polluters of the aquatic and other coastal environment. To put this into a positive light, aquaculture must be developed in a spirit of social harmony, environmental sustainability, and economic progress. This necessitates the spatial planning and integrated aquaculture technologies which need to follow an ecosystem approach. The rationale behind evaluation and recommendation of most of the integrated aquaculture systems involving crop, fish, animal, molluscs, shellfish, seaweeds, sea cucumber, crustaceans, etc. in different combinations for the coastal region stabilizes the production of, and decrease effluents from, cultures of high-valued species. This chapter highlights the importance of some of the integrated aquaculture systems suitable for coastal areas and ways to reduce the pollutant from aquaculture activities and improve the environmental conditions by responsible fishery practices.

Keywords

Integrated aquaculture · Rice-fish · Good practices · Traditional practices

22.1 Introduction

Aquaculture plays a major role in meeting the rising demand for fish products and supplying healthy and nutritious protein to human beings (Larsen and Roney 2013). For this reason the past 20 years have seen aquaculture evolve from a traditional

V. Ayyam et al., *Coastal Ecosystems of the Tropics - Adaptive Management*,
https://doi.org/10.1007/978-981-13-8926-9_22

practice to a science-based activity and grow into a significant food production sector, contributing more to national economies and providing better livelihoods for rural and farming families. Aquaculture is used to be regarded as an infant when compared with crop and livestock farming and capture fisheries. In most parts of the tropical Asian countries, the activity has matured to become a clearly defined economic sector that is better organized and characterized by state patronage and strong private sector participation (Kongkeo 2001).

This is evidenced from the growth of aquaculture more than 10% per annum globally, outpacing terrestrial livestock and capture fisheries. Excluding aquatic plants, 60% of production comes from inland and 40% from coastal or marine areas (Shearer et al. 1997). Farmers in the Asia-Pacific Region contribute over 80% of the world's aquaculture production (Edwards and Demaine 1997). At the same time, the evergrowing population in the tropical coastal countries and demand for coastal land for other uses and fresh water created enormous pressure on the limited natural resources. This resulted in the degradation of natural resources and loss of biodiversity. Proponents of aquaculture have called for increased fish production from mariculture in the nearshore areas to increase fish delivery for human consumption and to reduce the use of scarce natural resources such as fresh water (Duarte et al. 2009) and wild fish (FAO 2012). However, available space for new aquaculture and mariculture development in coastal zones is now becoming increasingly limited, because the complex interactions among users of coastal areas often leave little space for aquaculture, particularly since marine aquaculture requires coastal waters with specific environmental and water quality characteristics.

This trend can be expected to continue because, along with aquaculture, other human activities in coastal and oceanic waters will increase significantly in the coming years including tourism and other recreational activities (Hall 2001). Further use of offshore for renewable energy generation will also increase (Douvere and Ehler 2009). Traditional activities, such as capture fisheries, are deeply embedded in many coastal communities, and they will continue to play an important role in the use of nearshore coastal seascapes. The result is that suitable coastal areas for aquaculture are severely limited in many areas, and this has become a major barrier to expansion. This competition for space has resulted in negative interactions with traditional and new coastal users (Dempster and Sanchez-Jerez 2008) and negative effects of other activities on aquaculture such as agriculture and sewage discharges (Díaz et al. 2012).

This necessitates the spatial planning and integrated aquaculture technologies which needs to follow an ecosystem approach and to define socio-economic, environmental, and governance objectives that result in the integrated management of land, water, and living resources for the development and expansion of the coastal region in a sustainable and equitable way (FAO 2013). Further to ensure a balance between conserving ecosystem services and other legitimate uses of natural resources, good aquaculture practices have to be followed. This chapter highlights the significance of aquaculture practices particularly integrated systems to produce food by following ecological principles and the importance of following good aquacultural practices to minimize the environmental pollution and harmonious development of the coastal region.

22.2 Scope of Aquaculture

The main objective of promotion of aquaculture in the coastal areas is sustainable development of the coastal region. It means management and conservation of the natural resource base of the coastal region and the orientation of technological and institutional change to ensure continued supply of human needs for present and future generations. In other words, it is the harmonious development of the coastal region with conservation of coastal environment which is inclusive. Such sustainable development conserves land, water, and plant, and animal genetic material is environmentally non-degrading, technically appropriate, economically viable, and socially acceptable.

Aquaculture is the farming of aquatic organisms in inland and coastal areas, involving intervention in the rearing process to enhance production and the individual or corporate ownership of the stock being cultivated. In rural areas of poorer countries, agronomy, water management, aquaculture, and wild aquatic resource harvesting are often physically and functionally integrated. Thus aquaculture is an integral and indivisible part of the management of aquatic resources (Haylor and Bland 2001).

In contrast to the aquaculture of temperate region, coastal fish culture systems in the tropics, with the exception of the species used in traditional coastal fish farming (mullets and milkfish), have tended to focus on high market value species such as shrimp, grouper, and seabass. This provides opportunities for the integration of aquaculture into rural development by utilizing the diverse aquatic resources, viz. estuaries, bays, lagoon, coral reefs, mangroves, mudflats, and ponds, involving a wide range of stakeholders. Objectives may further range from food production, income generation, wild stock enhancement, or recreation (ornamental fish or sport). The scale may be intensive commercial operations through subsistence aquatic resource management within developed and less-developed economies (Haylor and Bland 2001). Such integrated and harmonious development provides not only livelihood to the coastal communities but ensures food production with much less harm to the environment and aids in conservation of coastal natural resources.

22.3 Traditional vs Modern Aquaculture

Traditionally in many coastal areas in the tropics, fishing communities have raised fish in natural areas with artificial enclosures. Such systems rely on natural pond production particularly for feed, enriched by nutrients in the incoming water from nearby mangroves and rice fields. Juvenile fish (grey mullet, sea bream, sea bass, eels, and flatfish) enter the pools during high tide and sometimes collected from nearby areas.

In the coastal areas of Indonesia, ponds (*tambak*) serve as a collection or cultivation basins for whatever species are washed into them with the tide (Hannig 1988). The ponds were stocked with fish fry naturally by tidal currents or artificially by gathering fish fry. Milkfish (*Chanos chanos*) is the major species cultivated, often in combination with Java carp (*Puntius gonionotus* L., now known as *Barbodes gonionotus*) or shrimp (*Penaeus monodon*) and with incidental harvests of mullet, barramundi (*Lates calcarifer* Block.), crabs, and various smaller shrimps. In West Africa another form of raising fish was devised by the Tofinnu people, who stuck bundles of branches in the bottom of shallow lagoons. In the short term, these *acadjas* serve as fish-attracting devices. The attracted fishes to the refuge increased productivity within the structure, and, in the long term, *acadjas* provide a habitat which allows fish to grow and reproduce (Laserre and Ruddle 1983). In Mauritius, ponds called *barachois* are built by closing off a lagoon with stone walls and stocking with juvenile fish (King 1993).

Such traditional systems are comparatively more sustainable than modern intensive aquaculture in that they provide livelihood to the local community, maintain the biodiversity of the local environment, use less inputs, and do not pollute surrounding waters with waste products by using little or no feed, low stocking densities, and no fertilizers.

In contrast, modern intensive aquaculture systems use much higher amounts of inputs such as water, feeds, fertilizers, and chemicals. Consequently these intensive farms discharge more wastes into the surrounding environment. Larsson et al. (1994) indicate that modern shrimp farming practices are, in a majority of cases, unsustainable in that they require large support areas (agricultural, marine, and mangrove) and resources (water, feed, fertilizers, chemicals, etc.) which are pumped in and out without recycling and released as wastes into the environment. Salmon farming requires an input of industrial energy (oil, coal, electricity) of 15,000–25,000 kcal per kg of reared salmon to produce net pens, automatic feeders, aerators, pumps, etc. and to run equipment, drive boats, raise juvenile salmon, and produce feed pellets. This amount is similar to capital- and energy-intensive offshore fisheries for salmon. Cage farming of salmon is among the most expensive food production system requiring 25–50 kcal of industrial energy per kcal of fish produced, comparable to broiler farming or pig raising (Folke and Kautsky 1992). Intensive aquaculture industries are also emerging for high-value warm water piscivorous fish, such as groupers and barramundi (in Hong Kong, Malaysia, Thailand, Indonesia, the Philippines, Taiwan, Singapore, and Australia).

The impacts of coastal aquaculture are directly from farm activities and from farm discharges (introduced species, wastes and faecal material, uneaten food, organic and inorganic fertilizers, drugs, disinfectants, pesticides, and antifoulants). Such resource-intensive methods of aquaculture production are aimed at export market or transported to urban centres. Generally it does not supply the food requirements of coastal communities and rarely improve their standard of living, which often result in social dislocation (Weeks and Pollnac 1992). The investment and running costs involved attract rich investors mainly from large city centres, resulting in the management of an area being taken away from local control. As a result modern aquaculture practices are considered as potential polluters of the aquatic environment and the cause of degradation of wetland areas.

Another crucial point that should be emphasized is that the countries in the tropical coastal region have the highest population densities in the world. There are many issues that aquaculture has to face, but the fundamental one is the widening gap between supply and demand. The critical concerns identified for resolution of this issue relate to social, economic, biological, environmental, and institutional questions, which must be resolved with a minimum of conflict and most efficient use of resources. Therefore, each practices or aquaculture method needs to be evaluated to reduce its negative impact on the environment and should contribute to increase in production.

22.4 Integrated Aquaculture Production

Although intensive aquaculture production has environmental concerns, it has huge production potential. For example, intensive shrimp farming methods allow an increase in stocking densities and survival which enables production to increase from 100 to 500 kg/ha/year in an extensive system to 4000–15,000 kg/ha/year under intensive culture. At the same time, the coastal zones in the tropics are facing increasing depletion of resources, pollution problems, and overcrowding. With dramatic population increases in the coastal areas, food requirements have increased many folds. As a result food production must increase to meet present and future requirements without using vast amounts of land.

One of the solution lies with increased aquaculture production in the coastal areas. But aquaculture systems are needed which are more integrated with the environment, can be more realistically community-based, and have less negative impact on the environment yet which can produce sufficient food for today's requirements in an efficient manner. Such an integrated approach applied to coastal resources management would reduce waste production in addition to producing additional food or cash crops. However, it requires understanding the trophic relationships and the nutrient cycling in culture systems—termed as ecological aquaculture (Mackay 1983). The stocking of fish with different feeding habits results in a more efficient utilization of the pond system, distributes grazing pressures among different feeding niches, and levels and uses waste of one as food for another. This form of polyculture is especially appropriate when there is a need for low-cost production and when feed is unavailable.

In general there are two types of integrated aquaculture system particularly fish farming, viz. agri-aquaculture and livestock-fish system. Agri-aquaculture-based system includes paddy-fish, horticulture-fish, mushroom-fish, etc. as major components, whereas other agriculture practices are taken as minor components. The livestock-fish system includes cattle-fish, pig-fish, poultry-fish, and duck-fish systems. The main objective is increasing farm productivity and maximizing synergies between these components.

22.4.1 Paddy-Fish System

Rice is the major crop and staple food for over half the world's population particularly in the tropical coastal areas. Over 90% of the rice produced in Asian countries is the livelihood for rural farmers. Collection of wild and indigenous fish from paddy field for home consumption among small farm holder family is an age-old practice as the rice cultivation by itself. At the same time, planned rice-fish system ensures higher productivity, increase in farm income, and employment with reduction of investment risk.

In paddy-fish systems, some physical modifications have been devised over the years in order to make the rice field better suited for fish culture. Most are common to many countries and may have been developed independently from each other (Fig. 22.1). In this system, paddy fields retain water for 3–8 months in a year. The

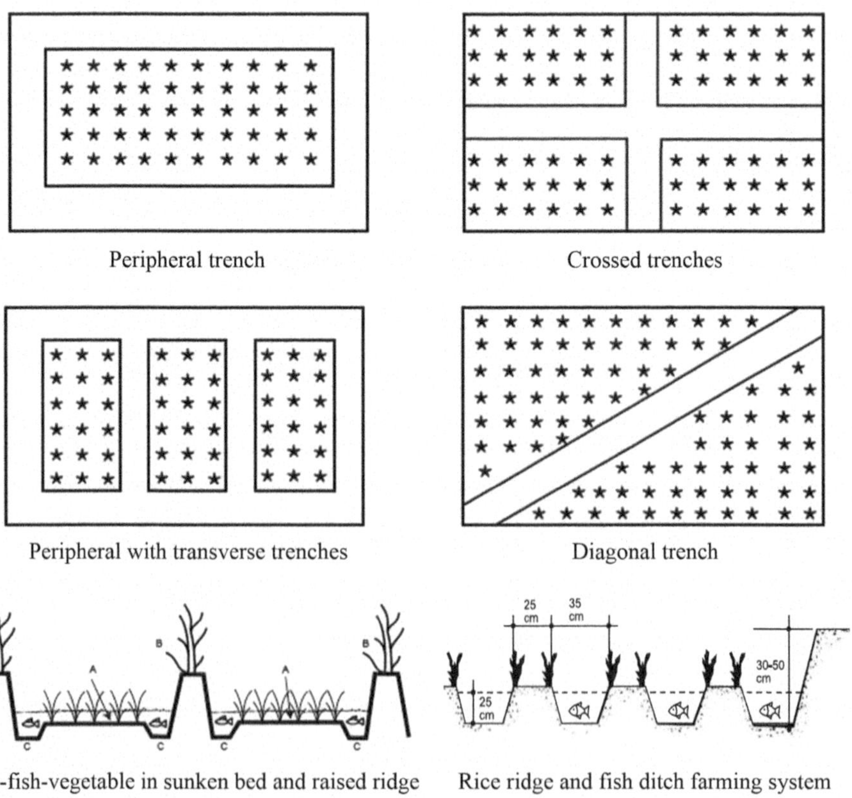

Fig. 22.1 Paddy-fish culture in different field designs

culture of selected species of fish in field which remains waterlogged even after the harvest of paddy crop provides an opportunity to grow fish and provide off-season occupation for farm families. The modern concept of paddy-fish integration, digging of peripheral trenches, construction of dykes, nutrient utilization of pond refuge, and sowing of improved varieties of rice and release of fish in the trenches helped to harvest better yield of rice and fish. Fish ponds receive the crop residues as pond input. The ecological benefits are weed control, consumption of some pest and molluscs, and bioturbation of soil-water interface. Green leafy vegetables on the pond dyke in aerial system and cash crop such as banana and papaya provide very good economic return to the farmers. In this system the major product yields are fish, rice, seasonal horticultural crops, etc.

In general paddy-fish production system can be grouped into three production types, viz. concurrent system, rotational system, and alternating system.

22.4.1.1 Concurrent Production

In concurrent production fish culture is taken up along with rice crop where the growing season of both rice and fish overlaps with each other. This is often referred in short as 'rice + fish' (Yunus et al. 1992; Roger 1996). However, some physical modifications are required to make a rice field 'fish-friendly'. The timing in stocking fingerlings is crucial since if stocked too soon after the rice is planted, some fish species are likely to damage the newly planted seedlings (Singh et al. 1980), and if too late, there may be a multitude of predator species in the fields. In coastal rice fields affected by saltwater, it is common for saltwater shrimps to enter the rice fields with the floodwater and grow among the rice plants. In Vietnam some farmers have been successful in growing shrimp together with a traditional tall rice crop in a brackish water environment. Supplementary feeding results in higher yields (Mai et al. 1992).

One constraint of the concurrent system is that the growing period of the fish is limited to that of rice, which is usually 100–150 days. Consequently the harvested fish are small, especially if early-yielding rice varieties are used. This can be partly remedied by the use of larger fingerlings, but there is a limit to this since large fish may be able to dislodge the rice seedlings. Another solution is to limit the production to that of large-size fingerlings for sale to farms growing table fish.

22.4.1.2 Rotational Production

In rotational culture rice and fish are grown at different times, in rotation. Rice growing season normally don't overlap with fish growth cycle. In this system, fish is used during the fallow period or as a winter crop to make use of the rice field when it otherwise would not be used (Ni and Wang 1995). In Indonesia, rotating fish with rice has been developed to a greater degree, which is called 'fallow-season crop'. The only physical modification required is the raising of the dike to hold water. Without the rice, the entire rice field can be operated and managed just like a regular fish pond from 3 to 6 months a year. It can be used for growing table fish or producing fingerlings. The production of two or three crops of fingerlings instead of one crop of table fish can also be practices to avoid problems of poaching or fish mortality due to infestation by predators such as snakes, birds, and water insects (Koesoemadinata and Costa-Pierce 1992). In rice-fish system due to the provision of trenches, whether peripheral or otherwise, the fish tend to remain in the trench during the ploughing and harrowing process.

For example, in the Javan system of rice-fish, rice field is flooded along with the standing rice stubble, either trodden down or cut off before or after the first flooding. Within 2 or 3 days, the water becomes putrid due to the decomposition of plant materials and is released and replaced with new water. Sufficient depth of water depth is maintained (30–80 cm). Here the stocking density of fingerlings is based on the magnitude of the rice harvest and the size of the fingerlings. The rule of thumb is to stock from 500 to 700 fingerlings (5–8 cm long) for 1 ton of unhusked rice harvested. Sometimes large fingerlings (100 g) are also stocked at the rate of 10% of the main stock. These larger fish keep the soil surface loose by their activities. Alternatively, 10-day-old carp fry may be stocked at the rate of 100,000 fry ha^{-1} for growing into fingerlings. This practice often results in high mortality but is apparently resorted to only if no other area is available as a nursery. Marketable fish are harvested in 40–60 days and fingerlings after only 4 weeks. There is enough time for a second, third, or even fourth crop of fish prior to the next rice planting season, depending on the availability of water. Table fish are harvested by draining the field, forcing the fish into trenches where they are picked by hand.

Another Indonesian system is called *penyelang* or 'intermediate crop' where farmers who practise double crop of rice with an adequate water supply can still raise fish in between the two rice crops. Since the seedbeds occupy only a very small portion of the rice field, the farmers can use the rest of the rice fields for growing fish during a period of 1–1.5 months sufficient to produce fingerlings. Some farmers let fish breeders use their rice fields during this period (Koesoemadinata and Costa-Pierce 1992).

22.4.1.3 Alternating Production System

Alternating production system is also some form of rotational culture, but it is distinct enough from the rotational and concurrent system in terms of overlap between rice and fish growth cycle and number of fish crop grown in a year. In this system rice takes 110–125 days to mature and in the remaining period depending on the water availability one filler crop of fish can be taken up. This means growing of fingerlings for 30–45 days for marketing for the next season. Therefore, by alternating between rice-fish and fish-only farming, rice fields can be productive throughout the year, and higher incomes can be realized from the available coastal paddy land.

Here also there are options depending on the farmers' economic conditions and water availability. A farmer may practise two rice crops and then a fish-only crop or two rice-fish crops followed by a fish-only crop. These systems are more popular in Southeast Asian countries (Koesoemadinata and Costa-Pierce 1992). In some coastal areas, even though rice is the main crop, fish are raised year-round in the rice field. In Java, farmers who practised two rice-fish crops followed by a fish-only system had a net return to input of 173% per year as against 127% for those practising a rice-rice-fish system and 115% for those practising rice-rice-fallow system (Yunus et al. 1992).

22.4.2 Aquatic Weed-Fish System

This system integrates various types of aquatic plants as feed source for herbivorous fishes. Grass carp directly consumes several weed species and *Azolla*. These floating plants with higher protein, low fat, low fibre, and lower lignin and cellulose have beneficial impact on digestibility for livestock and fish. In this system herbivorous fish (grass carp 50%) with other fishes (catla, rohu, mrigal, silver carp, common carp 10% each) in stocking density of 6000–8000 fry/ha fetches production over 5 tons/ha/year. The excreta of carp and decomposed plant / organic materials serve as manure for the pond. *Azolla* is a nitrogenous biofertilizer on the water surface in fishponds producing N, P, and K by trapping the solar energy. Due to high-protein content, use of *Azolla* as feed ingredients is of special significance in aquaculture system. In polyculture pond with water area of 1 ha, *Azolla* can provide over 100 kg of N, 25 kg of P, and 90 kg of K in addition to 1.5 tons of organic matter. Similarly inclusion of *Spirulina*, blue-green algae with high biomass can serve as a good feed supplement. Still there are gaps in identification of suitable aquatic weeds as fish feed but compatible in the aquatic environment.

22.4.3 Animal-Fish System

Judicious combination of animal component and fish culture in the coastal areas will ensure effective recycling of wastes and efficient use of available resources. The system also helps in farming intensification, nutrient recycling, effective use of bio-resources for optimum protein production, agricultural diversification, and environmental sustainability. Nutritive values of animal waste/excreta are given in Table 22.1. These wastes are converted into valuable food by fish grown in the pond. However, it is to be noted that the nutrient content of manures varies with type of animal, and the conversion efficiency varies with fish species and the surrounding environment. Some of the most important animal-fish systems suitable for coastal areas are cattle-fish, pig-fish, goat-fish, poultry-fish, and duck-fish.

Fish farming using cattle manure is one of the general practices all over the tropical world. A healthy cow weighing 400 kg excretes around 400–500 kg of dung and 3500–4000 l of urine on annual basis. These animal wastes contain easily available nutrients which are responsible for increase in natural food organisms, detritus, and beneficial bacteria in fish ponds. The biological oxygen value of cow manures is low because it is already semi-decomposed manure. A unit of five to six cows can provide adequate quantity of dung and urine to produce 3–4 tons of fish per ha per year. Additionally this system also produces milk and animal meat as well. Linkages and resource flow between different components of integrated aquaculture system are given in Fig. 22.2.

Table 22.1 Average nutrient status of different animal wastes

Animal	Excreta	Moisture (%)	Organic matter (%)	Nitrogen (%)	Phosphorus (P_2O_5 %)	Potash (%)
Cattle	Faeces	80–85	14.0	0.3	0.2	0.1
	Urine	92–95	2.3	1.0	0.1	1.4
Pig	Faeces	85	15	0.6	0.5	0.4
	Urine	97	2.5	0.4	0.1	0.7
Chicken	Faeces	78	25.5	1.4	0.8	0.6
Duck	Faeces	81	26.2	0.9	0.4	0.6
Rabbit	Faeces	10	37	2	1.3	1.2
Goat	Faeces	10	–	2.7	1.7	2.9

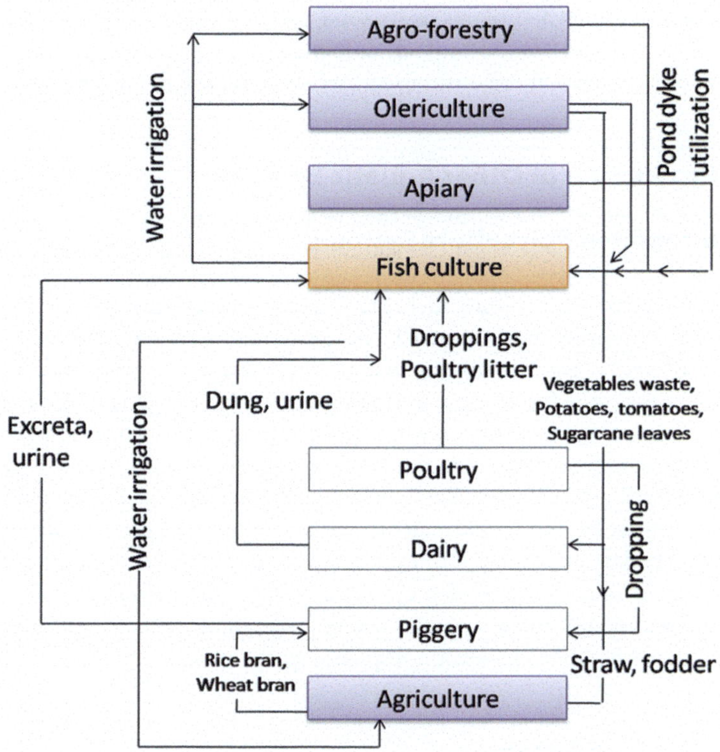

Fig. 22.2 Linkages and resource flow between different components of IFF systems

Pig-fish system produces meat and fish at a cheaper feeding cost, hence, better compared to other systems. Pig requires a floor space of 3–4 m^2 and can produce 75–120 kg depending on breed. The pig dung and urine are utilized for fertilization of culture ponds which is either channelized directly or partially decomposed before application in ponds. In fish pond, the application of pig wastes enhances nutrient content and produces dense bloom of planktons. Stocking of bottom feeder mrigal in higher quantity with focus on filter feeding fish, silver carp is advantageous in this type of integration. On an average fish harvest results in 3–4 tons/ha without any feed and fertilization in 12 months culture period at the stocking rate of 8000 fingerlings/ha.

In poultry-fish systems, poultry droppings and litter directly fall on the fish pond and act as a fertilizer source for production of fish. In this system, birds are housed in intensive, semi-intensive, and extensive systems. The house of the birds is built on the pond dyke or inside the pond areas. The poultry floor is covered with litter prepared with dropped straw, dry leaves, groundnut shells, or sawdust. All these wastes are recycled through fishes grown in the fish pond.

Similar to poultry, duck-fish system also helps in recycling of waste more effectively. In the pond environment, ducks roam freely during daytime and get several aquatic animals and plants as its food. Duck droppings and urine from the house used as nutrient source provide carbon, nitrogen, and phosphorus for the production of natural food organism used by fish in duck-fish integration.

22.5 Production of Other Aquatic Species

22.5.1 Integrated Polyculture Systems

Other than integrated fish culture, there are different polyculture systems relevant to coastal aquaculture that are in practice. In open-water fish cage culture, molluscs and seaweed can be used to take up excess feed and waste from the fish, while molluscs are a potential food source for the cultured fish. In Southeast Asia, molluscs and seaweed are raised together on lines, since they mutually enhance each other's growing environments. In pond culture, shrimp and/or fish can be grown in combination with molluscs, seaweed, and crabs in order to stabilize the growing environment, decrease food requirements, and clean up pond effluent which in turn increase production and profit (Brzeski and Newkirk 1997). In addition to this, aquaculturing of other species of food value (snails, sea cucumbers, sea urchins, etc.) and commercial value (pearl oysters, aquarium species, etc.) has been standardized and practised in several tropical coastal areas.

22.5.2 Mud Crab Fattening

Another option available for using the coastal lowland to provide livelihood to the fisher folk is in the form of mud crab fattening in mangrove ecosystem as they are widely found in most of the tropical coastal and islands. Juvenile crabs can be collected from estuaries, lakes, backwaters, creeks, and mangroves and grown in grow-out ponds constructed in tide-fed estuaries, backwaters, and creeks. The crab ponds can also be constructed by converting one portion of existing fish ponds and providing provision for brackish water inundation into that area. A pond of 0.1 ha area can be used for mud crab culture. With the stocking density of 500 numbers, ha^{-1} of 50–60 g size crab for a period of 6 months production of about 780 kg/ha can be achieved (Dam Roy et al. 2008).

22.5.3 Seaweed Cultivation

Seaweeds refer to any large marine benthic algae that are multicellular and macrothallic and thus differentiated from most algae that are of microscopic size (Smith 1944). Seaweed farming is the practice of cultivating and harvesting seaweeds which is largely carried out as a diversification activity in mariculture. Many of the rocky beaches, mudflats, estuaries, coral reefs, and lagoons provide ideal habitats for the growth of seaweeds. They form an important renewable resource in the marine

Table 22.2 Different types of marine algae cultivated in India and their use

Sl. no	Type of algae	Scientific name	Cultivation method	Use
1	Red algae	*Gracilaria edulis*, *G. crassa*, *G. foliifera*, and *G. verrucosa*	Longline ropes and nets by vegetative propagation	Agar manufacturing
		Gracilaria edulis	Single-rope floating	Hydrocolloids
			Raft technique	
		Gelidiella acerosa	Bottom-culture method using coral stone as a substratum	Hydrocolloids
		Kappaphycus alvarezii	Net bag and raft method	Source of food and carrageenan
2	Brown algae	*Sargassum* spp., *Turbinaria* spp., *and Cystoseira trinodis*	Collection and using nets	Production of alginates and liquid seaweed fertilizers

environment as evidenced from its annual production of about 7.0–8.0 million tons of wet seaweed along the coastal regions (McHugh 2003).

Seaweeds belonging to different genera are mainly used for edible and industrial purposes all over the world. The edible seaweed are algae that can be eaten and used in the preparation of food that belong to one of several groups of multicellular algae, viz. red algae, green algae, and brown algae. Alternatively, seaweeds are also harvested or cultivated for the industrial extraction of alginate, agar, and carrageenan substances collectively known as hydrocolloids or phycocolloids. Hydrocolloids have attained commercial significance, especially in food production as food additives. The food industry exploits the gelling, water retention, emulsifying, and other physical properties of these hydrocolloids. The sources of such materials are presented in Table 22.2.

22.6 Pollutant from Aquaculture

Wastewater from shrimp farms is high in nitrogen, phosphorous, carbon compounds, organic matter, shrimp excretory products, plankton, and chemical and antibiotic residues. Most of the shrimp farms in unregulated coastal areas discharge the drain water directly into public water courses of the sea. If the wastewater is discarded without dilution of treatment, it pollutes the environment. Therefore, in such areas the discharge of effluents has been found to deteriorate water quality in irrigation canals and coastal areas.

The discharge of dissolved organic matter from metabolites and decomposition of uneaten food and shrimp faeces into the sea from intensive shrimp farming operations have resulted in fish death and foul-smelling water in many wetland areas. Therefore, it is inferred that water treatment facilities are compulsory for farms.

Aquaculture effluents discharged into the marine environment are reported to threaten the survival of coastal resources. The effects on coastal habitats are mainly due to improper saline water released as pond effluents, which have a pronounced effect on both marine and freshwater organisms particularly in the coastal wetlands.

The salinization of surface water is another impact from intensive aquaculture farms to the receiving waters. It is reported that shrimp farm construction can potentially alter surface water flow patterns and water quality. Water flows were found to be affected by the temporary blockage of permanent division of canals. Discharge of untreated pond effluents into the natural system were known to hinder the quality of water. Further, seepage of saline water from the ponds into the ambient areas also leads to salinization of groundwater that will affect agricultural land when farmers unknowingly pump salinized water into farmland. Further, more pumping of saline water, in the case of bore water-dependent farms, and fresh water, in the case of coastal farms, lowers the groundwater table.

Another dimension of the salinization problems is its effect on drinking water sources of coastal villages. Shrimp farming causes dispersion of salt into land around the shrimp farms, thereby affecting paddy fields and other plantations. The soil salinization further devalues marginal agricultural land. Abstraction of fresh water from underground aquifers for intensive shrimp farming has also resulted in saltwater intrusion and salinization of freshwater aquifers.

Overall the rapid expansion of coastal aquaculture and skewed market forces has led to both environmental and social problems. Mangrove deforestation, land degradation, habitat loss, and disease all resulted from poorly planned development in this sector. Degradation of, and exclusion from, resources has marginalized many poor people who have not benefited from this growth (Haylor and Bland 2001).

22.7 Responsible Fisheries

Fisheries sector is increasingly contributing to the economic growth and nutritional security of nations particularly tropical coastal countries. Meanwhile efforts should be made to sustain the growth rate of fisheries including aquaculture and its contribution to the national GDP through responsible development of this sector, because there are lots of issues cropped up along with the booming aquaculture in the coastal areas which could be addressed by Code of Conduct for Responsible Fisheries.

The Code of Conduct for Responsible Fisheries (CCRF) was unanimously adopted by the FAO in October 1995. The CCRF or simply the 'Code' as it is popularly known defines the general principle that 'the right to fish carries with it the obligations to do in a responsible manner'. The Code sets out principles and standards of behaviour for such practices and aims at effective conservation, management, and development of living aquatic resources. The main objective is to ensure that the resources are utilized in a sustainable and responsible manner through adoption of sound national and international policies and programmes. The Code covers in its 12 articles not only the capture of fish and fishing operations but also the integration of fisheries into coastal area management, etc.

The Code provides the required framework for national and international efforts to ensure sustainable exploitation of aquatic living resources in harmony with environment. On fisheries management measures based on the best scientific evidence are available. The Code suggests setting up of mechanisms of monitoring, surveillance, control, and enforcement of fishing. The Code advocates optimum fishing fleet and gear to ensure sustainability and precautionary approach as the guiding principle for fishery management, and the regulation should facilitate resolution of conflict among fishers. The Code also advocates measures to minimize waste, discards, catch by lost or abandoned gear, catch of non-targeted species, etc. It also suggests integration of fisheries into coastal area management and urges an institutional framework and policy measures to facilitate sustainable use of coastal resources.

22.8 Aquaculture and Environmental Rehabilitation

Aquaculture systems, particularly the integrated systems, can provide many positive environmental benefits such as water storage on small-scale farms, efficient use of farm resources, and protection of wild and endangered species. Perhaps less well known and publicized is the fact that coastal aquaculture, including shrimp culture, can also contribute to environmental improvement. For example, seaweed and mollusc culture can contribute to the removal of nutrients and organic materials from coastal waters. It can also provide an alternative source of income and employment for people involved in environmentally destructive fishing practices and other forms of habitat destruction (Kongkeo 2001). Other activities of note include the following:

- Mixed aquaculture-mangrove systems could be used to restore mangrove habitats in South and Southeast Asian countries. This assumed more significance after December 2004 Indian Ocean tsunami to establish coastal bio-shield.
- Coral reef fish mariculture provides an effective alternative to destructive fishing practices in coral reef areas. Promotion of such practices can save coral reef environment in several lagoons and coral islands.
- The rehabilitation of fish populations through stock enhancement.
- Aquaculture itself is also a technique for the effective monitoring of environmental conditions and promotion of self-recruiting species available in the wild.
- The environmental assessment of an aquaculture area, rather than targeting individual projects, is now being used to identify opportunities for minimizing the impacts of the aquaculture sector within a particular area (e.g. in a bay, an estuary, or a watershed). Such applicable measures are promoted through a wide range of instruments and should be brought together within the framework of an aquaculture development plan, ideally as part of an Integrated Coastal Management Plan.

Thus there is a strong case for such interventions to be strategically planned, rather than reactive and uncoordinated. Planning performance is frequently disappointing, since the process is complex and significant institutional and legal changes are needed, which require time, resources, and continuity participation. A review of aquaculture in the Southeast Asian countries has suggested that aquaculture within saline inland areas can provide benefits through the use of previously degraded resources, contributing to environmental rehabilitation and poverty alleviation among communities whose livelihoods lie on such damaged resources (Haylor and Bland 2001).

22.9 Conclusion

Although there is considerable technology and relevant information are available to put integrated coastal aquaculture into practice, much of it is designed around commercial exploitation of economically valuable species and thus deals with two or three species at the most. Therefore, the analysis revealed the need for ecological aquaculture system-based approach. Perhaps research and efforts should concentrate on pond-based culture, as this system is more manageable and measurable with respect to physical and ecological parameters. The challenge is to develop systems than can be used in the extensive open-water areas of coastal zones throughout the tropical areas.

Further, to avoid conflict of interest and minimize environmental concerns, innovative approaches are required which should be inclusive to address the issues of different stakeholders. The implementation of zoning for different aquaculture and agricultural activities to regulate resource use in the coastal areas will address the needs of local communities better and minimize the environmental degradation. Some of the innovative and ecologically sustainable approaches such as integrated fisheries and polyculture should be adopted to minimize wastewater and maximize production. The analysis suggests that rice-fish and rice-polyculture systems hold great potential to enhance farm income and provide environmentally harmonious than intensive or rice monoculture in the coastal lowlands. In summary, in inland and coastal areas, improved aquatic resources management, including aquaculture integrated into existing farming systems, has the potential to enhance livelihoods, but considerable effort is required to include local people and support their management of sustainable development.

References

Brzeski V, Newkirk G (1997) Integrated coastal food production systems of current literature- a review. Ocean Coast Manag 34(1):55–71

Dam Roy S, George G, Sarma K, Chaturvedi CS, Sethi SN (2008) Mud crab culture. In: Zamir Ahmed SK, Srivastava RC (eds) Micro business module in agriculture and allied fields for livelihood options in A & N Islands. Central Island Agricultural Research Institute, Port Blair, pp 59–60

Dempster T, Sanchez-Jerez P (2008) Aquaculture and coastal space management in Europe: an ecological perspective. In: Holmert M, Black K, Duarte CM, Marbà N, Karakassis I (eds) Aquaculture in the ecosystem. Springer, Dordrecht, pp 87–116

Díaz R, Rabalais NN, Breitburg DL (2012) Agriculture's impact on aquaculture: hypoxia and eutrophication in marine waters. Organisation for Economic Co-operation and Development. www.oecd.org/tad/sustainableagriculture/49841630.pdf. Accessed 20 Feb 2014

Douvere F, Ehler CN (2009) New perspectives on sea use management: initial findings from European experience with marine spatial planning. J Environ Manag 90:77–88

Duarte CM, Holmer M, Olsen Y, Soto D et al (2009) Will the oceans help feed humanity? Bioscience 59:967–976

Edwards P, Demaine H (1997) Rural aquaculture: overview and framework for country reviews. FAO Regional Office for Asia and the Pacific (RAP), RAP Publ. 1997/36 RAP/FAO Bangkok. 61 pp

FAO (2012) The state of world fisheries and aquaculture 2012. FAO, Rome

FAO (2013) The FAO Fisheries and Aquaculture Department's efforts in implementing the recommendations of the past sessions of the COFI sub-committee on Aquaculture. Seventh session of the Sub-Committee on Aquaculture (SCA) of the FAO Committee on Fisheries (COFI), St. Petersburg, 7–11 Oct 2013

Folke C, Kautsky N (1992) Aquaculture with its environment: prospects for sustainability. Ocean Coast Manag 17:5–24

Hall CM (2001) Trends in ocean and coastal tourism: the end of the last frontier? Ocean Coast Manag 44:601–618

Hannig W (1988) Towards a blue revolution. Gadjah Mada University Press, Yogyakarta, 404 pp

Haylor G, Bland S (2001) Integrating aquaculture into rural development in coastal and inland areas. In: Subasinghe RP, Bueno P, Phillips MJ, Hough C, McGladdery SE, Arthur JR (eds) Aquaculture in the third millennium, Technical Proceedings of the Conference on Aquaculture in the Third Millennium, Bangkok, Thailand, 20–25 February 2000. NACA/FAO, Bangkok/Rome, p 73.81

King HR (1993) Aquaculture development and environmental issues in Africa. In: Pullin RSV, Rosenthal H, Maclean JL (eds) Environment and aquaculture in developing countries. ICLARM Conference Proceedings 31. pp 116–124

Koesoemadinata S, Costa-Pierce BA (1992) Development of rice-fish farming in Indonesia: past, present and future. In: De la Cruz CR, Lightfoot C, Costa-Pierce BA, Carangal VR, Bimbao MP (eds) Rice-fish research and development in Asia. ICLARM Conference Proceedings 24. pp 45–62. 457 p

Kongkeo H (2001) Current status and development trends of aquaculture in the Asian Region. In: Subasinghe RP, Bueno P, Phillips MJ, Hough C, McGladdery SE, Arthur JR (eds) Aquaculture in the third millennium, Technical Proceedings of the Conference on Aquaculture in the Third Millennium, Bangkok, Thailand, 20–25 February 2000. NACA/FAO, Bangkok/Rome, pp 267–293

Larsen J, Roney M (2013) Farmed fish production overtakes beef. Earth Policy Institute, Washington, DC. www.earthpolicy.org/plan_b_updates/2013/update114. Accessed 15 Feb 2018

Larsson J, Folke C, Kautsky N (1994) Ecological limitations and appropriation of ecosystem support by shrimp farming in Columbia. Environ Manag 18:663–676

Laserre P, Ruddle K (1983) Traditional knowledge and management of marine coastal systems. Report of the Ad Hoc Steering Group Organised by the International Association for Biological Oceanography, IABO/IUBS in cooperation with the UNESCO Division of Marine Sciences. Biology International Special Issue 4, 1983, 18pp

MacKay KT (1983) Ecological aquaculture, new approaches to aquaculture in North America. J World Mariculture Soc 14:704–713

Mai TD, Le TD, Dang KS, Pham NM, Nguyen ND (1992) Rice field aquaculture systems in the Mekong Delta, Vietnam: potential and reality. In: De la Cruz CR, Lightfoot C, Costa-Pierce

BA, Carangal VR, Bimbao MP (eds) Rice-fish research and development in Asia. ICLARM Conference Proceedings 24. 457 p
McHugh DJ (2003) A guide to the seaweed industry. FAO Fisheries Technical Paper, 441. p 105
Ni D, Wang J (1995) Material cycles and economic returns in a rice-fish ecosystem. In: MacKay KT (ed) Rice-fish culture in China. International Development Research Centre (IDRC), Ottawa, pp 177–182, 276 p
Roger PA (1996) Biology and management of the floodwater ecosystem in rice fields. International Rice Research Institute, Los Banos, 250 p
Shearer TR, Wagstaff SJ, Calow R, Stewart JA, Muir JF, Haylor GS, Brooks AC (1997) The potential for aquaculture in saline groundwater. Technical Report WC/97/58 Overseas Geology Survey DFID/NERC. 235 pp
Singh VP, Early AC, Wickham TH (1980) Rice agronomy in relation to fish culture. In: Pullin RSV, Shehadeh ZH (eds) Proceedings of the ICLARMSEARCA conference on integrated agriculture-aquaculture farming systems, 6–9 August 1979, Manila, Philippines. pp 15–34. 258 p
Smith GM (1944) Marine algae of the Monterey Peninsula, 2nd edn. Stanford University, Stanford
Weeks P, Pollnac RB (1992). Coastal aquaculture in developing countries: problems and perspectives) In: Pollnac RB, Weeks P (eds) Coastal aquaculture in developing countries: problems and perspectives. International Center for Marine Resource Development (ICRMD), Kingston, pp 1–13
Yunus M, Hardjamulia A, Syamsiah I, Suriapermana S (1992) Evaluation of rice-fish production systems in Indonesia. In: De la Cruz CR, Lightfoot C, Costa-Pierce BA, Carangal VR, Bimbao MP (eds) Rice-fish research and development in Asia. ICLARM Conference Proceedings 24. pp 131–138. 457 p

Mangroves and Sustainable Development of the Coastal Region

23

Abstract

Mangrove ecosystem represents one of the most productive natural capitals occurring in tropical regions capable of producing a wide range of goods and services for coastal communities and environment. Analysis of compiled data on mangrove area shows that more than 60% of the total mangrove area is found in just ten countries. The largest areas of mangroves are found in Asia and Africa, followed by North and Central America. Due to various reasons, loss of mangroves continues at alarming rates, but in the last two decades, the rate of net loss is showing signs of slowing down. Diversion of mangrove wetlands for aquaculture, agriculture, infrastructure, and tourism activities is among the major causes of the reported decrease in mangrove areas over time. Mangrove species involving agroforestry systems, agro-aqua farming, and providing alternate livelihood and community-based management of mangroves are indispensable component of mangrove management. Such efforts help in restoration and regeneration of mangrove, address the coastal land degradation, and provide livelihood support to the coastal population.

Keywords

Bio-shield · Coastal vegetation · Ecosystem services · Livelihood · Conservation

23.1 Introduction

Coastal zones of the tropics are endowed with diverse ecosystems that include a wide range of mangroves, coral reefs, seagrasses, salt marshes, mud flats, estuaries, lagoons, and unique marine and coastal flora and fauna. Coastal fishing and the postharvest fisheries employ millions of people directly and indirectly and provide livelihood to other dependant population from the adjoining areas. Mangroves are diverse group of woody trees, palms, shrubs, vines, and ferns that share a common

V. Ayyam et al., *Coastal Ecosystems of the Tropics - Adaptive Management*,
https://doi.org/10.1007/978-981-13-8926-9_23

ability to live in waterlogged saline soils subjected to regular flooding which occurs in the intertidal zone. The term 'mangrove' describes both the ecosystem and the plant families that have developed specialized adaptations to live in this tidal environment (Tomlinson 1986).

The mangroves are bordering margins of the tropical coastlines and estuaries providing habitat for a vibrant mix of species of various life stages. As such, mangroves function as nursery habitats for commercially important crab, prawn, and fish species. Mangrove roots trap silt and buildup thick, shifting sediments and thus help in preventing soil erosion. Mangroves are also important to coastal and island regions for a variety of reasons, including aquaculture, agriculture, forestry, protection against shoreline erosion, as a source of firewood and building material, and other local subsistence use. The entangled root masses of mangroves dissipate the wave energy and guard the coastlines and hence are often referred to as wetland bio-shields.

Unlike other biotic ecosystem, mangrove communities comprise both biotic and abiotic components which are having diverse edaphic conditions. The biotic components include plants, animals, and microbial organisms that are highly adapted to intertidal environmental conditions (Jayatissa et al. 2002). However, none of these can be identified as a community in a mangrove ecosystem without the actual mangrove plants (trees and shrubs), meaning that the true mangrove vegetation is the major constituent of the ecosystem.

However, in spite of their ecological richness, significant ecosystem services and contribution to the livelihood of locals and national economy, areas under mangrove forests are decreasing in many tropical countries due to poor management and increasing demand for coastal lands for other developmental activities. Global mangrove distribution data indicated that mangroves are continuously declining at a speedy rate of 1–2% per year, however at a threatening level in the developing countries where they found in abundances (Duke et al. 2007). The projection based on current trend is highly alarming which showed loss of mangroves throughout the world may reach up to 60% by 2030 (Valiela et al. 2001; Alongi 2002). In India, about 34% of India's mangroves were destroyed during 1950–2000. Such rapid depletion and degradation, unless arrested, will impact the livelihood, health, and wellbeing of the coastal population, affecting in turn prospects for sustained economic growth of the tropical region. In this context, knowledge of the exact species composition of mangroves in any country is a basic and important prerequisite to understand all the aspects of structure and function of mangroves, as well as their biogeographical affinities and their conservation and sustainable utilization (Jayatissa et al. 2002; Wang et al. 2003). Therefore, management/protection of mangroves along the coastline should form part of the developmental strategy to protect the cultivated land as well as to provide livelihood support to the peoples of this region. In this chapter the distribution, composition, and sustainable utilization of mangroves in tropical countries are discussed.

23.2 Composition of Mangrove Communities

Mangroves occur only in the tropical and subtropical intertidal estuarine regions, sheltered coastlines, and creeks which are dominated by partly submerged sclerophyllous plant species. They occur in the intertidal region where the terrestrial limit for other plant community ends and the influence of marine ecosystem begins (Fig. 23.1). True mangroves are characterized by a number of adaptations for thriving in a saline, waterlogged, and anaerobic conditions which generally seen in roots, leaves, and fruits of the plants. Non-mangrove vegetation generally occurs at the landward margins of mangroves or at the freshwater influx side. Sometimes, it is very difficult to separate the plant species and habitat as in the case of the Sundarbans where mangrove flourish in small islands interspersed with sea water of varying salinity.

Mangrove forests occupy only 1% of the total tropical forest area and yet provide valuable ecosystem services along the coastal line of the tropics. They help in coastal protection acting as barriers against devastating tropical storms and waves. Mixed mangrove species stand at seafront and bay act as a bio-shield against the surging sea waves and tsunamis. Generally two categories of plants are recognized in mangroves, namely, 'plants which are restricted to mangrove habitats' (tropical intertidal habitats) and 'plants which are not restricted to mangrove habitats' (Polidoro et al. 2010). These are also referred as 'true mangroves' and 'mangrove associates' by some authors. These two groups have been separated in a more descriptive way by Tomlinson (1986) as 'major mangrove components (true or strict mangroves', 'minor mangrove components)', and 'mangrove associates'.

Mangrove plants widely range from trees (*Avicennia*, *Rhizophora*), to trunkless palm (*Nypa fruticans*), to shrubs (*Aegiceras*, *Aegialitis*, *Pemphis*, and *Conocarpus*), and ground fern (*Acrostichum*). Characteristically, trees and shrubs may be erect

Fig. 23.1 Occurrence of mangroves in the intertidal zones

Fig. 23.2 Mangrove habitat and landward changes in species composition

type (*Pelliciera rhizophorae*, *Bruguiera parviflora*) to spreading, sprawling (*Acanthus* spp., *Scyphiphora hydrophyllacea*), and multiple-stemmed (*Ceriops decandra*). In general, it is observed that plants at both waterfront and landward have more lower limbs and foliage, and their stems tend to sprawl more laterally rather than erect and straight. Different plant composition and structure can be seen in Fig. 23.2 which represents dense mangrove stand of Andaman Islands, India. There is also difference in their canopy cover as well which varies from combined closed canopies as in the case of *Avicennia marina*, *Rhizophora apiculata*, *Bruguiera parviflora*, *Bruguiera gymnorrhiza*, *Camptostemon schultzii*, and *Xylocarpus* spp., while others are commonly found as undercanopy plants beneath the closed canopy (*Aegiceras*, *Cynometra*, *Acanthus*, *Acrostichum*, and *Ceriops decandra*). Mangroves require some special structures to stand firmly on the muddy areas such as buttresses and sturdy prop roots as they occupy one of the most difficult habitats in the world. Root structures above ground include four types, viz. pneumatophores, knee roots, stilt roots, and buttresses. Roots are used by various fauna but most notably by burrowing shipworms, molluscs, termites, and crabs.

23.3 Distribution

Mangroves evolved from terrestrial or freshwater plant species rather than marine plants. It is believed that the break-up of continental land masses provided conditions, favourable for the development of mangroves in the fringe areas of tropics. In the distant past, these land plants adapted to brackish water due to fast growth, diversity, and evolutionary process and became the ‘core’ mangrove flora in

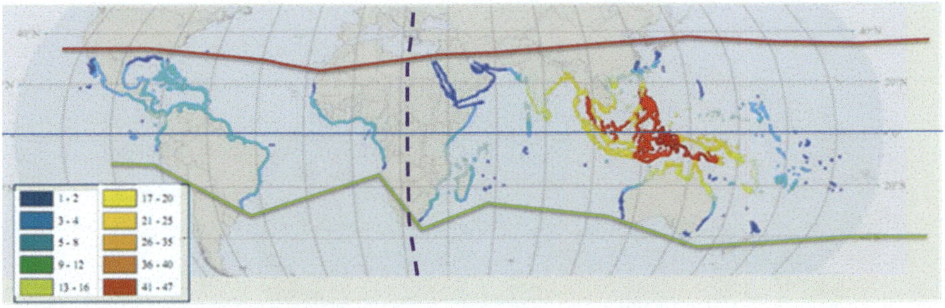

Atlantic East Pacific (AEP)   Indo West Pacific (IWP)

Created from Duke, N.C. (1992)

Fig. 23.3 World distribution of mangroves and species diversity. (Created from Duke, N.C. (1992))

the tropics. Therefore, the present distribution of individual mangrove species must be seen against this background of plate tectonics and continental drift (Duke 1992). Consequently mangroves are largely confined to the regions between 30° North and South of the equator, with notable extensions beyond this to the north in Bermuda (32°20′N) and Japan (31°22′N) and to the south in Australia, New Zealand, and the east coast of South Africa (Fig. 23.3). The widespread occurrence of mangrove vegetation and the floristic divergence between the 'old' and the 'new' world can only be explained by geological events (Spalding 1997). Mangroves have broader ranges along the warmer eastern coastlines of the Americas and Africa than along the cooler western coastlines. This difference in distribution is due to the presence of warm and cold oceanic currents.

As the mangroves are commonly found along sheltered coastlines, they fulfil important socio-economic and environmental functions in the tropics and subtropics. High population pressure in coastal areas has, however, led to the conversion of many mangrove areas to other uses, including infrastructure, aquaculture, rice, and salt production. The first attempt at estimating the total mangrove area in the world was undertaken as part of the FAO/UNEP Tropical Forest Resources Assessment in 1980, where the world total was estimated as 15.6 million hectares. More recent estimates range from 12 to 20 million ha (FAO 2003).

A total of 124 countries and areas were identified as containing one or more true mangrove species (Tomlinson 1986). The most extensive area of mangroves is found in Asia, followed by Africa and South America. Four countries (Indonesia, Brazil, Nigeria, and Australia) account for about 41% of all mangroves, while 60% of the total mangrove area of the world is found in ten countries (Fig. 23.4). The changes in mangrove area over time showed that large mangrove deforestation rates occurred in Asia followed by the Caribbean and Latin America; primarily it reflect large-scale conversion of mangroves for aquaculture and tourism infrastructure. Most countries have now banned the conversion of mangroves for aquaculture purposes and require environmental impact assessments prior to large-scale conversion of mangroves areas for other uses.

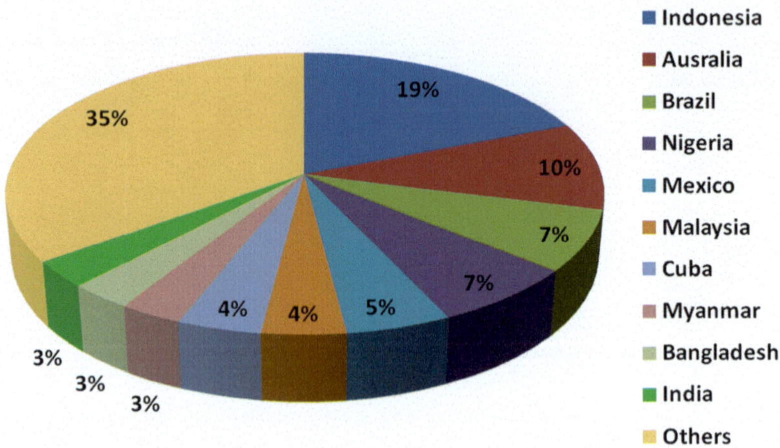

Fig. 23.4 Percentage distribution of mangrove areas in the world (2005)

23.4 Significance of Mangroves

Mangrove habitats exhibit community assemblage of specific species among mangroves and associations with some non-mangrove species as well. Mangrove ecosystems are generally rich in associated flora occurring in difficult conditions which includes bacteria, fungi, algae, sea grasses, lichen, and other vegetation. In addition, mangroves have tremendous ecological and social values. Most importantly mangroves also provide habitats for a large number of molluscs, crustaceans, birds, insects, monkeys, and reptiles. Other mangrove ecosystem services include the filtering and trapping of pollutants and the stabilization of coastal land by trapping sediment. In general the significance of mangrove ecosystem is grouped into three categories based on the basic services it offers, viz. provisioning, protective, and ecosystem services (Fig. 23.5). The details are discussed in the following subsections.

23.4.1 Integral Component of Coastal Ecosystem

Mangroves are the principal habitat for at least a part of the life cycle of a number of coastal biota, and it is connected with both land and marine ecosystems (Fig. 23.6). Mangroves could function as an important link in the chain of habitats that provide complementary resources and benefits. Mangroves, reefs, and fisheries often have a synergistic relationship, based on their connectivity (Mumby et al. 2004). Areas where mangroves benefit adjacent ecosystems by filtering sediments and pollutants or providing nursery habitats should be granted greater protection. Mangroves also stabilize sediments and trap heavy metals and nutrient-rich runoff, thus improving the water quality for sea grasses, corals, and fish communities. Mangroves and sea grasses filter freshwater discharge from land, maintaining necessary water clarity

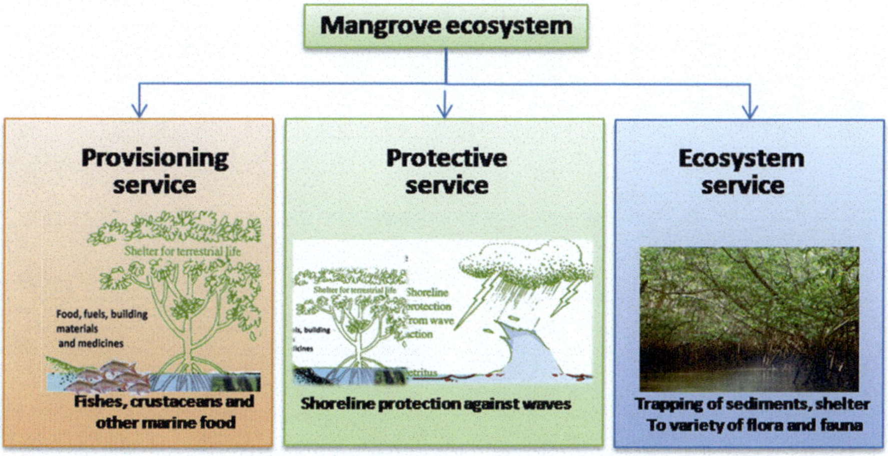

Fig. 23.5 Significance of mangrove ecosystem

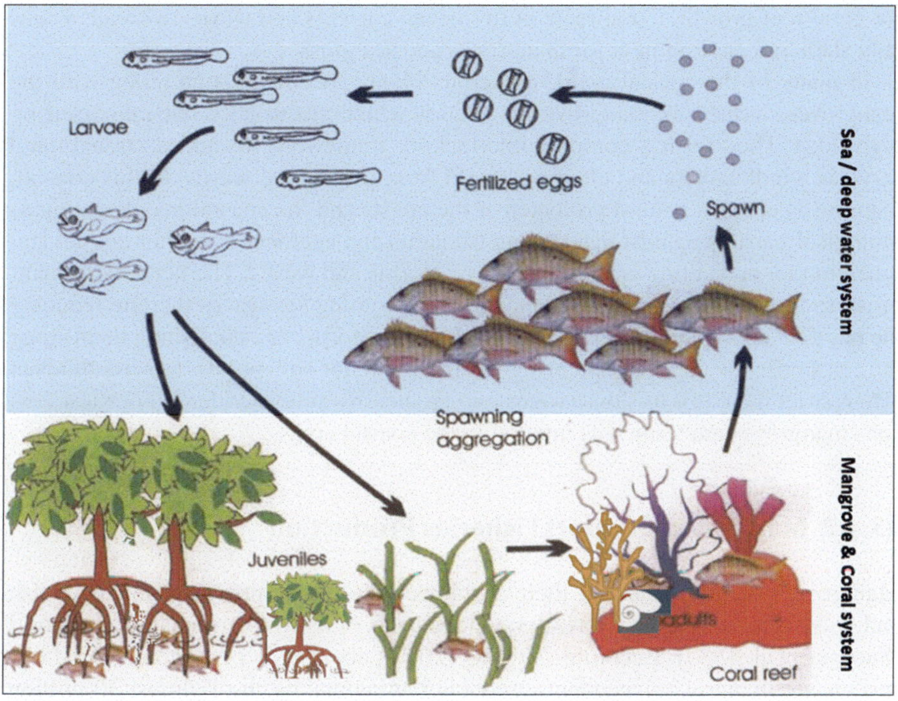

Fig. 23.6 Connectivity between mangrove, seagrass beds, and coral reefs

Fig. 23.7 Thick root network of mangroves

for coral reef growth. Coral reefs buffer ocean currents and waves to create a suitably sheltered environment for mangroves and sea grasses.

In many of the coastal areas, numerous plants are often found along with the mangroves, termed as mangrove associates, which are usually not immersed by high tides. They form a complex interlocking framework for added strength and serve as windbreakers and effective shield from destructive waves. In this context, mangroves act as a natural protector of the coasts and the coastal people acting as biological barriers against devastating tsunamis and cyclones and offer coastal stabilization by preventing erosion caused by storms and waves. The network of mangrove roots and poles offer resistance to the incoming sea waves thereby reducing the physical energy of the waves (Fig. 23.7). By restoring or establishing them along the coast or in the degraded waterlogged areas will not only protect the coastline but also restrict the entry of saline water into the nearby cultivated land. For these reasons mangroves are termed as integral to the coastal areas.

23.4.2 Mangroves Support Fisheries Production

Mangrove systems, including their contained waterways, provide habitat for fish and crustacean species that have formed the basis of fisheries throughout the world (Saenger et al. 1983). According to FAO (2007), about 30% of all commercial fish species are mangrove-dependent, producing an annual catch of almost 30 million tons in 2002 worldwide. For example:

- Fish catches in mangrove waterways of the Philippines ranged from 1.3 to 8.8 kg h^{-1} throughout the year, and the fish biomass of the total catch was positively correlated with the carbon content of the sediments and the mangrove litterfall (Pinto 1987).

- A study of three different mangrove areas over 2 years in southeast India has reported that luxuriant mangroves supported a catch of 11 kg of shellfish per hectare per day and 4.5 kg of finfish per hectare per day and that these catches are much greater than those from sparse mangroves in the same region (Kathiresan and Rajendran 2002).
- This is not confined to fish species, as crustaceans and shellfish species are also commonly harvested within and around mangroves. Several species of *penaeid* shrimps are dependent on mangrove forests for shelter during their juvenile stages (e.g. *Penaeus indicus*, *P. merguiensis*, *P. monodon*, *P. subtilis*, *P. brasiliensis*, and most species of *Metapenaeus*) (Kenyon et al. 2004).

In most of the tropical nations where coastal areas are lined with mangroves, collecting molluscs such as oysters (*Saccostrea* spp.) and cockles (*Anadara* spp.) is an important subsistence activity. It provides scope for improving the livelihood of people by conserving the mangroves, because mangroves are nursery grounds for innumerable aquatic lives and provide hiding places for juveniles. The entangled root masses along with the abundant food supply that comes from mangrove leaves make mangrove areas very good nurseries for many commercially important sea animals. Prop roots of *Rhizophora* spp. are known to provide an important structural refuge for hatchlings. Therefore, of all ecosystems, estuaries have the highest value per hectare making it significant for subsistence in many coastal communities. Further, estuarine habitats have been recognized as important drivers of nearshore fish productivity. At the same time, evidence for linkages between mangroves and offshore habitats by animal migration is still scarce but essential for management and conservation purposes. However, studies have shown that the fish assemblage in any mangrove system is usually unique because of these site-specific factors and differences in their relative importance.

But in recent times, due to over-exploitation of particular species and other resources, it has become a threat to the sustainability of mangrove ecosystem. In addition other anthropogenic and developmental activities accelerate the degradation of mangrove ecosystem. Therefore, conservation of this valuable ecosystem by involving the locals and creation of awareness assumes significance.

23.4.3 Protection Against Storm Surges

The crown and stem of mangroves by virtue of its dense network of roots and canopy in the intertidal zone serve as physical barrier against sea waves. The entangled root masses of mangroves dissipate the wave energy and guard the coastlines and hence are often referred to as bio-shields or natural sea defence. Their specialized roots trap and hold sediments and siltation from the uplands. Mangroves played a protective role in saving the lives of coastal dwellers by taking the brunt of destructive waves during the giant tsunami. Much of the ecological services of

Fig. 23.8 Bio-shield protects against the sea surges and storms

mangroves lie in protecting the coast from solar UV-B radiation, greenhouse effects, fury of cyclones, sea level rise, coastal erosion, and other natural threats in the coastline. Therefore, establishment of mangrove-based shield (Fig. 23.8) in the sea-front of the coastal land should be ideal choice to protect them from sea surges and tsunami-like incidents in the future.

23.4.4 Productivity of Estuaries and Coastal Waters

The leaves that fall from a mangrove tree break up and decompose into small pieces known as detritus which is further degraded by microorganisms that nourish marine animals. Mangroves contribute about 3.65 tons of litter/ha/year and contribute to 1.8–4.2 kg C m^{-2} $year^{-1}$. Besides as discussed previously, it also acts as a breeding ground for fishes and other aquatic life. They also support conservation and utilization of biological diversity—including a number of endangered mammals, reptiles, amphibians, and birds; protection of coral reefs, sea grass beds, and shipping lanes against siltation; and provision of habitat, spawning grounds, and nutrients for a variety of fish and shellfish, including many commercial species.

23.5 Mangroves and Livelihood

Many rural communities in the tropics have been deriving their livelihood by using the mangroves to produce honey, tannins, and traditional medicines besides harvesting of marine food which is the primary activity. Though the coastal population derives their livelihood from the mangroves and mangrove-based products, sometimes overexploitation leads to destruction of mangroves. Industrial exploitation and land diversion for other developmental activities have been not only destructive but triggers ecological consequences. Therefore, it is very important to make a balance between exploitation and conservation of mangroves. Alternative livelihood options and diverse income opportunities should be evolved for the mangrove-dependent communities which allow them to be flexible to adapt to social, political, and economic changes.

23.5.1 Support to the Rural Economy

Coastal populations in Asia have traditionally used mangroves as a source of wood and non-wood forest products for the production of charcoal, fuelwood, and timber and poles for houses, boats, and fish traps. But the level of exploitation was well within the natural regeneration capacity. In recent times, ecotourism activities are increasing in the Asian region, providing further sources of sustainable income to local populations and creating awareness of the services and benefits provided by mangroves. Alternatively mangrove leaves are often used as fodder for camels, goats, and sheep in the Near East, where mangroves are often the only coastal forest resource. However, overexploitation of the trees for this use may reduce mangrove growth and density and stunting trees and degrade stand quality. When the outer trees are overused, it is common practice to provide fresh, additional fodder for camels by opening the way to the inner mangroves. In some places in the Asian coastal region, this practice has affected the quality of the stand even if the mangrove area was not reduced significantly.

Similarly in Oceania, as in other regions, mangrove ecosystems are traditionally used for the collection of fish, shellfish, and crabs which are consumed by the local population. Mangroves have also traditionally been used for the production of fuel wood and as timber, posts, and poles for the construction of houses and boats, which in some cases has resulted in overexploitation of the resource. In Australia, on the other hand, where there is an abundant supply of hardwood timber from other forest types, the use of mangrove wood has been very limited, and several mangrove forests remain well conserved. However, limited areas of mangrove forest have been cleared to convert the land to shrimp farms, but this industry has not expanded as much in this region as in other parts of the world. In some Pacific countries (e.g. Fiji, the Northern Mariana Islands, and Vanuatu), mangrove areas have been converted to agriculture especially sugar production and rice paddies. Some other most important economic uses of true mangroves which are major source of livelihood are given below (Table 23.1).

23.5.2 Alternative Livelihood

Sustainable harvest and utilization of mangroves are possible only if local stewardship are encouraged to develop alternative livelihoods that are less destructive than overharvesting of mangroves or conversion to fish or shrimp ponds is a crucial step to mitigate mangrove deforestation (Hussain and Ahmed 1994). For example, alternative livelihoods include charcoal

production from coconut shells instead of from mangroves as well as traditional honey harvesting in mangroves, which encourages agroforestry and conservation of existing mangrove forests (Bandaranayake 1998). In Vietnam, seaweed farming and cage culture are tried as an income generating alternative to mangrove destruction

Table 23.1 Some economical species of mangroves and their uses

Sl. no.	Species	Uses
1	*Avicennia marina*	Fruits are edible. Leaves fed to livestock and used to treat burns
2	*Avicennia officinalis*	Leafy branches are used as cattle fodder. Timber is used for making boats. Fruits are plastered onto boils
3	*Excoecaria agallocha*	Timber is used as firewood and charcoal
4	*Bruguiera cylindrica*	Timber is an excellent source of firewood
	B. gymnorrhiza	
	B. parviflora	
	B. sexangula	
5	*Scyphiphora hydrophyllacea*	Leaf extracts are known to be helpful for stomach aches. Wood can be used to craft small objects and toys
6	*Cynometra iripa*	Timber is used as fuel wood and for boat building
7	*Ceriops tagal*	Best firewood source
8	*Rhizophora apiculata, R. mucronata*	The timber is used for fuel and charcoal making
9	*Rhizophora stylosa*	Used for firewood, charcoals, boat building
10	*Heritiera littoralis*	Used for boat building. Seeds edible. Seeds are used in cures for diarrhoea and dysentery. Timber is termite-resistant and used in making of poles for fishing
11	*Lumnitzera littorea*	Used as firewood. Has ornamental potential due to its scarlet flowers
12	*Lumnitzera racemosa*	Ornamental potential because of the beautiful white flowers
13	*Sonneratia alba*	Used as firewood
14	*Sonneratia caseolaris*	Used as timber
15	*Aegiceras corniculatum*	Seeds and bark contain a toxic substance named saponin which is used as a 'fish poison', i.e. to stun fish and scoop them up. Tender leaves are edible. The wood is used to make charcoal
16	*Xylocarpus granatum*	Used in boat building, furniture wood, and firewood
17	*Acanthus ebracteatus*	Pounded seeds are used to cure boils and dressing for burns
18	*Acanthus ilicifolius*	Believed to be a medicine for snake bites. Also used as fodder
19	*Nypa fruticans*	Fodder for pigs. Dried fronds for thatching. Woven into mats and baskets. Fruits are eaten by local people
20	*Phoenix paludosa*	Used in thatching

(Crawford 2002). In North Sulawesi, Indonesia, sustainably harvested bamboo provides an alternative to overharvesting of mangrove wood used in construction (Nugent 2003). Similarly local communities in Polynesian islands harvest and collect non-wood products from these forests, such as food, medicine, and tannin. The leaves of the nipa palm are traditionally used for thatching and weaving, and the sap is used for the production of alcohol. With these successes and wide scope for alternative livelihood, it is widely recognized that governance and management frameworks must be developed for mangrove-dominant ecosystem and dependent communities that include diverse patterns of resource use to maintain social and ecological resilience (Adger et al. 2005). Alternative livelihood options and diverse income opportunities allow communities to be flexible to adapt to social, political, and economic changes.

Fig. 23.9 A view of the agro-aqua model suitable for island condition with different components (duckery, horticultural crops, *azolla*, and fish)

23.5.3 Mangrove-Based Agro-aqua Farming

Human interface on mangroves cannot be avoided in a coastal agro-ecosystem as the available pressure on land area make people look into mangroves for alternate livelihood. Thus, the mangrove ecosystem can be conserved as a livelihood source for the residents of the coastal fisherman against their conventional perception of cutting the mangroves for wood and converting the lands to cultivable one. A mangrove-based agro-aqua farming model suitable for the local conditions can be developed and upscaled after its social acceptance (Fig. 23.9). One such model was developed for Andaman Islands, India, which has two ponds with provision for water exchange as per tide level and a mangrove nursery bordering the ponds. The ponds are stocked with mullet, *Mugil cephalus (300 Nos)*; scat, *Scatophagus argus (150 Nos)*; tilapia, *Oreochromis mossambica (200 Nos)*; and tiger prawn, *Penaeus monodon (200 Nos)* under polyculture system. Vegetables (bottle gourd, bitter gourd, asparagus, cucumber, snake gourd) and fruits (banana, papaya, guava, sapota, and mosambi) are raised on the bunds. A duck shed (for ten birds) is provided on the bund with a provision to wash off the duck waste directly into the ponds. Azolla is cultivated in a FRP tank on the bund which serves as a source of fish and duck feed. A mangrove nursery needs to be constructed bordering the ponds where the seedlings of *Bruguiera gymnorrhiza*, *Rhizophora apiculata*, and *Acanthus ilicifolius* are maintained. In this model, the ponds receive organic wastes from the duck shed and the nutrient-rich mangrove water during high tide; therefore, no supplementary feed was provided to the fishes. This mangrove-based agro-aqua integrated farming system model provides additional income of around Rs.125000–140,000 per year from 1 ha area besides nutritional security and generation of mandays. This model provided livelihood support to the coastal communities immediately after tsunami while it favoured regeneration of mangroves.

23.6 Loss of Mangrove Habitat

The value of mangrove ecosystem and its role in supporting the livelihood have been known to the coastal communities. Yet human pressure on coastal ecosystems throughout the tropics is often high, with land competition for aquaculture, agriculture, infrastructure, and tourism. Consequently mangroves and its associated species are being destroyed at alarming rate which also seriously endangered the existence of innumerable species thriving in the mangrove ecosystem. With the advent of modern tools like satellite-based observation, it is clearly evident that the balance has shifted against the mangroves resulting in gradual loss of mangrove areas besides degradation and change in the species structure. The factors responsible for such changes are many but vary for different countries.

23.6.1 Dynamics of Change

Over the last 50 years, about one third of the world's mangrove forests have been lost (Alongi 2002). The major cause of clearings and losses of mangrove areas over time is competition for land from alternate uses such as urban development, tourism, agriculture, shrimp pond, etc., which are major sources of economic activity in the developing tropical countries. According to trend analyses of the compiled data on mangroves in the world, nearly 15.2 million hectares of mangroves are estimated to exist in 2005 down from 18.8 million hectares in 1980. It means that during the past 25 years, about 3.6 million hectares have been lost which is about 20% of the global mangrove area in 1980 (Fig. 23.10). The maximum change has occurred during 1990s. At the regional level, Asia suffered the largest net loss of more than 1.9 million hectares since 1980, mainly due to changes in land use from 1980 to 1990. North and Central America and Africa also contributed significantly to the decrease in mangrove area at the global level, with losses of about 690,000 and 510,000 ha, respectively, over the last 25 years.

With the change in the data quality and reliability, it is now realized that large negative changes in the estimated mangrove area in Asia, the Caribbean, and Latin America in the 1980s have been caused primarily by large-scale conversion of these areas for aquaculture and tourism infrastructure. Shrimp aquaculture accounts for the loss of 20–50% of mangroves worldwide (Primavera 1997) which is being practised as mutually exclusive livelihood activity. At the country level, Indonesia, Papua New Guinea, Panama, and Mexico recorded the largest losses of mangroves during the 1980s. A total of about one million hectares were lost in these countries alone. In the 1990s Pakistan and Panama succeeded in reducing their rate of mangrove loss. On the other hand, Vietnam, Malaysia, and Madagascar suffered increased clearing and moved into the top five countries with major area losses in the 1990s and/or 2000–2005. Even though efforts are under way by national governments and international organizations to decrease mangrove loss, some 500,000 ha have still been lost worldwide within the period 2000–2005.

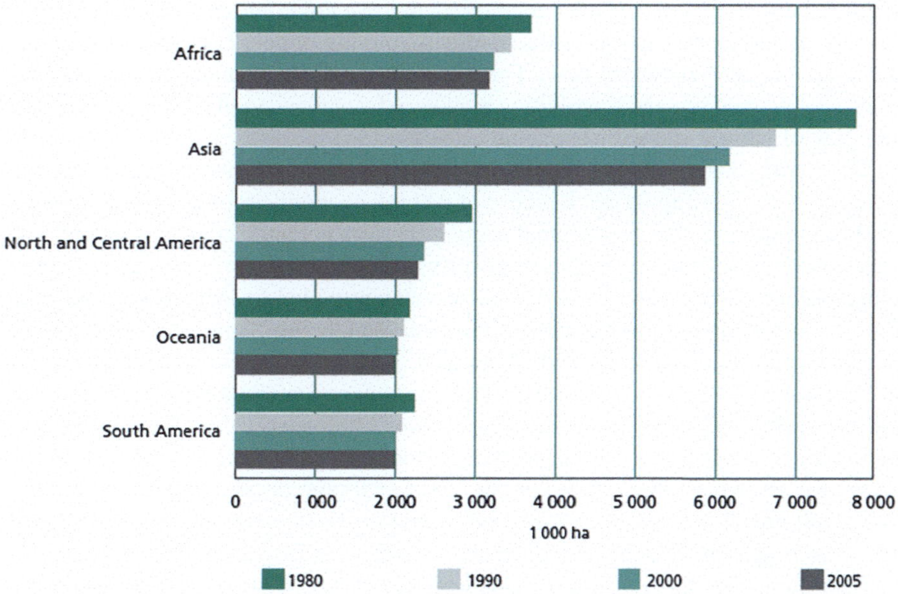

Fig. 23.10 Changes in world mangrove areas (1980–2005)

The consequent conversion of mangrove areas to other uses over the past decades has been alarming. Realizing the importance of mangrove stand in the coastal areas, most countries have now banned the clearing of mangrove areas for aquaculture and require environmental impact assessments prior to large-scale conversion of these areas to other uses.

23.6.2 Drivers of Change

The area under mangrove has been subjected to intense modification and decreasing over the period of time though at different rates in different regions of the world. In the last few decades, national and international efforts have changed the decreasing trend due to the realization of its importance and ecological services. In spite of these efforts, degradation and over-exploitation of mangroves are still continuing in several countries. The reasons are not solely attributable to anthropogenic causes, but natural factors also play prominent role in the modification of mangrove stand. Natural mangrove destruction is caused by cyclones, storms, occasional tsunamis, floods, intensive grazing, pest problems, and modification of surrounding land use. Besides natural factors, human threats to mangroves include the overexploitation of forest resources by local communities; conversion into large-scale development such as agriculture, forestry, aquaculture, salt extraction, urban development, and infrastructure; and diversion of fresh water for irrigation (UNEP 1994). It has been observed that these anthropogenic drivers of destruction play a predominant role. In addition human activities also change the mangrove environment such as dumping of waste material and garbage, effluent release, mining, fodder harvest, overgrazing, and diverting of freshwater flow. These modifying factors and their intensity of mangrove destruction and distribution may vary from place to place due to complex interplay of natural and anthropogenic causes.

The main causes of loss of area in Asia have been overexploitation and the development of shrimp farms. Because of its high economic return, shrimp farming has been promoted to boost national economies, as a potential source of income for local communities and as a means of poverty alleviation. However, this activity may cause loss of habitat and of ecosystem services provided by mangroves and may damage surrounding areas through pollution caused by the chemical products used. In some countries (e.g. India and Myanmar), government policies have promoted self-sufficiency in food production through the conversion of mangrove lands to rice paddies, while in other countries (e.g. China, Indonesia, the Philippines, and Vietnam), mangrove areas have been used for salt production. In addition, some mangrove areas have been clear-cut for urban or tourist development or converted to agriculture. The decrease in freshwater input or the interruption of flows caused by the construction of dams and diversion of water for irrigation may increase soil salinity, damaging the surrounding mangroves, as has occurred in Singapore and Pakistan. During the Vietnam War (1962–1972), mangrove forests often served as bases for military operations and were consequently highly damaged by chemical warfare (herbicides and napalm). Further losses have been caused by the conversion of mangrove land to other uses such as shrimp ponds, agriculture, salt pans, and human settlements.

In the Caribbean and Latin American region, mangrove loss has been caused primarily by large-scale conversion of these areas for aquaculture and tourism infrastructure. Similarly one of the primary causes of mangrove degradation and loss in the biodiversity rich Oceania region has been the development of tourism infrastructure. In most of these countries and territories, tourism is a main contributor to the national economy, but the construction of resorts and other tourism infrastructure has often been undertaken at the expense of coastal ecosystems, including mangroves. Mangrove areas have also been cleared to facilitate urban development. Currently, several mangrove areas and other coastal ecosystems are threatened by industrial pollution, as in New Caledonia, or by domestic and solid wastes and local pollution, as in Fiji, Kiribati, and Tuvalu (Maharaj 2002; USDA Forest Service 1998; United Nations 1998).

23.7 Conservation of Mangroves

23.7.1 Conservation Plan

Threats to and losses of mangroves in many countries are still a problem; at the same time, awareness of the importance of mangroves has greatly improved resulting in several afforestation and rehabilitation efforts are being undertaken, especially for protection and conservation purposes. In this context the terms restoration and rehabilitation are used to describe the conservation. The term 'restoration' has been adopted to specifically mean any activity that aims to return system to a preexisting condition, whereas the term 'rehabilitation' is applied more generally and is used to denote any activity including restoration and habitat creation that aims to convert a degraded system to a stable alternative. Further, these activities derive support from either natural regeneration or managed mangrove nurseries and sometimes together. At the same time, these efforts have clearly indicated the importance of peoples' participation for conservation and management of mangrove ecosystem. The management plan should include:

- Afforestation in new suitable areas
- Reforestation/regeneration of degraded mangrove areas
- Protective measures for conserving the mangrove cover
- Eco-development to derive sustainable benefit for the local communities

There is imperative need to formulate proper restoration practices for mangrove plantation in degraded coastal areas to protect the ecosystem and provide livelihood to the local people. Natural and artificial regeneration activities should be taken up as planned activity in mangrove-degraded areas besides protection of existing stand. Afforestation activities are taken up in a relatively new and suitable areas for mangrove establishment, while reforestation is carried out in a deforested area or area previously occupied by mangroves which has been cleared. Natural regeneration involves the natural process of establishment of seeds of mangroves. In this method the natural system is supported to reestablish itself by way of removing anthropogenic interference or proving best possible natural conditions. Artificial regeneration involves planting of seeds or seedlings in areas where there is inadequate availability of planting materials or the conditions are not favourable for natural establishments (Fig. 23.11). Under such conditions, fast-growing species (*Avicennia* and *Sonneratia*) are preferable. Further, it is also necessary to develop nursery stocks of high-quality species for their efficient performance in the field. While regenerating it is essential to have tidal inundation, but in field two different situations are experienced, viz. low and high tidal amplitude based on the coastal geomorphology.

The growing conditions for seedlings may become harsh due to high salinity of dry soils as a result of lack of regular tidal flushing (low tidal amplitude) and blocking of freshwater flow. The adverse condition can be managed by canal-bank planting technique (Ravishankar and Ramasubramanian 2004). In canal-bank method, canals are formed in a such a way that high saline soils gets regular tidal inundation, leaches out salts, and becomes suitable for mangrove regeneration. This involves formation of the feeder canal (3 m top, 1 m bottom, and 1 m deep) and the distribution canals (2 × 0.75 × 0.75 m) in the mud flats. Propagules are directly planted in the intertidal zone of the canals. Alternatively, in high tidal amplitude areas, direct seed sowing and planting of seedlings in the mud flats can be followed.

Fig. 23.11 Nursery raising for mangrove species and canal-bank method

23.7.2 Strategies to Promote Conservation

It is essential to protect mangroves from anthropogenic threats, because mangroves that are healthy will also be better able to adapt to global changes and in turn will provide livelihood and resilience to coastal communities against climate change. At the same time, anthropogenic threats to mangroves are expected to increase with climate change due to increased demand for food and developmental activities. This emphasizes the need for viable strategies to conserve the mangrove ecosystem which should be inclusive. In general this includes:

- Spread risk by identifying and protecting representative mangrove habitats replicates of these, and sources of seed to ensure replenishment following disasters.
- Biological methods should be supported with mechanical structures such as coastal seawalls or revetments in response to increased coastal erosion which also prevent coastal mangroves from retreating landward in response to sea level rise.
- Identify and protect critical areas that are naturally positioned to survive climate change and represent species diversity.
- Establish greenbelts and buffer zones to allow for mangrove migration and to reduce impacts from adjacent land use change and practices.

- Restore degraded areas that have demonstrated resistance or resilience to climate change, and hold large scope to provide livelihood to coastal communities.
- Understand and preserve connectivity between mangroves and sources of fresh water and sediment and between mangroves and their associated habitats like coral reefs and seagrasses.
- Establish baseline data of mangroves and monitor its response to changes.
- Implement adaptive strategies to compensate for changes in species ranges and environmental conditions.
- It is vital to develop alternative livelihoods for mangrove-dependent communities as a means to reduce mangrove destruction.

As discussed earlier in this chapter, conservation of mangrove won't be successful without the involvement of the local communities and strong government policy support as the government priority would be always pro-development. Therefore to attain a balance between sustainable use, societal expectations, and conservation, the following strategies are essential.

(i) Encouraging community regulation for long-term sustainability
(ii) Zoning mangrove ecosystems for proper management and utilization
(iii) Developing mangrove management plans for ensuring the stand and diversity
(iv) Reassessing the value of mangroves for efficient planning
(v) Improving community information
(vi) Rehabilitation of degraded mangroves for social and environmental benefit

23.7.3 Conservation Efforts in Tropical Countries

The analysis of the trend in mangrove area changes in the last 5 years (2000–2005) and shows a reduction of the rate of loss in all regions due to conservation efforts. On a positive note, a number of countries have registered an increase in mangrove area over time. In Bangladesh the Sundarban Reserved Forest (the Bangladeshi section of the Sundarban)—part of the largest mangrove area in the world—is well protected, and no major changes have occurred during the last few decades. In addition, successful efforts in coastal afforestation have contributed to an increase in the extent of mangrove area despite some losses outside the reserve. A long tradition of plantation programmes is found in other countries as well and has led to an increase in the extent of mangrove area. The recolonization and natural regeneration properties of these forests also help to increase their extent. In Ecuador, for example, the abandoning of ponds and structures for shrimp and salt production resulted in a recolonization of various sites. It should, however, be noted that in a few cases, a slight positive or negative trend may have been caused by changes in methodology or definitions or, as in the United Arab Emirates, by the discovery of new mangrove areas not included in previous assessments.

More than 20 countries are reported as having had no significant change in mangrove area over time. In some of these countries and areas, this may be caused by the limited quantitative information available at the time the study was prepared. In a few of them—Saudi Arabia and the Sudan, for example—although the extent did not appear to have changed significantly over the last decades, the quality of the stands may have severely deteriorated, due primarily to camel grazing, pollution, and/or oil spills.

Bangladesh, Malaysia, and Vietnam have very long traditions of sustainable management, plantation, and afforestation programmes in mangroves. Mangrove management and protection in Bangladesh, for example, started at the beginning of the twentieth century, with the preparation of forest management working plans and the creation of what is now called the Sundarban Reserved Forest. Coastal afforestation programmes started in 1966, involving several districts.

Experimental plantations have also been established successfully in extremely arid countries such as Bahrain, Kuwait, and the United Arab Emirates. Many countries in Asia have been increasingly promoting establishment and maintenance of mangrove greenbelts as protection against natural hazards, especially those countries affected by destructive typhoons each year (e.g. the Philippines and Vietnam). Other protection plantation activities are being undertaken in Bangladesh, Brunei Darussalam, India, Indonesia, and Vietnam.

Many countries have promulgated laws and regulations to protect remaining mangrove areas and mitigate widespread loss. Effective enforcement of this legislation is, however, often hampered by a lack of financial and human resources. Several Asian countries have ratified the Ramsar Convention on Wetlands (2004) and have designated mangrove areas as Ramsar sites or as national parks, reserves, and wildlife sanctuaries.

23.8 Conclusion

Mangrove communities comprise unique plant species that form the critical interface between terrestrial, estuarine, and nearshore marine ecosystems in tropical and subtropical regions. Mangroves fulfil many extremely important productive, protective, and social functions wherever it occurs. Mangroves are the major source of livelihood and integral component of social system for many of the coastal communities in the tropical coastal region. Similarly it also provides gainful employment opportunities and enhances the productivity of farmland and nearshore fishers by way of supplying fishlings. Yet, increased population pressures in coastal areas and lack of awareness have led to large-scale conversion of mangroves to other alternative uses which seriously undermines its ability to provide livelihood support and protect the coastal areas against several natural and human accelerated hazards. If deforestation of mangroves at the current rate were to continue, it could well lead to severe losses of biodiversity and livelihoods, in addition to salt intrusion in coastal zones and the siltation of coral reefs, ports, and shipping lanes. Consequently there would be losses of income from tourism and the loss of knowledge of mangroves and their use as recreational sites for coming generations.

Therefore, there is an imperative need to formulate proper rehabilitation programme for mangrove plantation in degraded coastal areas and conservation of existing stand to protect the coast against the sea surges and tides. This is a social necessity as the tropical coastal region is also known for large natural hazards and projected to get severely affected by climate change in the future. For these reasons the mangrove management plan should essentially include afforestation, regeneration of degraded mangrove areas, protective measures, and inclusive eco-development. When sustainably managed and the biodiversity is conserved without doubt, they can support the livelihoods of millions of coastal people in the tropical coastal region while storing globally significant quantities of carbon and reducing the vulnerability of coastal regions to storm surges and other threats.

References

Adger H, Hughes TP, Folke C, Carpenter SR, Rockstom J (2005) Social ecological resilience to coastal disasters. Science 309:1036–1039

Alongi DM (2002) Present state and future of the world's mangrove forests. Environ Conserv 29:331–349

Bandaranayake WM (1998) Traditional and medicinal uses of mangroves. Mangrove Salt Marshes 2(3):133–148

Crawford B (2002) Seaweed farming: an alternative livelihood for small-scale fishers? Coastal Resources Center, Narragansett

Duke NC (1992) Mangrove floristics and biogeography. In: Robertson AI, Alongi DM (eds) Tropical mangrove ecosystems. American Geophysical Union, Washington, DC, pp 63–100

Duke NC, Meynecke JO, Dittmann S, Ellison AM, Anger K et al (2007) World without mangroves? Science 317:41–42

FAO (2003) In: Wilkie ML, Fortuna S (eds) Status and trends in mangrove area extent worldwide, Forest resources assessment working paper no. 63. Forest Resources Division, FAO, Rome. http://www.fao.org/docrep/007/j1533e/J1533E00.htm#TopOfPage

FAO (2007) The world's mangroves: 1980–2005. FAO Forestry Paper 153, Rome. www.fao.org/forestry/site/mangrove/statistics

Hussain MZ, Ahmed I (1994) Management of forest resources. In: Hussain MZ, Acharya G (eds) Mangroves of the Sundarbans, vol 2. International Union for Conservation of Nature and Natural Resources, Bangladesh

Jayatissa LPF, Dahdouh-Guebas, Koedam N (2002) A review of the floral composition and distribution of mangroves in Sri Lanka. Bot J Linn Soc 138(1):29–43

Kathiresan K, Rajendran N (2002) Fishery resources and economic gain in three mangrove areas on the south-east coast of India. Fish Manag Ecol 9:277–283

Kenyon RA, Loneragan NR, Manson FJ, Vance DJ, Venables WN (2004) Allopatric distribution of juvenile red-legged banana prawns (*Penaeus indicus* H. Milne Edwards, 1837) and juvenile white banana prawns (*Penaeus merguiensis* De Man, 1888), and inferred extensive migration, in the Joseph Bonaparte Gulf, northwest Australia. J Exp Mar Biol Ecol 309:79–108

Maharaj RJ (2002) Pacific islands at risk: foreshore development and their vulnerability and implications for adaptation strategies to climate change. In: Mimura N, Yokoki H (eds) Proceedings of the APN/SURVAS/LOICZ joint conference on coastal impacts of climate change and adaptation in the Asia-Pacific Region, Kobe, Japan, 14–16. November. Asia-Pacific Network on Climate Change (AP-Net), Kobe, p 2000

Mumby PJ, Edwards AJ, Arias-Gonzalez JE, Lindeman KC, Blackwell PG, Gall A, Gorczynska MI, Harborne AR, Pescod CL, Renken H, Wabnitz CCC, Llewellyn G (2004) Mangroves enhance the biomass of coral reef fish communities in the Caribbean. Nature 427:533–536

Nugent S (2003) Mangrove action project: coastal communities resource centers take root. Earth Island J:18. 2 March 2006. www.earthisland.org/eijournal/new_articles.cfm?articleID=652&journalID=67

Pinto L (1987) Environmental factors influencing the occurrence of juvenile fish in the mangroves of Pagbilao, Philippines. Hydrobiologia 150:283–301

Polidoro BA, Carpenter KE, Collins L, Duke NC, Ellison AM, Ellison JC, Farnworth EJ, Fernando ES, Kathiresan K, Koedam NE, Livingstone SR, Miyagi T, Moore GE, Nam VN, Ong JE, Primavera JH, Salmo SG, Sanciangco JC, Sukardjo S, Wang Y, Yong YWH (2010) The loss of species: mangrove extinction risk and geographic areas of global concern. PLoS One 5(4):1–10

Primavera JH (1997) Socio-economic impacts of shrimp culture. Aquac Res 28:815–827

Ramsar Convention on Wetlands (2004) Annotated ramsar list: Australia. www.ramsar.org/profile/profiles_australia.htm

Ravishankar T, Ramasubramanian R (2004) Manual on mangrove nursery techniques. M.S. Swaminathan Research Foundation, Chennai, p 48

Saenger P, Hegerl EJ, Davie JDS (1983) Global status of mangrove ecosystems. Environmentalist 3(Suppl):1–88

Spalding M (1997) The global distribution and status of mangrove ecosystems. Int Newsl Coast Manag-Intercoast Netw Special edition 1:20–21

Tomlinson PB (1986) The botany of mangroves. Cambridge University Press, Cambridge

UN (1998) United Nations report on Tuvalu. http://islands.unep.ch/CLB.htm. Accessed on 8th Nov 2017

UNEP (1994) United Nations environment programme. Assessment and monitoring of climatic change impacts on mangrove ecosystems. UNEP Regional Seas Reports and Studies. Report no. 154

USDA Forest Service (1998) Mangroves of Kiribati: a priceless resource needs protection. In: Ward JD (ed) Supported by Northeastern Area State and Private Forestry; Pacific Islands Forests and Trees Support Programme; and Kiribati Ministry of Natural Resources Development, Division of Agriculture. United States Department of Agriculture, Forest Service, Washington, DC. Available at www.na.fs.fed.us/spfo/pubs/misc/mangroves/mangroves.htm

Valiela I, Bowen J, York J (2001) Mangrove forests: one of the world's threatened major tropical environments. Biosci 51:807–815

Wang BS, Liang SC, Zhang WY, Zan QC (2003) Mangrove flora of the world. Acta Bot Sin 45(3):644–653

Bioremediation: Key to Restore the Productivity of Coastal Areas

24

Abstract

Human activities particularly in the coastal areas resulted in pollution of different kinds. This seriously affected the entire coastal ecosystem. Most important pollutants are sediments, nutrients, heavy metals, and hydrocarbons. Removal of these pollutants and remediation of the contaminated/polluted soil and water assume greater significance to protect the environment and improve the productivity of the land. In this context, microbial assemblages, bioproducts, and higher plants are gaining increasing prominence for their potential in the remediation of contaminated coastal and marine ecosystems. The details of different methods used for bioremediation are discussed with case examples.

Keywords

Coastal pollution · Oil spills · Heavy metals · Phytoremediation · Biomolecules

24.1 Introduction

Anthropogenic activities associated with the economic advancement resulted not only in development and human wellbeing but also led to massive contamination of natural resources by industrialization, urbanization, and modern agriculture including aquaculture practices. The level of contamination and type and amount of pollutant are much severe in coastal areas as they harbour several industries, support tourism activities, and give space for urbanization; besides the coastal regions receive pollutants from faraway places through drainage systems. Such situations pose severe threat to human wellbeing and his future economic interest in the heavily populated coastal region. The combined effect of population growth, race for economic development, and the level of technological adoption is threatening the coastal ecosystems (Creel 2003).

V. Ayyam et al., *Coastal Ecosystems of the Tropics - Adaptive Management*,
https://doi.org/10.1007/978-981-13-8926-9_24

In general, over-exploitation of natural resources by burgeoning human population, together with destruction of forests on land and rapid urbanization, is leading to massive loads of sediments and pollutants in coastal and marine ecosystems, which represent a major concern for their potential detrimental consequences on ecosystem health and the provision of goods and services for human wellbeing (Spalding et al. 2001). Moreover the type and amount of pollutants accumulating in the coastal region may change in the future depending on the developmental path adopted by different governments, and hence huge uncertainty exists in the future degradation scenario.

Removal of these pollutants and remediation of the contaminated/polluted soil and water assume greater significance for two reasons. First, the contaminated sites, whether secured or not, pose serious threat to human wellbeing. Second, due to increase in population, the demand for land and water is also on the rise. Therefore, remediation of contaminated sited and removal of pollutants are essential to bring these resources into use. This is very pertinent to many of the tropical islands and tropical coastal countries where the scarce soil and water resources are getting contaminated with pollutant due to increased economic activities. The conventional techniques used for remediation have been to dig up contaminated soil and remove it to a landfill or to cap and contain the contaminated areas (Vidali 2001). This method has inherent weaknesses; hence, a better approach is required.

In this context, it is pertinent to mention the fact that contaminants such as metals and hydrocarbons undergo complex biogeochemical transformations which are largely dependent upon microbial activity. The microbial activity in particular prokaryotic activity influences the organic matter degradation, pH, and oxidation-reduction reactions in sediments and soils effecting metal speciation, partitioning, and bioavailability (Dell Anno et al. 2018). At the same time, prokaryotes can degrade hydrocarbons both under anoxic and oxic conditions, thus playing a key role in natural degradation processes, i.e. reduction of hydrocarbon contamination levels with time (Catania et al. 2015). In recent times metabolic products of microbial organisms are also known to have potential for decontamination of heavy metal and hydrocarbon pollutants from soil, sediments, and water resources. For these reasons, the exploitation of microbial assemblages, bioproducts, and even higher plants such as trees is gaining increasing prominence for their potential in the remediation of contaminated coastal and marine ecosystems. This chapter attempts to highlight the status of coastal pollution across different coastal countries and scope for bioremediation and different bioremediation methods to restore the productivity of coastal areas with case examples.

24.2 Coastal Pollution and Its Impact

The pollutants in the coastal areas come from either on-site activities or get carried from places away from the coastal region through water or intentional disposal of human beings. These are of different types and come out of different human activities. Presence of these pollutants in the environment causes degradation and negative effect on human health. Major sources and effects of different types of coastal pollution are summarized in Table 24.1.

Table 24.1 Major sources and effects of different types of coastal pollution

Type	Sources	Effects
a. Sediments		
Land based	Agricultural runoff, deforestation, construction sites, coastal mining, urbanization (domestic and municipal sewage)	Reduces light penetration reducing photosynthesis in corals and seagrasses. Impacts the visibility of predators, smothers coral polyps, benthic organism
Marine	Nearshore mining, dredging, offshore construction	
b. Nutrient enrichment		
Nitrogen and phosphorus	Untreated human and animal waste, industrial discharge, agriculture runoff, and coastal aquaculture	Algal bloom which may suppress the growth of seagrass and corals. Increased severity of coral diseases and reduces fertilization in corals. Jelly fish blooms, changing the structure of phytoplankton community
c. Toxic compounds		
Persistent organic pollutants (POPs)	Improper use and disposal of agrochemicals and industrial chemicals and the release of by-products leaching from landfills, atmospheric deposition from incineration	Bioaccumulation in marine organisms
Petroleum hydrocarbons or polycyclic aromatic compounds	Crude oil production and transportation, accidental oil spills, terrestrial runoff from vehicles and industries, burning of wastes and vegetation	Carcinogenic. Disrupts cellular and biochemical functions, reduction in population of sea birds
Heavy metals	Mining, smelting, oil refineries, chemical industry, ship yards, untreated sewage sludge and diffuse contamination, vehicular emissions, industrial effluents, sewage water, untreated human and agricultural wastes, fertilizers, pesticides, and batteries	Inhibit recruitment, fertilization, and development in various marine organisms including corals and crustaceans
Cyanide fishing	Live coral fish trade includes both ornamental and food fish trade	Coral bleaching and impaired calcification of skeletons. Death of various organisms in reef ecosystem

Modified from Todd et al. (2010)

24.2.1 Major Causes and Types of Pollution

Coastal areas are facing increasing pollution hazard from human activities which are either on-site or happening away from the polluted area and get carried to the coastal area through water. There are different human activities such as urbanization, industry, services, agriculture, and household which generate different types of pollutants. The major sources of coastal and marine pollution come from both land (through urbanization, industrialization, agriculture, mining, transport, etc.) and sea (through dredging, mining, dumping, and shipping). Non-point sources of pollution from urban areas include suspended solids, oil and grease, organic chemicals, inorganic compounds, pathogens, litter, and heavy metals which are brought down by runoff to the coastal areas from urban and even far-off sources and get accumulated in the coastal and marine ecosystem. In general these pollutants are grouped into organic, inorganic, physical, and biological materials which are inimical to human health and environment. These pollutants may get dispersed and dissolved or accumulate in soil, water bodies, and air in the coastal areas.

The basic reason for the coastal pollution arises from the resources and physical nature of coastal areas itself. The coastal areas are densely populated having a density of 80 persons per km^2 which is twice that of global average density due to the resource endowment and geographical nature of coastal areas facilitating trade and transport with outside the region and country. More than 50% of the population of Bangladesh, North Africa, and many of the Pacific island nations lives in coastal areas causing heavy damage to coastal resources through various activities. The coastal regions are increasingly urbanized, and 14 out of 17 megacities are located in coastal region, and 11 of these large cities are located in Asia. The rapid growth of the coastal urban centres in tropical coasts of Asia, Africa, and Latin America and the fast development of the coastal tourism sector produce vast quantities of pollution from untreated domestic sewage, posing a threat to the nearshore habitats such as coral reefs and also mangrove forests covering 25% of world's tropical coastline (Creel 2003). At the same time, increasing urbanization resulted in loss of mangrove forests for making fish and shrimp ponds, human settlements, and agricultural and industrial development. More than 70% of the mangroves were lost in Kenya, Liberia, the Philippines, and Puerto Rico. In these urban centres, the increased solid waste generation with inadequate disposal system results in release of nutrients and toxic chemicals to the environment.

24.2.2 Pollution and Its Impact

24.2.2.1 Urbanization and Industrialization

The urbanization and industrialization together resulted in greatest volume of waste discharges into the coastal and marine environment. This includes industrial waste, municipal wastes, animal remains and slaughterhouse wastes, and water and wastes from domestic usages. As many of the megacities are located in tropical coastal region, huge loads of such wastes are discharged into the nearby rivers or aquatic systems finally reaching the coastal waters. This is further accentuated by industrial pollution as major industrial development occurs in and around the highly populated cities. The untreated sewage effluents entering coastal waters contain a diverse array of polluting agents including pathogens, organic substances, heavy metals, and trace elements posing direct and indirect threats to coastal ecosystem. The heavy metals like lead, copper, chromium, arsenic, and cadmium are found in concentration up to 0.1–0.7 mg l^{-1} in sewage.

It is estimated that in the Gulf of Thailand, a BOD load of 659–34,376 tons is dumped annually generated from municipal and industrial wastes (Cheevaporn and Menasveta 2003). Gomez (1990) reported that over 80% of sewage originating from urban slums in Southeast Asia is deposited into the sea untreated. In Indonesia, organic pollution emissions into the South China Sea more than doubled between 1980 and 1993 (Chou 2009). The untreated sewage contains harmful human pathogens such as *Escherichia coli*, *Mycobacterium tuberculosis*, *Vibrio cholera*, *Citrobacter freundii*, *Salmonella* sp., etc., and they were also found to occur in marine mammals (Grillo et al. 2001). Besides, the dumping of sewage sludge at sea can smother the benthos, decrease species biodiversity, and increase heavy metal concentrations. Decaying organic matter and nutrients in sewage may lead to eutrophication resulting in alterations in ecosystem structure (Shahidul Islam and Tanaka 2004).

24.2.2.2 Agriculture and Allied Activities

Agriculture is one of the major contributors to coastal pollution accounting about 50% of the total pollution source of surface water. Agricultural activities pollute the coastal region by means of soil erosion leading to siltation, nutrient enrichment, pesticide contamination, and agro-industrial pollution. The increased and often indiscriminate use of agrochemicals like fertilizers and pesticides is a major source of aquatic pollution throughout the Caribbean region, Small Island Developing States (SIDS) in tropical region, and in coastal waters of many developing countries. The leaching of nitrate (NO_3) and ammonium (NH_4) from fertilizer sources results in nutrient enrichment of surface water sources (Rawlins et al. 1998).

The natural processes of soil erosion and the transport of sediment from the land and its discharge into coastal waters are exacerbated by human activities such as deforestation, urbanization, and agricultural activities in river-basin watersheds. The steep slopes, intensive rainfall, deforestation, and cultivation on steep slopes in many of the tropical islands result in severe erosion and subsequent siltation of coastal waters. In Jamaica, where hillside farming occurs on slopes of up to 70°, primitive agronomic practices have been responsible for estimated losses of 1300 tons of soil per year from 'cultivated steep slopes' (Eyre 1990).

The coastal aquaculture including prawn and fish farming releases large quantities of nutrients through faecal matter and wastage of animal feed adding substantial nutrient loadings to aquatic systems and subsequent discharge into receiving waters. In general, 85% of phosphorus, 80–88% of carbon, and 52–95% of nitrogen may be lost into the environment from marine fish culture systems (Wu 1995).

24.2.2.3 Mining and Exploration

The oil and gas exploration and mining of coastal sand for heavy metals exerted heavy damage to coastal and marine ecosystem. The activities related to oil and gas exploration and exploitation affected wetlands and coastal ecosystems at various stages of development such as exploration, site access, site preparation, drilling, production, pipeline installation, spill control and cleanup, and site closure. The natural hydrologic regime existing in coastal ecosystem is disrupted due to construction of artificial levees, canals, and impoundments and laying of pipelines for transporting oil and gas from the site of production to processing, i.e. from offshore production site to processing units. The depressurization during drilling also enhances the subsidence of land, and in many places it also causes seawater intrusion, thus affecting the groundwater resources (Ko and Day 2004). Similarly mining activities cause loss of protective vegetation cover, habitat degradation, and increase in heavy metals and sediment load in the river systems, all reaches the coastal waters. The loss of wetland area in deltaic regions decreases the value of the estuarine zone as a nursery ground for aquatic animals and also results in huge loss of biodiversity.

24.2.2.4 Pollution from Fossil Fuel and Related Activities

Oil pollution is a major threat, and huge volume of oil is added into the coastal and marine environment from various sources like drainage, routine maintenance, smoke, natural seeps, oil spills, onshore drilling, etc. Huge volume of oil is discharged into the sea each year. Approximately 1.3 million tons of petroleum enters the marine environment each year (Ventikos and Sotiropoulos 2014). The transportation of crude oil across the seas also poses a severe risk of marine pollution. For example, oil pollution is severe in the Straits of Malacca, South China Sea, Gulf of Thailand, and the Jakarta-Manila and Singapore-Bangkok routes (Lu 2003). The oil pollution also occurs from coastal refineries through leakages, spills, breakages, etc. Besides, the destruction of pipelines results in oil spillage in the coastal environment. The oil pollution poses serious adverse effects on coastal and marine ecosystem and the organisms at different levels of food web affecting the biodiversity in the estuarine zone, beaches, and ocean floor and kills benthic organisms such as crabs. It also coats birds, impairing their flight or reducing the insulative property of their feathers, thus making the birds more vulnerable to cold.

24.2.2.5 Persistent Organic Pollutants

The persistent organic pollutants (POPs) include polychlorinated biphenyls (PCBs) and organochlorine pesticides such as DDT, chlordane, and hexachlorocyclohexane (HCH). They are associated with waste discharge from industries, atmospheric deposition from incineration, and leaching from landfills. High concentrations of PCBs are found in industrialized areas and urban centres like Manila, Bangkok, and western Singapore in South Asian region. As the oceans are the major sink for PCBs, marine animals are much exposed to the toxicity of these compounds. They are found to affect productive capabilities and posing immunotoxic risks to marine mammals. The organochlorine pesticides such as DDT, HCH, and chlordane were used in malarial control, in timber industries, as well as in agriculture in many of the developing countries since the 1950s. DDT is still used for malarial control in the Philippines, Thailand, Papua New Guinea, and India though many countries have banned these compounds. The bioaccumulation and biomagnification of pesticides can interfere with hormonal, reproductive, and nervous systems in molluscs, crustaceans, fishes, shore birds, marine mammals, as well as humans (Todd et al. 2010). Other important toxics are butyltin compounds which are used as antifouling paint on ship hulls and also contribute to marine pollution proving detrimental to living organisms. They are known as endocrine disruptors, causing ovarian spermatogenesis in oysters and imposex in gastropods and bivalves. Occurrence of imposex has been recorded in many SE Asian countries including Singapore, Malaysia, Indonesia, and Thailand. The frequency of imposex in field populations shows a clear relation to environmental TBT levels, and sex imbalance causes a decline and species extinction in some natural populations (Cadee et al. 1995). They can bioaccumulate in molluscs, fishes, and polychaetes and be transferred to higher tropic levels including humans.

24.2.2.6 Heavy Metal Contamination

Metal contamination is ubiquitous in coastal ecosystems. They are the by-products of many industrial processes, contributing varying amounts of different metals and trace elements. In coastal and marine environment, the major sources are mining, leaching from landfills, industrial effluents, sewage, untreated human and agricultural wastes, fossil fuel combustion, and natural and vehicular emissions. Some of the common toxic metals which appear to be toxic to marine life are mercury (Hg), lead (Pb), cadmium (Cd), copper (Cu), chromium (Cr), manganese (Mn), zinc (Zn), silver (Ag), nickel(Ni), and selenium (Se), while some metalloids such as antimony (Sb) and arsenic (As) are also found to be toxic (Davies 1978; Raj et al. 2018). These heavy metals have serious effects on terrestrial and aquatic ecosystems, which increase physiological health risks (Pandey 2012). Major impacts on the aquatic organisms are inhibition of recruitment, fertilization, and development in various marine organisms including corals and crustaceans.

24.2.2.7 Tourism and Recreational Activities

Tourism is one of fastest growing economic activities in coastal areas and island nations worldwide. Tourism dominates the economy of Caribbean region and small island states. Ninety five (95%) percent of the economy of the Maldives depends on tourism, and it is the only source of hard currency. Though it results in economic development, it can also lead to unsustainable coastal development, vast quantities of waste generation, and discharge of solid wastes affecting the coastal resources (Creel 2003). In developing countries, more than 90% of wastewater and 70% of industrial wastes are discharged in coastal waters without proper treatment resulting in contamination.

24.3 Scope for Bioremediation

The use of alive or inactivated biological materials for removal and/or degradation of organic and inorganic contaminants has been increasingly used in recent times. This holds greater promise of reclamation/remediation of contaminated soil and water; in turn it provides scope for reuse of such remediated resources. The pollutants in the environment have varying degrees of biodegradability depending on the chemical composition and structure. The longer the persistence of these compounds in the environment, the more will be the effect on these inhabiting species. The living organisms especially microbes and higher plants have been enormously harnessed for biodegradation of toxic and xenobiotic compounds due to their rich diversity and their potential for adaptation to adverse conditions. Microorganisms are capable of performing detoxification reaction due to their diverse metabolic activities and accumulation of heavy metals in their body cells. Further, other adaptation strategies like cellular modifications to maintain the necessary biological functions, production of surface-active compounds, and use of efflux mechanisms to decrease the concentration of toxic compounds make the microorganisms an efficient decontamination tool. However all the chemical pollutants can't be remediated through bioremediation which may require combination of other methods, and sometimes it becomes uneconomical. Some of the contaminants potentially suitable for bioremediation are pesticides; heavy metals; many of the organic pollutants such as chlorinated solvents, chlorinated phenols, polyaromatic hydrocarbons, benzene, toluene, ethylbenzene, and xylene; and nutrients like nitrogen and phosphorus.

Similarly, higher plants typically cope up with pollutants by suitably changing the soil properties through their exudates and also their interaction with rhizosphere microorganisms (Mench et al. 2009). In contrast, some plants are capable of accumulating excessive concentrations of heavy metals in their tissues. For example, *Eleocharis acicularis* can accumulate arsenic, copper, and zinc. Such plants are exploited for removal of contaminants from the soil, water, and/or sediments. There are about 45 plant families and 500 plant species known to have the ability to accumulate heavy metals, and some of the important families are Brassicaceae, Euphorbiaceae, Asteraceae, etc. (Ullah et al. 2015). Plant growth-promoting bacteria enhance growth and root development, which in turn improves the tolerance of plants to various environmental stresses and promotes larger and healthier plants which are more competent in their ability to remediate both organic and inorganic contaminants (Ullah et al. 2015). In summary, bioremediation is an option available that offers the possibility to destroy or render various contaminants present in the coastal areas and is harmless using natural biological activity. Inclusion of successful and proven bioremediation methods in the developmental plan and coastal zone management will strengthen the resolve for sustainable development. This provides scope for using soil and water by removing the contaminant and improves the productivity.

24.4 Bioremediation

24.4.1 The Concept of Bioremediation

Bioremediation is a process which uses living organisms (both micro- and macroorganisms) and their products to remove contaminants from the soil. Bioremediation is defined as the process by which environmental contaminants are biologically degraded into non-toxic or less toxic materials by natural biological processes. In recent times the concept has progressively grown to include the productive utilization of living organisms (microbes and higher plants) to remove the environmental contaminants into naturally occurring compounds which are safe for human, animal, plant, and aquatic life or to render them harmless. The principle of bioremediation is based on the use of microbial populations which possess the ability to modify or decompose certain pollutants. The genetic diversity and metabolic versatility of the microorganisms are exploited for transformation of toxic substances or contaminants into less toxic products, which are then integrated into natural biogeochemical cycles (Alexander 1994). The maximum benefit of this process is the complete mineralization of compounds besides biomass production (Santos et al. 2011). Numerous mechanisms have been elucidated for the biodegradation such as degradation, detoxification, transformation, immobilization, or stabilization in which the toxic compounds are converted to harmless products or to safer levels or levels acceptable to regulatory authorities.

24.4.2 Favourable Conditions for Bioremediation

The successful biodegradation of any contaminants depends on the presence of specific microorganisms and adequate environmental conditions. Similarly for effective bioremediation, microorganisms must enzymatically attack the pollutants and convert them to harmless products. At the same time, the environmental conditions should permit the microbial growth and activity; therefore, its application often involves the manipulation of environmental condition to allow microbial growth and degradation to proceed to a logical end (Vidali 2001).

The factors favourable for bioremediation are the existence of a microbial population capable of degrading the pollutants, the availability of contaminants/pollutants and its access to the microbes/plants, and favourable environmental conditions. The main requirements for microbes are source of carbon and energy. Similarly the favourable environment for its effective performance is soil type, temperature, pH, presence of oxygen, electron acceptor, and nutrients.

24.5 Methods of Bioremediation

In bioremediation microorganisms, higher plants and microbial products such as enzymes, polysaccharides, and surfactants are being used. Based on the organisms involved, it can be classified as microbial remediation and phytoremediation. Each method can be further subdivided based on the process involved in bioremediation. The details are given in the flow chart depicting different bioremediation methods (Fig. 24.1).

24.5.1 Microbial Remediation

The potential of microorganisms for detoxification or degradation of toxic pollutants is well known since long time. The rich biodiversity and vast catabolic potential

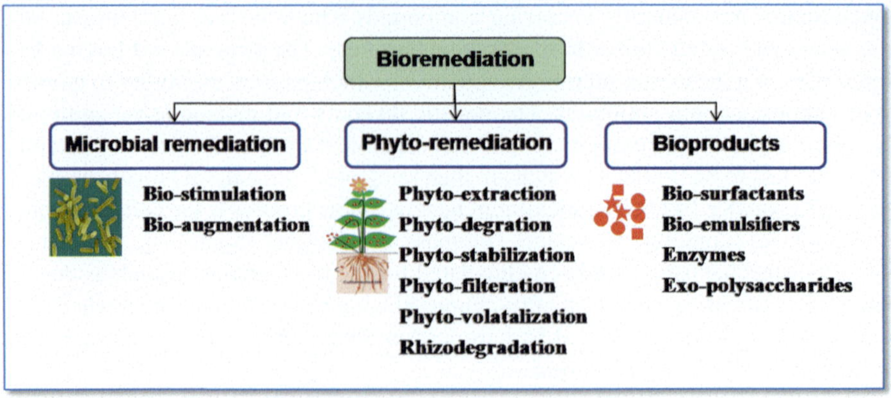

Fig. 24.1 Overview of different bioremediation methods

of microorganisms along with other adaptation strategies like ability to modify the cellular membrane to maintain essential biological functions, the production of surface-active compounds as biosurfactants, and the use of efflux pumps to decrease the concentration of toxic substances inside the cells made them as an economic and eco-friendly tool for the bioremediation of contaminated sites (Tyagi et al. 2010). Microorganisms such as bacteria, fungi, and algae are exploited for degradation of environmental contaminants as they have enzymes, viz. oxidoreductases, hydrolases, lyases, transferases, isomerases, and ligases which can use the contaminants as a food during metabolic reactions. In bioremediation these microbes are encouraged to work by supplying optimum levels of nutrients and other substances which are essential for their metabolism to degrade/detoxify the hazardous substances. The bioremediation can be done by addition of the microbes directly or by modifying the environment of the contaminated site so as to facilitate the native organisms to detoxify or decontaminate the site. Accordingly, the approaches involved in microbial remediation are classified into microbiological approach and microbial ecological approach (Ritter and Scarborough 1995). The microbiological approach involves augmentation of the contaminated site with one or more species of contaminant-specific degrading organisms, while microbial ecological approach involves adjusting certain physical and chemical factors at a site to enhance degradation by native organisms. As the bioremediation can be effective only where environmental conditions permit microbial growth and activity, its application often involves the manipulation of environmental parameters to allow microbial growth and degradation to proceed at a faster rate (Kumar et al. 2011). The bioremediation can be done either on-site (in situ remediation) or off-site (ex situ remediation), and based on the actual processes involved, the microbial remediation can be grouped into biostimulation and bioaugmentation.

24.5.1.1 Biostimulation

It is actually a microbial ecological approach whereby the environment is modified suitably to stimulate existing (native) microbes capable of biodegradation. It involves the stimulation of native microbial (bacteria, fungus) growth and accelerating the decontamination rate by addition of one or more rate limiting factors such as nutrients, electron acceptors and/or donors (phosphorus, oxygen), and energy source, i.e. carbon source, to enhance their metabolic activity (Prince 1997; Atagana 2008). Therefore, it is defined as a type of natural remediation that can improve pollutant degradation by optimizing conditions such as aeration, nutrient concentration, pH, and temperature control.

The stimulation is done by the correction of environmental factors, viz. pH, temperature, soil aeration, humidity, and moisture content (Santos et al. 2011). Thus the native microbial growth can be stimulated by supplying fertilizers, growth supplements, and trace minerals or by optimizing the environmental conditions like pH, temperature, and oxygen to fasten their metabolic rate and pathway (Adams et al. 2015). Sometimes, even trace amount of pollutant itself can act as stimulant by turning on the metabolic pathways necessary for production of enzymes involved in bioremediation. This method is widely used for decontamination of petroleum products and their derivatives (petroleum hydrocarbons). Some of the works related to stimulant and type of contaminant removed pertaining to tropical countries are listed in Table 24.2.

The addition of nutrients through organic and/or inorganic sources for degradation of petroleum hydrocarbons was well studied and documented by several researches. Contaminated soil containing more than 38,000 mg/kg TPH was remediated using sewage sludge and woodchip compost by Atagana (2008). Hamzah et al. (2014) used oil palm empty fruit bunch and sugarcane bagasses as a biostimulant for degrading crude oil and reported 100% decontamination for sugarcane bagasses and up to 97% for empty palm fruit. However, in a marine environment, the use of biostimulants is more critical, as they may not reach out to the microbes near the contaminants in the sea water and they may get diluted in water and drifted away from the pollutants by wave action. Moreover, higher concentrations of nutrient sources (N and P) can cause eutrophication resulting in enhanced algal growth and reduced oxygen supply in the water (Nikolopoulou and Kalogerakis 2009).

The major advantage of biostimulation is that well-adapted local or native microbes are exploited, and they are present spatially well distributed in both surface and subsurface layers. Hence, there is no concern regarding the establishment or stabilization of microbes as required when an organism is introduced in a new environment as in bioaugmentation. However, the major challenge is the application of additives (stimulants) covering the targeted area as it is often influenced by local lithology. For example, an impermeable soil layer may limit the flow of additives to subsurface layers, and also presence of fractures in the subsurface creates preferential pathways in the subsurface resulting in preferentially flow of additives, preventing their even distribution. Similarly addition of nutrients not only promotes the beneficial microbes but also promotes the growth of other heterotrophs as well, thus resulting in competition between the resident microflora (Adams et al. 2014) and

Table 24.2 Bioremediation of petroleum hydrocarbons (PHCs) through biostimulation

Nature of the stimulant	Type of contaminant	Percentage removed	Duration	Reference
Compost made from woodchips and sewage slugs	Crude oil	100% removal of two- to three-ringed PAHs within the first 3 months	570 days	Atagana (2008)
Poultry droppings	PHC contaminated marine sediments	95.35% for TPH and 98.92% for PAH	56 days	Chikere (2012)
Cow dung	Hydrocarbon polluted mangrove swamp	62.96%	70 days	Orji et al. (2012)
Tea leaf, soy cake, and potato skin	Diesel fuel	Between 25% and 82%	126 days	Dadrasinia and Agamuthu (2013)
Oil palm empty fruit bunch and sugarcane bagasses	Crude oil	100% for sugarcane bagasse and up to 97% for empty palm fruit bunch	20 days	Hamzah et al. (2014)
Biosolids obtained from treatment of domestic sewage	PHCs	96% enhanced degradation	–	Sarkar et al. (2005)
Phosphate, nitrate, or nitrate + phosphate amendment	Oil refinery sludge	46–55% attenuation of total petroleum hydrocarbon (TPH)	–	Roy et al. (2018)
Oleophilic fertilizer (containing urea, phosphoric esters, oleic acid, water, and a solvent)	Linear alkanes and aromatics	30% of linear alkanes and aromatics	60 days	Gallego et al. (2007)
Compost of the source-selected organic fraction of municipal solid waste (OFMSW) and rabbit food	Polycyclic aromatic hydrocarbons	89% degradation of the total PAHs	30 days	Sayara et al. (2011)
Fertilizer application and tilling	Crude oil	86% hydrocarbon loss	9 weeks	Odokuma and Dickson (2003)
Mature compost and bacterial consortia	Diesel	96% TPH degradation	120 days	Taccari et al. (2012)

reducing the efficiency. The 'nutritional imbalance' between the supplied carbon and the nitrogen and phosphorus concentrations needed for microbial growth is found to be a major limitation, and the rate of nutrient replenishment is insufficient to support the rapid microbial growth for degradation of oil, especially if found in high concentrations (Santos et al. 2011).

24.5.1.2 Bioaugmentation

It involves the addition of selected microorganisms to the contaminated sites in order to supplement native microbial population and hasten the degradation process. The addition of pollutant-degrading microorganisms (native/exotic/engineered) to augment the biodegradative capacity of indigenous microbial populations on the contaminated area is known as bioaugmentation. Since the 1970s, bioaugmentation has been proposed as an alternate strategy for the bioremediation of oil-contaminated environments. This strategy of bioaugmentation is used when native microbial populations are not capable of degrading the wide range of potential substrates present in complex mixtures such as petroleum (Leahy and Colwell 1990) or when the native organisms are under stress due to the recent exposure to the contaminants. Other conditions under which bioaugmentation may be considered are when the faster decomposition is the prime factor and native microbial load is low. Under such circumstances, seeding may reduce the lag period to start the bioremediation process. For this approach to be successful in the field, the seed microbial consortia should maintain genetic stability and viability during storage, survive in foreign and hostile environments, be able to degrade most petroleum components, effectively compete with native microorganisms, and move through the pores of the sediment to the contaminants (Adams et al. 2015).

Bioaugmentation involves the introduction of microorganisms isolated from the contaminated site, from a historical site or carefully selected and genetically modified to support the remediation of hydrocarbon substances based on the assumption that native organisms could not degrade the complex mixture of petroleum products due to sudden exposure to the contaminants. The success of this strategy depends on the use microbial strains or consortia well adapted to the contaminated site, and efficiency of such microbes depends on their ability to compete with native microorganisms and predators and various abiotic factors. While screening the microbes for application, the rate of proliferation of microorganisms used for augmentation, chemical structure and concentration of pollutants, the availability of the contaminant to the microorganisms, the size and nature of the microbial population, and the physical environment should be taken into consideration. The best approach for selecting suitable microbes should be based on the prior knowledge of the microbial communities inhabiting the targeted site (Thompson et al. 2005), and any successful bioremediation program requires site-specific strategies based on the nature of contaminants and also the environmental conditions of the decontamination site. Many studies have shown the advantages of using a microbial consortium rather than a pure culture for the bioremediation as it provides the metabolic diversity and robustness needed for field applications (Tyagi et al. 2010). It was reported that pure strains or single microbial culture can degrade low-molecular-weight hydrocarbons, while microbial consortia having different microbial population are necessary for the mobilization and biodegradation of high-molecular-weight PAHs (Santos et al. 2011).

Several studies demonstrated the potential of remediation of bioaugmentation for decontamination of organic pollutants, and some of the studies are presented in Table 24.3. Alisi et al. (2009) successfully obtained complete degradation of diesel oil and phenanthrene by using microbial consortia comprising native strains and

Table 24.3 Various techniques of phytoremediation

Sl. No	Techniques	Description
1	Phytoextraction	Also called phyto-accumulation. It involves the use of metal accumulating (hyperaccumulators) plants to remove metals from soil by concentrating them in harvestable plant parts. After a certain time period of growth, the plants are harvested and disposed or incinerated
2	Phytodegradation	Degradation of contaminants in plant tissues where they are metabolized or biotransformed by plant enzymes. The transformation can occur in any plant part like roots, stem, or leaves
3	Phytostabilization	Also called phyto-sequestration. It involves limiting the mobility and arresting the leaching of contaminants through the thick mat of adventurous roots and associated rhizosphere microbes. The different processes involved in phytostabilization are absorption by roots, adsorption to the surface of roots, or the production of biochemicals by the plant roots that are released into the soil or groundwater which can sequester, precipitate, or immobilize the contaminants
4	Phytofiltration or phytohydraulics	Sequestration/removal of pollutants by plants from contaminated water or aqueous streams, e.g. poplar trees were used to contain a groundwater plume of methyl-tert-butyl ether (MTBE). The various strategies involved in this technique are (a) rhizofiltration (the use of hydroponically cultivated plant roots), (b) constructed wetlands (CWs) and lagoons, and (c) bioadsorbent-based systems
5	Phytovolatilization	Useful for volatile compounds. The plants take up the volatile compounds through roots and transpire the same or its metabolites, thereby releasing them into the atmosphere
6	Rhizodegradation	Degradation of organic pollutants by rhizospheric microorganisms. Exudates from plants stimulate rhizospheric bacteria to enhance biodegradation of soil contaminants

found an overall reduction of about 75% of the total hydrocarbons in 42 days. A novel strain of *Serratia* sp. (BF40) isolated from crude oil-contaminated soil significantly reduced the TPH content (>60%) in saline soil due to its high surface activity and salt tolerance (Wu et al. 2012). Diaz-Ramirez et al. (2013) have studied hydrocarbon biodegradation potential of native and exogenous microbial inocula in Mexican tropical soils. The microbial consortium developed by Lin and Cai (2008) showed degradation capabilities ranging from 87% to 95% for different PAHs like pyrene, fluoranthene, phenanthrene, and fluoeno under laboratory conditions.

The microbial inoculum required for bioaugmentation is produced in bioreactors under optimum conditions. The factors which hamper the microbial growth may include fluctuations in temperature, soil moisture, pH, deficiency of nutrients, and also potentially toxic levels of pollutants in the contaminated soil (Gentry et al. 2004). Many instances the microbes performing well under controlled conditions failed in natural conditions might be due to problems concerning the adaptation of the inoculated microorganisms; insufficiency of substrate; competition between introduced and native organisms; use of other organic substrates in preference to the pollutant; and predation by other organisms like protozoa (Goldstein et al. 1985). Thus seeding alone is generally not enough and should be accompanied by suitable physical and environmental alterations (Leahy and Colwell 1990). So the combination strategy combining both bioaugmentation and biostimulation is found to be successful and used for decontamination of organic pollutants. Rosenberg et al. (1992) after optimizing conditions for bioremediation of crude oil using a combination of bioaugmentation and biostimulation technique in the laboratory successfully implemented the same for field and in situ beach remediation. To stimulate the growth of natural or introduced microbes in the soil, oxygen is supplied through direct air injection into residual contamination in soil by means of wells as in bioventing. It uses low air flow rates to provide only enough oxygen to sustain microbial activity.

Adsorbed fuel residuals are biodegraded, and volatile compounds also are biodegraded as vapours move slowly through biologically active soil (Lee et al. 2006; Agarry and Latinwo 2015).

Similarly, a combination of bioaugmentation (using native hydrocarbon-degrading bacterial consortium) and biostimulation using agricultural fertilizers (15:15:15 NPK) was employed as an option for the recovery of the contaminated environment over 20 weeks in a mangrove located on the New Calabar River, Nigeria. However, the results indicated that the combination of nutrient addition and the degrading bacterial consortium did not result in a significant increase in biodegradation. As the use of nutrients in the form of fertilizers is cheaper than the large-scale cultivation of native microbial populations, biostimulation alone can degrade the oil pollutants in tropical soils (Odokuma and Dickson 2003b). The hydrocarbonoclastic bacteria called marine bacteria which degrade hydrocarbons degrade the hydrocarbons present in the polluted sites of marine environments. Hydrocarbonoclastic bacterial consortium has a wide range of degradation capabilities on both aliphatic and aromatic fractions of crude oil (Karlapudi et al. 2018).

24.5.2 Phytoremediation

Contamination of land and water resources in coastal ecosystems of tropical region poses a greater challenge by affecting ecosystem quality, human health, and biodiversity. In recent times, plant-based environmental remediation has been widely pursued as a favourable cleanup technology. Plant-mediated decontamination/detoxification processes are commonly referred to as 'phytoremediation'. It has been proposed as an alternative method to remove pollutants from air, soil, and water or to render pollutants harmless and does not affect soil biological activity, structure, and fertility (Rajkumar et al. 2012). A schematic representation of phytoremediation is given in Fig. 24.2.

Phytoremediation is an emerging green technology which signifies the use of plants to contain, sequester, remove, or degrade inorganic and organic contaminants in soil, sediment, surface, and groundwater, thereby restoring the contaminated sites to a relatively clean, non-toxic environment. Even though it is relatively new, it has

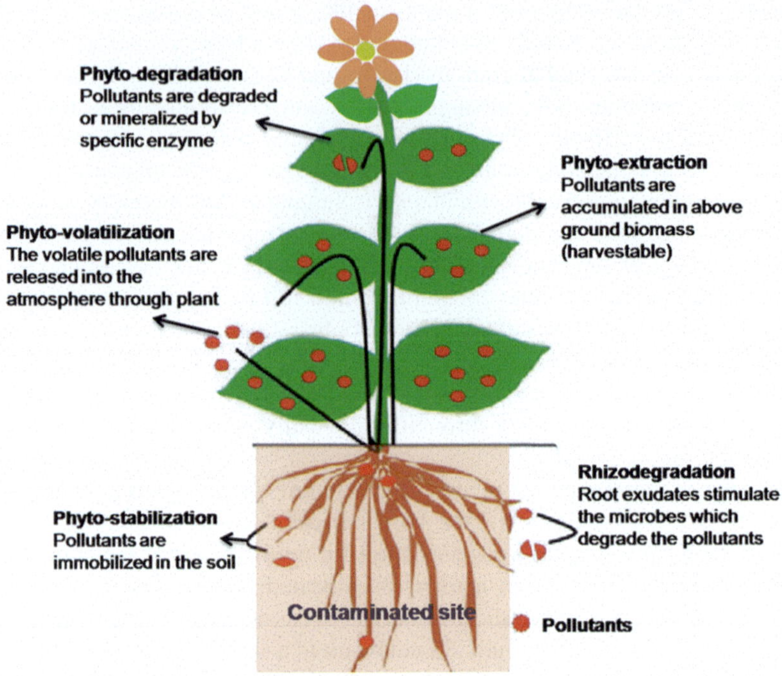

Fig. 24.2 Phytoremediation of contaminated sites

greater potential as a green alternative for the management of pollutants, especially in tropical areas where rich plant diversity offers greater scope for handling different contaminants (Salamanca et al. 2015). A variety of polluted waters can be phytoremediated, including sewage and municipal wastewater, agricultural runoff/drainage water, industrial wastewater, coal pile runoff, landfill leachate, mine drainage, and groundwater plumes. Similarly it is used for decontamination of soils polluted with heavy metals. The various techniques of phytoremediation are summarized in Table 24.3.

The phytoremediation can be carried out by using hyperaccumulators; they are the plants which can absorb high levels of contaminants with their roots and accumulate up to 1–5% of their total dry weight (Lee 2013). Currently about 500 species of plant species have been recognized to accumulate high amounts of heavy metals (Jaffré et al. 2013). The most effective plants used for removing heavy metals are *Thlaspi caerulescens* (alpine pennycress) and *Alyssum bertolonii*. The potential of plants for phytoremediation depends on their ability to accumulate excessive concentration of contaminants, a faster growth rate, high biomass production, resistance to pests and diseases, and tolerance to one or more metals (Lee 2013). Fast-growing trees like poplar can be most suitable for phytoremediation because of their high biomass and wide-ranging root system (Glick 2010). The harvested biomass could be processed by drying, ashing, or composting, and even some metals can be reclaimed from the ash. Even the plant biomass can be converted into bioenergy, e.g. biodiesel/biofuels, biogas, etc., through various energy recovery techniques (Van Ginneken et al. 2007).

The normal plants can also be used in phytoremediation together with the manipulation of soil conditions, to increase either the bioavailability or the mobilization of metals. The soil amendments like chelates will increase the bioavailability of metals, thereby increasing the plant uptake. In induced phytoextraction, EDTA is added to the soil, shortly before harvesting the plants. The application of EDTA to soil enhanced the accumulation of Pb in the shoots of plants such as *Zea mays* and *Brassica juncea* (Lee 2013). Similarly to enhance the biomass production by hyperaccumulating plants, plant growth-promoting microbes are used in recent times, and a wide range of bacteria has been recognized that successfully aid in phytoremediation (Glick 2010). Jing et al. (2014) isolated metal-resistant and growth-promoting bacteria from *Polygonum pubescens* grown in metal-contaminated soil and found to increase the plant growth and the accumulation of Cd, Pb, and Zn by *B. napus*.

The use of plants in association with plant growth-promoting bacteria for bioremediation of heavy and trace metals is well documented by several authors (Table 24.4). The major mechanisms in such interaction are the enhanced biomass production and nutrient uptake by plants primarily facilitated by microbes through the production of growth-promoting substances like phytohormones (auxins, cytokinins, and gibberellins); decreased production of ethylene in plants, thereby alleviating the stress and enhanced plant growth and reducing the metal phytotoxicity by biosorption and bioaccumulation mechanisms; and enhanced mobilization and availability of nutrients (N,P, and K) and metals by production of siderophores, organic acids, biosurfactants, etc. The metabolites released by such microbes can alter the uptake of heavy metals by plants through their effects on plant growth dynamics and directly through acidification, chelation, precipitation, immobilization, and oxidation-reduction reactions in the rhizosphere.

24.5.3 Use of Microbial Products

The microorganisms are known to produce bioactive compounds like biosurfactants, bioemulsifiers, enzymes, and exopolysaccharides (EPS) which have greater potential for bioremediation of hydrocarbons, pesticides, herbicides, heavy metals, etc. These compounds were produced by microbes to mediate solubilization of hydrophobic compounds present in their environment which are later used as substrates (Margesin and Schinner 2001). They are either extracellular compounds released into the environment or intracellular compounds localized on cell surfaces which become integrated into cell membrane. The presence of surface-active molecules on the microbial cell surface increases the hydrophobicity of the cell which enables them to survive in a hydrophobic environment (Satpute et al. 2010). A summary of use of different bioactive compounds to remove heavy metals and organic pollutants is given in Table 24.5.

The major problem concerning the bioremediation of contaminated marine sediments with polycyclic aromatic hydrocarbons (PAHs) is their limited bioavailability

Table 24.4 Examples of phytoremediation of heavy metals

Plant host	PGP bacteria	Special features of plant growth promotion	Effects on plants	Reference
Alyssum murale	*Microbacterium oxydans*	Mobilization of metals (Ni)	Increased uptake of Ni to the extent of 27 to 39%	Abou-Shanab et al. (2006)
Brassica juncea	*Pseudomonas* sp., *Bacillus* sp.	Production of ACC deaminase, siderophore, IAA, solubilization of P, K, and N fixation, mobilization and bioaccumulation of metals	Enhanced plant growth and above ground biomass, increased metal uptake Ni, Zn, Cr, and Cd	Zaidi et al. (2006) and Kumar et al. (2008)
	Azotobacter chroococcum			
	Bacillus megaterium			
	Enterobacter sp.			
	Pseudomonas aeruginosa			
	Enterobacter aerogenes			
	Psychrobacter sp.			
Brassica napus	*Bacillus* sp.	IAA, ACC deaminase, siderophore, metal mobilization	Root elongation, shoot and root dry weight, increased plant uptake of metals Cd, Pb, Cu	Sheng and Xia (2006)
	Pseudomonas tolaasii			
	P. fluorescens			
	Mycobacterium sp.	P solubilization		
	Arthrobacter sp.			
Ricinus communis	*Pseudomonas* sp.	Biosorption, mobilization, ACC deaminase, IAA, siderophore	Increased metal extraction (Zn)	Rajkumar and Freitas (2008)
	P. jessenii			
Lycopersicon esculentum	*Methylobacterium oryzae*	ACCD, phytohormone production, siderophore, biosorption of Ni and Cd, and mobilization of Cd and Pb	Increased plant growth, decreased uptake and translocation of Ni and Cd (phytostabilization)	Madhaiyan et al. (2007)
Oryza sativa	*Burkholderia* sp.			He et al. (2009)
	Pseudomonas sp.		Increased uptake and enhanced biomass production (phytoextraction)	

Salix caprea	*Pseudomonas* sp.	Siderophore, IAA, immobilization of metals	Increased plant leaf biomass, reduced uptake of Cd and Zn (phytostabilization)	Kuffner et al. (2008)
	Janthinobacterium lividum			
	Serratia marcescens			
	Flavobacterium sp., *Streptomyces* sp.			
Polygonum pubescens	*Enterobacter* sp.	IAA, siderophores, ACC deaminase, and P solubilization	Solubilization and increased uptake of Cd, Pb, and Zn	Jing et al. (2014)
	Klebsiella sp.			
Solanum nigrum	*Pseudomonas* sp.	Biosurfactants, siderophores, organic acids	Enhanced uptake of Cd, Zn, and Cu	Chen et al. (2014)
Zea mays	*Burkholderia* sp.	ACCD, siderophore, IAA, P solubilization	Increased above- and belowground biomass, increased uptake of Pb, Cd (phytoextraction)	Jiang et al. (2008)
Eichornia crassipes	Effluents of shrimp and Nile tilapia farms	P and N removal	82% removal of P and 42.7% removal of N	
Zizaniopsis bonariensis	Municipal wastewater	Nutrient removal	68% N and 79–81% P removed	
P. stratiotes, *E. crassipes*, *Salvinia rotundifolia*, *Cyperus alternifolius*,	Wastewater from metallurgic industry	Removal of heavy metals	Cr, 86%; Ni, 67%; and Fe, -95%	
P. elephantipes, *Thalia geniculata*, *Polygonum punctatum*, *Pontederia cordata*, *Pontederia rotundifolia*,				
Typha domingensis, *Aechmea distichantha*				
E. crassipes, *T. domingensis*, and *Pontederia cordata* L.	Metallurgic plant wastewater	Removal of heavy metals	Removal, Cr 88%, Ni 93%, Zn 98%	

Table 24.5 Bioremediation of heavy metals and organic pollutants by biomolecules

Type of contaminant	Microbial source	Type of biomolecule	Bioremediation yield	References
Crude oil	*Pseudomonas aeruginosa*	Biosurfactant—rhamnolipids	10–20% for various constituents	Nikolopoulou et al. (2013)
Crude oil	*Candida bombicola*	Biosurfactant—glycolipids	72–88%	Kang et al. (2010)
Heavy metals	*Pseudomonas aeruginosa*	Biosurfactant—di-rhamnolipids	78–92%	Juwarkar et al. (2008)
Heavy metals and total petroleum hydrocarbons (TPH)	*Bacillus subtilis*	Biosurfactant—lipopeptide	26 to 44% for HM and 64.5% for TPH	Singh and Cameotra (2013)
Heavy metals	*Microbial consortia*	Biosurfactants—lipopeptide, rhamnolipid, sophorolipid	16–60%	Mulligan et al. (2001)
Polycyclic and aromatic hydrocarbons and heavy metals	Plant-derived biosurfactant	Biosurfactant—saponin	Phetratrene (88%) Cd (76%)	Song et al. (2008)
Heavy metals	*Bacillus firm*, *Mycobacterium* sp.	Exopolysaccharides	Chromium, arsenic, cadmium	Raj et al. (2018)

due to low solubility of these compounds in water, while high-molecular-weight hydrocarbons tend to be absorbed on sediment particles reducing their bioavailability (Duran and Cravo-Laureau, 2016). Hence, surfactants are used for bioremediation of the hydrocarbons to transfer the hydrophobic compounds to the aqueous phase and make them available for the microorganisms to degrade. Thus the surfactants could increase mobility as well as bioavailability of the hydrocarbons. As the synthetic surfactants may cause environmental pollution, are harmful to the organisms, have low biodegradability, several efforts have been undertaken to identify biosurfactants for the enhancement of bioremediation processes in highly polluted environments (Radmann et al. 2015).

Biosurfactants produced by microorganisms are the biological active surface molecules with vast applications due to their specific versatile properties, minute toxicity, and biological acceptability (Karlapudi et al. 2018). Biosurfactants have higher surface activity and can withstand extreme conditions like acidity, salinity, temperature, etc. They are biodegradable and have demulsifying-emulsifying ability, anti-inflammatory potential, and antimicrobial activity. Biosurfactants are amphiphilic compounds consisting of hydrophilic polar moiety (as carbohydrates and proteins) and the hydrophobic moiety (alcohols or fatty acids). Because of the amphiphilic structure, biosurfactants not only have the ability to increase the surface area of the hydrophobic substances but also have the ability to change the property of cell surface of the microorganisms. The properties of biosurfactants such as substrate specificity, lower toxicity, higher biodegradability, better environmental compatibility, higher foaming, high selectivity, and ability to withstand extreme conditions make them ideal for bioremediation (Shekhar et al. 2015) of industrial wastes such as heavy metals, pesticides, petroleum hydrocarbons, etc.

Oil spills cause devastating effect on aquatic life in coastal and marine ecosystem. As the chemically synthesized surfactants are reported for their toxicity on aquatic organisms, it is unsuitable for remediation. The biosurfactants emulsify the hydrocarbons in water to form various mixtures and make them water soluble. Lecithins, rhamnolipids, and sophorolipids are the few surfactants which are found to be successful in the remediation of the oil contamination (Kang et al. 2010). The rhamnolipids were found to be more effective in the biodegradation of petroleum hydrocarbons (Dell'Anno et al. 2018). Kim et al. (1997) isolated a bacterium from a crude oil sample which produced a biosurfactant that had good emulsifying properties on crude oil and paraffin. It is found that biosurfactants produced from marine bacterium were capable of destroying the oil slicks floating on the surface of water and cause the dispersion of oil in water by forming a stable emulsion, thereby enhancing the rate of biodegradation. This ability of biosurfactants can be exploited for cleaning up the oil spills on shorelines and in the sea. Moreover, the biosurfactants produced by oil-degrading bacteria can enhance the assimilation of the hydrocarbons as well as the nutrients available in the environment. Like microbial consortia, the use of mixture of the biosurfactants including synthetic surfactants can exert a synergistic effect in stimulating the rate of degradation of hydrocarbons (PAHs) from the contaminated soil (Dell'Anno et al. 2018).

Some groups of microorganisms synthesize emulsifying agents that could help in hydrocarbon degradation; hence emulsifiers have been used for cleaning up the oil (Kim et al. 1997). Bioemulsifiers are primarily composed of mixtures of heteropolysaccharides, lipopolysaccharides, lipoproteins, and proteins that generally have a higher-molecular-weight compared to biosurfactants. As the bioemulsifiers mainly show emulsifying capability but no surface activity, their primary role is to create and stabilize the emulsions due to the high number of reactive groups exposed in their structures which allow an enhanced emulsion kinetic stability (Dell Anno et al. 2018). Alasan is an emulsifying agent produced by *A. radioresistens* KA53 by allowing aggregation of PAH oligomers and significantly increasing their solubility (Navon-Venezia et al. 1995). Similarly another compound 'emulsion' was isolated from common marine organism *Acinetobacter calcoaceticus* which has greater potential for bioremediation of hydrocarbon-polluted sites. Several marine bacterial strains are known to produce emulsifying agents giving greater scope for bioremediation.

The biosurfactants are also used for bioremediation of heavy metal contamination in soil, water, and sediments. The biosurfactants most commonly used in metal bioremediation include molecules with an electric charge, i.e. cationic and anionic biosurfactants that bind metals having opposite charge and removing them through desorption. Rhamnolipids are the group of biosurfactants extensively used to desorb metals such as cadmium, copper, lanthanum, lead, zinc, chromium, and nickel from contaminated sediments (Dell'Anno et al. 2018), and they preferentially complex toxic metals such as cadmium and lead, rather than cations such as calcium and magnesium (Frazer 2000). Other compounds like saponin, lipopeptide, etc. are also used for removal of heavy metals from soil and sediments. Few studies also indicated the potential of biosurfactants in removing both organic and inorganic pollutants from the contaminated soil and sediments. The biosurfactant 'saponin' is reported to be effective for the removal of cadmium (76%) and phenanthrene (88%) from contaminated soils (Song et al., 2008). Singh and Cameotra (2013) reported the washing of contaminated sediments with a 50 CMC biosurfactant solution removed 30–40% of metals and 65% of total petroleum hydrocarbons present in the sediments. Nowadays, extracellular polysaccharides or exopolysaccharides (EPS) are bacterial by-products with high-molecular-weight compounds which are extensively used for bioremediation of water and soil tainted with heavy metals; EPS comprises of polysaccharides, proteins, nucleic acids, lipids, uronic acid, and organic and inorganic compounds that promote the attachment of bacterial cells to solid surfaces. They are acidic in nature due to the presence of uronic acid, pyruvic acid, and/or sulphate and phosphate ions. Because of their acidic nature, these substances tolerate heavy metal stress and are preferable to heavy metal remediation (Raj et al. 2018).

24.6 Advantages of Bioremediation and the Future

When compared to the conventional approaches of landfills and cap and contain method, bioremediation is more efficient and suitable for large areas. Similarly bioremediation has clear advantages than the chemical or physical methods which are suitable for only small sites. From the compilation of works carried out elsewhere in the world, the following advantages are reported for bioremediation methods at the present state of research on this aspect.

- Bioremediation is a natural process as an acceptable waste treatment process for contaminated material such as soil. Comparatively this takes lesser time to clean up the contaminated sites.
- The microbes used for microbial remediation are able to degrade the contaminants and increase in numbers when the contaminant is present. When the contaminant is degraded, the biodegradative population starts declining. The residues for the treatment are usually harmless product including water, carbon dioxide, and cell biomass.
- Bioremediation requires less effort and can be carried out on-site, often without causing a major disruption of normal activities. This also eliminates the need to transport quantities of waste off-site and the potential threats to human health and the environment that can arise during transportation.
- Bioremediation helps in complete destruction of the pollutants, many of the hazardous compounds can be transformed to harmless products, and this feature also

eliminates the chance of future liability associated with treatment and disposal of contaminated material.

- At the same time, the bioremediation methods do not use any dangerous chemicals. Nutrients especially fertilizers are added to make active and fast microbial growth and commonly used on lawns and gardens. Because bioremediation changes harmful chemicals into water and harmless gases, the harmful chemicals are completely destroyed.
- It is applied in a cost-effective process as it costs less than the other conventional methods that are used for cleanup of hazardous waste.
- The method is simple, less labour intensive, eco-friendly, nonintrusive, potentially allowing for continued site use, and sustainable. The methods can also be used by coastal communities and marginal farmers for cleaning up their contaminated land and water.

In spite of their advantages, there are some concerns that the products of biodegradation may be more persistent or toxic than the parent compound. This needs to be evaluated on a long-term basis and continuously monitored for its performance or toxic effect on the environment. It has been often observed that biological processes are highly specific. Important site factors required for success include the presence of metabolically capable microbial populations, suitable environmental growth conditions, and appropriate levels of nutrients and contaminants. The same condition needs not to be found everywhere; therefore, it is difficult to extrapolate from bench- and pilot-scale studies to full-scale field operations. This aspect needs some research attention, and the alternate option for large-scale application should be explored to harness the benefit of bioremediation methods.

Moreover, research is needed to develop and engineer bioremediation technologies that are appropriate for sites with complex mixtures of contaminants that are not evenly dispersed in the environment because contaminants may be present as solids, liquids, and gases. Another important feature of bioremediation is the uncertain regulatory provision for its production and use on large scale. Therefore, it should be defined clearly what constitute bioremediation and clean environment besides quantifiable standards for different pollutants and residues of microbial products released into the environment. An independent institution/authority needs to be established to arrive at a policy decision based on the research outcome.

24.7 Conclusion

Human activities have increased the level of contamination and type and amount of pollutant in coastal areas as they harbour several industries, support tourism activities, and give space for urbanization. Further, studies have shown that the coastal regions receive pollutants from faraway places through drainage systems. Such situations pose severe threat to coastal ecosystems, human wellbeing, and his future socio-economic interest in the region. Bioremediation offers greater scope to reclaim and utilize the land and get rid of the environment from the pollutants. From the literature, bioremediation is understood as a process which uses living organisms (both micro and macro organisms) and their products to remove contaminants from the soil, water, and the surroundings.

In general, microorganisms, higher plants, and microbial products such as enzymes, polysaccharides, and surfactants are used in bioremediation. These methods are more eco-friendly, cost-effective, green, and efficient technology when compared to traditional methods of chemical treatment or physical removal of hazardous compounds from contaminated site. Most importantly the bioaugmentation and biostimulation can be combined and could be effectively utilized for in situ remediation of oil spills and chronically contaminated sites. When compared to the conventional approaches of landfills and cap and contain method, bioremediation is more efficient and suitable for large areas. But it can be directly upscaled to large areas.

In order to improve the efficiency of bioremediation, there should be regulatory mechanism with supervision to ensure safety of use and to get desired benefit; bioremediation should be followed by restoration process in coastal mines and quarry spoils; research should address the research gap and the evolving challenges in the future.

References

Abou-Shanab RAI, Angle JS, Chaney RL (2006) Bacterial inoculants affecting nickel uptake by Alyssum murale from low, moderate and high Ni soils. Soil Biol Biochem 38:2882–2889

Adams GO, Tawari-Fufeyin P, Igelenyah E (2014) Bioremediation of spent oil contaminated soils using poultry litter. Res J Eng Appl Sci 3(2):124–130

Adams GO, Tawari Fufeyin P, Okoro SE, Igelenyah E (2015) Bioremediation, biostimulation and Bioaugmention: a review. Int J Environ Bioremediation Biodegrad 3(1):28–39

Agarry S, Latinwo GK (2015) Biodegradation of diesel oil in soil and its enhancement by application of bioventing and amendment with brewery waste effluents as biostimulation – bioaugmentation agents. J Ecol Eng 16:82–91

Alexander M (1994) Biodegradation and bioremediation. Academic, San Diego

Alisi C, Musella R, Tasso F, Ubaldi C, Manzo S, Cremisini C, Sprocati AR (2009) Bioremediation of diesel oil in a co-contaminated soil by bioaugmentation with a microbial formula tailored with native strains selected for heavy metals resistance. Sci Total Environ 407:3024–3032. https://doi.org/10.1016/j.scitotenv.2009.01.011

Atagana HI (2008) Compost bioremediation of hydrocarboncontaminated soil inoculated with organic manure. Afr J Biotechnol 7(10):1516–1525

Cadee GC, Boon JP, Fischer CV, Mensink BP, Tjabbes CC (1995) Why the whelk Buccinum undatum has become extinct in the Dutch Wadden Sea. Neth J Sea Res 34:337–339

Catania V, Santisi S, Signa G, Vizzini S, Mazzola A, Cappello S, Yakimov MM, Quatrini P (2015) Intrinsic bioremediation potential of a chronically polluted marine coastal area. Mar Pollut Bull 99:138–149

Cheevaporn V, Menasveta P (2003) Water pollution and habitat degradation in the Gulf of Thailand. Mar Pollut Bull 47(1–6):43–51

Chen L, Luo S, Li X, Wan Y, Chen J, Liu C (2014) Interaction of Cd hyperaccumulator Solanum nigrum L. and functional endophyte Pseudomonas sp. Lk9 on soil heavy metals uptake. Soil Biol Biochemist 68:300–308

Chikere CB (2012) Culture-independent analysis of bacterial community composition during bioremediation of crude oil-polluted soil. Br Microbiol Res J 2(3):187–211

Chou LM (2009) The east Asian seas state of the marine environment report. UNEP EAS/RCU, Bangkok

Creel L (2003) Ripple effects: population and coastal regions, Making the Link, Population reference Bureau. https://www.prb.org/wp-ontent/uploads/2003/09/RippleEffectsEng.pdf. Accessed on 16 July 2018

Dadrasnia A, Agamuthu P (2013) Potential biowastes to remediate diesel contaminated soils. Global NEST J 15(4):474–484

Davies AG (1978) Pollution studies with marine plankton; Part II. Heavy metals. Adv Mar Biol 15:381–508

Dell' Anno F, Sansone C, Ianora A, Dell Anno A (2018) Biosurfactant-induced remediation of contaminated marine sediments: current knowledge and future perspectives. Mar Environ Res 137:196–205

Díaz-Ramírez I, Escalante-Espinosa E, Schroeder RA, FócilMonterrubio R, Hugo Ramírez-Saad (2013) Hydrocarbon biodegradation potential of native and exogenous microbial inocula in Mexican tropical soils. Biodegradation of hazardous and special products. https://doi.org/10.5772/56233

Duran R, Cravo-Laureau C (2016) Role of environmental factors and microorganisms in determining the fate of polycyclic aromatic hydrocarbons in the marine environment. FEMS Microbiol Rev 40:814–830

Eyre LA (1990) Forestry and watershed management, mineral development planning and production national strategy consultation workshop, 25–27 April 1990, Kingston, Jamaica

Frazer L (2000) Lipid lather removes metals. Environ Health Perspect 108:320–323

Gallego JR, Fernandez JR, Diez-Sanz F, Ordonez S, Sastre H, Gonzalez-Rojas E, Pelaez AI, Sanchez J (2007) Bioremediation for shoreline cleanup: in situ vs. on-site treatments. Environ Eng Sci 24:493–504. https://doi.org/10.1089/ees

Gentry TJ, Rensing C, Pepper IL (2004) New approaches for bioaugmentation as a remediation technology. Crit Rev Environ Sci Technol 34:447–494

Glick BR (2010) Using soil bacteria to facilitate phytoremediation. Biotechnol Adv 28:367–374

Goldstein RM, Mallory LM, Alexander M (1985) Reasons for possible failure of inoculation to enhance biodegradation. Appl Environ Microbiol 50:977–983

Gomez ED (1990) State of the marine environment in the east Asian seas region. UNEP Regional Seas Reports and Studies No 126

Grillo V, Parsons ECM, Shrimpton JH (2001) A review of sewage pollution and cetaceans: a Scottish perspective. Paper presented to the Scientific Committee at the 53rd Meeting of the International Whaling Commission, 3–16 July 2001, London

Hamzah A, Chia-Wei P, Pek-Hoon Y, Nurul H (2014) Oil palm empty fruit bunch and sugarcane bagasse enhance the bioremediation of soil artificially polluted by crude oil. Soil Sediment Contam Int J 23(7):751–762

He LY, Chen ZJ, Ren GD, Zhang YF, Qian M, Sheng XF (2009) Increased cadmium and lead uptake of a cadmium hyperaccumulator tomato by cadmium-resistant bacteria. Exotoxicol Environ Saf 72:1343–1348

Islam S, Tanaka M (2004) Impacts of pollution on coastal and marine ecosystems including coastal and marine fisheries and approach for management: a review and synthesis. Mar Pollut Bull 48:624–649

Jaffré T, Pillon Y, Thomine S, Merlot S (2013) The metal hyperaccumulators from New Caledonia can broaden our understanding of nickel accumulation in plants. Front Plant Sci 4:1–7

Jiang CY, Sheng XF, Qian M, Wang QY (2008) Isolation and characterization of a heavy metal resistant Burkholderia sp. from heavy metal-contaminated paddy field soil and its potential in promoting plant growth and heavy metal accumulation in metal polluted soil. Chemosphere 72:157–164

Jing YX, Yan JL, He HD, Yang DJ, Xiao L, Zhong T, Yuan M, Cai XD, Li SB (2014) Characterization of bacteria in the rhizosphere soils of *polygonum pubescens* and their

potential in promoting growth and Cd Pb, Zn uptake by *Brassica napus*. Int J Phytorem 16:321–333
Juwarkar AA, Dubey KV, Nair A, Singh SK (2008) Bioremediation of multi-metal contaminated soil using biosurfactant a novel approach. Indian J Microbiol 48:142–146
Kang SW, Kim YB, Shin JD, Kim EK (2010) Enhanced biodegradation of hydrocarbons in soil by microbial biosurfactant. Sophorolipid App Biochem Biotechnol 160:780–790
Karlapudi AP, Venkateswarulu TC, Tammineedi J, Kanumuri L, Ravuru BK, Dirisala VR, Kodali VP (2018) Role of biosurfactants in bioremediation of oil pollution-a review. Petroleum 4:241–249
Kim PD, Oh K, Kim SY, Kim JH (1997) Relationship between emulsifying activity and carbohydrate backbone structure of emulsan from Acinetobacter calcoaceticus RAG-1. Biotechnol Lett 19:457–459
Ko JY, Day JW (2004) A review of ecological impacts of oil and gas development on coastal ecosystems in the Mississippi delta. Ocean Coast Manag 47:597–623
Kuffner M, Puschenreiter M, Wieshammer G, Gorfer M, Sessitsch A (2008) Rhizosphere bacteria affect growth and metal uptake of heavy metal accumulating willows. Plant Soil 304:35–44
Kumar K, Singh N, Behlh HM, Srivastava S (2008) Influence of plant growth promoting bacteria and its mutant on heavy metal toxicity in Brassica juncea grown in fly ash amended soil. Chemosphere 72:678–683
Kumar A, Bisht BS, Joshi VD, Dhewa T (2011) Review on bioremediation of polluted environment: a management tool. Int J Environ Sci 1:1079–1093. https://goo.gl/P6Xeqc
Leahy JG, Colwell RR (1990) Microbial-degradation of hydrocarbons in the environment. Microbiol Rev 54:305–315
Lee JH (2013) An overview of phytoremediation as a potentially promising technology for environmental pollution control. Biotechnol Bioprocess Eng 18:431–439
Lee TH, Byun IG, Kim YO, Hwang IS, Park TJ (2006) Monitoring biodegradation of diesel fuel in bioventing processes using in situ respiration rate. Water Sci Technol 53:263–272
Lin Y, Cai LX (2008) PAH-degrading microbial consortium and its pyrenedegrading plasmids from mangrove sediment samples in Huian. Chin Mar Poll Bull 57:703–706
Lu J (2003) Marine oil spill detection, statistics and mapping with ERS SAR imagery in South-East Asia. Int J Remote Sens 24:3013–3032
Madhaiyan M, Poonguzhali S, Sa T (2007) Metal tolerating methylotrophic bacteria reduces nickel and cadmium toxicity and promotes plant growth of tomato (Lycopersicon esculentum L.). Chemosphere 69:220–228
Margesin R, Schinner F (2001) Bioremediation (natural attenuation and biostimulation) of diesel-oil-contaminated soil in an alpine glacier skiing area. Appl Environ Microbiol 67:3127–3133
Mench M, Schwitzguebel JP, Schroeder P, Bert V, Gawronski S, Gupta S (2009) Assessment of successful experiments and limitations of phytotechnologies: contaminant uptake detoxification and sequestration, and consequences for food safety. Environ Sci Pollut Res 16:876–900
Mulligan CN, Yong RN, Gibbs BF (2001) An evaluation of technologies for the heavy metal remediation of dredged sediments. J Hazard Mater 85:145–163
Navon-Venezia S, Zosim Z, Gottlieb A, Legman R, Carmeli S, Ron EZ, Rosenberg R (1995) Alasan, a new bioemulsi er fromfi Acinetobacter radioresistens. Appl Environ Microbiol 61:3240–3244
Nikolopoulou M, Kalogerakis N (2009) Biostimulation strategies for fresh and chronically polluted marine environments with petroleum hydrocarbons. J Chem Technol Biotechnol 84:802–807. https://doi.org/10.1002/jctb.2182
Nikolopoulou M, Pasadakis N, Norf H, Kalogerakisa N (2013) Enhanced ex situ bioremediation of crude oil contaminated beach sand by supplementation with nutrients and rhamnolipids. Mar Pollut Bull 77:37–44
Odokuma LO, Dickson AA (2003a) Bioremediation of a crude oil polluted tropical rain forest soil. Glob J Environ Sci 2(1):29–40
Odokuma LO, Dickson AA (2003b) Bioremediation of a crude oil polluted tropical mangrove environment. J Appl Sci Environ Manag 7:23–29

Orji FA, Abiye AI, Dike EN (2012) Laboratory scale bioremediation of petroleum hydrocarbon – polluted mangrove swamps in the Niger Delta using cow dung. Malays J Microbiol 8(4):219–228

Pandey VC (2012) Phytoremediation of heavy metals from fly ash pond by *Azolla caroliniana*. Ecotox Environ Safe 82:8–12

Prince RC (1997) Bioremediation of marine oil spills. Trends Biotechnol 15:158–160

Radmann EM, de Morais EG, de Oliveira CF, Zanfonato K, Vieira Costa JA (2015) Microalgae cultivation for biosurfactant production. Afr J Microbiol Res 9:2283–2289

Raj KK, Sardar RU, Bhargavi E, Devi I, Bhunia B, Tiwari ON (2018) Advances in exopolysaccharides based bioremediation of heavy metals in soil and water. A critical review. Carbohydrate Polymers 199:353–364

Rajkumar M, Freitas H (2008) Influence of metal resistant-plant growth-promoting bacteria on the growth of Ricinus communis in soil contaminated with heavy metals. Chemosphere 71:834–842

Rajkumar M, Sandhya S, Prasad MNV, Freitas H (2012) Perspectives of plant-associated microbes in heavy metal phytoremediation. Biotechnol Adv 30:1562–1574

Rawlins BG, Ferguson AJ, Chilton PJ, Arthurtons RS, Rees JG, Baldock JW (1998) Review of agricultural pollution in the Caribbean with particular emphasis on small island developing states. Mar Pollut Bull 36(9):658–668

Ritter WF, Scarborough RW (1995) A review of bioremediation of contaminated soils and groundwater. J Environ Sci Health A: Environ Sci Eng Toxicol 30(2):333–357. https://doi.org/10.1080/10934529509376204

Rosenberg E, Legmann R, Kushmaro A, Taube R, Adler E, Ron EZ (1992) Petroleum bioremediation—a multiphase problem. Biodegradation 3:337–350

Roy A, Dutta A, Pal S, Gupta A, Sarkar J, Chatterjee A, Saha A, Sarkar P, Sar P, Kazy SK (2018) Biostimulation and bioaugmentation of native microbial community accelerated bioremediation of oil refinery sludge. Bioresour Technol 253:22–32

Salamanca EJP, Madera-Parra CA, Avila-Williams CA, Rengifo-Gallego AL, Ríos DA (2015) Phytoremediation using terrestrial plants. In: Ansari A, Gill S, Gill R, Lanza G, Newman L (eds) Phytoremediation. Springer, Cham. https://doi.org/10.1007/978-3-319-10969-5_25

Santos HF, Carmo FL, Paes JES, Rosado AS, Peixoto RS (2011) Bioremediation of mangroves impacted by petroleum. Water Air Soil Pollut 216:329–350

Sarkar D, Ferguson M, Datta R, Birnbaum S (2005) Bioremediation of petroleum hydrocarbons in contaminated soils: comparison of biosolids addition, carbon supplementation, and monitored natural attenuation. Environ Pollut 136:187–195. https://doi.org/10.1016/j.envpol.2004.09.025

Satpute SK, Banat IM, Dhakephalkar PK, Banpurkar AG, Chopade BA (2010) Biosurfactants, bioemulsifiers and exopolysaccharides from marine microorganisms. Biotechnol Adv 28:436–450

Sayara T, Borràs E, Caminal G, Sarrà M, Sánchez A (2011) Bioremediation of PAHs-contaminated soil through composting: influence of bioaugmentation and biostimulation on contaminant biodegradation. Int Biodeterior Biodegradation 65(6):859–865

Shekhar S, Sundaramanickam A, Balasubramania T (2015) Biosurfactant producing microbes and their potential applications: a review. Crit Rev Environ Sci Technol 45:1522–1554

Sheng XF, Xia JJ (2006) Improvement of rape (Brassica napus) plant growth and cadmium uptake by cadmium-resistant bacteria. Chemosphere 64:1036–1042

Singh AK, Cameotra SS (2013) Efficiency of lipopeptide biosurfactants in removal of petroleum hydrocarbons and heavy metals from contaminated soil. Environ Sci Pollut Res 20:7367–7376

Song SS, Zhu LZ, Zhou WJ (2008) Simultaneous removal of phenanthrene and cadmium from contaminated soils by saponin, a plant-derived biosurfactant. Environ Pol 156:1368–1370

Spalding MD, Ravilious C, Green EP (2001) World atlas of coral reefs. UNEP-WCMC and the University of California Press, Berkley

Taccari M, Milanovic V, Comitini F, Casucci C, Ciani M (2012) Effects of biostimulation and bioaugmentation on diesel removal and bacterial community. Int Biodeterior Biodegradation 66(1):39–46

Thompson IP, van der Gast CJ, Ciric L, Singer AC (2005) Bioaugmentation for bioremediation: the challenge of strain selection. Environ Microbiol 7:909–915

Todd AP, Ong X, Chou LM (2010) Impacts of pollution on marine life in south east Asia. Biodivers Conserv 19:1063–1082

Tyagi M, da Fonseca MMR, de Carvalho CCCR (2010) Bioaugmentation and biostimulation strategies to improve the effectiveness of bioremediation processes. Biodegradation. https://doi.org/10.1007/s10532-010-9394-4

Ullah A, Heng S, Munis MFH, Yang X, Fahad S (2015) Phytoremediation of heavy metals assisted by plant growth promoting (PGP) bacteria: a review. Environ Expt Bot 117:28–40

Van Ginneken L, Meers E, Guisson R, Ruttens A, Elst K, Tack FM, Vangronsveld J, Diels L, Dejonghe W (2007) Phytoremediation for heavy metal contaminated soils combined with bioenergy production. J Environ Eng Landsc Manag 15:227–236

Ventikos NP, Sotiropoulos FS (2014) Disutility analysis of oil spills: graphs and trends. Mar Pollut Bull 81:116–123. https://doi.org/10.1016/j.marpolbul.2014.02.007

Vidali M (2001) Bioremediatin. An overview. Pure Appl Chem 73(7):1163–1172

Wu RSS (1995) The environmental impact of marine fish culture: towards a sustainable future. Mar Pollut Bull 31:159–166

Wu T, Xie WJ, Yi L, Li XB, Yang BH, Wang J (2012) Surface activity of salt-tolerant Serratia spp. and crude oil biodegradation in saline soil. Plant Soil Environ 58(9):412–416

Zaidi S, Usmani S, Singh BR, Musarrat J (2006) Significance of Bacillus subtilis strain SJ-101 as a bioinoculant for concurrent plant growth promotion and nickel accumulation in *Brassica juncea*. Chemosphere 64:991–997

Balancing Development and Environmental Impact in the Coastal Regions

25

Abstract

Coastal regions are vital component of strategic plan to accelerate economic growth in many of the tropical coastal countries. Nevertheless coastal development is often accompanied by increasing competition between different activities for space within the coastal environment. Consequently the coastal resources are exploited unsustainably causing environmental degradation, loss of biodiversity, pollution of soil and water bodies, loss of recreational values, and increase in the vulnerability of coastal areas. This necessitates balancing developmental activities with the normal functioning of coastal ecosystems to promote sustainable development. In this context, interaction among various sectors and activities may be categorized as being synergistic, complementary, competitive, or antagonistic in the developmental process based on the environmental standards and ecological sustainability. The harmonizing strategy involving development and environmental quality aims at maximizing synergistic and complementary interactions, and minimizing competitive and antagonistic ones at the same time regulates different developmental activities to minimize its environmental impact. This chapter discusses these aspects in details with reference to primary, secondary, and service activities taking place in the tropical coastal region.

Keywords

Developmental activities · Coastal degradation · Promote synergy · Regulation

25.1 Introduction

The coastal regions all over the world have been used extensively and increasingly for large number of developmental and other economic activities. This includes human settlement, agriculture, fisheries, forestry, trade, industry, recreation, and other services. These multiple uses are not always compatible and may result in a

V. Ayyam et al., *Coastal Ecosystems of the Tropics - Adaptive Management*,
https://doi.org/10.1007/978-981-13-8926-9_25

wide array of problems for resource users and decision-makers. The coastal sea presents problems related to transport, fishing, dumping, mining, etc., stemming from an intensification and diversification of activities in the coastal and inland region. There is over-exploitation of fishery resources, pollution by dumping, and an array of conflicts due to the interaction among user groups in congested marine areas (Clark 1991).

In addition, coastal regions receive and concentrate pollutants and other negative consequences of development activities taking place in the hinterlands. In particular, the flow of rivers has often been severely modified for irrigation purposes, and they carry to the coastal areas the pollutants coming from inland agricultural and industrial activities. Such unplanned and unsustainable development often result in over-exploitation of resources, pollution, alteration of natural systems and conflicts between the users for limited coastal and shoreline space. At the same time, the potential of coastal region for providing livelihood to coastal communities and accelerate economic growth in the tropical coastal countries cannot be downgraded. Therefore, it is imperative to understand the problems of resource use for various developmental activities and its impacts on the coastal environment to take suitable remedial measures. This may lead to balancing developmental activities in tandem with the normal functioning of different coastal ecosystems and minimal impact on the environment. There should be general guiding principles to promote various developmental activities, while suitable environmental regulations should be in place to restrict the detrimental activities.

25.2 Drivers of Change

The natural coastal environment is being modified to serve the human needs, while some of the natural events pose serious hazards by virtue of geographical location and terrain condition of the coastal regions. There are three major drivers of environmental change taking place in the coastal areas, viz. economic growth, population growth, and climate change. These drivers exert range of pressures on the coastal environment. Events associated with variations in climate have become major pressures in the tropical coastal areas, and concern about preparing for possible future impacts of climate change and variability has been a strong driver of developmental planning inclusive of adaptation responses. Concern about changes in the size and composition of coastal populations has also been growing for several decades but not addressed adequately. Urbanization and coastal development for farming, services, and industry are a major pressure on terrestrial and marine biodiversity and environmental quality, water resources, air quality, and cultural and natural heritage. These drivers are associated with the specific developmental activity and its consequential effect on the environment. But the magnitude and type may vary with locations, regions, and countries as the physical and development environments are different.

25.3 Developmental Footprints on the Coastal Environment

Coastal ecosystems are extremely sensitive to environmental perturbations particularly human-induced changes. Nearly two third of the human population in the tropics lives in coastal areas, and their activities exert significant pressure on coastal ecosystems. This happens primarily in the form of land use change and pollution. This affects the ecosystem process, services, and natural resources in the coastal region. Sometimes this becomes irreversible and beyond the recovery of natural ecosystem resulting in reduced ecosystem goods and services. Habitat loss due to land use change affects the biodiversity in terms of its number and composition. Pollution has much serious consequences in the aquatic life. Contaminants of major concerns include persistent organic pollutants, nutrients, oils, radio nuclides, heavy metals, pathogens, sediments, litters, debris, etc. (Williams 1996). The effects are exaggerated by the climate change and other accelerated natural process to produce significant negative impact on the coastal ecosystem (Fig. 25.1). In general, pollution directly and indirectly impacts the fish and other aquatic animal populations resulting in economic loss to both fisherman and other stakeholders. The causes of ecosystem or natural resource damage in coastal areas are documented and analysed globally as developmental footprints on ecosystem process, natural resources, and critical habitats. These are discussed in the following subsections.

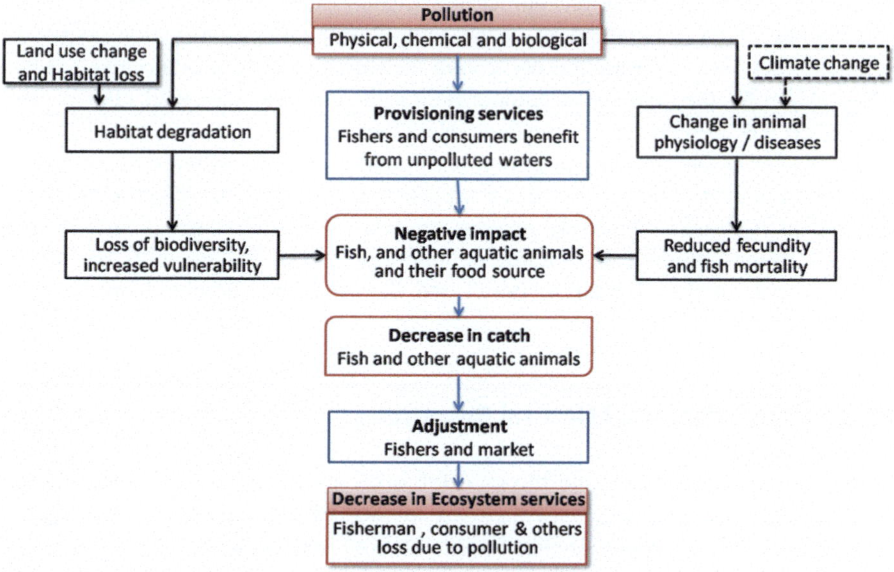

Fig. 25.1 Schematic representation showing effect of pollution and land use change getting translated into reduced ecosystem services

25.3.1 Impact on Coastal Natural Process

In the coastal areas, there is a dynamic equilibrium between different erosional and accretion process which are conditioned by the prevailing climatic factors. Similarly there is a flow of energy and nutrient between and within different ecosystems and its components. But the dynamics of coastal landscapes and the process that determine the nutrient and energy flows in coastal areas are increasingly being modified by human activities, in particular those that affect water flows (dams, increased water extraction, deviation of rivers) and erosion (deforestation). This prevents or slows down vertical accretion, thus aggravating saltwater intrusion and impairing drainage conditions in riverine, delta, or estuarine areas. It reduces or blocks sediment supply to the coast itself, which may give rise to the retreat of the coastline through wave erosion (FAO 1998).

Coastal areas are also prone to threats from natural causes such as tidal surges and sea level rise. The relative change of sea and land is the main factor deciding the impact of sea waves. Many cities, for instance, suffer land subsidence as a result of excessive groundwater withdrawal. This may be compounded by sea level rise, when the rates of subsidence exceed the rate of sea level rise. The effect is felt on the erosion patterns and damage to coastal infrastructure, salinization of wells, suboptimal functioning of the sewage system of coastal cities, loss of littoral ecosystems, and loss of biotic resources. In coastal areas particularly in islands, factors such as modified ocean circulation patterns, climate change in the catchment basin, and change in coastal climate should be taken into account while planning for developmental activities. These processes are also affected by the pollution of water bodies. Knowledge of the pollution sources and impacts on ecosystems is important not only for the better understanding of the ecosystem responses to various types of pollutants but also to formulate preventive measures (Shahidul Islam and Tanaka 2004). Many of the sources of aquatic pollutions are generally well known, and huge effort has been devoted to the issue. However, new concepts and ideas on environmental pollution are emerging (e.g. biological pollution) with a corresponding need for an update of the knowledge. For these reasons coastal developmental strategy and plan should effectively address the environmental issues arising out of human activities, while it should have collaborative efforts to cope with the natural process accelerated by human activities and the predicted climate change having global impact.

25.3.2 Degradation of Natural Resources

In the coastal areas, biotic and physical (abiotic) resources both of renewable or nonrenewable are extensively being used to meet the human needs. Biological resources include all economically important flora and fauna in the coastal areas which may be abundant or endangered. These resources are linked in a complex way through which the flow of energy and matter occurs. Therefore, degradation or over-exploitation of one resource will have its consequential impact on the environment and the other affecting the flow pattern. The most important coastal resources affected by human activities are land, water, and biodiversity; in turn it will degrade or affect the protective, provisioning, and regulatory services of the coastal ecosystems.

25.3.2.1 Land Use Change and Modifications

The impact of human activities on coastal land is felt by way of diversion and degradation. This includes urban expansion, recreational activities, mining, aquaculture, industries, landfills, and other developmental activities. For the nearshore environment, issues relevant to coasts include modification of coastal habitats by processes such as coastal urban development, catchment development, marinas, breakwaters, island reclamation projects, coastal and nearshore mining and dredging, harbours, and shipping channels. A particular concern is the incremental nature of coastal development, which reduces the abundance of native vegetation and breaks down connectivity among remnant habitat patches. Land modification results in change of river flow and enhanced erosion with a consequential impact on the availability of fresh water, wetlands, sediment and nutrient flow to estuaries and coastal environments.

25.3.2.2 Water Quality and Pollution

The impacts of human activities on coastal water are very prominent which include pollution of coastal rivers, estuary, nearshore, and other wetland water bodies, desalination, seawater intrusion, and impacts of water abstraction (removal) on flora and fauna. However, moisture stress/drought and excessive exploitation of groundwater have increased tensions over water use, and this is likely to be an important consideration for coastal management in the future. The discharge of toxic effluents from industries located far away from the coast into the water bodies affects the quality of water in the natural environment. More fertilizers are being used in the coastal agricultural lands to allow farmers to grow crops more quickly. When rainwater carries these fertilizers into the oceans, they can stimulate algal blooms, also known as red tides. These blooms can be toxic to fish, shellfish, and humans.

In Bangladesh about 9000 metric tons of different pesticides and more than 2 million metric tons of fertilizers are used annually, and about 1800 tons year^{-1} of pesticide residues are added to the coastal waters every year (Shahidul Islam and Tanaka 2004). Large volume of animal manure are discharged from intensive livestock farms located near urban areas of Hanoi, Bangkok, and Manila with little or no treatment (Gerber and Menzi 2006). It was estimated that inputs from various sources into rivers in Cambodia, Malaysia, Thailand, and Vietnam result in the release of around 0.3 million t of nitrogen per year. In the Gulf of Mexico, due to more loading of nutrients, algal blooms (red tide effect) produced a secondary growth of bacteria that depleted the water of oxygen, killing many aquatic lives (UNEP/GRA 2006).

In Asia, Latin America, and Africa, 75–80% feed loss was recorded depending on the aquaculture systems and the degree of management applied. Intense cultivation of fish, shellfish, and finfish in coastal waters can be a source of environmental disturbance associated with unnaturally high concentrations and deposition of organic matter that alters sedimentary processes and oxygen concentrations in the nearshore and other wetlands (Shahidul Islam and Tanaka 2004).

25.3.2.3 Biodiversity

For biodiversity in general, many species of plants and animals in the coastal areas are threatened by developmental and recreational activities. One of the major problems that occur in an impacted coastal ecosystem is the loss of biodiversity and the changes in the composition of species living in the coastal ecosystem. For example, there are 75 threatened species of fish and other aquatic organisms that are reported due to pollution around the Thailand coast (Chavalit and Siraprapha 2002). Destruction of coastal habitat often results in decrease in population and species diversity besides interference in breeding and migratory nature of birds. In some cases, changes to the species composition allow for the invasion of non-native or intrusive species that compete with the local populations for limited resources. Changes to the species composition also have the potential to alter the food supply and the balance of competition between species in an ecosystem. Often it results in the introduction of new weeds and pest species besides pathogens that disturb the ecosystem composition and functioning with the consequential economic loss though reduction in ecosystem productivity.

25.3.2.4 Loss of Protective Cover

The impact of human activity on vegetation in the coastal strip is highly variable from extraction of minor forest products and wood to removal of entire vegetation cover. This becomes detrimental beyond the natural replenishing capacity affecting the ecosystem services provided by the coastal vegetation. The loss of mangrove and littoral species resulted in the removal or reduction of protection function of these vegetations against sea surges and tsunamis. This was witnessed during the December 2004 Indian Ocean tsunami which devastated the South and Southeast Asian countries. Loss of protective cover also results in biodiversity and habitat loss and enhanced the sediment flow into the nearshore area affecting the coral ecosystem. In the absence of protective cover, storms not only damage the physical environment but also displace animals and disrupt the natural food web. Storms associated with El Niño are blamed for decreases in fish populations in some coastal regions of the Pacific Ocean. Coral bleaching is also attributed to storm action on some reefs. Therefore, considering the climate change and sea level rise, loss of protective cover significantly enhances the vulnerability of coastal communities.

25.3.2.5 Pollution from Fossil Fuel and Related Activities

Every year large volume of fossil fuel (oil) is added into the coastal and marine environment from various sources like oil spills, onshore drilling, discharge, transport, drainage, routine maintenance, smoke, natural seeps, etc. Approximately 1.3 million tons of petroleum enters the marine environment each year (Ventikos and Sotiropoulos 2014). Niger is one of the largest oil producer in Africa, but due to mismanagement and accidents, it was estimated that around 2.3 million barrels of oil spilled into the environment over the period of 1976–1996 with the largest spill of around 8.4 million US gallons of oil from an offshore well which blew out in January 1980 that spilled into the Atlantic Ocean damaging 340 ha of mangrove (Nwilo and Badejo 2005).

The transportation of crude oil across the seas also poses a severe risk of marine pollution. For instance, the South Asian Seas (SAS) region is situated in the middle of east-west shipping route in the Indian Ocean, through which a total of 525 million tons of crude oil pass through in a year, which accounts for about 25% of total world movement of crude oil by sea (SACEP 2017). The oil pollution is found in the Straits of Malacca, South China Sea, Gulf of Thailand, and the Jakarta-Manila and Singapore-Bangkok routes (Lu 2003). Oil pollution also occurs from deliberate operational discharge, automobiles, and street dust, and various industries contribute via terrestrial runoff and atmospheric deposition (Todd et al. 2010). Many substances in crude oil are toxic, and of a particular concern are PAHs, which are persistent in mangrove ecosystem due to the high organic content which enhances adsorption of these compounds. The oil pollution also occurs from coastal refineries through leakages, spills, breakages, etc. Besides, the destruction of pipelines results in oil spillage in the coastal environment. The other sources of anthropogenic oil pollution are due to incomplete fossil fuel combustion, the incorrect disposal of industrial wastes, diesel engine exhausts, aviation exhaust, and oil spills caused by oil exploration sites, oil tankers and pipelines (underwater and underground), spent marine lubricants, and storage tanks. Oil endangers fish hatcheries in coastal waters and contaminates the commercially valuable fish. The mortality and elimination of sea stars and sea urchins were reported by Smith 1970 due to diesel oil pollution.

25.3.2.6 Loss of Heritage

Human activities affect the natural and cultural heritage of the coastal region in several countries. It includes many important wetlands, places of importance in the traditional culture and practices of Indigenous people, buildings associated with areas of earlier civilization/settlements, historically important shipwrecks, threatened species and communities, and other places of natural heritage significance.

25.3.3 Effect of Primary Activities Supporting Livelihood

Over the years the coastal areas have witnessed increase in human activities either for livelihood or large industry and widened to exploit every natural resource. In the 'dry zone' of the coastal areas (landwards), agriculture, fisheries (aquaculture), and forestry are the major activities, and captured fisheries is the major activity in the 'wet zone' affecting the coastal ecosystems. Table 25.1 illustrates the major economic activities in agriculture, forestry, and fisheries subsectors that have an impact on environmental and social concerns within or outside the subsector in the coastal areas. The main consequences of resource use and its mismanagement in general are felt in ecosystem health, productivity, and sociocultural values. Some of these issues are intrasectoral, such as depletive/destructive resource uses (FAO 1998),

Table 25.1 Impact of agriculture, forestry, fisheries, and other activities on coastal areas

Area of impact/ subsector	Use or activity	Degradation process/ environmental impact	Impact of social/economic concern
Estuary, harbour, and inshore water quality impacts			
Agriculture, urban settlement, and others	Diversion of rivers for irrigation, industry, and urban use	Reduced water flow in rivers, increased estuary salinity, decreased estuary circulation	Decreased fish yields and loss of habitat
Agriculture	High use of pesticides	Toxic residues enter the estuaries and inshore waters	Decreased fish yields and death of critical fauna
	High use of fertilizers	Increased amount of nutrients in the discharge results in eutrophication	Decreased fish yields and stock
Agriculture and animal husbandry	Intensive cropping or grazing in the coastal and catchment	Erosion, river turbidity, sedimentation of habitat, floodplain deposition, and beaches covered with sediment	Silting of navigation channels, increased flood hazard, and decreased tourism value
Groundwater quality/quantity			
Agriculture/ coastal aquaculture	Withdrawal of groundwater at a greater rate than natural recharge	Saltwater intrusion of aquifer leading to increased salinity, contamination of groundwater	Reduction in the water available for use, risk to human health
Mangrove and other coastal wetland impacts			
Agriculture/ coastal aquaculture	Reclamation of mangrove for rice and aquaculture	Destruction of mangrove and wetlands	Reduced filtration capability, increased risk of shore erosion, flooding and storm damage, reduction in fish yields
Forestry	Mangrove harvesting for building materials and fuelwood	Harvesting at a level greater than the sustainable yield	Reduction or loss of rare or endangered species, reduction in fishery yield and non-timber forest products used by local communities
Agriculture	Draining of salt marsh and coastal wetlands	Lowering of land surface and accelerated rise in sea level	Increased frequency and extent of beach erosion, flooding, increased salinity in coastal areas
Coral reef and atoll impacts			
Agriculture/ forestry	Irresponsible agricultural, tourism, and/or forestry practices in coastal areas	Watershed erosion, turbidity, and siltation of coral reefs	Coral bleaching and decreased fish yields and recreational value

(continued)

Table 25.1 (continued)

Area of impact/ subsector	Use or activity	Degradation process/ environmental impact	Impact of social/economic concern
Fisheries	Fishing with dynamite, muroami fishing	Coral reef destruction	Decreased fish yields and recreational value
	Intensive localized fishing effort	Harvesting targeted species at a level greater than sustainable yield	Decreased yield of associated fish species
Beach, dune, and delta impacts			
Animal husbandry	Grazing of livestock	Destruction or removal of dune stabilizing vegetation	Dune migration on to agricultural and inland, risk of sea surges, and erosion increases
Agriculture and urban centres	Diversion of rivers for irrigation, drinking, recreation, and other activities	Decreased supply of beach material to the shoreline	Initiation or increase of shoreline erosion, decreased productivity
Social impacts			
Aquaculture	Commercial shrimp farm development	Acquisition of land used by peasant/locals, unfair contract arrangements	Creation of landless people, lack of equity, social conflict
Agriculture	High use of pesticides	Toxic pollution of inshore and wetlands	Potential human consumption of toxic fish due to biomagnifications
	Intensive use of nutrients	Eutrophication	Reduction in recreation attraction and human habitats

Modified from FAO (1998)

while others are often trans-sectoral, such as loss of habitat and environmental degradation. Some of the major human activities are fishing, aquaculture, agriculture, mining, oil and gas exploitation, exploitation of shell, sediment and mineral nodules, waste disposal, industries, recreation, power generation, and shipping. Further, urban settlements and reclamation of marshy areas have reduced the critical habitats of many of the coastal flora and fauna. All these activities are location specific but produce consequential response from the coastal ecosystems due to interlinks and multiplying effects.

25.3.4 Effect on Critical Coastal Habitat

In general, habitat (which is Latin for 'it inhabits') is defined as type of natural environment in which a particular species of organism lives. It is characterized by both physical and biological features. A species' habitat is very important where it can find food, shelter, protection, and opportunity for reproduction. In the coastal areas, both on-site human activities and activities of humankind elsewhere in the world have significantly affected the coastal habitats. For example, fragmentation and destruction of coastal habitats happen by on-site human activities, whereas sea level rise and consequential submergence of land are a global phenomenon. In recent times oil pollution has been receiving increasing attention with the increase in tanker operations, oil use, and frequent marine tanker collisions and accidents resulting in oil spills. Millions of oils are being added into the coastal and marine environments from variety of sources affecting the habitats. Degradation of coastal habitats and natural resources are reported in African coast by Federico and Giovanni (2000) and loss of nesting habitats and population decline of leatherback turtles along Malaysian coast and degradation of nursery ground of fishes by Siddeek et al. (1999). Most important coastal habitats severely affected by human activities are mangroves, seagrass beds, coral reef, lagoons, and estuaries.

Mangroves and other coastal wetlands are a key resource in many areas since they produce nutrients for most tropical shrimp and many fish species; provide physical habitat for a number of fish species especially in their juvenile phase; and protect the coastline from natural hazards such as tidal waves. Mangroves can be affected in a number of ways including excessive siltation or sedimentation and major oil spills. A particular problem has been the clearing of mangroves to reclaim land for other uses such as aquaculture, agriculture, or housing and other infrastructure.

Seagrass systems play a similar role to mangroves in providing nutrients and shelter for fish and shellfish. Again they may play a particularly important role as nursery areas. They may be affected by excessive siltation and water pollution. They may also be damaged by trawling, an example of an intrasectoral effect, demonstrating the difficulty of defining sectors appropriately from a management viewpoint.

Coral reefs, which are living organisms, contribute to fisheries in three ways, viz. reef fishing itself, fishing in shallow coastal waters where the reef forms an essential part of the food web, and offshore fisheries which depend in part on the reef's productivity. It has been estimated that one third of the world's fish species live on coral reefs (WRI 1986). Many artisanal fisheries also depend on coral reefs. Such fisheries represent 90% of fish production in Indonesia and 55% in the Philippines (Clark 1991). Consequently there tends to be a high level of conflict over coral reef usage, especially between fishing, tourism, and coral mining. Regrettably, all three activities may contribute to the perturbation of reef ecosystems. Clearly, coral mining leads directly to physical degradation as do some fishing methods, notably muroami. Recreational visits may also cause damage, e.g. anchoring. Reefs are also subject to a variety of natural disasters, including hurricanes, reef-destroying animals (crown-of-thorns starfish), and disease.

Lagoons and estuaries are also important habitats similar to the other habitats, notably the provision of organic materials to outside waters via tidal activity and the provision of habitat to fish species, especially in their juvenile phase. They may be seriously affected through waste disposal, residential development (via land reclamation), construction that changes water circulation (harbours), and oil and chemical pollution. For example, decline in populations and colonies of seabirds are reported by Pablo et al. (1999) and Dunnet et al. (1990) and fishes and crustaceans by Reeves et al. (1997).

25.4 Balancing Development and Environmental Concern

There are many anthropogenic activities and interests that shape the coastal land use in any coastal countries. Developmental activities are vital for the progress of any country which is very much pertinent to the tropical region developing countries. At the same time, in the race for development, there are potential chances to have incompatible activities adjacent to each other. Further, policy decisions and local people's interest about coastal land development can also impact on the coastal and nearby marine area.

This created environmental concern often seriously affecting livelihood of coastal communities and negative impact on the coastal ecosystems compounded by natural causes. The main intention of activities such as urban and rural development, food production, transport, infrastructure, energy generation and transmission, recreation, services, and mineral extraction is without doubt focused on development. However, when these activities become incompatible or not managed proactively and effectively, it creates more environmental problems and conflicts than social externalities. Further there can be loss of property, public values, and damage to important infrastructure. For example, traditional swimming areas may be made unsafe by adverse impacts such as the discharge of contaminants and other recreational activities such as jet skiing and surfing. Consequently this may restrict public access to varying degrees; impact on natural character and coastal landscapes, increasing the risks from natural hazards; affect coastal historic heritage; and impact on coastal biodiversity and the special relationship of social value of the locals with the coastal environment.

Concerned with the negative consequences of developmental activities, governments across the tropical coastal countries and other stakeholders have certainly made efforts to reduce the negative effects. But such efforts and actions are not significant and failed to protect the natural environment. Further, coastal land is affected by the dynamic influence of the sea and weather which also have some linkages with human activities. Strong desire to have better quality of life is resulted into the collision of developmental activities with natural environment. Under such circumstances carrying capacity of environment provides the base for development activities and acts as guideline for national government to approach the developmental activities. Therefore critical evaluation of our activities is a must to avoid collision that our present course will bring about (Fig. 25.2).

Therefore, coastal space needs to be managed so that conflicts are avoided or minimized, amenity values are maintained and enhanced, and safety and navigation requirements are met. Managing conflict will often involve determining what activity takes priority in each locations with due considerations for socio-economic

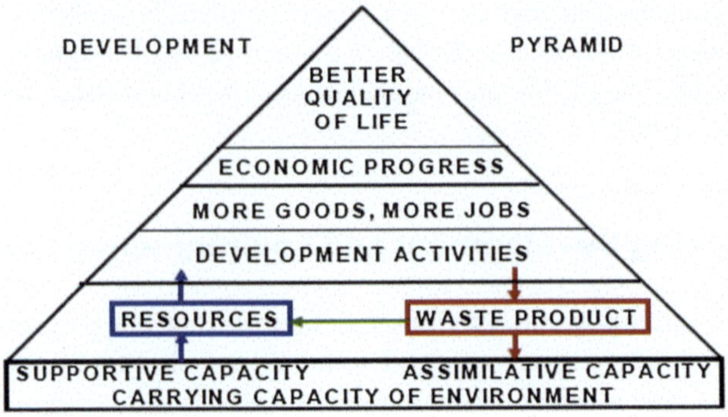

Fig. 25.2 Development pyramid illustrating the need for balancing

aspects. This requires strong efforts to balance these conflicting activities; at the same time, priority should be given to address the environmental concerns. Some of the harmonizing management options, activities, and general guidelines are discussed in this subsection.

25.4.1 General Guideline for Coastal Development

- The developmental activities in the coastal areas must follow an integrated coastal zone management plan drawn based on sound ecological principles, addressing environmental concern, provide livelihood to local as priority, and meet the national developmental goals.
- Give priority to locate and design sites to avoid vulnerable coastal ecosystems and habitats besides conflicts.
- Designate critical habitats and species for special protection or status in the coastal region particularly areas close to developmental hubs.
- Locate development that is not coastal-dependent away from important coastal ecosystems areas.
- Incorporate buffer zones whenever development close to vulnerable ecosystems cannot be avoided.
- In addition to avoiding unnecessary negative impacts on coastal ecosystems, establish shoreline setbacks or buffer zones aimed at avoiding damage to development from coastal and riverine natural hazards.
- Avoid breeding and nesting seasons while undertaking construction work; in other words there should be developmental plan along with activity chart for different months.
- Design infrastructures and land uses so as to avoid blocking water circulation.
- Dispose of effluents, runoff, and discharges after proper treatment in a manner that avoids pollution of ecosystems.
- Require compensation for ecological losses caused by development; wherever possible, provide for replacement of lost habitat or restoration of degraded one.

25.4.2 Harmonizing Strategy

The healthy state of coastal environment and natural resources is interdependent and interrelated and has direct and indirect connections with inland resource systems as well. For this reason it is imperative to have a holistic developmental plan. Further, each developmental activity if not managed sustainably will have negative consequences in the other system. Inept economic development in coastal areas is likely to create serious problems related to water pollution, degradation of critical habitats, depletion of natural resource stocks, and other effects. When this occurs, the much publicized benefits of growth such as increased employment and rising nominal incomes will be undermined by increased costs in the areas of health, productivity, and aesthetics.

Therefore, it is vital to have a balanced coastal developmental strategy than any other region. For instance, production of fish may be dependent on the habitat for juveniles provided by mangrove swamps, and the health of a coral reef may be related to the filtering properties of the mangrove ensuring that only clear water reaches the reef. On the other hand, a coral reef may die as a result of being covered with silt due to soil erosion, perhaps occurring in the upstream away from the coral belt, caused by inappropriate forestry or agricultural practices.

Within each demarcated zone, harmonious activities should be selected which are essential to (1) maximize the desired benefits from the developmental activity, (2) limit/minimize the negative interactions between the main uses, and (3) capitalize on potential synergies between different activities (Table 25.2). Best practices comprise, inter alia, the conventional, site-specific management of pollution, coastal development and tourism, fisheries and aquaculture, and biodiversity conservation.

For example, zones dedicated to biodiversity conservation will usually be most effective away from urban centres, whereas aquaculture should be located as close to urban markets as water quality permits. Food production from small-scale subsistence and artisanal fisheries will be optimized by providing fishers with access to most coastal areas and by closing their fishing grounds to larger-scale, commercial fisheries. Water quality is fundamental for coastal biodiversity, habitat and ecosystem health, and coastal activities such as swimming, fishing, marine farming, and shellfish gathering. Therefore strict imposition of water quality and effluent standards through regulations or promoting good practices will improve the synergies between different uses of water in the coastal areas.

At the same time, these harmonizing strategies should also ensure participation of different stakeholders and cooperation at local, regional, national, and international level, because often it is observed that many coastal structures and natural habitats cross the mean high water springs boundary and cross the jurisdictional boundary of territorial authorities. For example, jetties are usually located both on land and within the coastal marine area to provide easy access to the coast. Natural habitats of fishes and other endangered aquatic life are located across national boundaries and in high seas. Therefore, integrated and harmonious management is imperative for managing cross-boundary issues and the effects of the occupation of these types of structures or facilities within the coastal environment.

Table 25.2 Balancing activities to promote development and environmental conservation

Developmental benefit	Minimize negative effect	Improve the synergies	References
Aquaculture			
Locate aquaculture, and maintain water quality using ICM plan	Select sites, and specify density of farms to avoid benthic anoxic conditions	Modify aquaculture facilities, and avoid habitat for wild juvenile fish	Cochrane et al. (2009) and FAO (1998)
Promote and provide incentives for investment in aquaculture	Stock species (bivalves, sea cucumbers) to assimilate waste products from sea cages	Extract nutrients by farming seaweeds and harvesting them for fuel and fertilizer	FAO (1998)
Partition use of water column to maximize opportunities for co-culture/polyculture	Use sterile animals to avoid outbreeding of escapees with wild stocks	If possible and need arises, produce cultured juveniles to rebuild wild stocks	Bell et al. (2005)
Introduce suitable species for food production with low invasive potential	Reduce disease risks, and improve sanitation and monitoring	Explore microbes in the wild to promote probiotic effect	Lorenzen et al. (2010)
Ports/shipping			
Add subsurface structural complexity to wharfs and breakwaters to improve shelter for fishes around that place	Ensure water and sediment quality and strict monitoring of effluent discharge and vessel movement	Dedicate areas for aquaculture of species favoured by higher nutrient levels	Hair et al. (1994), McNeill et al. (1992), and Sale et al. 2008
Use social and financial capital of urban centres to clean up habitats of aquatic life and ecosystem management	Minimize risks of toxic algal blooms and invasive species in ballast water	Maximize opportunities for fishing from wharfs, jetties, etc.	Bax et al. (2003) and Anderson et al. (2002)
Fisheries			
Use management frameworks and approaches for small-scale fisheries	Prohibit or ban fishing methods that damage fish habitats, or prevent habitat recovery by suitable regulatory tools	Implement restocking, stock enhancement, and sea ranching along with management	Pomeroy and Andrew (2011) and Bell et al. (2005)
Promote fishing rights as means of livelihood and human rights of local communities	Specify fishing exclusion areas around aquaculture operations	Identify source-sink areas, and harvest appropriately	Charles (2011) and Kritzer and Sale (2006)
Protect fish habitat by regulating coastal transport and other activities	Use fishery reserves to protect spawning aggregations of valuable species		FAO (1998) and Bell et al. (2005)
Tourism			
Manage visitor activities to prevent damage to fish habitats	Ensure effluent from resorts does not promote harmful algal blooms	Transplant harvested or cultured corals to enhance reef structure to support more fish	
Create protected areas for recreation	Restrict tourism activities within ecological capacity		

25.4.3 Regulatory Activities

The coastal areas across tropical coastal nations are increasingly and extensively used for a large number of developmental activities. These multiple uses are not always compatible with the natural process resulting in wide array of problems for resource users, decision-makers, and natural environment. Coastal land is particularly used for human settlement, agriculture, trade, industry, and other amenities. At the same time, the coastal sea presents problems related to transport, fishing, dumping, mining, sedimentation, pollution, etc., stemming from an intensification and diversification of ocean uses. Since the twentieth century, crude oil is purified and processed in refineries to produce a variety of fuels, lubricants, and solvents. During these operations, continuous small-scale pollution occurs through leakages, spills, breakages, etc. Water is used in many processes and inevitably become contaminated with oil and derivatives and, when discharged, carries appreciable oil loads. This needs to be regulated at every level beginning from transport, storage, processing, and use (Shahidul Islam and Tanaka 2004).

There is also over-exploitation of coastal biological resources and an array of conflicts due to the interaction among user groups in congested marine areas (Clark 1991). All these contrasting interests and environmental concern require suitable conflict solving mechanism and other legal arrangements. Therefore, wide varieties of legal instruments have been developed and implemented by several coastal countries to promote harmonious development of the coastal areas which are envisioned in the integrated coastal zone management plan. However, identifying appropriate, practical legal mechanisms to give effect to the concepts and principles such as sustainable development that underlie ICAM remains one of the major challenges facing legislative drafters. This necessitates evolving and implementing certain legal mechanisms for the successful development and implementation of coastal area management programmes (FAO 1992) for harmonious development. Some of the most important regulatory mechanisms based on the experienced problems are on pollution control, creating protected areas, critical areas, biodiversity, natural hazards, restoration, environmental and socio-economic assessment, and conflict resolution. These regulations should provide for a central authority to regulate or control the identified problems at different level. In addition such regulatory activities should be in conformity with the national commitments to the international community and other multilateral agreements.

25.5 Conclusion

The coastal region is important to most tropical coastal countries due to the concentration of population, resource endowment, and potential to accelerate the economic growth. In this region traditional resource-based activities, such as coastal fisheries, aquaculture, forestry, and agriculture, are found side by side with activities such as industry, shipping, services, and tourism. Owing to these importance, there is growing concern for its future, particularly regarding the status of its natural resources which are affected by the competition for space among different unsustainable activities and to certain extent failure of policies and regulations. Therefore it is imperative to have balance between development and environmental concerns, to conserve the natural resources, and continue to provide livelihood to the coastal communities and derive economic benefit for the coastal nations.

In the same context, the potential for economic opportunities in coastal cities is a strong attractive force, fuelling immigration, often from economically depressed rural areas. As a result, in the future much larger, younger populations can be expected in the coastal areas of developing countries. These future coastal residents will demand employment, housing, energy, food, water, and other goods and services, thus presenting a substantial development challenge. Coastal areas are also important ecologically, as they provide a number of environmental goods and services. The peculiar characteristic of coastal environments is their dynamic nature which results from the transfer of matter, energy, and living organisms between land and sea systems. For this reason preservation of coastal biological diversity and its habitats assumes significance which emphasises on encouraging development that minimizes negative impacts on living natural resources and its surroundings.

The coastal systems are also affected by the changing climate systems particularly sea level rise which demand strong adaptation response from the coastal areas. This would be one of the strong determining factors of resource use and developmental objectives in the future. Therefore integrated coastal region management aimed at sustainable resource use is not only aimed to conserve the resources and ecosystems at present, but it is a continuous and dynamic process to continue.

References

Anderson DM, Glibert PM, Burkholder JM (2002) Harmful algal blooms and eutrophication: nutrient sources, composition, and consequences. Estuaries 25:704–726

Bax N, Williamson A, Aguero M, Gonzalez E, Geeves W (2003) Marine invasive alien species: a threat to global biodiversity. Mar Policy 27:313–323

Bell JD, Rothlisberg PC, Munro JL, Loneragan NR, Nash WJ, Ward RD, Andrew NL (2005) Restocking and stock enhancement of marine invertebrate fisheries. Adv Mar Biol 49:1–370

Charles A (2011) Human rights and fishery rights in small-scale fisheries management. In: Pomeroy RS, Andrew NL (eds) Small-scale fisheries management: frameworks and approaches for the developing world. CAB International, Oxfordshire, pp 59–74

Chavalit V, Sirapraphа P (2002) The status of estuarine fish diversity in Thailand. Mar Freshw Res 53:471–478. https://doi.org/10.1071/MF01122

Clark JR (1991) The status of integrated coastal zone management: a global assessment. CAMPNET, University of Miami/RSMAS, Miami. 118p

Cochrane K, De Young C, Soto D, Bahri T (2009) Climate change implications for fisheries and aquaculture: overview of current scientific knowledge, Food and Agriculture Organization of the United Nations fisheries and aquaculture technical paper 530. FAO, Rome

Dunnet GM, Rurness RW, Tasker ML, Becker PH (1990) Seabird ecology in the North Sea. Neth J Sea Res 26(2–4):387–426

FAO (1992) Integrated management of coastal zones, FAO fisheries technical paper 327. Food and Agriculture Organization of the United Nations, Rome. http://www.fao.org/docrep/003/T0708E/T0708E01.htm#ch1

FAO (1998) Integrated coastal area management and agriculture, forestry and fisheries. FAO guidelines, Food and Agriculture Organization of the United Nations, Rome. http://www.fao.org/docrep/W8440e/W8440e00.htm#TopOfPage

Federico C, Giovanni A (2000) The Indian Ocean coast of Somalia. Mar Pollut Bull 41(1–6):141–159

Gerber PJ, Menzi H (2006) Nitrogen losses from intensive livestock farming systems in Southeast Asia: a review of current trends and mitigation options. Int Congr Ser 1293:253–261. https://doi.org/10.1016/j.ics.2006.01.028

Hair CA, Bell JD, Kingsford MJ (1994) Effects of position in the water column, vertical movement and shade on settlement of fish to artificial habitats. Bull Mar Sci 55:434–444

Kritzer JP, Sale PF (2006) Marine metapopulations. Academic, New York, p 576

Lorenzen K, Leber KM, Blankenship HL (2010) Responsible approach to marine stock enhancement: an update. Rev Fish Sci 18:189–210

Lu J (2003) Marine oil spill detection, statistics and mapping with ERS SAR imagery in south-East Asia. Int J Remote Sens 24:3013–3032

McNeill SE, Worthington DG, Ferrell DJ, Bell JD (1992) Consistently outstanding recruitment of five species of fish to a seagrass bed in Botany Bay, NSW. Aust J Ecol 17:359–365

Nwilo PC, Badejo OT (2005) Oil spill problems and management in the Niger Delta. In: International oil spill conference, Miami

Pablo Y, Esteban F, Patricia G, William C (1999) Status and conservation of seabirds breeding in Argentina. Bird Conserv Int 9(4):299–314

Pomeroy RS, Andrew NL (2011) Small-scale fisheries management: frameworks and approaches for the developing world. CAB International, Oxfordshire

Reeves RR, Wang JY, Leatherwood S (1997) The finless porpoise, Neophocaena phocaenoides (G. Cuvier, 1829) a summary of current knowledge and recommendations for conservation action. Asian Mar Biol 14:111–143

SACEP (2017) Marine environment pollution challenges and response in line with SDG-14 in South Asian Seas Region. South Asia Co-operative Environment Programme, Sri Lanka. 24pp

Sale PF, Butler MJ IV, Hooten AJ, Kritzer JP, Lindeman KC, Sadovy de Mitcheson YJ, Steneck RS, van Lavieren H (2008) Stemming decline of the coastal ocean: rethinking environmental management. UNU-INWEH, Hamilton, p 42

Shahidul Islam M, Tanaka M (2004) Impacts of pollution on coastal and marine ecosystems including coastal and marine fisheries and approach for management: a review and synthesis. Mar Pollut Bull 48:624–649

Siddeek MSM, Fouda MM, Hermosa GV Jr (1999) Demersal fisheries of the Arabian Sea, the Gulf of Oman and the Arabian Gulf. Estuar Coast Shelf Sci 49(Suppl. A):87–97

Todd PA, Ong X, Chou LM (2010) Impacts of pollution on marine life in Southeast Asia. Biodivers Conserv 19(4):1063–1082

UNEP/GRA (2006) The state of the marine environment: trends and processes. United Nations environment Programme-global plan of action, The Hague

Ventikos NP, Sotiropoulos FS (2014) Disutility analysis of oil spills: graphs and trends. Mar Pollut Bull 81(1):116. https://doi.org/10.1016/j.marpolbul.2014.02.007

Williams C (1996) Combating marine pollution from land-based activities: Australian initiatives. Ocean Coast Manag 33(1–3):87–112

WRI (1986) World resources 1986. World Resources Institute, Washington, DC, p 353

The manufacturer's authorised representative in the EU is Springer Nature Customer Service Centre GmbH, Europaplatz 3, 69115 Heidelberg, Germany. If you have any concerns regarding our products, please contact ProductSafety@springernature.com

Printed and bound by CPI Group (UK) Ltd, Croydon, CR0 4YY
30/09/2026
02225232-0011